AAPS Advances in the Pharmaceutical Sciences Series

The AAPS Advances in the Pharmaceutical Sciences Series, published in partnership with the American Association of Pharmaceutical Scientists, is designed to deliver well written volumes authored by opinion leaders and authoritarians from around the globe, addressing innovations in drug research and development, and best practice for scientists and industry professionals in the pharma and biotech industries. For more details and to see a list of titles in the Series please visit http://www.springer.com/series/8825

More information about this series at
http://www.springer.com/series/8825

AAPS Advances in the Pharmaceutical Sciences Series

Allen C. Templeton • Stephen R. Byrn •
Roy J. Haskell • Thomas E. Prisinzano
Editors

Discovering and Developing Molecules with Optimal Drug-Like Properties

aapspress

Springer

Editors
Allen C. Templeton
Analytical Sciences
Merck & Co.
Summit, New Jersey, USA

Stephen R. Byrn
Department of Industrial and Physical Pharmacy
Purdue University
West Lafayette, Indiana, USA

Roy J. Haskell
Discovery Pharmaceutics
Bristol-Myers Squibb
Wallingford, Connecticut, USA

Thomas E. Prisinzano
Department of Medicinal Chemistry
University of Kansas School of Pharmacy
Lawrence, Kansas, USA

ISSN 2210-7371 ISSN 2210-738X (electronic)
ISBN 978-1-4939-1398-5 ISBN 978-1-4939-1399-2 (eBook)
DOI 10.1007/978-1-4939-1399-2
Springer New York Heidelberg Dordrecht London

Library of Congress Control Number: 2014950891

Printed on acid-free paper

Springer is part of Springer Science+Business Media (www.springer.com)

Contents

About the Editors

Allen C. Templeton is executive director of the Analytical Sciences organization within Merck Research Laboratories. He is responsible for the managing staff in the pursuit of scientific problem-solving for pharmaceutical product development. Before assuming his current position, Dr. Templeton held positions of increasing responsibility within Merck, including leadership roles in preformulation and formulation. Dr. Templeton earned his Ph.D. in analytical chemistry from the University of North Carolina at Chapel Hill. His research experience has been in the area of analytical and materials chemistry. He has published more than 50 articles, served as co-inventor on 11 patents, and authored more than 120 presentations in the area of pharmaceutical analysis. He has organized a number of symposia and training courses on diverse topics within the field of pharmaceutical characterization. Dr. Templeton is an active member in a number of professional organizations, including the American Association of Pharmaceutical Scientists (AAPS) and the American Chemical Society (ACS). He has served in a number of roles for AAPS and is most recently the secretary/treasurer of the Physical Pharmacy and Biopharmaceutics section. He was elected to the United States Pharmacopeia (USP) expert committee on physical analysis and has worked to revise a number of USP standard chapters. He is also currently serving on the Analytical Leadership Group for the International Consortium for Innovation and Quality in Pharmaceutical Development (IQ). He is on the editorial advisory boards for the *Journal of Pharmaceutical Sciences*, *American Pharmaceutical Review*, and *Current Drug Delivery*.

Stephen R. Byrn is Charles B. Jordan Professor of Medicinal Chemistry in the Department of Industrial and Physical Pharmacy, Purdue University. Dr. Byrn set in motion the development of the field of solid state chemistry of drugs with his books and papers on the subject. He has also taught more than 100 short courses on solid state chemistry and pharmaceutical solids and has educated more than 50 Ph. D. students and postdoctoral fellows. Dr. Byrn has had numerous grants, including one of the first 13 from NIH Centers for AIDS Research. Dr. Byrn is cofounder of Purdue's graduate programs in regulatory and quality compliance. He is also cofounder of the Purdue-Kilimanjaro School of Pharmacy graduate certificate program in industrial pharmacy and manufacturing in Moshi, Tanzania. Dr. Byrn has served as chair of the Pharmaceutical Sciences Advisory Committee to the FDA and chaired several USP committees. Dr. Byrn is also cofounder of SSCI, Inc. (Solid State Chemical Information), a cGMP research and information company now owned by Aptuit. Dr. Byrn is an elected Fellow of the American Association of Pharmaceutical Scientists (AAPS) and has received several awards for his research and entrepreneurial activities including the first AAPS David Grant Award for Research Achievement in Physical Pharmacy. The *Journal of Pharmaceutical Sciences* has a special issue dedicated to Dr. Byrn. His current research interests include strategies for accelerated drug development and the use of synchrotron X-rays for pharmaceutical research.

Roy J. Haskell is a Research Fellow in the Discovery Pharmaceutics group of Bristol-Myers Squibb, where he is engaged in discovery support as well as the design and characterization of novel formulations. He received a Ph.D. in analytical chemistry from the University of Wisconsin–Madison. He joined The Upjohn Company and worked in the area of protein biophysics. As part of Pharmacia, his focus transitioned to formulating poorly soluble molecules and characterizing the formulations by which such compounds are delivered. Dr. Haskell's research interests include design and characterization of colloidal formulations, predictive modeling of physicochemical properties, sizing submicron particles, aggregation/precipitation mechanisms, the role of solubility and supersaturation in oral absorption, and the use nanotechnology in drug discovery and development. Dr. Haskell is the author of three patents and more than 30 publications and presentations.

Thomas E. Prisinzano received his Ph.D. in pharmaceutical sciences from the School of Pharmacy, Virginia Commonwealth University, in Richmond, Virginia. He was an Intramural Research Training Award Fellow in the National Institute of Diabetes and Digestive and Kidney Diseases in Bethesda, Maryland, and is currently Professor and Chair of the Department of Medicinal Chemistry at the University of Kansas. His research focuses on the development of novel agents to treat pain, substance abuse, and other CNS disorders through the identification, structure elucidation, and optimization of natural products. Dr. Prisinzano has received a number of awards including the D. John Faulkner Travel Award from the American Society of Pharmacognosy, the Jack L. Beal Award from the *Journal of Natural Products*, the Matt Suffness (Young Investigator) Award from the American Society of Pharmacognosy, the Joseph Cochin Young Investigator Award from the College on Problems of Drug Dependence, and the David W. Robertson Award for Excellence in Medicinal Chemistry from the American Chemical Society.

List of Contributors

John Baumann Bend Research, Bend, OR, USA

David M. Bender Small Molecule Design and Development, Eli Lilly and Company, Indianapolis, IN, USA

C. J. Benmore X-ray Science Division, Argonne National Laboratory, Lemont, IL, USA

Shobha N. Bhattachar Small Molecule Design and Development, Eli Lilly and Company, Indianapolis, IN, USA

Nathan Boersen Celgene Corporation, Summit, NJ, USA

Chad Brown Formulation Sciences, Merck & Co., West Point, PA, USA

Stephen R. Byrn Department of Industrial and Physical Pharmacy, Purdue University, West Lafayette, IN, USA

James DiNunzio Merck & Co., Summit, NJ, USA

Brett Duersch Analytical Development-Commercialization, Merck & Co., West Point, PA, USA

Kimberly A. Foster Discovery Pharmaceutics, Bristol-Myers Squibb, Lawrenceville, NJ, USA

Sudhakar Garad Discovery Pharmaceutics, Cubist Pharmaceuticals, Lexington, MA, USA

Roy J. Haskell Discovery Pharmaceutics, Bristol-Myers Squibb, Wallingford, CT, USA

John Higgins Discovery Pharmaceutical Sciences, Merck & Co., West Point, PA, USA

Grace Ilevbare Discovery Pharmaceutical Sciences, Merck & Co., Kenilworth, NJ, USA

Akash Jain Discovery Pharmaceutics, Cubist Pharmaceuticals, Lexington, MA, USA

Susan M. Jenkins Metabolism and Pharmacokinetics, Bristol-Myers Squibb, Wallingford, CT, USA

David Johnson Pharmaceutical Commercialization, Merck & Co., West Point, PA, USA

Filippos Kesisoglou Biopharmaceutics, Merck & Co., West Point, PA, USA

Michael H. Kress Process Chemistry, Merck & Co., Rahway, NJ, USA

Michael Lowinger Discovery Pharmaceutical Sciences, Merck & Co., Rahway, NJ, USA

William Marinaro Formulation Sciences, Merck & Co., Summit, NJ, USA

Patrick Marsac Formulation Sciences, Merck & Co., West Point, PA, USA

Craig McKelvey Merck & Co., Inc., Formulation Sciences, West Point, PA, USA

Michael McNevin Formulation Sciences, Merck & Co., Summit, NJ, USA

Elaine Merisko-Liversidge Pharmaceutical Research & Development, Alkermes, Waltham, MA, USA

Robert Meyer Pharmaceutical Commercialization, Merck & Co., West Point, PA, USA

Amitava Mitra Biopharmaceutics, Merck & Co., West Point, PA, USA

Michael Morgen Bend Research, Bend, OR, USA

Justin Moser Formulation Sciences, Merck & Co., West Point, PA, USA

Dawn D. Parker Discovery Pharmaceutics, Bristol-Myers Squibb, Wallingford, CT, USA

Thomas E. Prisinzano Department of Medicinal Chemistry, University of Kansas, Lawrence, KS, USA

Alfred C. F. Rumondor Formulation Sciences, Merck & Co., Summit, NJ, USA

Joyce Stellabott Formulation Sciences, Merck & Co., West Point, PA, USA

Stephanie A. Sweetana Small Molecule Design and Development, Eli Lilly and Company, Indianapolis, IN, USA

Allen C. Templeton Analytical Sciences, Merck & Co., Summit, NJ, USA

Karen Thompson Formulation Sciences, Merck & Co., West Point, PA, USA

Ching Kim Tye Discovery Pharmaceutics, Bristol-Myers Squibb, Lawrenceville, NJ, USA

David T. Vodak Bend Research, Bend, OR, USA

James A. Wesley Small Molecule Design and Development, Eli Lilly and Company, Indianapolis, IN, USA

Dan Zhang Formulation Sciences, Merck & Co., Summit, NJ, USA

Part I
Strategy and Tactics that Enable Discovery

Chapter 1
Developability Assessment and Risk Management During Drug Discovery

Sudhakar Garad and Akash Jain

1.1 Introduction

The pharmaceutical industry is seeing insufficient revenue and profit growth due to high development costs, long development and approval times, fewer new product launches, lack of rich pipelines, and loss of revenues to generics due to patent expiration (Cockburn 2004; Frank and Seiguer 2003). The lack of rich pipeline in the industry is caused by the high attrition rates due to inadequate physicochemical and biopharmaceutical attributes, acceptable safety, and sub-marginal efficacy in preclinical and clinical studies, all of which also contribute to increasing drug development costs (Subramaniam 2003; Rosiello et al. 2013). Because of rapidly rising drug development costs, there is an enormous pressure to cut cost and streamline development (Watkins 2002; Paul et al. 2010). While different companies are dealing with these challenges differently, the industry can benefit substantially by enhancing the discovery–development interface. The knowledge gained during the candidate selection phase will be useful in fast development of optimal formulation and clinical approaches for clinical studies, and consequently will help shorten the development timelines for new chemical entities (NCEs). This will directly benefit not only the industry, but also the society, as new and better lifesaving drugs can reach patients sooner (Venkatesh and Lipper 2000).

In such a scenario, the reduction of attrition rate and thus reduction in the development cost and timeline for a product pipeline can be achieved by strengthening the efficacy and toxicity screens, and establishing a developability screen to enable selection of developable compounds to the clinic.

Efficacy and toxicity screens involve dosing the drug in a variety of animal models. A general schematic of the most common in vivo studies performed during

S. Garad (✉) • A. Jain
Discovery Pharmaceutics, Cubist Pharmaceuticals, 65 Hayden Avenue, Lexington, MA 02421, USA
e-mail: sudhakar.garad@cubist.com

A.C. Templeton et al. (eds.), *Discovering and Developing Molecules with Optimal Drug-Like Properties*, AAPS Advances in the Pharmaceutical Sciences Series 15,
DOI 10.1007/978-1-4939-1399-2_1

different discovery stages is shown in Fig. 1.1. Selection of the right formulation principles is critical for the proper functioning of efficacy and toxicity screens, especially for NCEs with poor solubility and/or with potential for variable systemic exposure. For such NCEs, an inadequate formulation principle may limit their systemic exposure through limited solubilization or precipitation in the gastrointestinal tract or in the blood stream (from a solubilized formulation), thus leading to low and variable exposure and false safety interpretation. Additionally, sub-marginal exposure may also lead to poor efficacy, and thus could lead to rejection of a potential blockbuster. For example, blockbuster drug products such as Cyclosporine (Neoral®), Telaprevir (Incivek®) and Lopinavir/Ritonavir (Kaletra®) are excellent case studies where the poor solubility of respective APIs (active pharmaceutical ingredients) was successfully overcome by selecting the right enabling formulation technology. Therefore, the success of such efficacy and toxicity screens for NCEs depends a lot on proper selection of suitable formulations, which eventually will lead to objective go/no-go decisions before the NCEs enter into full development.

Developability screens are equally important as the efficacy and toxicity screens. Developability assessment typically involves physicochemical characterization of NCEs, characterization and selection of the most suitable/stable solid form of API, development of formulations for robust PK, efficacy and safety assessments in preclinical species, biopharmaceutics (e.g., factors limiting absorption and bioavailability, food effects, etc.) and drug delivery options for further development of NCEs. The data package based on in silico, in vitro and in vivo evaluation of multiple formulations/delivery technologies allows the selection of a biopharmaceutically optimized formulation/delivery strategy, and consequently, yields more reliable information on developability of NCEs moving into clinic and commercialization.

1.2 Developability Assessment Group (DAG)

1.2.1 Background

The organizational structure of the function performing developability assessment varies significantly from one pharmaceutical company to another (Kerns and Di 2002; Fiese 2003; Balbach and Korn 2004; Sun et al. 2004; Balani et al. 2005; Singh 2006; Maas et al. 2007). Developability Assessment Groups (DAGs) could reside in either discovery or development parts of an organization. In either case, the key enabler for their success is strong cross-functional collaborations, especially with medicinal chemists, biologists, pharmacologists and ADME/PK/Tox groups on the discovery side and technical development teams (API and Drug Product) and clinicians on the development side. Other key enablers for DAGs include in silico modeling tools, state-of-the-art analytical capabilities, solid-state characterization

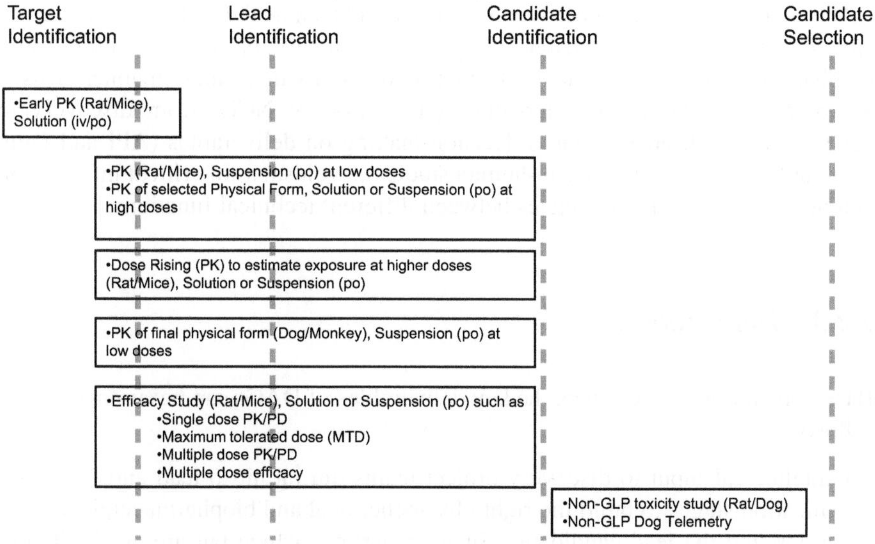

Fig. 1.1 General and most common sequence of in vivo studies during candidate selection. Reprinted from Saxena et al. (2009) with permission from John Wiley & Sons

technologies, automation and high-throughput platforms and strong understanding of biopharmaceutics and drug delivery.

1.2.2 Roles and Responsibilities

In principle, the DAGs can take a two-pronged approach on NCEs depending on complexity of biological target and chemical modification space available to medicinal chemists:

Pharmaceutics—Build right developable properties in molecule (e.g., p*K*a, solubility, stability, permeability, etc.), identify developable solid forms (e.g., salts, stable polymorphs, co-crystals) and selection of optimal delivery strategy (e.g., route of administration, formulation principle).

Enabling Technologies—Solubility and bioavailability enhancement (multiple technologies for parenteral and oral delivery), modified release (improve therapeutic index), and targeted drug delivery (local, site of inflammation/infection, etc.).

In scenarios where both of the above-mentioned approaches are unsuccessful in identifying NCEs with desired profile, a timely feedback can be provided to discovery teams to consider terminating further efforts on such targets. This is a great mechanism to derisk NCEs as soon as possible based on their biopharmaceutical, technical, and preclinical safety evaluation. Another key responsibility of

DAGs is to act as a liaison between discovery and technical development functions and ensure seamless transfer of physicochemical properties, solid form, biopharmaceutics and technical information on NCEs in a timely manner. This is an extremely value-adding step during transition of NCEs from discovery to development as it enables rapid decision-making on deliverables (API and Drug Product for GLP tox and first-in-human studies), avoids loss of critical information and helps minimize redundancies between different technical functions.

1.2.3 Deliverables

Based on authors' experience, the key deliverables of DAGs can be summarized as follows:

(a) Intellectual input to discovery project teams during hit-to-lead and lead optimization stages to build the right physiochemical and biopharmaceutical properties into the lead candidates. Such properties include but are not limited to p*K*a, log *P*/*D* and addition of functional groups to improve solubility and/or to reduce very high lattice energies due to strong intermolecular interactions.
(b) Physicochemical and solid-state characterization of NCEs (melting point, solubility, p*K*a, log *P*/*D*, permeability, etc.).
(c) Formulation development for pharmacokinetic, pharmacological and toxicological studies using conventional approaches such as solutions, suspensions, etc., or using non-conventional delivery approaches such as spray-dried dispersions, nanosuspensions, etc.
(d) Generate in silico predictions of biopharmaceutical performance using tools such as GastroPlus™, maximum absorbable dose, etc.
(e) Recommendation on stable physical form and clinical formulation.
(f) Recommendation to terminate non-developable candidates as soon as possible thereby enabling research colleagues to move on to new scaffolds/targets.
(g) Risk assessment with mitigation plan for the selected development candidates.
(h) Enabling technologies for NCEs (e.g., targeted delivery, modified release, etc.).

The creation and implementation of DAGs in multiple organizations, over the last decade or so, has yielded a new paradigm for transition of NCEs from discovery into development. The above listed deliverables traditionally, once a candidate was handed over from discovery to development, would take approximately 6–9 months for completion and utilize ~30–100 g of API. In addition, if any developability issues were identified, it was almost too late to provide the feedback to discovery teams. With formation of DAGs, similar deliverables can now be achieved in approximately 2–4 months with ~2–10 g of API (Table 1.1). These significant savings in time and material have been made possible by development of automated and high-throughput platforms. The early intervention of DAGs in discovery programs helps provide timely feedback on developability issues and suggests appropriate measures well before a candidate is nominated for full development.

Table 1.1 Key activities and deliverables of Developability Assessment Groups (DAGs)

Hit to lead (5–20 mg)	Lead optimization (50–500 mg)	Candidate selection phase (2–10 g)
p*K*a	Weight-based purity	Batch characterization
Solubility (pH) and simulated fluids	Solid-state characterization (DSC, TGA, XRPD, DVS)	Solid form selection salt/co-crystal, polymorph selection
Stability (pH)	Preliminary salt/crystallization profiling	Formulation and condition-of-use stability for DC-enabling studies
PAMPA (GI, BBB)	In-process analytical method development	Analytical method development and validation
Formulation for PK/acute tolerability	Formulation for efficacy/toxicity studies	Impurity characterization
HPLC purity	Dosing/delivery strategies	Establish standards and specification
	Biopharmaceutics	Initiate GMP and GLP validation
		Clinical formulation recommendation

For a DAG to be successful and effective, it is very important to have the right balance of expertise in the group. Scientists in such groups generally have background in areas such as pharmaceutics and biopharmaceutics, solid state and material science, analytical science and drug delivery. In addition to the technical expertise, competencies in cross-functional communication, working in team environment, and strong dedication and passion to bring new development candidates into clinic for unmet medical needs are critical in the role of a developability assessment scientist.

While the authors believe that the most optimal organizational structure is to have a DAG integrated within the discovery organization, several other organizational structures can function equally well, as long as there is close cooperation between a DAG and discovery teams. The DAG is not only responsible for building and selecting developable molecules but it is also responsible for identifying enabling drug delivery technologies for clinical and commercial products. This broad spectrum role of DAGs adds tremendous value to an organization in increasing or maximizing the potential of safe and efficacious candidates, especially, when it is very difficult to find NCEs for novel mechanisms. The emphasis in this chapter is not so much on the organizational structure, but on the valuable information that is generated by DAGs at various stages of drug discovery and development process with automated tools, minimum amount of material, resources and utilization of cutting edge technologies.

1.3 Developability Assessment of New Chemical Entities (NCEs): Methods and Best Practices

1.3.1 Biopharmaceutics-Based Molecular Design

Biopharmaceutical properties are usually secondary for any given molecule during early discovery stages as most time and effort is spent in identifying molecules with the right balance of potency and safety. However, based on lessons learned from failures and delays in advancing candidates with poor biopharmaceutical profile through clinical development, major pharmaceutical organizations are looking to build biopharmaceutical properties into molecules during early discovery stages (Saxena et al. 2009, Zheng et al. 2012). A few important properties are discussed in this section.

1.3.1.1 Ionization

Ionization constant, p*K*a, is a useful thermodynamic parameter to monitor the charge state of NCEs. Based on their p*K*a, all NCEs can be classified into four major classes—acid, base, neutral, and Zwitterion. Each of these classes of molecules represents different challenges and opportunities in their biopharmaceutical performance. Molecules with acidic p*K*a (p*K*a 3.5–6.0) predominantly exist in their ionized and solubilized form at physiological intestinal pH (5.5–7.4). However, it has been reported and well documented that ionized form of a molecule tends to have poor passive permeability (Thomayant et al. 1998). As both the ionized and unionized form of a molecule exist in a dynamic/equilibrium process in vivo, the authors believe that a sufficient fraction absorbed (Fa) is achieved in vivo from acidic molecules. Basic molecules demonstrate a much larger range of p*K*a (2.0–9.0) depending on the type of functional groups. Strongly basic molecules (p*K*a above 6.0) are considered ideal for improving solubility and dissolution rates and to minimize in vivo precipitation (and related complications) in the physiological pH range for both i.v. and oral delivery. Such molecules readily form salts with acidic counter ions and remain solubilized at the site of absorption. On the other hand, weakly basic molecules (p*K*a below 6.0) require stronger acidic counter ions for salt formation and demonstrate improved solubility/dissolution rates only at pH 4.0 and below. At physiological pH for absorption (pH 5–7), the weakly basic molecules rapidly dissociate from their salt forms and convert to poorly soluble neutral forms resulting in poor and/or variable absorption and bioavailability. Weakly basic molecules also pose a significant challenge in developing stable dosing solutions for parenteral delivery due to risk of precipitation upon injection (Jain et al. 2010). Zwitterionic molecules are the most complex in their physicochemical and biopharmaceutical behavior. Such molecules typically demonstrate high variability in oral absorption and bioavailability (Thomas et al. 2006). In addition, their p*K*a values and solubility profile are highly sensitive to presence of additional functional groups in vicinity of the ionizable groups. For instance, the presence of a strongly electron-withdrawing group next to the ionizable group can lead to a considerable shift in its p*K*a and results in poor solubility/dissolution. Given a preference, it is better to avoid Zwitterions especially for development, unless they demonstrate high intrinsic solubility across the entire pH range or high potency such that absolutely minimal doses are required in preclinical as well as clinical studies.

1.3.1.2 Intermolecular Interactions

Intermolecular interactions of NCEs (e.g., charge–charge, charge–dipole, dipole–dipole, van der Waal's dispersion, π–π stacking, hydrogen bonding, etc.) can significantly impact their crystallinity and solid-state properties which in turn can have a profound impact on their solubility, dissolution, precipitation, recrystallization behavior and ultimately their biopharmaceutical performance (Thomoyant et al. 1998). It is therefore advisable to minimize exploration of chemical scaffolds with very high lattice energies (e.g., π–π stacking of planar aromatic rings). Solid-state properties such as melting point and/or heat of fusion can serve as useful surrogates to estimate the lattice energies.

To summarize, it is always worthwhile to spend the extra time and effort during lead optimization stages in building some of the key biopharmaceutical properties into a molecule. This investment during discovery stages eventually pays off in development as it lays the foundation for design of a robust drug product strategy that can help clinical teams accelerate and shorten development timelines.

1.3.2 Physicochemical Characterization

A thorough understanding of physicochemical properties of a NCE is a prerequisite to prediction of biopharmaceutical behavior as well as screening and development of solid form and formulations for preclinical and clinical studies. Developability assessment teams are able to generate a robust package of physicochemical characterization data using minimal amounts of time and material, largely due to availability and constant evolvement of in silico prediction tools and high-throughput screening technologies. This section will describe the methods and best practices in characterization of physicochemical properties such as p*K*a, log *P*/*D*, solubility/dissolution rate, permeability, and in vitro metabolism.

1.3.2.1 Ionization Constant (p*K*a)

As noted above, the ionization constant, p*K*a, is a commonly used thermodynamic parameter in drug discovery to understand binding mechanisms and to predict ADME properties of NCEs due to the pH gradient of 1.0–7.0 in the human GI tract (Avdeef 2003). A number of important properties such as lipophilicity, solubility pH profile, permeability, and human Ether-à-go-go-Related Gene (hERG) binding affinity are modulated greatly by p*K*a. While p*K*a can be successfully predicted by in silico tools owing to the high dependence on the molecular structure of NCEs, this prediction may not be accurate unless the corresponding model has been parameterized to account for the novel chemical space spanned by many discovery programs, thereby justifying the need for in vitro determination. A range of experimental approaches with varying throughput, cycle time, sample requirement, and costs are available for p*K*a determination (Wan and Ulander 2006; Wang & Faller 2007; Wang et al. 2007). Some of the commonly used methods are potentiometric titration (Avdeef 2003), capillary electrophoresis (CE) (Cleveland et al. 1993; Ishihama et al. 2002), Spectral Gradient Analyzer (Box et al. 2003), and Sirius T3 (Sirius Analytical).

Potentiometric Titration

In the potentiometric titration method, a potentiometer records the pH changes with a glass electrode, caused by introducing a known volume of titrant to the well-mixed solution of a NCE. This methodology is tedious as it requires a lengthy process due to long and repetitive equilibrium steps after titrant additions. However, the p*K*a values obtained are reliable and therefore, this methodology is considered to be the gold standard despite a number of limitations. This methodology is only suitable for compounds with good solubility as potentiometric titration requires concentrations in the range of 0.1–1 mM. Although cosolvents have been used to circumvent poor solubility, it still requires several titrations to extrapolate the data from different water and cosolvent mixtures to aqueous solution (zero cosolvent concentration). An additional problem with potentiometric titration is the increasing numbers of NCEs in early discovery are delivered as salt forms with protogenic counter ions, like acetate, fumarate, titrate, etc. Finally, this methodology requires materials in milligram scale, and therefore, it is more useful in late discovery.

Spectral Gradient Analyzer

This assay works effectively with poorly soluble NCEs by using cosolvent in the media. The p*K*a data measured using this "rapid-mixing" approach correlate well with those from the potentiometric titration method. In addition, this method is useful for measuring p*K*a for early discovery compounds as it requires a small amount of material, and allows for high throughput/automation. The limitation is that it is suitable for ionizable compounds, which induce a change in the UV spectra scan. In other words, not only is a UV chromophore required, but also the chromophore may have to be located close enough to the ionization center within an NCE. This technique is based on a continuously flowing pH gradient and a UV-DAD (diode array detector). It was developed recently and allows for a much higher throughput (Box et al. 2003). This method establishes a stable time-dependent pH gradient by rapidly mixing acidic and basic buffers, during which drug candidates pre-dissolved in organic solvent are introduced at different pH conditions. This fast method allows full characterization of ionizable groups for 70 % of the discovery output and usually gives high quality data that are consistent with those obtained by potentiometric titration (Garad et al. 2009). Neither this "rapid mixing" nor the potentiometric titration approach is compound-specific, so compounds with purity or stability issues or counter ions containing similar UV chromophore may be problematic for this approach.

1.3.2.2 Lipophilicity (log *P*/*D*)

Lipophilicity, as expressed by the logarithm of partition coefficient or distribution coefficient ($\log P$ or $\log D$) of NCEs between a lipophilic phase (e.g., octanol) and aqueous phase, is a valuable physicochemical property. Compounds with very high ($\log P/D > 6$) or very low lipophilicity ($\log P/D < 1$) can present significant developability challenges such as poor formulability, poor permeability and absorption, accumulation in tissues/organs, etc. Therefore, a right balance of lipophilicity in any compound is essential prior to its selection as a candidate for development. Multiple techniques can be used to determine lipophilicity of the compound. While shake-flask is the conventional method for $\log P$ (or $\log D$) determination, the dual-phase potentiometric titration approach is also widely accepted, particularly during late drug discovery phase (Avdeef, 2003). For NCEs lacking an ionizable group, HPLC $\log P$ technique, also known as eLog*P* can be applied (Lombardo et al. 2000). A variety of techniques that are suitable for early discovery include liposome chromatography, immobilized artificial membrane (IAM) chromatography, capillary electrophoresis (CE) (Avdeef, 2003), and artificial membrane preparations (Wohnsland & Faller 2001).

1.3.2.3 Solubility/Dissolution

Solubility of NCEs is one of the most important physicochemical properties that govern its ADME, pharmacological and biopharmaceutical profile. Poor solubility and dissolution rates are the most common rate-limiting factors for oral drug absorption, especially for NCEs with high permeability (Ku 2008). A variety of solubility and dissolution measurements are performed throughout the discovery and development stages of a NCE (Hariharan et al. 2003; Vertzoni et al. 2005; Kibbey et al. 2001; Glomme et al. 2005; Avdeef 2007; Galia et al. 1998; Ingels et al. 2002; Lind et al. 2007). The objectives and methods for such measurements differ widely depending on the stage of the project and intended use of the data. A few examples of typical solubility experiments include high-throughput measurement in two or three buffer system during early discovery stages to rank order

Table 1.2 Solubility and dissolution experiments in discovery and early development of NCEs

Experiment	Application (s)
HT-solubility	Guide SAR during early discovery stages
Shake-flask solubility	pH solubility profile
Solubility in biorelevant media	Prediction of maximum absorbable dose, GastroPlus™ simulations, build IVIVC, food effect prediction
Organic solvents	Salt and polymorph screening, crystallization process development
Solubility in cosolvent, surfactant, complexing agents, lipids	Develop solubilized formulations for preclinical and clinical studies; develop dissolution media for release testing
Intrinsic dissolution rate	Rank-order solid forms (e.g., salts, co-crystals, polymorphs)
Non-sink dissolution	Rank-order of prototype solid forms and formulations (e.g., solid dispersions, salts, etc.)
Dissolution (type II apparatus)	Release testing for solid dosage forms; biorelevant testing and food effect prediction for solid dosage forms

compounds and develop SAR, solubility measurement of an API in organic solvents to develop a crystallization process or measuring kinetic solubility of API in biorelevant media to predict maximum absorbable dose. Solubility experiments can be performed under a variety of conditions such as equilibration at room temperature (e.g., 24–48 h), kinetic solubility at 37 °C (e.g., 1–4 h), temperature cycling in organic solvents or two-step solubility screen in biorelevant media to identify supersaturation-enhancing excipients. A detailed list of solubility and dissolution experiments typically performed during discovery and early development stages is shown in Table 1.2.

An important consideration for developability of a NCE is its tendency to precipitate under physiological conditions. Depending on its solid form and pH-dependent solubility, a compound may precipitate out in stomach fluids (pH ~2) or in small intestinal fluids (pH ~5–7) during its transit in the gastrointestinal tract. The precipitated compound may dissolve differently in the gastrointestinal fluids than the original form. Occasionally, for compounds with pH-dependent solubility, certain forms/salts of a compound may dissolve in stomach fluids, and upon gastric emptying in the intestine may remain dissolved, thus yielding supersaturated solutions. In such cases, the compound may show much higher bioavailability. Thus, it is critical to evaluate dissolution characteristics of various solid forms before selecting a final solid form for further development. Dissolution studies should be conducted either with suspension formulations or with compounds filled in capsule with or without formulation excipients. The physiological conditions can be simulated by using buffers (pH 1, 2, 4.5, and 6.8) and/or simulated fed/fasted gastric and intestinal media (Jantratid et al. 2008) at 37 °C either as single step or pH-shift method (Mathias et al. 2013). The presence of formulation excipients could help simulate a solid dosage form and can reflect the effects of formulation excipients on dissolution rate. For compounds with poor solubility, dissolutions studies should be conducted in simulated gastric and intestinal media to determine the potential for food effect.

Solubility and dissolution data are often used to select and/or rank-order NCEs during lead optimization stages. However, due to unknown efficacious clinical doses at such early stages of discovery, it is difficult to drop or terminate NCEs based on poor solubility values. In addition, for poorly soluble NCEs (typically with aqueous solubility <100 μg/ml), there are a number of formulation technologies (see Sect. 1.3.4.4) that could be explored to improve the solubility/dissolution rates and eventually their in vivo absorption and bioavailability. Early screening of such formulation technologies in combination with human PK/PD and dose predictions is essential for a successful developability assessment of poorly soluble NCEs.

1.3.2.4 Permeability

Permeability plays a vital role in the absorption of orally delivered NCEs. Permeability is a complex phenomenon and it involves multiple mechanisms across the GI mucosa (Artursson & Tavelin 2003). Hence, it is important that a developability assessment scientist identify permeability behavior of NCEs early during lead optimization stages and provide timely feedback to project teams, especially when the teams are considering an oral route of administration. Unlike poor solubility, there are only few formulation or delivery technologies that can overcome poor permeability of NCEs and provide significant improvement in their oral absorption or bioavailability (Maag 2012; Maher and Brayden 2012).

One of the simple parameters that correlate well with permeability is the molecular weight of any given NCE. As per Lipinski's rule of 5, any NCE with molecular weight <500 typically demonstrates medium to high passive permeability and NCE with molecular weight >500 usually tend to exhibit poor to medium or almost no passive permeability. Another in silico model is the one developed by Egan and coworkers (2000). Their absorption model is based on polar surface area (PSA) and calculated $\log P$ ($C \log P$). Based on their analysis, it was shown that majority of highly permeable marketed drugs were populated in an egg-shaped zone of the absorption model and therefore, specified as "good" molecules. On the other hand, drugs with poor permeability or those considered as substrates for efflux transporters were scattered outside the egg-shaped zone and denoted as "poor" in the model. Interestingly, an exception to this egg-shaped model is the blockbuster drug—Lipitor.

In addition to the in silico methods, a number of in vitro models were developed to predict permeability as well as to assess the contributions of active transporters in the permeation process (Hämäläinen & Frostell-Karlsson 2004; Balimane et al. 2006). A few important assays, commonly being used in the last few decades, for measuring permeability of NCEs with high speed and reliability/reproducibility are described as follows.

PAMPA

Parallel artificial membrane permeability assay (PAMPA), pioneered by Kansy et al. (1998), utilizes a chemical membrane immobilized on a 96-well filter plate and samples are analyzed by a UV plate reader. Distinct membrane models were established by Kansy et al. (1998), Avdeef et al. (2001), Faller (2001) and Sugano and coworkers (2001) to mimic passive diffusion of NCEs across the GI tract (Garad et al. 2009). Avdeef (2003) introduced the "double-sink" model that simulates the concentration and pH gradient across the GI membrane. Cosolvents (Ruell et al. 2004) or excipients (Kansy et al. 1998; Sugano et al. 2001; Liu et al. 2003; Bendels et al. 2006) were employed in PAMPA to overcome the low solubility issues frequently encountered in early discovery. Quantification using LCMS (Balimane et al. 2005; Wang & Faller 2007; Mensch et al. 2007; Cai et al. 2012) or HPLC (Liu et al. 2003) drastically improved the sensitivity and robustness of PAMPA by extending the limit of detection for NCEs with low solubility. It also prevents interference originating from impurities with high solubility and/or strong UV chromophore. PAMPA ultimately offers a fast and relatively cost-effective method to estimate permeability for NCEs absorbed by passive diffusion mechanisms.

Caco-2

The human colon adenocarcinoma (Caco-2) cell permeability model exhibits morphological (e.g., tight junction and brush-border) and functional similarities (e.g., multiple transport mechanisms) to human intestinal enterocytes (Ungell & Karlsson 2003; Englund et al. 2006), thereby serving as the gold standard for in vitro permeability assessment in industry (Artursson & Tavelin 2003; Lennernas and Lundgren 2004). Caco-2 cells, which extensively express a variety of transport systems beyond P-glycoprotein (Pgp), are amenable to investigate the interplay among different transport systems and differentiate the relative contributions from passive and active transport mechanisms to the overall permeability across the human GI tract (Ungell 2004; Steffansen et al. 2004). While the conventional protocol (e.g., 21-day cell culturing) is essential to assure the full expression of transporters, the Caco-2 models using accelerated cell culturing procedures (e.g., 3–7 days) are not suitable for studying active transport of NCEs due to inadequate expression of transporters (Liang et al. 2000; Alsenz & Haenel 2003; Lakeram et al. 2007). Scientists (Yamashita et al. 2002; Miret et al. 2004) are still debating whether the short-term Caco-2 culturing system is appropriate to rank permeability, with concern about its poor correlation with human absorption data and inability to differentiate medium permeability compounds (Miret et al. 2004; Sugano et al 2010).

Today, the cumbersome long culturing procedure can be readily handled by automation (Saunders 2004; Wang & Faller 2007) and therefore Caco-2 has been successfully validated and widely implemented in 24- or 96-well plate formats to assess the permeability and drug interaction with GI-related transporters (Ungell &

Karlsson 2003; Kerns et al. 2004; Marino et al. 2005; Balimane et al. 2004). Similarly to PAMPA, precaution should be taken when dealing with low solubility compounds in discovery to select agents inert to the permeability and transport process of the Caco-2 model (Yamashita et al. 2000). Introduction of bovine serum albumin (BSA) to the basolateral compartment is useful to mimic the in vivo sink condition and to help minimize the non-specific binding of NCEs to the cells and labware but the effect appeared to vary greatly upon the mechanism by which NCEs are transported across the monolayer (Saha & Kou 2002; Neuhoff et al. 2006).

The "bi-directional" approach is typically utilized to assess the transport mechanism in the Caco-2 model where permeability is measured from "apical" to "basolateral" compartments [absorptive permeability, P_{app}(A–B)] and in the reverse direction [secretory permeability, P_{app}(B–A)] (Artursson & Tavelin 2003; Lennernas & Lundgren 2004; Ungell 2004; Steffansen et al. 2004; Hochman et al. 2002; Varma et al. 2006). NCEs that function as substrates for efflux transporters are one of the major concerns in discovery as they may significantly limit molecules from absorption into the enterocytes and GI membrane and eventually retard the exposure (Varma et al. 2006). Historically, efflux ratio (ER) has been utilized to identify NCEs with potential efflux issues

$$\text{ER} = \frac{P_{\text{app}}(\text{B} - \text{A})}{P_{\text{app}}(\text{A} - \text{B})} \tag{1.1}$$

While the classification boundary may vary from lab to lab, NCEs with $\text{ER} \gg 1$ are characteristic of potential efflux substrates and those with $\text{ER} \approx 1$ are dominated by passive mechanism(s). Once oral absorption of a drug candidate or scaffold is limited by efflux-dependent GI permeability, Caco-2 mechanistic studies may help establish SAR, allowing for dialing out the efflux issue by optimization (Hochman et al. 2002). Experimentally, one can identify the transporters (e.g., Pgp) that NCEs may serve as substrates for and also the potential enhancement in oral absorption when primary transporters are inhibited. In addition, ER and P_{app}(A–B) can be assessed at elevated NCE concentrations to investigate the potential of saturated transporters at high dose and eventually build IVIVC for highly soluble or formulated NCEs (Hochman et al. 2002).

As Pgp is, by far, the most prevailing one among transporters in human intestinal enterocytes, one frequently initiates the first transporter mechanistic study by applying potent Pgp inhibitors such as Cyclosporin (first generation), Verapamil and Valspodar (secondary generation), Zosuiquidar/LY335979, and Elacridar/GF120918 (third generation) (Kuppens et al. 2005; Nobili et al. 2006). Troutman and Thakker found that ER is unable to properly characterize the Pgp-inhibition-mediated enhancement of absorptive permeability due to the asymmetric behavior of Pgp substrates in absorptive and secretory transport (Troutman and Thakker 2003a). Instead, absorption quotient (AQ) is recommended to better predict how Pgp-facilitated efflux activity attenuates intestinal permeability in vivo.

$$AQ = \frac{P_{Pgp}(A-B)}{P_{PD}(A-B)} = \frac{P_{PD}(A-B) - P_{app}(A-B)}{P_{PD}(A-B)} \quad (1.2)$$

where P_{Pgp}(A–B) is absorptive permeability (or apparent permeability from apical to basolateral direction) attributed to Pgp activity and P_{PD}(A–B) is absorptive permeability measured in the presence of Pgp inhibitor(s).

AQ differentiates the absorptive permeability from secretory transport and offers a more relevant approach to quantify the functional activity of transporters such as Pgp observed during absorptive permeability. For a potent transporter (e.g., Pgp) substrate, inhibition of transporter (or Pgp) activity usually leads to drastically enhanced absorptive permeability [$P_{PD}(A–B) \gg P_{app}(A–B)$] and thereby $AQ \approx 1$. On the other hand, comparable absorptive permeability values are anticipated (in the presence or absence of transporter inhibitor) for a weak transport substrate [$P_{PD}(A–B) \approx P_{app}(A–B)$] or $AQ \approx 0$.

MDRI-MDCK

The Madin-Darby canine kidney (MDCK) model originating from dog kidney with different expression of transporters than human intestine (Balimane et al. 2000; Irvine et al. 1999) has commonly been utilized for permeability evaluation of NCEs using the passive diffusion mechanisms (Ungell & Karlsson 2003). Recent developments using Pgp-transfected multidrug resistance-1 (MDRI)-MDCK model (Bohets et al. 2001) allows for estimating the contributions of efflux transporters with reduced cell culturing cycle time (3–5 days). By regulating the level of Pgp expression, the sensitivity of NCEs to Pgp in the MDCK assay may be amplified, although its relevance to GI physiology has yet to be established.

Given the advantages and limitations of each approach, the latest consensus appears to favor a strategy that combines all three approaches with in silico models to ensure high quality assessment of permeability in early discovery (Kerns et al. 2004; Balimane et al. 2006; Faller et al. 2007). PAMPA should serve as a fast and high-throughput permeability ranking tool in particular for scaffolds using passive diffusion mechanisms. The Caco-2 model should be applied to challenging scaffolds involving active transport mechanisms or with higher molecular weight (e.g., >600). The former (potential substrates/inhibitors for efflux transporters), most likely, will exhibit "medium" to "high" permeability in PAMPA but poor in vitro–in vivo correlation. Caco-2 mechanistic studies are valuable to identify the major transporters such as Pgp, MRP2 and BCRP and to appraise the impact of shutting-down active transporters via either inhibitory (Varma et al. 2003) or saturation mechanisms (Bourdet & Thakker 2006). MDCK, expressed with a specific transporter, may be ideal to tackle the impact of an individual transporter subsequent to Caco-2 transporter assays (Varma et al. 2005).

1.3.2.5 Oxidative Metabolism

Metabolism is an important intrinsic property that drives the clearance or elimination of NCEs. Two of the major sites of metabolism for orally administered compounds are in the gastrointestinal tract (sometimes referred to as pre-systemic metabolism) and the liver (also known as first-pass metabolism). Finding the right balance of metabolic stability is critical during developability assessment of NCEs as high metabolism (or rapid clearance) could result in poor bioavailability and poor efficacy whereas very low metabolism, along with entero-hepatic circulation, could lead to prolonged half-life and accumulation of NCEs in the body resulting in undesirable side-effects. Metabolism is a highly species-dependent phenomenon. Metabolic rates can vary significantly among different species, due to the presence of unique metabolizing enzymes in each species, strain and gender (Martignoni et al. 2006). For example, CYP3A4 is the most important metabolizing enzyme in humans and current literature suggests that more than 50 % of the marketed drugs are metabolized by CYP3A4 enzyme. However, CYP3A4 is not found in any animal species other than humans and monkeys as described in Table 1.3 (Martignoni et al. 2006; Emoto et al 2013). Typically, rodents have a higher metabolic rate than dogs, monkeys, and humans. However, the rate and extent of metabolism and ranking of species can vary significantly from one chemical structure series to another. In particular, for compounds that exhibit significant species differences in their metabolism, it is very difficult to extrapolate the PK parameters in human and obtain any reasonable estimate of human dose. In such instances, it is advisable to conduct micro dosing or exploratory study in human as soon as possible. Screening of metabolic stability in multiple animal species early in drug discovery is very useful to guide structural modification and selection of compounds for in vivo studies. Metabolite identification is also very helpful for bioanalytical and pharmaceutical scientists to understand metabolically labile as well as chemically labile sites in a NCE. Metabolism is a predominant factor that differentiates oral absorption from oral bioavailability. For instance, a compound could have complete absorption (~100 % Fa or fraction absorbed) through gastrointestinal tract upon oral dosing based on its high solubility and permeability. However, the oral bioavailability for same compound could be only ~20–30 % due to its high first-pass metabolism prior to reaching systemic circulation. This is an extremely important consideration while developing formulations for poorly soluble NCEs. Majority of the solubility enhancing formulation technologies can potentially maximize the absorption (or fraction absorbed Fa%) of a given NCE, but not necessarily their oral bioavailability. The collection and use of data relevant to metabolism is more thoroughly covered in Chap. 4.

Table 1.3 Species-dependent CYP3A enzymes (*m* male specific, *f* female specific)

Human	Mouse	Rat	Dog	Monkey
3A4	3A11	3A1/3A23	3A12	3A4
3A5	3A13	3A2m	3A26	
3A7	3A16	3A9f		
3A43	3A25	3A18m		
	3A41	3A62		
	3A44			

1.3.3 Solid-State Characterization

Solid-state characterization assays and technologies (Table 1.4) are well known to most scientists responsible for solid form selection in support of CMC development activities. Thermogravimetric analysis (TGA), Differential Scanning Calorimetry (DSC), Powder and single-crystal X-ray diffraction are commonly used tools to characterize the solid forms of any given NCE using small amounts of material. Advanced calorimetric methods such as modulated DSC (mDSC) provide higher sensitivity and can measure the heat of fusion of crystalline solids as well as glass transition events in amorphous solids. Hot stage and infrared microscopy are useful tools to study solid form conversions (e.g., polymorphs). Dynamic vapor sorption (DVS) is a routinely used instrument to measure hygroscopicity behavior of solid forms and provides useful information for solid form selection activities.

1.3.3.1 Polymorphism

Polymorphism studies require special attention during developability assessment of NCEs because changes in physical form during development could result in significant delays and often, it is very difficult to reproduce the original form with optimal physicochemical and biopharmaceutical performance. The authors' experience suggests that the larger the scale of a crystallization step and longer the processing time, greater is the likelihood that such process will generate the thermodynamically preferred physical form. The thermodynamically preferred polymorphic form can be ascertained from a simple bridging experiment where two polymorphic forms are placed on a microscope slide in contact with a common solvent. In this case, the low-energy form grows at the expense of the high-energy form. Hot-stage microscopy and IR microscopy are useful instruments in monitoring these conversions. In some cases, the low-energy form may be obtained only through the use of hazardous solvents and/or procedures and hence may not be practical for large-scale production. Therefore, a thorough evaluation of polymorphic space and its impact on physicochemical, biopharmaceutical and processability of a given NCE is an important component of developability assessment.

Table 1.4 Solid-state characterization assays and technologies

1	Crystallinity: XRPD, DSC, Tg, TGA, IR, and microscopy
2	Single-crystal X-ray: structure/unit cell
3	Topography: scanning electron microscopy
4	Particle size analysis: by laser light scattering
5	Polymorphism and salt/co-crystal screening: manual, HTS
6	Bulk density, elasticity
7	Rationale for selecting the preferred crystalline form
8	Particle size recommendation for biopharmaceutics performance or manufacturability of a dosage form

1.3.3.2 Amorphous Solids

Amorphous materials have become more prevalent in the development pipeline as a result of the "hydrophobic" nature of NCEs originating from discovery. Such compounds are neither hydrophilic nor lipophilic and present unique challenges in their formulation development, especially, overcoming their poor solubility. Generation and stabilization of amorphous solid forms, using polymeric matrices in solid dispersion for example, provides an attractive formulation approach to improve the dissolution, solubility and oral bioavailability of these hydrophobic NCEs. APIs can also be developed as amorphous forms if supported by detailed understanding of the amorphous system and robust scalable process for manufacturing these forms. This approach becomes inevitable especially when NCEs are difficult to crystallize due to molecular complexity or presence of trace impurities that act as crystallization inhibitors. Amorphous materials, with increased dissolution rate and aqueous solubility are chemically reactive and more hygroscopic than crystalline material (Byrn et al. 1999; Hancock and Parks 2000; Hancock and Zografi 1997). Amorphous materials exist in either the glassy state below their glass-transition temperature (Tg) or as a super-cooled liquid above their Tg. Although the physical properties differ between each amorphous state, Arrhenius relationships are applicable below the Tg and allow for extrapolation to ambient storage/handling conditions. The ratio of Tg, in Kelvin, to the melting point of a crystalline material is a constant of 0.72 (range 0.59–0.84). This is a useful rule of thumb to estimate feasibility and likelihood of developing an amorphous API or formulation for a given NCE. Analysis of amorphous forms of drug candidates should include, in particular, the measurement of Tg (most commonly using mDSC) and any changes in water solubility, hygroscopicity, and solid-state stability relative to the crystalline form. Water solubility may be the most difficult parameter to measure for an amorphous material because of rapid crystallization (Hancock and Parks 2000). In summary, the know-how and experience in development and characterization of amorphous APIs as well as solid dispersion formulations has progressed significantly in last decade and more than a dozen products have been launched in recent past using these technologies. These are discussed in Chap. 3.

1.3.3.3 Hygroscopicity

A detailed evaluation of hygroscopicity of NCEs and various solid forms is essential for optimal physicochemical behavior (e.g., solubility, physical and chemical stability) as well as processability and manufacturing (e.g., control and reproducibility of desired polymorphic form). Hygroscopicity, simply termed as the study of moisture uptake as a function of percent relative humidity, can be measured in an automated manner with dynamic moisture sorption analyzers that quickly assess the hygroscopicity of material in a closed system at controlled temperature and ambient or controlled pressure. These instruments allow the measurement of the weight change kinetics and equilibration for small samples exposed to a stepwise change in humidity. The authors would like to highlight two critical points to consider during DVS measurements: (1) drying the sample at the start of the measurement cycle and (2) failing to achieve equilibrium at each humidity condition, particularly at extreme humidity. For example, drying of a hydrated material at the beginning of DVS experiment could result in form conversion to anhydrous state and provide a hygroscopicity profile that is not representative of the hydrated form under ambient conditions of storage and handling. Equilibration of the system at each humidity condition can be confirmed by analysis of the sample weight versus time. This aspect is often overlooked, and can lead to confusion when samples from the same bulk lot appear to absorb different amounts of moisture, usually in different labs, under the same humidity stress. Hygroscopicity evaluation should start with an independent determination of the initial moisture content (TGA, Karl Fischer, etc.). Analysis of powder X-ray diffraction patterns of solid forms before and after DVS measurement also provides very useful information in detecting solid form transitions especially, the commonly observed transitions of dehydration and/or hydrate formation.

1.3.3.4 Particle Size

Particle size and distribution is an important solid-state property that heavily impacts the dissolution behavior, flowability, and processability of APIs. Low-magnification scanning electron microscopy (SEM) provides a simple record of particle size and crystal shape. Often a change in the crystallization process results in a change in crystal morphology with concurrent changes in powder flow. Low-magnification SEM can readily reveal morphological changes and alert solid-state/crystallization expert and the formulator. A number of automated particle-sizing methods are now available, each with its inherent shape limitations. Usually, the particle sizing methods calculate particle size distributions by normalizing the shape to an equivalent spherical particle. Based on actual morphology of particles (e.g., needle-shaped long crystals), such measurements can be limited in their accuracy. Nonetheless, these methods allow the counting of many particles in a short period of time and provide good quality control feedback on the reproducibility of a manufacturing process. Surface area analysis methods, such as Brunaur–Emmett–Teller (BET) also provide useful insight into changes in available surface area due to changes in chemical processing. The particle size recommendation for development is derived from the type of dosage form and impact of particle size on fraction absorbed/bioavailability. The particle size recommendation should assure a homogeneous blend for an oral dosage form, particularly for low strength tablets or capsule dosage forms. For poorly absorbed compounds due to low permeability, reducing the particle size may have no effect on the percent of dose absorbed, but may be necessary from manufacturing stand point such as blend homogeneity and content uniformity of the formulated product. The most significant effect of particle size on absorption is typically observed for low dose–low solubility compounds. A number of in silico methods (e.g., Noyes–Whitney equation, Dissolution number using GastroPlus™) can be applied during discovery stages and identify need for particle size reduction for improving dissolution and bioavailability. Such early guidance can be extremely useful to API and formulation development teams so that appropriate particle size reduction technologies can be incorporated in the manufacturing processes as early as IND-enabling or Phase I stages. However, it should be noted that particle size recommendations may change as the clinical dose is refined based on outcome of early clinical studies (Phase I/IIa). Typically for discovery stage compounds, flow properties are usually poor and therefore, the measurement of flow properties during early stages does not add much value to the developability assessment package. Tapped bulk density is a key physical measurement that can be performed during solid form screening activities in order to rank-order and select preferred crystalline forms as well as support the development of capsule dosage form.

Overall, solid-state properties of any given NCE are very critical for their biopharmaceutical performance and manufacturability into a suitable dosage form. Solid-state characterization is an important activity that should be performed during lead optimization stages especially on few selected advanced compounds. The results and recommendations from this evaluation should be shared with research and development colleagues in a timely manner and added to the overall developability assessment package prior to candidate selection. Although, sub-optimal solid-state properties by themselves do not create a no-go scenario for NCEs, it is critical for research and development teams to realize that selecting a candidate with poor solid-state properties may require upfront investment of time and resources in selecting an optimal solid form for development and more often than not, result in higher risk and longer development timelines.

1.3.3.5 Solid Form Screening and Selection

Screening and selection of developable solid form is an essential component of developability assessment. Due to availability and advancement of high-throughput screening technologies, it is now possible to screen several hundreds and potentially thousands of solid forms including salts, polymorphs, co-crystals, etc., using milligram amounts of API (Balbach and Korn 2004). Results from these early screens are typically followed up with more manual and scalable screening methods before final selection of suitable solid form for early development phase. In an ideal situation, it is expected that the solid form selected during early phase is the final form and should be used throughout the NCEs development cycle up to commercialization. Although this might be the case for some NCEs, more realistically, the solid form activities follow a continuum of increasing form space knowledge as an NCE progresses from discovery to early, mid and late development stages (Singhal and Curatalo 2004). Some of the key objectives of solid form activities during discovery and early development include crystallization of amorphous APIs, producing crystalline material to enable purification and isolation, solubility/bioavailability enhancement to support PK/tox studies, discovery and characterization of important crystalline forms (e.g., hydrates, anhydrous, solvates, etc.), understanding form-relationships (e.g., kinetic and thermodynamic) and optimization of physicochemical attributes such as stability, hygroscopicity, dosage form compatibility, etc.

1.3.4 Formulation Approaches for Preclinical Studies

1.3.4.1 Formulation for PK Studies

Low-dose pharmacokinetic studies in rodents (rats and mice) are usually performed during early discovery stages to understand ADME properties of NCEs, to guide SAR and to support advancement of candidates into efficacy and tolerability testing. An example of a decision tree to select NCEs for further profiling based on low-dose pharmacokinetic studies in rodents is highlighted in Fig. 1.2. Due to the large number of such studies performed for each discovery project, limited amounts of compound availability (~1–5 mg) and requirement for a rapid turnaround on formulation requests (~24–48 h), it is unreasonable to expect a formulator to design the most optimized dosing formulation for each and every compound. Therefore, it is a common practice in pharmaceutical companies to identify a set of ~3–4 pre-defined formulations containing different combinations of cosolvents (e.g., dimethyl acetamide, *N*-methyl pyrrolidone, ethanol, propylene glycol, polyethylene glycols, etc.), surfactants (e.g., Tween-80, Cremophor EL, Solutol HS 15, etc.) and aqueous diluents (e.g., buffers, saline, dextrose, etc.) that could be readily used to develop solution formulations for low-dose PK studies via multiple routes of administration including i.v, p.o., s.c., i.m. and i.p. The topic of formulations is treated in greater detail in Chap. 2.

One of the important considerations while developing a solution formulation is the physical and chemical stability of NCE during dosing as well as under physiological conditions. A quick dilution test of solution formulations in simulated physiological media (e.g., 1:1, 1:5 and 1:10 dilution in phosphate buffered saline, simulated gastric and intestinal fluids) and lack of any visible precipitation typically

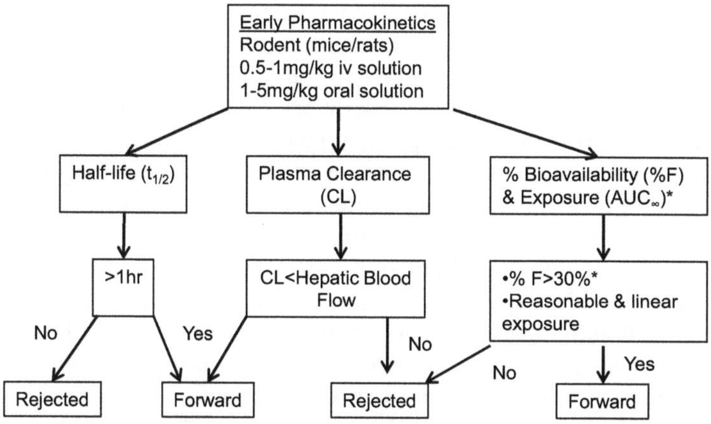

*Oral bioavailability >30% in mice is advisable, however, these values might differ based on disease areas and in case of highly potent molecules.

• AUC$_\infty$ - Area Under the Curve (till infinite time point)

Hepatic blood flow (ml/min/kg)	
Mouse (20-25g)	90
Rat (200-300g)	85
Rabbit (2.5-3kg)	71
Dog (10-12kg)	31
Monkey (4-5kg)	44

Fig. 1.2 Example NCEs selection process based on early PK studies in rodent species. Reprinted from Saxena et al. (2009) with permission from John Wiley & Sons

provides a formulator enough confidence to advance these formulations into PK study. If the PK study involves administration of a poorly soluble compound using i.v. infusion, a more detailed investigation of precipitation kinetics, in both static and dynamic systems might be required. It is important that early PK studies (i.v. and p.o.) are performed using a completely solubilized formulation. This approach ensures that physicochemical properties such as crystallinity, particle size, dissolution kinetics, etc., will not be the limiting factors in absorption of NCE and root-cause for poor absorption and/or poor bioavailability can be attributed to other important parameters such as permeability, metabolism, etc. This information is very useful for medicinal chemists to build SAR, enable rapid decision-making (selection or termination) and to incorporate right biopharmaceutical properties in the NCEs from the very beginning. Chapter 4 addresses these issues in more detail.

1.3.4.2 Formulation for Pharmacology Studies

In comparison to low-dose PK studies, pharmacology studies are more diverse in their study designs, dose requirements, duration and selection of animal species.

Depending on target indication, pharmacology studies may require single dose or multiple doses up to ~2 weeks. Dosing concentrations are typically higher than those required for low-dose PK studies and are predominantly driven by the potency, PK properties (e.g., rate of absorption, clearance, etc.) and maximum tolerable dose of a given NCE. The primary goal of initial pharmacology studies is to identify a minimal efficacious dose in a given animal model. More detailed studies to establish PK/PD relationships, dose–response, etc., are usually performed during later discovery stages. The types of dosage forms used in pharmacology studies could vary significantly from study to study. It is not uncommon for a scientist in developability assessment group to develop formulations ranging from a simple solution in buffers or cosolvents or an oral suspension of API to enabling formulations such as spray–dried dispersions or microemulsion pre-concentrates or high organic mixtures to be incorporated into subcutaneous osmotic pump devices. In addition to the solubility, dilution behavior, condition-of-use stability and tolerability, another important aspect to consider during development of pharmacology formulations is the likelihood of interfering or masking the pharmacodynamic response or efficacy of an NCE due to presence of certain excipients in the dosing vehicle. For example, it is advisable to avoid the use of sugars or lipid-based vehicles in animal models for diabetes and metabolic disease indications or to avoid certain surfactant-based excipients in animal models of pain. Another important formulation consideration for pharmacology studies is the route of administration. For example, a poorly soluble NCE may not demonstrate desired PD/efficacy response from oral dosing of a simple suspension formulation due to lack of sufficient systemic exposure and may therefore, require the use of enabling technologies (e.g., solid dispersion, microemulsions or nanosuspensions) to achieve desired outcome in animal study. In such instances, it is very important that the organization has right expertise to enable the molecule and validate the biological target in preclinical species. It is highly likely that the enabling formulation developed for pharmacology studies could very well be used for longer-term PD/efficacy studies as well as acute and chronic toxicity studies.

1.3.4.3 Formulation for Toxicology Studies

Toxicology studies in rodents and non-rodents are required by regulatory agencies before any NCE can be evaluated in a clinical study. Typically, such studies are performed in a step-wise manner from candidate selection to IND-enabling stages. Early toxicology studies during discovery stages could include a single-dose acute tolerability assessment to support pharmacology studies or repeat-dose study for ~4 or 7 days to address some specific safety concerns for a particular target or lead candidate. Non-GLP studies (typical duration ~7 days) in rodents and non-rodents are generally performed on ~1–3 candidates prior to development candidate nomination. IND-enabling studies are performed under GLP conditions and require repeat-dosing anywhere from 7 to 28 days depending on intended clinical indication. One of the primary goals of toxicology studies is to identify potential safety concerns and define the no adverse effect level (NOAEL) for any given NCE and therefore, such studies require administration of very high doses and high systemic exposures from selected formulation. In addition, it is also desirable to provide a benign formulation as these studies involve repeat-dosing for longer durations. For NCEs being developed for oral administration, a simple suspension formulation of a stable crystalline form (salt or free base) is the most attractive option. However, as many discovery compounds are poorly soluble, a simple suspension may not provide adequate systemic exposures especially at high doses, and therefore, such compounds will require a solubilized or enabling formulation for their toxicology evaluation. As previously noted, if a compound with sub-optimal physicochemical or biopharmaceutical properties requires an enabling formulation for its pharmacology and/or toxicology evaluation in preclinical models, it is highly likely that similar formulation principle will be required for Phase I clinical study. Single and multiple ascending dose studies in human, in principle, have similar objectives, which is to understand safety and PK of a NCE by administering high doses of a compound in ascending fashion and achieve high systemic exposures to widen the therapeutic index and support dose selection for subsequent clinical studies. In author's experience, this approach of using same enabling formulation principle from preclinical to clinical study can be quite useful in accelerating early development of poorly soluble NCEs and get faster read-out on their PK, safety and in some cases, proof-of-concept for target engagement or efficacy in humans. It should be noted that early involvement of clinical team is essential for the success of this approach as they can proactively identify appropriate study sites and modify study protocols to support the preparation and dosing of enabling formulations.

1.3.4.4 Enabling Formulation Approaches

As noted earlier, development of preclinical formulations requires consideration of several factors including highest dose required, duration of study, animal species, tolerability of excipients, etc. Based on a compound's intrinsic properties, a number of formulation approaches are available to develop robust toxicology formulations and are described below and in Chaps. 2 and 3.

Conventional Approaches

The majority of compounds with high aqueous solubility are formulated in physiological buffers or saline or dextrose for oral as well as IV administration at 1 mg/ml and higher concentrations. NCEs with intrinsic solubility less than 0.1 mg/ml require cosolvents and surfactants and are formulated as a stock solution at 5 or 10 mg/ml and further diluted to the desired concentration using saline or dextrose or buffers (Strickley 2004). For a single dose study, higher percentages of cosolvent/surfactant mixtures are well tolerated (confirm LD50 for cosolvents). However, for repeat-dose studies, it is preferable to have 70 % of aqueous phase (e.g., buffer, saline, dextrose, etc.) in the solution formulation. Higher percentages and/or amounts of cosolvents or surfactants for repeat dosing could have agonist/additive effect on toxicity (Brewster et al. 2007).

For an oral suspension formulation, the concentration of compound is not a limiting factor, the only exception being very high viscosity of dosing suspensions at very high concentrations that could result in technical hurdles while dosing smaller animals especially rodents. In such scenarios, it is likely that the viscosity of suspensions will potentially define the maximum feasible dose that can be achieved in a tox study. Typically, compounds can be formulated in suspensions at concentrations ranging anywhere from 1 to 300 mg/ml. Commonly used suspending media include low concentrations of polymers (e.g., 0.1–1 % of celluloses such as Na CMC, MC, HPC and low viscosity HPMC) along with small amounts of surfactants (e.g., 0.1–0.5 % of Tween-80, SLS, etc.). The most important parameters for a suspension formulation include pH, particle size and distribution, stability of physical form (amorphous vs. crystalline) during preparation and dosing and solubilized fraction of NCE in suspension which in turn helps to understand its absorption and systemic exposure in vivo.

Salts and Co-crystals

One of the primary reasons to conduct salt and co-crystal profiling for a lead compound is to identify solid forms with improved dissolution rates and higher solubility in order to overcome the poor biopharmaceutical properties of the original free acid or base form. A quick and effective way to identify the need for salt or co-crystal form profiling is to perform a pharmacokinetic study in rodents comparing a solution vs. a suspension formulation of the crystalline free acid or base form of a NCE. If the suspension arm demonstrates approximately twofold or less reduction in systemic exposure as compared to the solution arm, such compound is a good candidate for salt/co-crystal profiling. In cases where lead NCE is available in amorphous form, it is highly recommended to perform an early salt/crystallization screening to identify suitable crystalline forms prior to the pharmacokinetic evaluation of a solution vs. suspension formulation. Another important reason for identifying a salt or co-crystal form is to improve physiochemical properties of the original free form. This leads to lower likelihood of encountering complications during their processing, manufacturing, and storage and thus providing convenience/suitability for development purposes.

An early profiling of the alternative salt forms for lead compounds is done using a few common acidic or basic counter ions selected based on the p*K*a of the compound (Morris et al 1994). The goal is to achieve a salt form, which is more soluble and dissolves more rapidly in the physiological pH environment, crystalline, less hygroscopic, and more stable under accelerated degradation conditions than the original form of the compound (Serajuddin 2007). Once the different salt forms of a compound are synthesized, they are evaluated for solubility, stability, and physical characterization. The salt form that shows the most desired physical characteristics, increased solubility/dissolution rate, and stability is selected for PK studies. PK studies typically involve measurement of oral bioavailability (%F) from oral suspensions of selected salt forms vs. suspensions of free form at a high dose, followed by dose linearity study to demonstrate the likelihood of achieving enough exposures for safety assessment of NCE. If two or more salt forms have similar physicochemical attributes, a bridging PK study may be performed to select the better salt candidate. Eventually, the solid form that gives dose linear exposure is scalable and demonstrates suitable physicochemical stability is selected for further development.

Nanosuspensions

Nanosuspensions are widely used as an enabling formulation technology to improve dissolution rate/bioavailability of poorly soluble compounds when given orally. However, in recent times, nanosuspensions have been administered via i.v. bolus or i.v. infusion to enable parenteral delivery of poorly soluble candidates in certain disease areas (Wong et al. 2008). The authors' experience on nanosuspension technology for oral as well as i.v. is described as follows.

For Oral route of administration, compounds which have been developed as nanosuspensions historically share some common physicochemical properties such as neutral, Zwitterionic or very weakly basic functional groups (pKa < 3.0), melting point typically less than 200 °C or enthalpy less than 50–70 J/g. Compounds with higher melting points (greater than 200 °C) or higher enthalpies (greater than 80 J/g) did not show dose linear exposures from nanosuspension as eventually nanoparticles need to dissolve and get into solution prior to absorption at physiological conditions. These observations are purely based on the authors' experience on more than 1,000 molecules coming from research for PK/PD studies every year; however, there are always exceptions to above findings. For example, one exception to the above pattern was observed with a compound that in addition to its high melting point and high enthalpy was highly lipophilic and demonstrated higher solubility in physiological media (e.g., fed-state intestinal media and bile salts). As a result, higher exposure was observed in vivo for this compound when dosed as a nanosuspension (Junghanns and Müller 2008). Compounds with stronger basicity or acidity may not be good candidates for developing oral nanosuspensions as desired improvement in dissolution and biopharmaceutical performance can be achieved by simply micronization or salt formation. Compounds with basic functional groups when dosed as nanosuspension will most likely dissolve/solubilize in gastric pH of stomach and reprecipitate under intestinal pH conditions with a completely different particle size distribution. Similarly, acidic compounds will dissolve readily under intestinal pH conditions and may not benefit significantly from nanosized formulation. In summary, the physicochemical nature of a NCE, its thermodynamic properties and solubility behavior in simulated physiological media are the most important factors while selecting a nanosuspension formulation for oral dosing.

For parenteral nanosuspensions, compounds with high plasma protein binding (>99.9) are typically good candidates as they can be readily dissolved in the blood stream and chances of recrystallization during their distribution to site of action are minimal. If not dissolved readily, nanocrystalline particles may circulate for longer durations in plasma and as a result, may accumulate in major organs such as liver, lungs, kidneys, etc., which could be detrimental especially with higher doses (greater than 100 mg/kg in preclinical and greater than 500 mg in human via parenteral route). The generation of nanosuspensions appropriate to the discovery environment is described in Chap. 3.

Solid Dispersions

Another widely used enabling formulation technology for oral absorption enhancement is the solid dispersion technology, wherein a crystalline API is converted into an amorphous form and stabilized in a polymeric matrix using a variety of processing techniques (e.g., solvent evaporation, spray drying, melt extrusion, etc.). The generation and stabilization of metastable amorphous forms in turn leads to improved dissolution/solubility and maintenance of supersaturated state for desired length of time in GI tract (typically 1–2 h) to ensure complete absorption. Solid dispersions are usually administered in a solid dosage form (e.g., tablet or capsule) in clinical setting, however, such dosage forms are not technically feasible to be used in rodents, especially for repeat-dose toxicity studies. In such instances, administering the solid dispersion as a suspension is the only feasible alternative approach. A suitable suspending agent for dosing a solid dispersion should maintain a high active concentration for a reasonable period of time (approximately 2 h for animal dosing) without precipitation of crystalline drug. Commonly used vehicles for preparing suspensions of solid dispersion formulations include purified water, 0.5 % w/v methylcellulose (MC), 0.5 % w/v MC/0.5 % Tween-80, 0.5 % w/v polyvinyl pyrrolidone (PVP), 0.5 % w/v Tween-80, and 0.5 % w/v SLS. The presence of metastable amorphous forms of API in solid dispersions and their long-term physical stability (or lack of) is a major concern for further development. Therefore, it is highly recommended to perform a detailed investigation of their physical as well as chemical stability under ambient and accelerated conditions using HPLC-MS, XRPD, and DSC. As a general guideline, a NCE should maintain its amorphous form in the solid dispersion after storage at 40 °C/75 % RH for 3 and 12 months at RT in a desiccator. It is also important to note that solid dispersions might absorb a significant amount of water due to the hygroscopicity of the polymer and form a gel. In order to confirm the physical stability of solid dispersions, the mDSC profile should demonstrate minimal changes in the glass transition temperature and absence of any melting endotherms. Such a profile is indicative of a completely miscible system without any phase separation. In addition, chemical stability of NCEs in solid dispersion should be demonstrated under studied conditions by HPLC-MS assay. Final selection of solid dispersion formulation for in vivo studies should be based on the higher dissolution rate/kinetic solubility up to 2–4 h in physiological pH as well as its promising physicochemical properties such as stability under accelerated conditions. The authors' experience suggests that NCEs formulated using solid dispersion technology are usually recommended as powder in a bottle for Phase I studies as doses are high and filling solid dispersion into a capsule could be problematic due to a large number of capsules to be administered orally. Also, water content in gelatin/HPMC capsules (4–12 %) could cause long-term stability issues. For later stages of development and commercialization, tablets are the preferred dosage forms for NCEs formulated as solid dispersions. The manufacture and characterization of amorphous dispersions in discovery is thoroughly discussed in Chap. 3.

Lipid-Based Formulations

Lipid-based formulations for toxicology studies are mostly developed as microemulsion pre-concentrates (MEPC), also known as self-emulsifying drug delivery systems (SEDDS). In general, MEPC or SEDDS contain ternary mixtures of oils, surfactants, and cosolvents in which high amounts of poorly soluble API can be dissolved and maintained in solution. Once administered in vivo, the pre-concentrates come in contact with aqueous fluids in the GI tract and undergo spontaneous emulsification to generate nano- or micro-sized colloidal particles that enable rapid and complete absorption. For compounds that undergo rapid re-crystallization in vivo, supersaturated SEDDS formulations have been identified wherein a suitable crystallization inhibitor (such as PVP or HPMC) can be incorporated in the SEDDS mixture to extend in vivo supersaturation and achieve higher absorption and bioavailability (Gao and Shi 2012). Typically, NCEs with $C \log P$ values >5 are considered as good candidates for MEPC or SEDDS. NCEs with lower lipophilicity can be formulated as MEPC or SEDDS; however high drug loading may not be achieved. Chemical stability of a NCE in the liquid pre-concentrate vehicles and their individual components is another critical parameter to determine feasibility of MEPC or SEDDS approach.

In general to identify the components of MEPC, solubility screening of a molecule is done in GRAS excipients especially oils, glycerides (mono, di, and tri), surfactants and cosolvents. These excipients are listed in Table 1.5. Based on the solubility in single component systems and acceptable excipient amounts for preclinical and clinical studies, several permutations and combinations need to be screened. A few most promising MEPCs/SEDDS based on solubility/stability screen should be selected for further studies. The selection of final formulation for toxicity studies should be based on drug solubility, particle size distribution, and physicochemical stability of the diluted MEPC/SEDDS systems in water and simulated gastric fluids (typically 1:9 dilution). For example, a poorly soluble API when formulated in a MEPC/SEDDS formulation resulted in a translucent micro emulsion upon dilution with water (droplet size ~20 nm) and demonstrated dose linear exposures in preclinical studies (Zheng et al. 2012).

Table 1.5 Commonly used excipients in microemulsion pre-concentrates (MEPC) or self-emulsifying drug delivery systems (SEDDS)

Number	Excipients
1	Cremophor EL
2	Tween 80 (polysorbate 80)
3	Labrasol
4	Brij 98 50 % w/w in ethanol
5	Gelucire 44/14 50 % w/w in ethanol
6	Solutol HS 15 50 % w/w in ethanol
7	Vitamin E TPGS 50 % w/w in ethanol
8	Ethanol
9	Propylene glycol
10	PEG 400
11	Transcutol HP
12	Arlasolve (dimethyl isosorbide)
13	Corn oil
14	Maisoel glyceride
15	Olive oil
16	Sesame oil
17	Soybean oil
18	Miglyol 812
19	Peceol (glyceryl monooleate)
20	Maisine 35-1 (glyceryl monolinoleate)
21	Capmul MCM
22	Lauroglycol FCC
23	Labrafil M-1944 CS
24	Labrafil M-2125 CS
25	Triacetin (glycerol triacetate)

The authors' experience suggests that for Phase I/II studies MEPC/SEDDS are recommended as a drink solution in a bottle. After confirmation of the efficacious dose, MEPC/SEDDS could be formulated into a soft gelatin capsule with a volume of 0.8–1.2 ml. Hence, it is essential that very high solubility and drug loading can be achieved in the selected MEPC/SEDDS combinations. If efficacious doses selected for later stages of development or commercialization are higher than 100 mg, the MEPC/SEDDS in soft gelatin capsule approach can become less attractive option as multiple units may be required to support the dosing regimen. Despite these limitations, MEPC/SEDDS provide a viable formulation strategy for development and commercialization, especially for NCEs with the potential to be first-in-class therapy for lifesaving indications.

1.3.4.5 High-throughput Screening (HTS)

A discussion on enabling solid form and formulation technologies for toxicology and early clinical studies will not be complete without highlighting the contributions of high-throughput screening technologies, automation, and miniaturization. High-throughput screening has been routinely used in drug discovery over several decades. However, its utility in areas of solid form screening (e.g., salts, polymorphs) and formulation screening (e.g., solid dispersion) has been realized more recently (in past 10–15 years). Several fast-pace HTS platforms along with state-of-the-art analytical tools are now available to the developability assessment scientists that can provide valuable information with very small amounts of material in a short duration of time (Korn and Balbach 2004, 2014). Some of the major HTS suppliers for liquid and solid handling are Zinsser Analytics, Tecan, Emerald Biosystems, Biodot, Freeslate (former Symyx), and Chemspeed. Use of robots in conjunction with solid and liquid dispensers can significantly enhance the efficiency/productivity of developability assessment groups and enable the scientists to explore vast experimental spaces that would not be practically possible using traditional manual approach. For example, identification of crystalline solid form can be a daunting task for a discovery stage NCE and may require significant amounts of labor-intensive steps and utilize several grams of API. However, with the availability of HTS platforms (96-well format), it is now possible to screen hundreds and potentially thousands of experimental conditions (e.g., solvents, temperature, etc.) using milligram amounts of API and increase the likelihood of success in identifying desired crystalline form (Zheng et al. 2012). An example of analytical equipment miniaturization is the micro-dissolution apparatus (from pION) which reduces the amount of material needed for assessment of how a compound will dissolve in the gastrointestinal environment. Miniaturized dissolution helps in selecting and rank-ordering different salts or co-crystal forms based on their respective dissolution rates in simulated physiological conditions. It also helps understand the rate of precipitation of a compound from different solid forms and formulations under physiological conditions which can then be used in simulation models (e.g., Gastro Plus™) to predict the fraction absorbed/bioavailability (Mathias et al. 2013).

Another noteworthy application of HTS is the design of solid dispersion formulation by solvent evaporation. Using HTS platforms, scientists can simultaneously screen up to 96 combinations of polymers and surfactants with minimal consumption of the compound. For example, Zheng et al. demonstrated successful application of HTS in developing a solid dispersion formulation for toxicology studies. In their experimental design, a total of 64 variants containing 7 polymers and 7 surfactants were evaluated with a 10–50 % drug load of a compound using a 96-well block. Dissolution rate/solubility of each variant was measured at two time points (30 and 120 min) in physiologically biorelevant media to assess the potential of solid dispersions to achieve a high concentration and the ability to maintain that supersaturation for a length of time that would allow rapid absorption and increased bioavailability. The most promising variants from dissolution and solubility data were selected and scaled up via solvent evaporation using a rotary evaporator or mini-spray dryer (Zheng et al. 2012).

1.3.5 *Recommendations for First-In-Human Studies*

1.3.5.1 Solid Form

As already noted earlier, one of the most important functions of developability assessment group is to recommend a developable solid form for Phase I studies and possibly, for later stages of development and commercialization. The timing and decision-making process for solid form selection could vary from one organization to the other or from one NCE to the other. In the authors' experience, identification of a stable, developable solid form for a given NCE prior to initiation of IND-enabling studies is a value-adding investment as it can significantly accelerate the technical development timelines for API and drug product and at the same time, minimize risks of failure and surprises during later stages of development. Due to advancement and recent development of high-throughput technologies, the selection or recommendation of a stable solid form no longer requires 30–50 g of material and 3–4 months of time. Such activities can now be completed with approximately 2–5 g of material and 4–6 weeks of time (Korn and Balbach 2004, 2014; Zheng et al 2012; Saxena et al. 2009). The following key attributes should be considered and incorporated in the solid form recommendation/selection process while nominating a particular solid form for clinical studies:

(a) Crystalline.
(b) Non-hygroscopic (less than 2 % of moisture uptake up to 90%RH), especially for an oral dosage form, for parenterals, hygroscopicity up to 5–10 % is acceptable.
(c) Reproducible at manufacturing scale from mg to kg scale
(d) Reproducible crystal habit.
(e) Corrosivity: Few strong counter ions could lead to corrosive salts especially halogenated salt. Hence, selection of a non-corrosive salt form is a must.
(f) Hydrate/solvates: stable, reproducible at manufacturing scale.
(g) Physically and chemically stable at accelerated conditions with minimum 2 years of shelf life at room temperature.

1.3.5.2 Biopharmaceutics

As building the right biopharmaceutical properties in NCEs is highly desirable during early discovery stages, it is equally desirable to consider biopharmaceutical properties while selecting the final solid form and formulation principle for clinical studies. With advancement in simulation/modeling software, it is easy to simulate the physiochemical properties of a NCE based on its chemical structure and predict the in vivo absorption/exposure in humans as a function of dose based on solubility, permeability, dissolution, and PK data in animals. These are prophylactic approaches, which could save a lot of time and valuable resources spent in actual experimentation. Moreover, these techniques could help identify the potential issues early on. An example of such simulation software is GastroPlus™ which is useful in predicting the effect of particle size on absorption as a function of dose (Kuentz et al. 2006). This could be useful, especially for compounds with poor solubility and high permeability, where size reduction could lead to enhanced dissolution rate/solubility and oral bioavailability. Thus, such simulation tools could prove to be useful in guiding the formulator in evaluating the right candidate for particle size reduction techniques such as nano/micro-milling. Presently, several important simulation software are at different stages of their evolution and validation. Constant use and understanding of simulation techniques can eventually lead to higher productivity, better utilization of resources, faster screening, and better prediction of possible issues in humans.

In addition to the more sophisticated modeling tools, simple calculations such as maximum absorbable dose (MAD) are very helpful tools for a developability assessment scientist to predict risk of poor oral absorption in human and define formulation goals to overcome such issues. The MAD equation, as shown below, takes into account the permeability, converted to an absorption rate constant (Ka, min^{-1}) (Hilgers et al. 2003), solubility (typically, measured in biorelevant media such as simulated intestinal fluid at 1 h, 37 °C), GI residence time (270 min in human), and GI volume (250 ml in human). Although the original MAD equation was developed to predict oral absorption limitations in human, similar predictions can be made in different animal species based on their respective physiologies especially GI residence time and GI volume. A list of such parameters for different animal species is shown in Table 1.6. Burton and other MAD references (Hilgers et al. 2003; Curatolo 1998; Gu et al. 2007; Johnson and Swindell 1996) cite approximately 270 min for a typical drug absorption window, and this has been maintained here as most T_{max} values in the present study are within 2–3 h. Intestinal volume can be selected based on animal species used to calculate MAD (McConnell et al. 2008). The final maximum absorbable dose is normalized to an average weight of animal models in kilograms, and therefore MAD units are reported as milligrams of compound absorbed per kilograms of animal body weight (mg/kg). Wistar Han rats of ~0.3 kg , ~10 kg beagle dogs, and ~4 kg cyno monkey were utilized for the majority of studies ($N = 3$ per in vivo test was typical).

$$\mathrm{MAD} = \mathrm{Ka}\left(\mathrm{min}^{-1}\right) \times S(\mathrm{mg/ml}) \times \mathrm{SITT}(\mathrm{min}) \times \mathrm{SIWV}(\mathrm{ml}) \qquad (1.3)$$

where MAD = maximum absorbable dose with units of mg (human) or mg/kg (animal species)

(*Note*: *To calculate MAD for animal species, divide the MAD obtained using Eq.* (1.3) *by animal body weight* (*kg*) *to obtain MAD in units of mg/kg*)

Ka = absorption rate constant (0.05 for medium to highly permeable compounds)

S = Solubility (kinetic solubility is more relevant for absorption)

SIWV: Small intestine volume (250 ml in human)

SITT: Small intestine transit time (270 min in human)

Table 1.6 Physiological parameters in multiple species to calculate maximum absorbable dose (MAD)

Parameters	Mouse	Rat	Dog	Monkey	Human
Intestinal volume (ml)	1.5	11.23	480	230	250 (small intestine)
Transit time (min)	90	88	111	180	275
Average body weight (kg)	0.04	0.3	10	5	70

Table 1.7 Maximum absorbable dose (in milligrams) in preclinical species and human

Solubility (mg/ml)	Mouse	Rat	Dog	Monkey	Human
0.0001	0.02	0.02	0.03	0.04	0.3
0.005	0.8	0.8	1.3	2.1	17.2
0.01	1.7	1.6	2.7	4.1	34.4
0.015	2.5	2.5	4.0	6.2	51.6
0.5	84.4	82.4	133.2	207.0	1,718.8
0.75	126.6	123.5	199.8	310.5	2,578.1
1	168.8	164.7	266.4	414.0	3,437.5

Maximum absorbable dose as a function of compound solubility for preclinical species and human is shown in Table 1.7.

1.3.5.3 Formulation Principle

Prior to the selection of a clinical formulation, a number of enabling formulation approaches are usually explored for developing toxicology formulations especially for NCEs with poor solubility/dissolution rate (Sect. 1.3.4.4). However, preclinical tox doses are very high and depending on clinical SAD/MAD doses, the decision needs to be taken proactively what formulation principles needs to be selected for the clinical studies. A simple approach based on pharmacokinetic evaluation of a compound at pharmacological dose with a solution and suspension formulation can serve as a guiding tool in selecting a clinical formulation principle (Table 1.8).

One critical factor that needs to be considered while recommending a clinical formulation is whether the NCE is being developed as a *first-in-class indication* or *best-in-class*. For first-in-class target or indication, it is more important to validate the biological target in preclinical species as well as clinical subjects as soon as possible. Hence one needs to select/recommend the formulation principle which will not be a limiting factor to in vivo exposure/bioavailability. Simple dosage forms such as powder in bottle or drug in capsule are typically recommended in such scenarios. These dosage forms are simple to develop and offer fastest development path to get into the clinic. However, for NCEs anticipated to be *best in class*, as the biological target is validated and the dose is very well understood from preclinical to clinical species, in vitro/in vivo data can be used for prediction of the efficacious dose in human. Hence, the authors' experience indicates that for *best in*

Table 1.8 Prediction of clinical formulation principle based on oral bioavailability comparison from solution vs. suspension formulation of a NCE at pharmacological dose in animal species

Parameter	Scenario 1	Scenario 2	Scenario 3	Scenario 4
Oral bioavailability from solution	High (>80 %)	High (>80 %)	High (>80 %)	Low (<30 %)
Oral bioavailability from suspension	High (>80 %)	Low (<30 %)	Medium (40–60 %)	Low (<30 %)
Conclusion	Easy to develop	Exposure limited by solubility	Exposure limited by dissolution rate	Permeability, efflux, metabolism
Clinical formulation principle (also consider maximum absorbable dose, food effects)	Powder-in-bottle, capsule or tablet	MEPC (microemulsion pre-concentrate), solid solution	Salt, co-crystal, dispersions, micro or nanoparticulate systems	Low likelihood of improving exposure using formulation technologies
%NCE/BCS	5–15 (I/III)	60–75 (II)		<5–10 (IV)

class NCEs, it is recommended to develop a robust formulation for Phase I studies, which could be ultimately developed as a commercialized product.

1.3.5.4 Risks and Mitigation Plans

Along with providing clear recommendations for developable solid form and formulation strategy, developability assessment groups are also responsible for highlighting risks associated with the developability of any given NCE and identify necessary steps to mitigate those risks. Based on the authors' experience, this exercise of risk identification and mitigation planning is best done in a cross-functional team environment comprising of all key contributors (e.g., technical development functions, DMPK/safety groups, etc.). A comprehensive risk assessment can also provide useful guidance to research teams on back-up programs and prioritization.

Current data on discovery and development pipelines in pharmaceutical industry suggest that more than 70 % of NCEs get terminated due to safety issues, 20 % due to efficacy, 5 % due to poor pharmacokinetics, and 5 % due to biopharmaceutical properties/formulation principle (Kola and Landis 2004, Khanna 2012). Often due to complexity of the biological target or presence of on-target side effects, it is very difficult to overcome these issues even with enabling technologies. Firstly, an organization needs to think multiple times before selecting a high risk compound for development especially if the high risk is driven by safety, efficacy, and poor biopharmaceutical performance in human. For any given *first-in-class* target/molecule, the development risk is usually very high from a target validation perspective as well as commercialization success perspective. Hence, in such instances, it is important for the developability assessment group to give feedback to the research team to move on to new scaffolds or explore totally new targets for similar indication. This, of course, should be a team decision and rather than wasting further resources and time, it is better to move on to other promising leads or targets rapidly. This mechanism helps a lot because the sooner the team terminates a molecule or a program; the better it is for the organization to take on new challenges (i.e., for the effective utilization of resources within the individual departments as well as organization). Despite the drawbacks highlighted above in selecting a high-risk candidate, the authors' admit that a certain degree of risk-taking is essential for any organization that thrives to be innovative and wants to develop patient centric therapies and add value in patients' lives.

A number of mitigation approaches can be put in place to de-risk promising NCEs especially the ones intended to be *first-in-class* for unmet medical needs. For example, it is very important to identify a suitable animal model for developing a robust PK/PD relationship which could translate into human. In addition, it is equally important to develop the right biomarkers for target validation in humans. Sometimes, due to novelty and complexity of a particular target or NCE, if none of the above two approaches are feasible, it is advisable to either move away from such targets or if absolutely critical, proceed into the clinic with two to three promising lead candidates using a pharmacodynamic read-out and in parallel invest time and efforts in developing new biomarkers. Some of the commonly observed biopharmaceutical risks for NCEs include lack of a reliable human dose prediction and food effects, especially for NCEs with narrow therapeutic index (TI). Efficacious human doses are usually difficult to predict when NCEs demonstrate significant species differences in their absorption, metabolism and elimination profiles or due to lack of suitable animal models that can demonstrate a PK/PD/efficacy relationship. As a result, it becomes very challenging for technical development functions to develop robust formulation and dosing strategy for early clinical studies. In such instances, it is recommended to proceed faster into the clinic with a simple formulation principle such as powder in bottle, or drink solution in bottle and determine the fate of the molecule sooner rather than later instead of resolving the above issues/liabilities over several years. Food effects on pharmacokinetics of orally administered NCEs can pose significant development hurdles for compounds with narrow TI (Li et al. 2002, Jones et al. 2006, Gu et al. 2007, Lentz 2008). It is always advisable to select a compound or use an enabling technology that can minimize the impact of food on pharmacokinetics and safety that could be detrimental to patients and avoid any potential black-box warnings on commercial drug product, which could lead to a significant competitive disadvantage. If early in vitro and in vivo assessment in preclinical studies suggests the likelihood of food effects, it is recommended that a fed/fasted investigation should be carried out in Phase I studies at predicted efficacious doses.

Lastly, risks associated with the solid form such as unknown or poorly investigated polymorphic landscape should be highlighted and mitigation plans should be put in place to evaluate the polymorphs in a timely manner. For example, a final polymorphic form should be selected prior to pivotal clinical trials and its impact on exposure evaluated in a bridging clinical PK study. Poor physicochemical properties of a molecule such as density, particle size, shape, flow properties, and compressibility, etc., can bring significant challenges to scale up and reproducibility of drug product manufacturing processes. A detailed analysis of the risks and identification of corresponding mitigation plans is warranted as part of developability assessment.

1.4 Enabling Delivery Technologies: Maximizing the Potential of NCEs

It is now a well-known fact that discovery of NCEs with the right balance of potency, safety, physicochemical properties, and biopharmaceutics is a highly challenging and difficult task. Poor physicochemical properties such as solubility or dissolution can be improved using a variety of approaches as noted in earlier sections. In addition to these approaches, a number of enabling delivery technologies are now available with the potential to overcome challenges of poor safety and/or efficacy. The development and commercialization of enabling delivery technologies is a game-changing phenomenon for the pharmaceutical and biotech industry especially in times when organizations continue to struggle in populating their pipeline with developable NCEs. Enabling delivery approaches include but are not limited to targeted delivery, controlled release applications, alternate routes of administration and localized delivery.

The fundamental aim of enabling delivery technologies is to overcome poor pharmacokinetics, efficacy and/or safety profile of an otherwise promising NCE and maximize its potential for intended therapeutic benefit. For example, a NCE with very high potency (e.g., clinical dose <1 mg) but poor oral bioavailability due to hepatic first pass effects could be a great candidate for transdermal delivery. Similarly, a NCE that is delivered intravenously could benefit from an inhaled route of delivery if targeted site of action is in the lungs and high local concentrations are required to maximize its efficacy or TI. It is, however, critical for a project team to address whether a revised target product profile with new route of administration would still be compatible with the organization's strategy and ultimately benefit the patient population. A number of controlled release technologies are available for both oral and parenteral delivery to overcome safety issues (e.g., Cmax driven toxicities), reduce medical care costs, improve patient compliance by reduced frequency of dosing and ease of application. Significant advances have been made in the field of localized delivery (e.g., ocular implants, intra-articular delivery for joint disease and postoperative pain, etc.). Targeted delivery approaches based on nanoparticles has been an area of extensive research for last couple of decades especially in the field of oncology. For example, Doxil® is a pegylated liposomal formulation of doxorubicin that was developed to improve the safety profile by targeting doxorubicin at the site of action.

Nanostructured biomaterials, nanoparticles in particular, have unique physicochemical properties such as ultra-small and controllable size (typically in the 1–100 nm range) (Zhang et al. 2008), large surface area to mass ratio, high reactivity, and functionalizable structure. These properties can be applied to facilitate the administration and, thereby overcome some of the limitations of traditional therapeutics such as anticancer and antimicrobial agents. In recent years, encapsulation of antimicrobial or anticancer drugs in nanoparticle systems has emerged as an innovative and promising alternative that enhances therapeutic effectiveness and minimizes undesirable side effects of NCEs.

By loading NCEs into nanoparticles through physical encapsulation, adsorption, or chemical conjugation, their pharmacokinetics and therapeutic index can be significantly improved. Many advantages of nanoparticle-based drug delivery have been recognized, including improved serum solubility of NCEs, prolonged systemic circulation lifetime, drug release at a sustained and controlled manner, preferential delivery to the tissues and cells of interest, and concurrent delivery of multiple therapeutic agents to the same cells for combination therapy (Zhang et al. 2010; Davis et al. 2008; Peer et al. 2007; Wagner et al. 2006). Moreover, drug-loaded nanoparticles can enter host cells through endocytosis and then release drug payloads to treat infected or cancerous cells. As a result, a number of nanoparticle-based drug delivery systems have been approved for clinical uses to treat a variety of diseases and many other therapeutic nanoparticle formulations are currently under various stages of clinical tests (Zhang et al. 2010; Wagner et al. 2006). Knowing the vast scope of nanoparticle drug delivery, the authors would like to focus on the development and application of nanoparticles for anticancer and antimicrobial drug delivery.

An interesting example of nanoparticle-based targeted delivery for anticancer agents is the intravenous injection of magnetic particles (i.e., ferrofluids) bound to anticancer agents that can be concentrated in the desired area (e.g., the tumor) by applying an external magnetic field. In one study, experimental VX-2 squamous cell carcinoma was implanted in the median portion of the hind limb of New Zealand White rabbits ($n = 26$). When the tumor had reached a volume of approximately 3,500 mm^3, FF-MTX (methotrexate) was injected intraarterially (i.a.; femoral artery) or i.v. (ear vein), whereas an external magnetic field was focused on the tumor. FF-MTX i.a. application with the external magnetic field resulted in a significant ($p < 0.05$), complete, and permanent remission of the squamous cell carcinoma compared with the control group (no treatment) and the i.v. FF-MTX group with no signs of toxicity. The intratumoral accumulation of FFs was visualized both histologically and by magnetic resonance imaging. Thus, it suggests that i.a. application of FF-MTX is successful in treating experimental squamous cell carcinoma. This "magnetic drug targeting" offers a unique opportunity to treat malignant tumors locoregionally without systemic toxicity (Alexiou et al. 2000). Lipid-based nanocarriers have proven successful in the delivery of mainly chemotherapeutic agents, and currently they are being applied clinically in the treatment of various types of cancer. These drug delivery systems achieve increased therapeutic efficacy by altering the pharmacokinetics and biodistribution of encapsulated drugs, resulting in decreased drug toxicity and enhanced accumulation in tumor tissue. This increased accumulation is due to the relatively leaky immature vasculature of the tumor. More advanced targeted nanocarriers in general have clearly shown their potential in various animal tumor models and await clinical application. The development of targeted nanocarriers in which therapeutic and imaging agents are merged into a single carrier will certainly be of importance in the near future. Indeed, scientists active in the field of imaging (e.g., nuclear and magnetic resonance imaging) have already started to exploit nanocarriers for molecular imaging. Image-guided drug delivery using these multifunctional nanocarriers containing therapeutic and imaging agents will ultimately allow for online monitoring of tumor location, tumor targeting levels as seen in previous examples, intratumoral localization and drug release kinetics prior and during radio- and/or chemotherapeutic treatment (Koning and Krijger 2007).

Several nanoparticle-based approaches such as liposomes, polymeric nanoparticles, solid lipid nanoparticles and dendrimers have been widely investigated as antimicrobial drug delivery platforms, of which several products have been introduced into the pharmaceutical market (Zhang et al. 2010).

A few examples of each of these delivery platforms for antimicrobials are listed in Table 1.9.

In summary, there are more than dozens of targeted delivery technologies in preclinical and clinical development. For example, there were over 1,500 injectable delivery technologies including targeted delivery, depot systems, and technologies for water-insoluble candidates at various stages of development and commercialization (*Source*: Pharmacircle® 2013). It is estimated that by 2020, the commercial value of products utilizing enabling technologies would be close to $20 billion. Considering the significant advancements, projected commercial growth and most importantly, the opportunity to develop breakthrough therapies for unmet medical needs, it is almost inevitable for pharmaceutical and biotech organizations to get heavily involved in the area of enabling delivery technologies (through internal capacity build-up or partnering with external companies) and maximize the potential of their NCEs as well as their existing clinical or commercial candidates.

1.5 Summary/Concluding Remarks

The process of drug discovery and identification of developable NCEs is a multifaceted approach. Although, the ultimate goal is to identify a candidate that meets all the desired criteria for an intended target product profile (e.g., potency, physicochemical, biopharmaceutics, efficacy, safety, etc.), realistically it is not always feasible to achieve the right balance of all desired properties. It is therefore essential for organizations to have, at least in some capacity, a developability assessment group in place that can add significant value to the candidate selection process. The roles and responsibilities of developability assessment group are much broader than only preparing dosing formulations for preclinical studies. It is essential that a

Table 1.9 Nanoparticle-based delivery systems for antimicrobial drugs

Drug	Delivery technology	Targeted microorganism	Activity	Reference
Polymyxin B	Liposomes	*Pseudomonas aeruginosa*	Decreased bacteria count in lungs, increased bioavailability	Omri et al. (2002)
Amphotericin B	Liposomes	*Aspergillus fumigatus*	Delivery at infection site	Takemoto et al. (2004)
Amphotericin B	Polymeric nanoparticles	*Candida albicans*	Showed lower in vivo toxicity due to reduced accumulation in kidney and liver	Espuelas et al. (2003)
Rifampicin, isoniazid, pyrazinamide, and ethambutol	Polymeric nanoparticles	*Mycobacterium tuberculosis*	Improved pharmacokinetic and showed high therapeutic efficacy	Ahmad et al. (2006)
Tobramycin	Solid lipid nano particles	*Pseudomonas aeruginosa*	Increased drug bioavailability	Cavalli et al. (2002)
Clotrimazole	Solid lipid nano particles	Yeast, *Aspergilli*, *dermatophytes*	Showed prolonged drug release, improved physical stability and high encapsulation efficiency	Souto et al. (2004)
Sulfamethoxazole	Polyamidoamine (PAMAM) dendrimers	*Staphylococcus aureus*	Showed sustained drug release and increased antibacterial activity	Abeylath et al. (2008)
Artemether	PEGylated lysine based copolymeric dendrimer	*Plasmodium falciparum*	Increased drug stability, enhanced solubility of artemether and prolonged drug circulation half-life	Bhadra et al. (2005)

developability assessment scientist provides timely feedback to research project teams on incorporating the right physicochemical and biopharmaceutical properties in their lead candidates or scaffolds, as well as performs detailed risk assessment on solid form and formulation strategies. A number of in silico and high-throughput, material saving in vitro technologies are now available for a developability assessment group to perform the above-mentioned activities in a short period of time while utilizing minimal amounts of material. However, the availability of tools and technologies alone does not guarantee a successful outcome from the developability assessment function. What makes a developability assessment group ultimately successful is the right choice of personnel who can strike the perfect balance of technical expertise, excellence in cross-functional communication, multi-tasking abilities and a can-do attitude. In addition to the significant value-adding contributions in selecting developable candidates, the developability assessment group should be heavily involved in maintaining most up-to-date knowledge base on enabling delivery technologies and bring new ideas or new product opportunities to the attention of project teams on an on-going basis. In summary, developability assessment groups have proven to be a valuable asset for pharmaceutical organizations involved in the discovery and development of NCEs and will continue to reap more benefits in the future.

Acknowledgments The authors would like to sincerely thank Cubist colleagues for their constant support on innovation, especially, You-Seok Hwang, Yannick Fillon, Lisa Duong, Mingjin Fan, Amy Bombardier, Dev Prasad, Sandra O'Connor, and Cubist Senior Leadership Team - Ronnie Farquhar, Bertrand Le Bourdonnec, Rick Oleson, Jared Silverman, and Scott Coleman. The authors would also like to thank ex-colleagues from Novartis, Rick Panicucci, Yatindra Joshi, Abu Serrajuddin, Tony Tong, Madhu Pudipeddi, Colleen Ruegger, Dimitris Papoutsakis, Hyungchul Kim, Weijia Zheng, Vishal Saxena, Lipa Shah, Stephanie Dodd, Andrew Phimister, Vijay Sethuraman, Alex Aycinena and Anand Subramony. Last but not least, our GURUS, Dr. Robin Bogner, Univ. of Connecticut and Prof. Samuel Yalkowsky, Univ. of Arizona.

References

Abeylath SC, Turos E, Dickey S, Lim DV (2008) Glyconanobiotics: novel carbohydrated nanoparticle antibiotics for MRSA and Bacillus anthracis. Bioorg Med Chem 16:2412–2418

Ahmad Z, Pandey R, Sharma S, Khuller GK (2006) Alginate nanoparticles as antituberculosis drug carriers: formulation development, pharmacokinetics and therapeutic potential. Indian J Chest Dis Allied Sci 48:171–176

Alexiou C, Arnold W, Klein RJ, Parak FG, Hulin P, Bergemann C, Erhardt W, Wagenpfeil S, Lübbe AS (2000) Locoregional cancer treatment with magnetic drug targeting. Cancer Res 60 (23):6641–6648

Alsenz J, Haenel E (2003) Development of a 7-day, 96-well Caco-2 permeability assay with high-throughput direct UV compound analysis. Pharm Res 20(12):1961–1969

Artursson P, Tavelin S (2003) Caco-2 and emerging alternatives for prediction of intestinal drug transport: a general overview. In: Lennernas H, Artursson P, van de Waterbeemd H (eds) Drug bioavailability. Wiley, Weinheim, pp 72–89

Avdeef A, Strafford M, Block E, Balogh MP, Chambliss W, Khan I (2001) Drug absorption in vitro model: filter-immobilized artificial membranes. 2. Studies of the permeability properties of lactones in piper methysticum forst. Eur J Pharm Sci 14:271–280

Avdeef A (2003) Absorption and drug development, solubility, permeability and charge state. Wiley-Interscience, Hoboken

Avdeef A (2007) Solubility of sparingly-soluble ionizable drugs. Adv Drug Deliv Rev 59:568–590

Balani SK, Miwa GT, Gan LS, Wu JT, Lee FW (2005) Strategy of utilizing in vitro and in vivo ADME tools for lead optimization and drug candidate selection. Curr Top Med Chem 5:1033–1038

Balbach S, Korn C (2004) Pharmaceutical evaluation of early development candidates "the 100 mg approach". Int J Pharm 275:1–12

Balimane PV, Chong S, Morrison RA (2000) Current methodologies used for evaluation of intestinal permeability and absorption. J Pharmacol Toxicol 44:301–312

Balimane PV, Patel K, Marino A, Chong S (2004) Utility of 96 well Caco-2 cell system for increasing throughput of Pgp screening in drug discovery. Eur J Pharm Biopharm 58:99–105

Balimane PV, Pace E, Chong S, Zhu M, Jemal M, Van Pelt CK (2005) A novel high-throughput automated chip-based nanoelectrospray tandem mass spectrometric method for PAMPA sample analysis. J Pharm Biomed Anal 39(1–2):8–16

Balimane PV, Han YH, Chong S (2006) Current industrial practices of assessing permeability and P-glycoprotein interaction. AAPS J 8(1):E1–E13

Bendels S, Tsinman O, Wagner B, Lipp D, Parrilla I, Kansy M, Avdeef A (2006) PAMPA—excipient classification gradient map. Pharm Res 23(11):2525–2535

Bhadra D, Bhadra S, Jain NK (2005) Pegylated lysine based copolymeric dendritic micelles for solubilization and delivery of artemether. J Pharm Sci 8:467–482

Bohets H, Annaert P, Mannens G, Beijsterveldt VL, Anciaux K, Verboven P, Meuldermans W, Lavrijsen K (2001) Strategies for absorption screening in drug discovery and development. Curr Top Med Chem 1:367–383

Bourdet DL, Thakker DR (2006) Saturable absorptive transport of the hydrophilic organic cation ranitidine in Caco-2 cells: role of pH-dependent organic cation uptake system and P-glycoprotein. Pharm Res 23(6):1165–1177

Box K, Bevan C, Comer J, Hill A, Allen R, Reynolds D (2003) High-throughput measurement of pKa values in a mixed-buffer linear pH gradient system. Anal Chem 75:883–892

Brewster ME, Mackie C, Noppe M, Lampo A, Loftsson T (2007) The use of solubilizing excipients and approaches to generate toxicology vehicles for contemporary drug pipelines. In: Biotechnology: pharmaceutical aspects. Springer, New York, pp. 221–256

Byrn SR, Pfeiffer RR, Stowell JG (eds) (1999) Amorphous solids. Solid-state chemistry of drugs, 2nd edn. SSCI, Inc., West Lafayette, pp 249–258

Cai X et al (2012) Approach to improve compound recovery in a high-throughput Caco-2 permeability assay supported by liquid chromatography–tandem mass spectrometry. J Pharm Sci 101(8):2755–2762

Cavalli R, Gasco MR, Chetoni P, Burgalassi S, Saettone MF (2002) Solid lipid nanoparticles (SLN) as ocular delivery system for tobramycin. Int J Pharm 238:241–245

Cleveland JA Jr, Benko MH, Gluck SJ, Walbroehl YM (1993) Automated pKa determination at low solute concentrations by capillary electrophoresis. J Chromatogr A 652:301–308

Cockburn IM (2004) The changing structure of the pharmaceutical industry. Health Aff 23:10–22

Curatolo W (1998) Physical chemical properties of oral drug candidates in the discovery and exploratory development settings. Pharm Sci Technol Today 1:387–393

Davis ME, Chen ZG, Shin DM (2008) Nanoparticle therapeutics: an emerging treatment modality for cancer. Nat Rev Drug Discov 7:771–782

Egan WJ, Merz KM, Baldwin JJ (2000) Prediction of drug absorption using multivariate statistics. J Med Chem 43:3867–3877

Emoto C, Yoda N, Uno Y, Iwasaki K, Umehara K, Kashiyama E, Yamazaki H (2013) Comparison of p450 enzymes between cynomolgus monkeys and humans: p450 identities, protein contents, kinetic parameters, and potential for inhibitory profiles. Curr Drug Metab 14:239–252

Englund G, Rorsman F, Roennblom A, Karlbom U, Lazorova L, Grasjoe J, Kindmark A, Artursson P (2006) Regional levels of drug transporters along the human intestinal tract: co-expression of ABC and SLC transporters and comparison with Caco-2 cells. Eur J Pharm Sci 29 (3–4):269–277

Espuelas MS, Legrand P, Campanero MA, Appel M, Cheron M, Gamazo C, Barratt G, Irache JM (2003) Polymeric carriers for amphotericin B: in vitro activity, toxicity and therapeutic efficacy against systemic candidiasis in neutropenic mice. J Antimicrob Chemother 52:419–427

Faller B, Ertl P (2007) Computational approaches to determine drug solubility. Adv Drug Deliv Rev 59(7):533–545

Fiese EFG (2003) General pharmaceutics—the new physical pharmacy. J Pharm Sci 92:1131–1342

Frank RG, Seiguer E (2003) Generic drug competition in Business Briefings. Pharmageneric. pp. 66–70

Galia E, Nicolaides E, Horter D, Lobenberg R, Reppas C, Dressman JB (1998) Evaluation of various dissolution media for predicting in vivo performance of class I and II drugs. Pharm Res 15:698–705

Gao P, Shi Y (2012) Characterization of supersaturatable formulations for improved absorption of poorly soluble drugs. AAPS J 14(4):703–713

Garad S, Wang J, Joshi Y, Panicucci R (2009) Preclinical development for suspensions. In: Kulshreshtha AK et al (eds) Pharmaceutical suspensions: from formulation development to manufacturing. Springer, New York, pp 127–176

Glomme A, Marz J, Dressman JB (2005) Comparison of a miniaturized shake-flask solubility method with automated potentiometric acid/base titrations and calculated solubilities. J Pharm Sci 94:1–16

Gu CH, Li H, Levons J, Lentz K, Ganghi RB, Raghavan K, Smith RL (2007) Predicting effect of food on extent of drug absorption based on physicochemical properties. Pharm Res 24:1118–1130

Hämäläinen MD, Frostell-Karlsson A (2004) Predicting the intestinal absorption potential of hits and leads. Drug Discov Today Technol 1:397–406

Hancock BC, Zografi G (1997) Characteristics and significance of the amorphous state in pharmaceutical systems. J Pharm Sci 86:1–12

Hancock BC, Parks M (2000) What is the true solubility advantage for amorphous pharmaceuticals? Pharm Res 17:397–404

Hariharan M, Ganorkar LD, Amidon GE, Cavallo A, Gatti P, Hageman MJ, Miller JL, Shah UL (2003) Reducing the time to develop and manufacture formulations for first oral dose in humans. Pharm Technol 27:68–84

Hilgers AR, Smith DP, Biermacher JJ, Day JS, Jensen JL, Sims SM, Adams WJ, Fiis JM, Palandra J, Hosley JD, Shobe EM, Burton PS (2003) Predicting oral absorption of drugs: a case study with a novel class of antimicrobial agents. Pharm Res 20:1149–1155

Hochman JH, Yamazaki M, Ohe T, Lin JH (2002) Evaluation of drug interactions with P-glycoprotein in drug discovery: in vitro assessment of the potential for drug-drug interactions with P-glycoprotein. Curr Drug Metab 3(3):257–273

Ingels F, Deferme S, Destexhe E, Oth M, Van den Mooter G, Augustijns P (2002) Simulated intestinal fluid as transport medium in the Caco-2 cell culture model. Int J Pharm 232:183–192

Irvine JD, Takahashi L, Lockhart K, Cheong J, Tolan JW, Selick HE, Grove JR (1999) MDCK (Madin-Darby canine kidney) cells: a tool for membrane permeability screening. J Pharm Sci 88:28–33

Ishihama Y, Nakamura M, Miwa T, Kajima T, Asakawa N (2002) A rapid method for pKa determination of drugs using pressure-assisted capillary electrophoresis with photodiode array detection in drug discovery. J Pharm Sci 91:933–942

Jain A, Zheng W, Weaver M, Garad S, Panicucci R (2010) Importance of early characterization of physicochemical properties in developing high dose intravenous infusion regimens for poorly water-soluble compounds. PDA J Pharm Sci Technol 64:517–526

Jantratid E, Janssen N, Reppas C, Dressman JB (2008) Dissolution media simulating conditions in the proximal human gastrointestinal tract: an update. Pharm Res 25:1663–1676. doi:10.1007/s11095-008-9569-4

Johnson KC, Swindell AC (1996) Guidance in the setting of drug particle size specifications to minimize variability in absorption. Pharm Res 13:1795–1798

Jones HM, Parrott N, Ohlenbusch G, Lavé T (2006) Predicting pharmacokinetic food effects using biorelevant solubility media and physiologically based modelling. Clin Pharmacokinet 45 (12):1213–1226

Junghanns J-UAH, Müller RH (2008) Nanocrystal technology, drug delivery and clinical applications. Int J Nanomedicine 3(3):295–310

Kansy M, Senner F, Gubernator K (1998) Physicochemical high throughput screening: parallel artificial membrane permeation assay in the description of passive absorption processes. J Med Chem 41:1007–1010

Kerns EH, Di L (2002) Multivariate pharmaceutical profiling for drug discovery. Curr Top Med Chem 2:87–98

Kerns EH, Di L, Petusky S, Farris M, Ley R, Jupp P (2004) Combined application of parallel artificial membrane permeability assay and Caco-2 permeability assays in drug discovery. J Pharm Sci 93(6):1440–1453

Khanna I (2012) Drug discovery in pharmaceutical industry: productivity challenges and trends. Drug Discov Today 17(19/20):1088–1102

Kibbey CE, Poole SK, Robinson B, Jackson JD, Durham D (2001) An integrated process for measuring the physicochemical properties of drug candidates in a preclinical discovery environment. J Pharm Sci 90:1164–1175

Kola I, Landis J (2004) Can the pharmaceutical industry reduce attrition rates? Nat Rev Drug Discov 3:711–715

Koning GA, Krijger GC (2007) Targeted multifunctional lipid-based nanocarriers for image-guided drug delivery. Anticancer Agents Med Chem 7(4):425–440

Korn C, Balbach S (2004) Pharmaceutical evaluation of early development candidates "the 100 mg-approach". Int J Pharm 275(1–2):1–12

Korn C, Balbach S (2014) Compound selection for development—is salt formation the ultimate answer? Experiences with an extended concept of the "100 mg approach". Eur J Pharm Sci 57:257–263

Kuentz M, Nick S, Parrott N, Rothlisberger D (2006) A strategy for preclinical formulation development using GastroPlus™ as pharmacokinetic simulation tool and a statistical screening design applied to a dog study. Eur J Pharm Sci 27:91–99

Kuppens IELM, Breedveld P, Beijnen JH, Schellens JHM (2005) Modulation of oral drug bioavailability: from preclinical mechanism to therapeutic application. Cancer Invest 23 (5):443–464

Lakeram M, Lockley DJ, Sanders DJ, Pendlington R, Forbes B (2007) Paraben transport and metabolism in the biomimetic artificial membrane permeability assay (BAMPA) and 3-day and 21-day Caco-2 cell systems. J Biomol Screen 12(1):84–91

Lennernas H, Lundgren E (2004) Intestinal and blood-brain drug transport: beyond involvement of a single transport function. Drug Discov Today Technol 1:417–422

Lentz KA (2008) Current methods for predicting human food effect. AAPS J 10(2):282–288

Li Z, Vachharajanib NN, Krishnab R (2002) On the assessment of effects of food on the pharmacokinetics of drugs in early development. Biopharm Drug Dispos 23:165–171

Liang E, Chessic K, Yazdanian M (2000) Evaluation of an accelerated Caco-2 cell permeability model. J Pharm Sci 89(3):336–345

Lind ML, Jacobsen J, Holm R, Mullertz A (2007) Development of simulated intestinal fluids containing nutrients as transport media in the Caco-2 cell culture model: assessment of cell viability, monolayer integrity and transport of a poorly aqueous soluble drug and a substrate of efflux mechanisms. Eur J Pharm Sci 32:261–270

Liu H, Sabus C, Carter GT, Du C, Avdeef A, Tischler M (2003) In vitro permeability of poorly aqueous soluble compounds using different solubilizers in the PAMPA assay with liquid chromatography/mass spectrometry detection. Pharm Res 20(11):1820–1826

Lombardo F, Shalaeva M, Tupper KA, Gao F, Abraham MJ (2000) ElogPoct: a tool for lipophilicity determination in drug discovery. J Med Chem 43:2922–2928

Ku MS (2008) Use of the biopharmaceutical classification system in early drug development. AAPS J 10(1):208–212

Maag H (2012) Overcoming poor permeability—the role of prodrugs for oral drug delivery. Drug Discov Today Technol 9(2):e121–e130

Maas J, Kamm W, Hauck G (2007) An integrated early formulation strategy—from hit evaluation to preclinical candidate profiling. Eur J Pharm Biopharm 66:1–10

Maher S, Brayden DJ (2012) Overcoming poor permeability: translating permeation enhancers for oral peptide delivery. Drug Discov Today Technol 9(2):e113–e119

Marino AM, Yarde M, Patel H, Chong S, Balimane PV (2005) Validation of the 96 well Caco-2 cell culture model for high throughput permeability assessment of discovery compounds. Int J Pharm 297(1–2):235–241

Martignoni M, Groothuis GMM, de Kanter R (2006) Species differences between mouse, rat, dog, monkey and human CYP-mediated drug metabolism, inhibition and induction. Expert Opin Drug Metab Toxicol 2:875–894

Mathias NR et al (2013) Assessing the risk of pH-dependent absorption for new molecular entities: a novel in vitro dissolution test, physicochemical analysis, and risk assessment strategy. Mol Pharm 10:4063–4073

McConnell EL, Basit AW, Murdan S (2008) Measurements of rat and mouse gastrointestinal pH, fluid, lymphoid tissue, and implications for *invivo* experiments. J Pharm Pharmacol 60:63–70

Mensch J, Noppe M, Adriaensen J, Melis A, Mackie C, Augustijns P, Brewster ME (2007) Novel generic UPLC/MS/MS method for high throughput analysis applied to permeability assessment in early drug Discovery. J Chromatogr B Analyt Technol Biomed Life Sci 847 (2):182–187

Miret S, Abrahamse L, de Groene EM (2004) Comparison of in vitro models for the prediction of compound absorption across the human intestinal mucosa. J Biomol Screen 9(7):598–606

Morris KR, Fakes MG, Thakur AB, Newman AW, Singh AK, Venit JJ, Spagnuolo CJ, Serajuddin ATM (1994) An integrated approach to the selection of optimal salt form for a new drug candidate. Int J Pharm 105:209–217

Neuhoff S, Artursson P, Zamora I, Ungell A-L (2006) Impact of extracellular protein binding on passive and active drug transport across Caco-2 cells. Pharm Res 23(2):350–359

Nobili S, Landini I, Giglioni B, Mini E (2006) Pharmacological strategies for overcoming multidrug resistance. Curr Drug Targets 7(7):861–879

Omri A, Suntres ZE, Shek PN (2002) Enhanced activity of liposomal polymyxin B against Pseudomonas aeruginosa in a rat model of lung infection. Biochem Pharmacol 64:1407–1413

Paul SM, Mytelka DS, Dunwiddie CT, Persinger CC, Munos BH, Lindborg SR, Schacht AL (2010) How to improve R&D productivity: the pharmaceutical industry's grand challenge. Nat Rev Drug Discov 9:203–214

Peer D, Karp JM, Hong S, Farokhzad OC, Margalit R, Langer R (2007) Nanocarriers as an emerging platform for cancer therapy. Nat Nanotechnol 2:751–760

Prueksaritanont T et al (1998) In vitro and in vivo evaluation of intestinal barriers for the zwitter ion L-767,679 and its carboxyl ester prodrug L-775,318. Drug Metab Dispos 26(6):520–527

Rosiello A, Dimitri N, Fiorini F (2013) A new approach to assess drug development performance. Drug Discov Today 18(9–10):420–427

Ruell JA, Tsinman O, Avdeef A (2004) PAMPA—a drug absorption in vitro model. 12. Cosolvent method for permeability assays of amiodarone, itraconazole, tamoxifen, terfenadine, and other very insoluble molecules. Chem Pharm Bull 52:561–565

Saha P, Kou JH (2002) Effect of bovine serum albumin on drug permeability estimation across Caco-2 monolayers. Eur J Pharm Biopharm 54(3):319–324

Saunders KC (2004) Automation and robotics in ADME screening. Drug Discov Today Technol 1:373–380

Saxena V, Panicucci R, Joshi Y, Garad S (2009) Developability assessment in pharmaceutical industry: an integrated group approach in selecting developable candidates. J Pharm Sci 98 (6):1962–1979

Serajuddin ATM (2007) Salt formation to improve drug solubility. Adv Drug Deliv Rev 59:603–616

Singh SS (2006) Preclinical Pharmacokinetics: an approach towards safer and efficacious drugs. Curr Drug Metab 7:165–182

Singhal D, Curatalo W (2004) Drug polymorphism and dosage form design: a practical perspective. Adv Drug Deliv Rev 56:335–347

Souto EB, Wissing SA, Barbosa CM, Muller RH (2004) Development of a controlled release formulation based on SLN and NLC for topical clotrimazole delivery. Int J Pharm 278:71–77

Steffansen B, Nielsen CU, Brodin B, Eriksson AH, Andersen R, Frokjaer S (2004) Intestinal solute carriers: an overview of trends and strategies for improving oral drug absorption. Eur J Pharm Sci 21(1):3–16

Strickley RG (2004) Solubilizing excipients in oral and injectables formulations. Pharm Res 21 (2):201–230

Subramaniam S (2003) Productivity and attrition: key challenges for biotech and pharma. Drug Discov Today 8:513–515

Sugano K, Hamada H, Machida M, Ushio H, Saitoh K, Terada K (2001) Optimized conditions of bio-mimetic artificial membrane permeability assay. Int J Pharm 228:181–188

Sugano K, Kansy M, Artursson P, Avdeef A, Bendels S, Di L, Ecker GF, Faller B, Fischer H, Gerebtzoff G, Lennernaes H, Senner F (2010) Coexistence of passive and carrier-mediated processes in drug transport. Nat Rev Drug Discov 9:597–614

Sun D, Yu LX, Hussain MA, Wall DA, Smith RL, Amidon GL (2004) *In vitro* testing of drug absorption for drug 'developability' assessment: forming an interface between in vitro pre-clinical data and clinical outcome. Curr Opin Drug Discov Devel 7:75–85

Takemoto K, Yamamoto Y, Ueda Y, Sumita Y, Yoshida K, Niki Y (2004) Comparative studies on the efficacy of Am Bisome and Fungizone in a mouse model of disseminated aspergillosis. J Antimicrob Chemother 53:311–317

Thomas VH, Bhattachar S, Hitchingham L, Zocharski P, Naath M, Narayanan S, Chad LS, El-Kattan A (2006) The road map to oral bioavailability: an industrial perspective. Expert Opin Drug Metab Toxicol 2(4):591–608

Troutman MD, Thakker DR (2003) Efflux ratio cannot assess P-glycoprotein-mediated attenuation of absorptive transport: asymmetric effect of P-glycoprotein on absorptive and secretory transport across caco-2 cell monolayers. Pharm Res 20:1200–1209

Ungell AL, Karlsson J (2003) Cell culture in drug discovery: an industrial perspective. In: van de Waterbeemd H, Lennernas H, Artursson P (eds) Drug bioavailability. Wiley, Weinheim, pp 90–131

Ungell A-LB (2004) Caco-2 replace or refine? Drug Discov Today Technol 1:423–430

Varma MVS, Ashokraj Y, Chinmoy SD, Panchagnula R (2003) P-glycoprotein inhibitors and their screening: a perspective from bioavailability enhancement. Pharmacol Res 48(4):347–359

Varma MVS, Sateesh K, Panchagnula R (2005) Functional role of P-glycoprotein in limiting intestinal absorption of drugs: contribution of passive permeability to P-glycoprotein mediated efflux transport. Mol Pharm 2(1):12–21

Varma MVS, Perumal OP, Panchagnula R (2006) Functional role of P-glycoprotein in limiting peroral drug absorption: optimizing drug delivery. Curr Opin Chem Biol 10(4):367–373

Venkatesh S, Lipper RA (2000) Role of the development scientist in compound lead selection and optimization. J Pharm Sci 89:145–154

Vertzoni M, Dressman J, Butler J, Hempenstall J, Reppas C (2005) Simulation of fasting gastric conditions and its importance for the in vivo dissolution of lipophilic compounds. Eur J Pharm Biopharm 60:413–417

Wagner V, Dullaart A, Bock AK, Zweck A (2006) The emerging nanomedicine landscape. Nat Biotechnol 24:1211–1217

Wan H, Ulander J (2006) High-throughput pKa screening and prediction amenable for ADME profiling. Expert Opin Drug Metab Toxicol 2(1):139–155

Wang J, Urban L, Bojanic D (2007) Maximising use of in vitro ADMET tools to predict in vivo bioavailability and safety. Expert Opin Drug Metab Toxicol 3(5):641–665

Wang J, Faller B (2007) Progress in bioanalytics and automation robotics for ADME screening. In: Testa B, van de Waterbeemd H (eds) Comprehensive medicinal chemistry, vol 5, 2nd edn, ADME-Tox approaches. Elsevier Ltd., Oxford, pp 341–356

Watkins KJ (2002) Fighting the clock. Chem Engl News 80:27–33

Wohnsland F, Faller B (2001) High-throughput permeability pH profile and high-throughput alkane/water LogP with artificial membranes. J Med Chem 44:923–930

Wong J, Brugger A, Khare A, Chaubal M, Papadopoulos P, Rabinow B, Kipp J, Ning J (2008) Suspensions for intravenous (IV) injection: a review of development, preclinical and clinical aspects. Adv Drug Deliv Rev 60:939–954

Yamashita S, Furubayashi T, Kataoka M, Sakane T, Sezaki H, Tokuda H (2000) Optimized conditions for prediction of intestinal drug permeability using Caco-2 cells. Eur J Pharm Sci 10(3):195–204

Yamashita S, Konishi K, Yamazaki Y, Taki Y, Sakane T, Sezaki H, Furuyama Y (2002) New and better protocols for a short-term Caco-2 cell culture system. J Pharm Sci 91(3):669–679

Zhang L, Pornpattananangkul D, Hu C-MJ et al (2010) Development of nanoparticles for antimicrobial drug delivery. Curr Med Chem 17:585–594

Zhang L, Gu FX, Chan JM, Wang AZ, Langer RS, Farokhzad OC (2008) Nanoparticles in medicine: therapeutic applications and developments. Clin Pharmacol Ther 83:761–769

Zheng W, Jain A, Papoutsakis D, Dannenfelser R, Garad S, Panicucci R (2012) Selection of oral bioavailability enhancing formulations during drug discovery. Drug Dev Ind Pharm 38 (2):235–247

Chapter 2
Discovery Formulations: Approaches and Practices in Early Preclinical Development

Shobha N. Bhattachar, David M. Bender, Stephanie A. Sweetana, and James A. Wesley

2.1 Introduction

Technological innovations in biology, chemistry, and medicine have provided the pharmaceutical industry a wealth of targets and molecules with the potential to treat diseases once thought intractable to drug therapy. These advances have brought about a renaissance in the industry and current estimates suggest there are more than 5,000 potential new medicines in human testing, a high percentage of which would be considered "first in class" (Long and Works 2013). It has been suggested that pharmaceutical portfolios have shifted from commercially crowded therapeutic areas where the probability of approval is high to less crowded areas with novel targets and subsequent lower approval rates (Scannell et al. 2012). Additionally, there is a growing recognition that modulation of multiple targets (e.g., magic shotguns) rather than a single target (e.g., magic bullet) by a drug may provide greater therapeutic benefit to the patient (Roth et al. 2004; Morphy 2010; Gleeson et al. 2011). These transformations have resulted in a decline in new drug approvals and more importantly, a gradual but significant shift out of conventional druggable chemical space (Pammolli et al. 2011). The consequential increase in complexity, both in terms of the molecules and their biological targets, combined with the increasing need to work in an efficient and cost-constrained environment has necessitated an evolution in the role of pharmaceutical sciences in discovery support.

Traditionally, the pharmaceutical scientist participated on discovery teams only in the later phases of lead development or in the lead optimization phase, and their role was largely to assess the development risks (developability) of the molecule advancing to clinical dosing (Venkatesh and Lipper 2000). These activities, while

S.N. Bhattachar • D.M. Bender • S.A. Sweetana • J.A. Wesley (✉)
Small Molecule Design and Development, Eli Lilly and Company, 893 S Delaware St, Indianapolis, IN 46285, USA
e-mail: wesley_james_anthony@lilly.com

A.C. Templeton et al. (eds.), *Discovering and Developing Molecules with Optimal Drug-Like Properties*, AAPS Advances in the Pharmaceutical Sciences Series 15,
DOI 10.1007/978-1-4939-1399-2_2

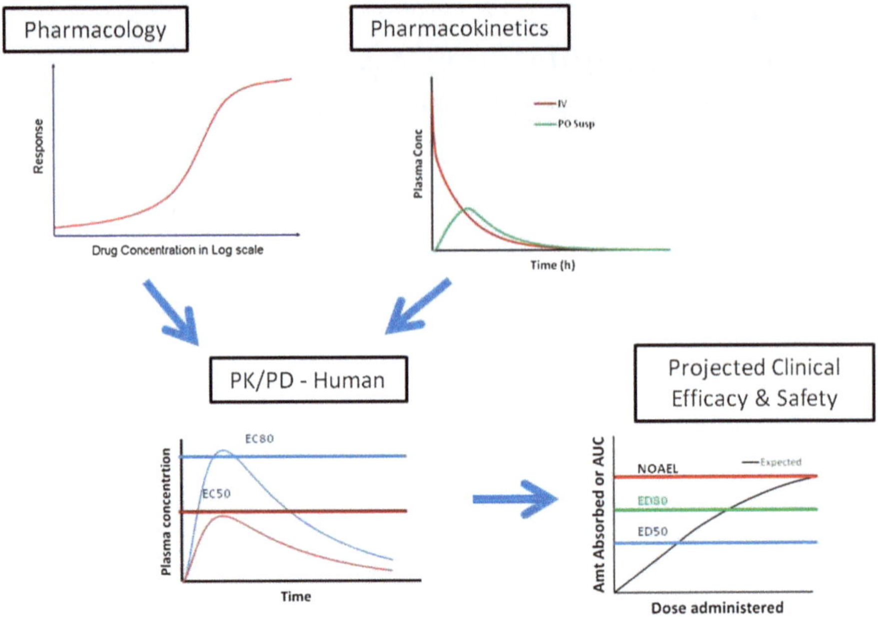

Fig. 2.1 Illustration of connectivity between pharmacology, pharmacokinetics, pharmacodynamics, toxicology and clinical dose range, and the relevance of discovery formulations and impact on clinical studies

important, have been augmented to include early discovery formulation support related to building a basic understanding of biology through in vivo target validation and demonstration of proof of mechanism (Neervannan 2006; Li and Zhao 2007; Shah and Agnihotri 2011). In addition, the desire to shorten development timelines while placing greater emphasis on patient centered design and delivery has brought about the need for development strategy discussions to start to take place earlier in preclinical development. Pharmaceutical scientists are ideally positioned to provide this type of support to project teams, given their knowledge of the physicochemical properties of compounds and training in formulation development (Hageman 2006). Formulations can profoundly impact drug release, absorption, and metabolism, which influence the resulting pharmacokinetic (PK) profile and the associated pharmacodynamic response. Thus, formulation and drug delivery technologies play an important role in in vivo discovery efforts.

The in vivo studies performed in the preclinical setting can broadly be classified as pharmacology, pharmacokinetic, and toxicology studies. The goals and challenges of these studies are diverse. Connectivity of key data collected from these studies, their impact on clinical formulation development, and ultimately on the in vivo clinical performance is illustrated in Fig. 2.1. The main output from pharmacology studies is the pharmacologic response (in the form of pharmacodynamic outcome, receptor occupancy, etc.) as it relates to in vivo plasma concentration or exposure of the compound. The primary outputs from a pharmacokinetic

Table 2.1 Preclinical formulations for in vivo studies in discovery: goals of studies and role of formulations

	Target to hit	Hit to lead	Lead optimization and clinical candidate selection
Pharmacology	Proof of concept/ target validation studies.	Compounds of interest from cell-based potency screens tested for in vivo activity. Multiple scaffolds not uncommon.	Nonoral routes of administration less common (if intent is oral dosing of clinical candidate).
		Wide range of concentrations to test for activity, target selectivity, and durability.	Studies focused on a thorough assessment of in vivo pharmacology for selection of clinical candidate.
		Time course and dose range assessments to understand on-target and off-target effects. Some studies done using nonoral formulations (e.g., IP or SC route).	
		Most formulation recommendation is based on assessment of selected compounds from each scaffold and vehicle effect considerations.	
	Mostly limited to 1 or 2 tool compounds, generally with poor druggability.	Formulations for expensive studies are based on compound-specific (or lot specific) assessments. Physical and chemical stability data generated as needed.	
	Single dose studies. Frequently high concentrations needed at the target.		
	Nonoral routes of dosing commonly used.		
PK/ADME	Limited to assessment of ADME properties of tool compounds in rodents	PK assessment in rodents to get basic understanding of clearance mechanisms and PK properties as they relate to scaffolds. Goal is to assist in selection of lead scaffold.	Rodent and nonrodent PK studies. Dogs are the most common nonrodent species.
			Oral absorption and metabolism parameters must be acceptable for oral dosing of clinical compound.
		Scaffold-wise formulation recommendation based on physicochemical properties of	Solubility and solid state data on material going

(continued)

Table 2.1 (continued)

	Target to hit	Hit to lead	Lead optimization and clinical candidate selection
		representative compounds. Basic crystallinity data on compounds of interest to inform formulation properties and/or absorption modeling.	into dog studies is essential. High emphasis on the absorbable dose in humans and potential need for enabled formulations.
			Formulation or study design options for overcoming PK variability associated with dog gastric pH may be used.
Toxicology	No in vivo studies	Short-term rat toxicology study on one or two compounds to support lead declaration. Doses up to 1,000 mg/kg not uncommon depending on potency data from pharmacology studies.	Short-term rat toxicology studies, followed by longer term or pilot toxicology studies to support clinical candidate selection.
			Rodent and nonrodent species.
Developability considerations	None	Preliminary assessment of developability to guide SAR.	Definitive assessment of commercial developability and understanding of associated risks.

study are the absorption, distribution, and clearance parameters for that compound as it relates to the species used in the study. These data feed into the generation of the pharmacokinetic–pharmacodynamic (PK-PD) model which describes the dose–concentration–effect relationship. Plasma exposures pertaining to the safety of the compound come from toxicology studies and provide a no observed adverse effect level (NOAEL), a level of exposure where there is no biologically significant increase in adverse effects compared to control. With appropriate scaling between species and the expected in vivo performance of the clinical formulation, the projected absorption and plasma exposures in the clinical dose range are established. Thus, while the formulations used for the various in vivo studies may be different, the outcomes of the studies are highly connected and have an important bearing on the design and execution of early clinical studies. Analysis of early clinical data enables further refinement of the models for next generation discovery efforts. In addition, the availability of exposure data from human studies allows for assessment of the performance of the drug product and provides a context for computational simulations of modified delivery systems, should the human pharmacokinetic profile suggest they are needed.

Table 2.1 summarizes the distinctive features and goals of preclinical in vivo studies based on the general type of the studies and the discovery phase during which they are conducted. As noted in the table, formulations used in the early phases of discovery are geared toward target validation and/or proof of concept, with little or no developability considerations for the compounds or the formulations tested. As discovery programs progress toward lead declaration and subsequent optimization however, developability considerations take on increasing importance and the formulations used must be selected accordingly.

Strategies used in the development and assessment of preclinical formulations, and their application in the different types of in vivo studies will be discussed in the sections that follow.

2.2 Discovery and Preclinical Formulation Approaches

Formulation approaches to deliver molecules in the preclinical setting include suspensions, solutions, and amorphous dispersions administered as solids or in aqueous vehicles and each is discussed briefly below. These general approaches to formulation development, particularly related to solubilization, have been extensively reviewed and therefore, emphasis in this chapter is placed on application in the preclinical setting. The development of an overall formulation strategy to support in vivo studies should be considered carefully as it can reduce cycle time and resources. This strategy must be comprehensive, encompassing early studies designed to identify and validate drug targets, to long-term toxicology studies and ultimately, to support clinical studies in man. A focus on developing these types of strategies is presented in the next section, followed by a detailed discussion around practical considerations and examples for various types of studies, including pharmacology, ADME, toxicology, and alternate drug delivery.

2.2.1 Suspension Formulations and Nanosuspensions

Suspension formulations are the most widely used formulations in the discovery phase, owing to their ease of preparation and applicability to a wide variety of chemical platforms. In general, suspensions may serve as surrogates for exposure predictions from a standard human dosage form (capsule or tablet), provided the solid state properties of the compound are reflective of the API form to be developed. Standard suspension vehicles include a 1–10 % mixture in water of a cellulose polymer, such as methylcellulose, hydroxyethylcellulose, or acacia, accompanied by low levels (0.1–0.2 % w/v) of a nonionic surfactant such as polysorbate 80 to facilitate wetting. Use of nonionic components can minimize agglomeration due to charge interactions with ionizable drugs. Common practice in the discovery setting is to reduce the particle size of the suspension using ultrasonic probe sonication, thereby creating more favorable properties for dissolution and absorption. With appropriate equipment configuration, a mean particle size diameter of 10 μm can routinely be achieved. One of the challenges with suspensions is that early lots of material often have less than ideal physical properties, which can include amorphous material or mixtures of amorphous and crystalline forms. This batch-to-batch variability can confound interpretation of in vivo results if consistent characterization of material in the dose preparation is not conducted. Additionally, the physical stability of the suspension must be monitored to ensure no form changes are occurring which may impact exposure. Daily preparation may help to avoid the need for this testing.

The use of nanoparticle formulations has much precedence in the discovery setting (Rabinow 2004). Nanoparticles are submicron (<1 μm) solid colloidal systems in which the drug is in a colloidal state of subdivision. This is in contrast to micronized drug where particles in the 2–5 μm range are typically achieved. Nanoparticles have a greater total surface area than the same mass of larger particles, resulting in increased dissolution rate. At very small particle sizes (e.g., <200 nm), saturation solubility may increase, however, the predicted increase is small, approximately 10–15 % at a particle size of 100 nm, according to the Freundlich–Ostwald equation (Kesisoglou et al. 2007). Nanoparticulate suspension formulations can be prepared on a small scale and in a short time frame using a variety of techniques (Merisko-Liversidge and Liversidge 2008) and can be administered by multiple routes (oral, intranasal, intraperitoneal, or intravenous). Key considerations in formulating nanoparticulate suspensions include appropriate choices of: (a) wetting agents for suspension formation and (b) polymers to coat the particles in order to create adequate steric hindrance for prevention of aggregation and particle growth. The nanomilling process should be optimized to achieve a very tight range for particle size distribution in the final product. This minimizes Oswald ripening and consequent particle growth in the suspension (Van Eerdenbrugh et al. 2008). Characterization should be performed in appropriate in vitro screens under biorelevant conditions, to ensure that the particles do not agglomerate upon dosing (Kesisoglou and Mitra 2012). The physical stability of the formulation should be monitored to ensure that no crystal form changes or undesired particle growth occurs during the intended shelf life. Formulating nanosuspensions for discovery is covered in Chap. 3.

2.2.2 pH Adjustment

Ionizable compounds with p*K*a's in the biorelevant range can often be formulated as pH adjusted solutions. The Henderson–Hasselbalch equation describes the relationship between the pH, p*K*a, and relative concentrations of the ionized and unionized forms of the compound in solution. The solubility of the ionized form is generally much greater than the solubility of the neutral form, and so the pH is

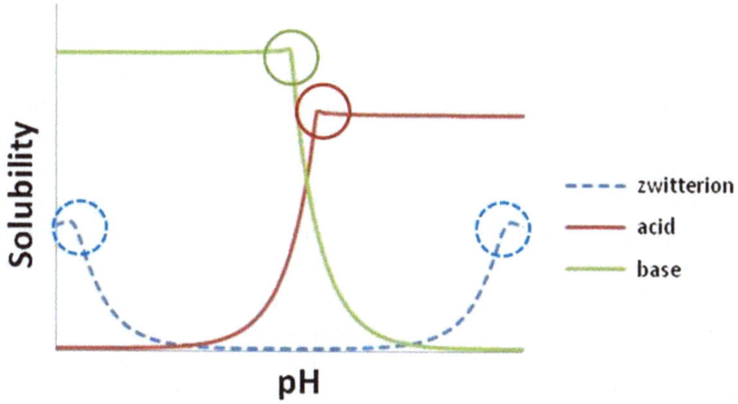

Fig. 2.2 pH–solubility relationships for different types of ionizable compounds. *Circles* indicate pH_{max}

modified in the direction of greater ionization as shown in Fig. 2.2 in order to achieve solubilization. Dilute solutions of sodium hydroxide or hydrochloric acid are most commonly used for pH adjustment. The final pH of the formulation should be monitored to ensure it is within an acceptable range for administration to laboratory animals, typically between pH 2 and 9 for oral administration. Chemical stability of the compound at the desired pH should also be verified. For compounds with high pH-dependent solubility in the pH range of the gut (pH 2–7), buffered systems may be used for short-term studies to minimize the risk of precipitation within the gastrointestinal tract. For parenterally administered formulations with a pH outside the range of approximately pH 6–8, it is important to make sure that the buffer capacity is sufficiently low (generally solution molarity should be ~25 mM or less) so as not to cause venous irritation.

Although pH adjustment increases the total amount of drug in solution, it should be recognized from the standpoint of oral absorption, it is the nonionized species that is more readily absorbed across the intestinal mucosa. In order to appreciate the effects of pH modification on solubility, it is necessary to consider the behavior of the various species present in equilibrium. The pH at which the ionized and unionized species are both saturated is referred to as the pH_{max} (Fig. 2.2). For basic compounds passing from the region of pH_{max} to regions of higher pH during GI transit, the equilibrium solubility at the corresponding pH could induce precipitation. However, depending on the intrinsic properties of the compound, or the composition of the local environment, it is fairly common for compounds to remain in a metastable state known as supersaturation for a significant duration of time. This increases the thermodynamic activity in the GI tract leading to enhanced flux and subsequent exposures (Pole 2008). However, supersaturation can also lead to precipitation of an acidic drug in the stomach, and a basic drug in the intestine, possibly leading to low and/or variable exposure. Precipitation can sometimes be inhibited or reduced by the use of polymers added to the vehicle (Warren et al. 2010; Xu and Dai 2013).

Another effective method for obtaining a solution formulation via pH adjustment is the formation of an in situ salt, which is accomplished by adding molar equivalents of an acidic or basic counterion to the free form of an ionizable compound. This technique takes advantage of the different equilibrium constants (K_{sp}'s) that arise when different ions are present in a saturated solution of an ionic compound (Tong and Whitesell 1998). It is also useful when compounds prove difficult to formulate due to solid state issues encountered with the free form, such as poor suspendability or stickiness of the material in the formulation media. While the selection of an appropriate salt form is an important aspect of drug development, it is not usually practical to conduct a traditional salt screen in the early discovery phase. This challenge is readily overcome by the formation of an in situ salt. Typically, the p*K*a of the counterion selected should be approximately 2 p*K*a units away from that of the free form. For a basic compound, the counterion should have a p*K*a that is at least 2 units lower than the free form while for an acidic compound, the p*K*a should be at least 2 units higher than the free form. Due to the small amounts of material required, a number of different counterions can be screened rapidly in order to select the most desirable for use. It should be emphasized that since the primary goal of this approach is to find an aqueous formulation that offers the best solubility advantage, no consideration is given as to the long-term viability of the salt form identified.

2.2.3 Cyclodextrin Complexation

Cyclodextrins are cyclic sugar oligomers and their use in drug solubilization has been reviewed extensively (Stella and He 2008; Loftsson and Brewster 2010; Kurkov and Loftsson 2013). Cyclodextrins possess a hydrophilic exterior and a hydrophobic core, and therefore the primary mechanism of solubilization is due to the ability of these agents to form noncovalent inclusion complexes with lipophilic drugs. If the cyclodextrin–drug complex results from a 1:1 interaction, solubility increases linearly as a function of cyclodextrin concentration as shown in Fig. 2.3. The primary advantage this offers is low risk of drug precipitation upon dilution. Upon administration, dilution and competitive binding with plasma components are the major driving forces for dissociation of the complex (Stella et al. 1999; Kurkov et al. 2012). In most cases, dissociation is complete, providing rapid release of drug. However in a few cases, where the drug–cyclodextrin binding constant (K) is reported to be very high ($>1 \times 10^5\ M^{-1}$), an effect on drug disposition has been observed (Charman et al. 2006). The importance of the binding constant has been detailed in an excellent review (Carrier et al. 2007) and these authors state that most poorly soluble drugs will have increased oral bioavailability when dosed as a cyclodextrin complex provided the binding constant is below $1 \times 10^4\ M^{-1}$. Equally important is consideration of the drug and cyclodextrin concentrations in the

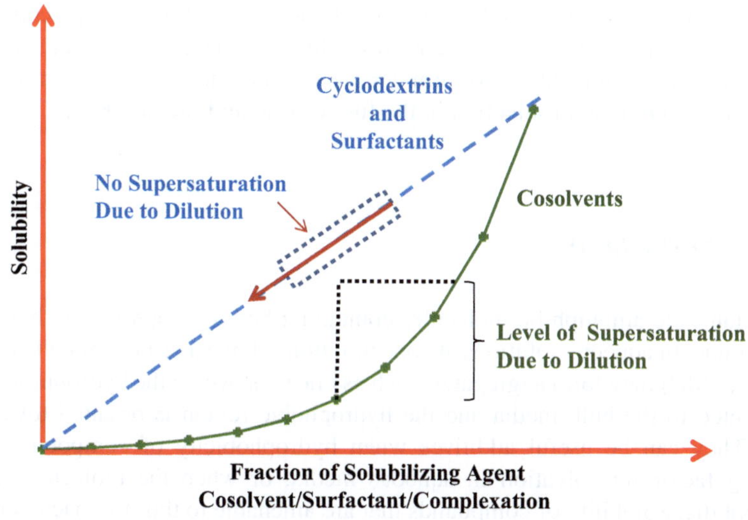

Fig. 2.3 Solubilization techniques—solubility vs. dilution. The dilution of cosolvent systems can result in supersaturation followed by precipitation. Cyclodextrins and surfactants typically increase solubility in a linear manner and therefore dilution does not result in supersaturation

formulation since these will influence the complexation equilibrium. In vivo performance may be impacted in cases where dilution is minimal (e.g., intranasal, intramuscular) or if concentrated, large volumes of cyclodextrins are administered orally (Stella and He 2008). This may be due to decreased uptake of the drug through biological barriers containing unstirred water layer (UWL) that exists between the membrane and bulk water (Loftsson and Brewster 2011) and a decrease in the free fraction available for permeability (Miller and Dahan 2012). The cyclodextrins most commonly used in discovery and development are 2-hydroxypropyl-β-cyclodextrin (HPβCD) and sulfobutylether-β-cyclodextrin (SBEβCD). These cyclodextrins are highly water soluble with solubility >500 mg/mL and have been extensively characterized with regard to safety profile and material properties. While both cyclodextrins are found in US marketed parenteral formulations, preclinical data suggests SBEβCD may be preferred for parenteral administration due to lower in vitro hemolysis compared with HPβCD (Shiotani et al. 1995; Luke et al. 2010).

Typical cyclodextrin use concentrations are approximately 10–20 % w/v. For example, the amount of cyclodextrin required for solubilization, given a target drug solubility of 10 mg/mL, a molecular weight of 500, and a 1:1 complex formulation, is 5 % w/v SBEβCD (average MW 2,163) or 2.8 % w/v HPβCD (average MW 1,400). Cyclodextrin complexation is often used along with complementary approaches such as pH adjustment and low levels of polymers to improve the extent of solubilization. In particular, SBEβCD carries a negative charge at physiological pHs due to the low p*K*a of the sulfonic acid groups. As a result, through charge attraction, cations may bind better than the neutral form to SBEβCD.

Thus, solubility screens include both neutral and ionized forms of the compound, where appropriate. In silico methods to predict binding constants, and therefore those compounds most likely to form complexes, are under development and may further guide compound selection in the future (Rao and Stella 2003).

2.2.4 Surfactants

Surfactants are amphiphilic molecules containing both hydrophilic and hydrophobic regions. In aqueous solutions at concentrations above a critical micelle concentration (CMC) they form aggregates, such as micelles, where the hydrophilic region is oriented to the bulk media and the hydrophobic region is oriented toward the core. They can be useful additives when hydrophobicity of compounds is the limiting factor for solvation in aqueous media, or when the molecule itself is amphiphilic. Solubility of compounds that are amenable to this approach generally increases in a linear manner with increasing surfactant concentration. As a result, the risk of precipitation upon dilution of surfactant-based formulations is minimal as depicted in Fig. 2.3. Several authors have published lists of common surfactants and typical formulation concentrations for both oral and IV use (Neervannan 2006; Li and Zhao 2007; Strickley 2008). Examples of commonly used surfactants include polysorbates (e.g., Tween 80) and polyoxyl castor oil (Cremophor EL). In general, the primary challenge with the use of surfactants in preclinical formulations is the large amounts of these excipients typically required for solubilization, which are associated with tolerability issues. Hypersensitivity reactions have been well documented following intravenous administration of certain surfactants such as Cremophor EL and polysorbate 80 to sensitive animals (particularly dogs) and humans (Lorenz et al. 1977; 1982; Weiss et al. 1990; ten Tije et al. 2003). Changes in plasma clearance with the use of surfactants have been reported both in vitro and in vivo (ten Tije et al. 2003; Bravo et al. 2004; Bittner et al. 2005). Inhibitory effects on intestinal drug transport processes such as P-glycoprotein (P-gp) efflux are well established for numerous surfactants and details have been compiled in a review (Williams et al. 2013). Thus, care must be taken in interpretation of pharmacokinetic and pharmacodynamic data generated in the presence of high concentrations of surfactants. The use of Vitamin E-TPGS to increase the bioavailability of poorly soluble drugs is well document (Li et al. 2012). However, oral administration of Vitamin E-TPGS in longer term studies may result in absorption of D-α-Tocopheryl (Vitamin E) by the hydrolysis of the TPGS moiety (Traber et al. 1986; Dimitrov et al. 1996; Jacquemin et al. 2009). This can lead to exposures to Vitamin E that potentially could complicate interpretation of pharmacology or safety studies. Therefore, the nature of the surfactant and level of use should be carefully considered in designing formulations.

2.2.5 Cosolvents

Organic cosolvent systems can be powerful solubilizing agents for molecules in the discovery setting. Cosolvents alter the polarity of aqueous systems to provide a more favorable solubilization environment for nonpolar solutes. Most cosolvents are characterized by hydrogen bond donor and acceptor groups that interact strongly with water and help ensure mutual miscibility in practically any proportion. They also have small hydrocarbon regions that do not interact strongly with water. These hydrocarbon regions reduce the ability of the aqueous system to squeeze out nonpolar solutes. As a result, cosolvency is a highly versatile and powerful means of solubilizing nonpolar solutes in aqueous media (Yalkowsky 1999). Typical cosolvents include *N*-methyl pyrrolidone, 2-pyrrolidone, dimethylsulfoxide, polyethylene glycol 400, and dimethylacetamide. The solubility profile as indicated in Fig. 2.3 follows a log-linear relationship with cosolvent concentration. Therefore, cosolvent-based formulations, if formulated near their solubility maximum, will supersaturate upon dilution during parenteral or oral administration. With supersaturation comes the possibility of precipitation in the in vivo environment causing low or variable exposure. Orally, this may sometimes be mitigated using small amounts of polymers or surfactants (Gao et al. 2004; Xu and Dai 2013). Toxicity and tolerability of cosolvents should be a consideration and often limit their use for studies with high dose requirements (necessitating larger volumes of cosolvents to be dosed) or studies with long duration (such as toxicology testing). When used in parenteral formulations, hemolysis may occur, which can cause pain due to the release of hemoglobin from erythrocytes into the plasma. An in vitro screening approach, as described by Reed and Yalkowsky (1985) may be used to evaluate the hemolytic potential of formulations. Additionally, cosolvents (particularly PEG400) have been shown to interfere in mass spectrometry based bio-analytical methods due to ion suppression, in which the analytical response for the compound of interest is reduced due to the coelution of an excipient (Larger, Breda et al. 2005). Once identified as an issue, this problem can generally be overcome, either by altering the HPLC method or by reducing or eliminating the excipient responsible for interference.

2.2.6 Lipids

Lipid-based formulations can be an attractive formulation approach for molecules with high $\log P$ (>4). These formulations can include simple oils to emulsions, microemulsions, self-emulsifying and self micro-emulsifying drug delivery systems (SEDDS/SMEDDS). SEDDS and SMEDDS are mixtures of lipids, surfactants, and cosolvents that disperse in aqueous media to form emulsions or microemulsions and have been effectively used to increase exposure of highly lipophilic molecules and been reviewed extensively in the literature (Porter et al. 2008; Pouton and Porter 2008; Williams et al. 2013). The in vivo performance of these formulations depends on how they are processed in the gastrointestinal tract. For example, formulations composed of long chain triglycerides undergo lipolysis and the digestion products are further solubilized by bile salt–lecithin complexes, resulting in the formation of fine colloidal dispersions that bypass first-pass metabolism and are predominantly absorbed through the intestinal lymphatic system. Thus, bioavailability of compounds formulated in this manner can be greater than what might be achieved when solubilized compound is absorbed through the standard mechanism via the portal system. For example, lipid-based formulations make it possible to effectively deliver testosterone derivatives via the oral route by targeting the lymphatic system and reducing first pass liver exposure (Dudley 2011; Yin et al. 2012). In order for compounds to be amenable to these formulations, their solubility in the lipid system should be sufficiently high to support the dose requirements for animal studies. This is often a limitation to the use of this formulation approach. Additionally, chemical and physical stability of the compound in the vehicle/dosage form can sometimes be a major hurdle for long-term use and should be studied carefully (Pouton and Porter 2008). The impact of lipid-based components on the clinical pharmacological parameters being assessed in the study, and their safety and tolerability also need to be assessed as they can significantly restrict the amounts used and the duration of the studies. Despite the barriers, it is possible to leverage the numerous advantages offered by lipid-based formulations. A recent review by Chen et al. describes an effective strategy for incorporating lipid-based formulations into discovery flow schemes, such that the properties of the chemistry templates can be appropriately influenced in order to make them viable candidates for lipid-based formulations. This is especially valuable when the intrinsic properties of the biological targets do not lend themselves to ligands that can be delivered through conventional formulation approaches (Chen et al. 2012).

2.2.7 Solid Dispersions and Supersaturation

Amorphous solid dispersions are enabled oral formulations that have received a great deal of attention in the discovery phase. This is primarily due to the observed increases in exposure in animal and human testing, the small scale in which solid dispersions can be manufactured, and relative safety of the excipients used. Amorphous solid dispersion formulations are dispersions of amorphous drug in a carrier matrix (usually a polymer). They form supersaturated solutions upon dosing, thereby increasing the flux across the intestinal membrane. With appropriate choice of the polymer, it is often possible to sustain the duration of supersaturation for several hours, thereby overcoming absorption limitations due to low equilibrium solubility. Polymers and other excipients used to make amorphous solid dispersions generally have greater acceptable daily intake (ADI) amounts compared to the excipients used to make the simpler formulations described in the previous paragraphs, which makes solid dispersions an attractive option for longer term studies. A number of innovative products have reached the market in recent years which have been developed as solid dispersions in order to overcome solubility limitations of the crystalline forms of the drugs that were found to negatively impact the performance of the drug product (Vo et al. 2013). The use of amorphous solid dispersion formulations in discovery has also been described (Verreck et al. 2003; Vasconcelos et al. 2007; Bikiaris 2011). In our experience at Lilly, amorphous solid dispersions have been successfully applied for oral dosing in both toxicology and clinical studies and have resulted in significant improvements in plasma exposure and decreases in variability compared to conventional formulations with crystalline material.

Solid dispersions present a greater level of complexity when compared to the other formulation approaches that have been described. They require greater resources for formulation development and preparation of supplies for in vivo studies. In addition, the chemical and physical stability of the solid dispersion formulation must be carefully evaluated to ensure that it possesses sufficient handling and storage characteristics for use in the desired study. Briefly, the development process includes small-scale experiments to select the drug–polymer combination that results in the best dissolution profile, followed by a slightly larger, but still milligram scale set of experiments, to assess thermal properties and physical/chemical stability of the formulation (Six et al. 2004; Vandecruys et al. 2007; Qian et al. 2010). Owing to the nature of the manufacturing processes for amorphous solid dispersions, adequate overages need to be built in to material estimates to cover for loss during production and handling. This option is therefore utilized only for compounds that do not lend themselves to any other options, or when the specific needs of the in vivo studies preclude the use of excipients that would be required with the other available formulation approaches.

2.3 Formulation Strategy

The previous section describes *Formulation Approaches* for drug solubilization using various techniques (aqueous and cosolvent) or the use of particulate/solid dispersion systems. The question remains as to a viable *Formulation Strategy* to determine which approach is best suited to the molecule of interest. A number of recent publications have presented flow charts or high throughput screening paradigms as a means to "zero in" on formulations most amenable to the compound being tested (Chaubal 2004; Li and Zhao 2007; Maas et al. 2007; Saxena et al. 2009; Balazs 2011). These flow charts tend to be a linear progression of various in vitro assays, evaluating a broad range of excipient classes and solubilization methods. Logical in design, these approaches ultimately require an iterative process of in vivo testing and reformulation in order to identify a formulation capable of producing the desired exposures. This may not be conducive to the

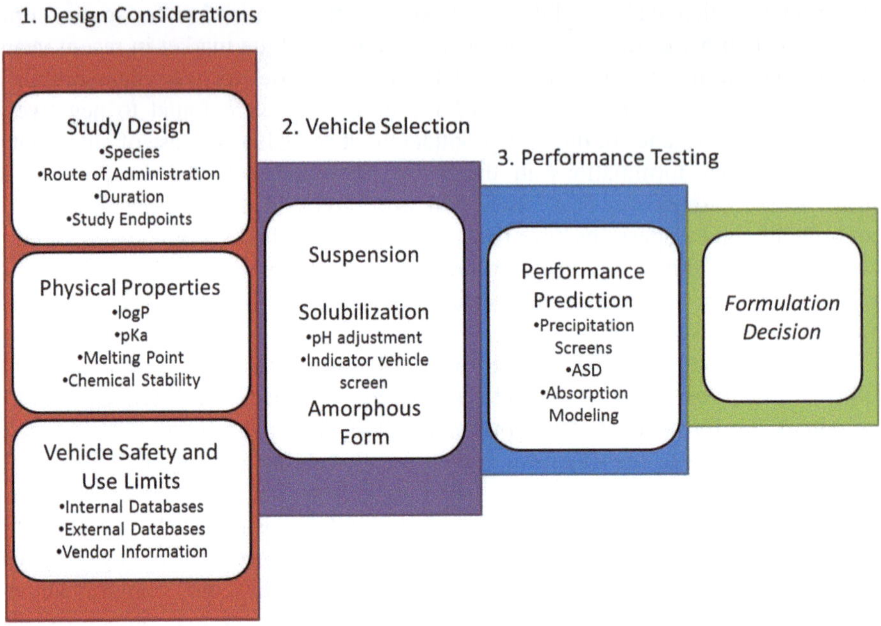

Fig. 2.4 Streamlined paradigm for the selection and evaluation of preclinical formulations

speed required in the discovery phase because iterations can be time consuming and require relatively large quantities of drug.

An alternative strategy for identifying preclinical formulations in a more streamlined manner is outlined in Fig. 2.4. This strategy relies on three integrated steps, each with its own unique set of models and tools. The first step consists of an assessment of the physicochemical properties of the compound combined with a careful evaluation of the in vivo study parameters. The physicochemical properties for the compound of interest are obtained through in vitro measurement (e.g., solubility, pH stability, permeability) or through the use of in silico models. This allows a molecule to be described by a few fundamental properties that can be tied to potential methods of solubilization, such as ionization potential (p*K*a) and lipophilicity (log *P*). This must then be considered within the context of requirements for the in vivo study such as species, dose, route of administration, duration, etc. Based on these data, an initial list of preferred vehicles is generated. The list of vehicles is further narrowed by incorporation of important excipient data related to safety and potential pharmacokinetic/pharmacologic interference. These data are derived from a number of different sources, including external data from the literature, vendors, and from compilations of extensive internal in vivo study data. At the completion of this evaluation, an initial hypothesis is generated as to the general types of vehicles that would likely be successful in meeting the requirements of the study as well as what methods of solubilization would best take advantage of the inherent structural properties of the molecule. In doing so, the relatively exhaustive list of possible excipients can be narrowed simply by eliminating those vehicles that are not compatible with the functional groups present in the compound or the design of the in vivo study.

The next step in this strategy centers on the use of in vitro screening tools to determine if solubility targets are achieved. As shown in the figure, vehicles are subdivided into three broad categories: suspensions, solubilized formulations, and stabilized amorphous formulations. From the analysis conducted previously, one or more of these classes would have been identified as an appropriate starting point for formulation development, based on the type of compound and study. Studies where only a single dose is to be administered, such as a pharmacokinetic study, we have found that an appropriate vehicle can be selected from a defined list of "standard" vehicles, which has been developed using institutional knowledge. These vehicles, as well as recommended characterization to facilitate interpretation of in vivo data are summarized in Table 2.2. Rather than screening all possible excipients and

Table 2.2 Standard vehicles for use in single dose studies

Route	Formulation	Minimum characterization
Oral suspensions	•HEC 1 % w/v with PS80 0.25 % v/v and simethicone 0.05 % v/v •Acacia 10 % w/v with simethicone 0.05 % v/v	Visual, microscopy
Oral enabled formulations	•Cyclodextrin (20 % w/v) with or without pH adjustment	Microscopy of solid forms
	•PEG 600 90 % v/v, Solutol® HS 15 10 % v/v	
	•Soybean Oil 80 % v/v, Capmul® PG8 20 % v/v (if clogP>5)	
	•Gelucire® 44/14 100 %	
	•Solid dispersion (30–50 % w/w polymer)	
	•Nanosuspension (PVP 2 % w/v and SLS 0.15 % w/v for steric stabilization and wetting)	
PK Intravenous dosing (single dose only, 1 mL/kg)	•SBEβCD 20 % w/v, 25 mM pH 2 or 8 $NaPO_4$ buffer	In vitro plasma precipitation screen (if cosolvent)
	•Microemulsion 20 % water	
	•DMA 10 % v/v, EtOH 15 % v/v, PG 30 % v/v in 25 mM pH 2 or 8 $NaPO_4$ buffer	
	•DMA 25 % v/v, EtOH 15 % v/v, PG 10 % v/v, 2-pyrrolidone 25 % v/v (last resort!)	
Intraperitoneal/subcutaneous dosing	•SBEβCD 20 % w/v in 25 mM pH 3 or pH 8 $NaPO_4$ buffer	Visual
	•NMP 10 % v/v, Captex 300 or soybean oil 90 % v/v (Last resort!)	

Solutions/suspensions prepared using deionized water

HEC=hydroxyethyl cellulose, PS80=polysorbate 80, PEG=polyethylene glycol, PVP=polyvinylpyrrolidone, DMA=dimethylacetamide, NMP=*N*-methylpyrrolidone, PG=propylene glycol

combinations thereof, a vehicle is selected from this abbreviated list which contains vehicles for each route of administration that we have found have the highest success rate for the types of study. In cases where solubilization has been proposed as a means to achieve the desired exposure target, we have devised a means of screening a broad range of solubilization methods using a very limited set of representative vehicles as opposed to conducting solubility determinations in large numbers of individual excipients. If the results point to intractable solubility space where it would be necessary to use vehicle(s) not amenable to the study design (e.g., aggressive cosolvents in long-term toxicology testing) then alternate technologies such as solid dispersions or adjustment in dose regimen are considered.

As will be discussed later in this chapter, toxicology studies are somewhat unique from most other in vivo studies conducted in the discovery setting. The need to achieve high exposures, either by administering large doses of compound or through the use of enabling formulations, can make the identification of a suitable vehicle challenging. For short investigative toxicology studies, a relatively wide range of vehicles may be used, as excipient toxicity should be minimized due to the limited duration. However, from the standpoint of formulation development, toxicology studies conducted late in the discovery framework are primarily designed to identify acceptable vehicles for use in longer duration FHD-enabling studies in early clinical development. As a result, greater consideration must be given to both the safety profile of the excipients used in these formulations as well as the complexity of manufacturing and formulation stability that will be required.

Prediction of the potential in vivo performance of formulations makes up the final step in designing and implementing an efficient formulation strategy. This is of particular utility when there are several viable formulation options to choose from or when considering high dose administration typically encountered in toxicology studies. As discussed previously, the use of in vitro systems is preferred as they allow for evaluation of formulations without the need for large numbers of costly and time-consuming in vivo studies. A particularly useful tool for this purpose is the Artificial Stomach Duodenum model (ASD) (Carino et al. 2006), which is a dynamic dissolution system that simulates the pH and mass transfer of the stomach and duodenal compartments. By comparing of the duodenal dissolution profiles of various test formulations, the relative supersaturation of each may be evaluated, which theoretically correlates with the rank order of absorption of compounds displaying solubility limited absorption. This type of system has been further simplified by Gao and coworkers (2010) who described a pH-dilution method which mimics the relevant pH, volumes, and transit times in the gastrointestinal tract of the rat. Much like the ASD, this simple method has been used to estimate regional changes in drug concentration along the GI tract for various formulations. Results from these types of in vitro systems are often used in conjunction with absorption modeling, which will be discussed in more detail later in this chapter. Absorption modeling is used to predict the relative in vivo performance of formulations by simulating plasma exposure profiles, which makes it possible to explore many different hypotheses simply by modifying the simulation parameters.

In the discovery phase, the formulator is often asked to develop formulations that would be generally acceptable for any compound within a given scaffold of interest. This enables compounds to be progressed rapidly through in vivo assays without incurring delays associated with developing novel formulations for each compound one at a time. As has been outlined in the aforementioned strategy, scaffold-wise formulation recommendations must also be based on careful assessment of an adequate number of compounds, and a proper understanding of the relationship between specific structural motifs, compound properties, and the underlying approach to the formulation. Above all, it is essential that there is a well-defined feedback loop within the discovery team, so the preformulation/developability scientist is aware of the performance of the formulations, and any unusual observations with regard to the physical appearance of the formulations, or the in vivo response and/or exposure.

While many different approaches to formulation strategy may exist across the pharmaceutical industry, they share a common goal of working to identify formulations that support the progression of new molecules through the discovery pipeline in a rapid and efficient manner. Given the highly complex nature of the process of drug discovery, it is clear that formulation development plays a significant role in the overall success or attrition of discovery projects.

In the sections that follow, additional discussion and examples of formulation development for various types of studies are presented.

2.4 Pharmacology Formulations

It should be appreciated that preclinical formulations strongly influence the link between pharmacology, pharmacokinetics, and pharmacodynamics. The biological targets being explored today are far more complex than those of a decade ago (Hopkins and Groom 2002), and the cost of typical pharmacology studies can exceed \$40,000–50,000 due to the highly sophisticated nature of the design (e.g., in vivo efficacy studies in xenograft models) as well as long lead times due to study preparation. Finally, as a result of the recent emphasis on translational research and the development of biomarkers, there is a much greater focus on the identification of both outcome and mechanism biomarkers early in discovery (Kwong et al. 2011). It is therefore critical that the formulations used in pharmacology studies are designed to perform consistently and reproducibly in order to meet the needs of these studies and to drive the right decisions.

Formulation needs for pharmacology studies gradually shift over the different phases of discovery efforts as presented in Table 2.1. In the early phase of most projects, the primary goal is to test a biological hypothesis for a mechanism of action or to validate a novel biological target. Many experiments are run with compounds that have not yet been optimized and therefore, have poor druggability properties, including incomplete absorption upon oral dosing and/or rapid clearance. In addition, there is rarely enough information available at this stage

Table 2.3 Summary of the various routes of administration used in in vivo studies

Route of administration	Typical reasons for choice	Comments
IV (bolus or infusion)	1. Maximize exposure by avoiding first-pass metabolism, e.g., cell cycle targets in oncology and in acute invasive studies in metabolic disorders.	Solubility could be a limiting factor for amount that can be delivered through this route.
	2. To overcome oral absorption limitations such as permeability and first-pass metabolism.	
	3. IV is intended route in clinical development.	Injection site irritation potential may also limit the use of this route (Turner et al. 2011).
	4. Infusions are used to achieve sustained exposures to assess PD response at longer time points.	
SC	1. Similar to IV, but this route is also amenable to suspension formulations for bolus dosing. This is useful when solubility limitations preclude use of the IV route (neurodegenerative, behavioral pharmacology, diabetes targets).	Dose volume, solubility (for solution-based formulations), or suspendability for (suspensions) may be limiting factors.
	2. To overcome oral absorption limitations such as permeability and first-pass metabolism.	Injection site irritation may limit the use of this route.
	3. Use of osmotic pumps for sustained target engagement, especially with rapidly cleared compounds.	In addition, in pain studies, e.g., pain due to injection can obscure the efficacy of the compound.
IP	1. Compounds with absorption limitations due to low permeability across intestinal mucosa.	Compounds will enter the portal vein immediately upon dosing and be subject to first-pass extraction similar to oral route (Lukas et al. 1971). The ability of the lymphatic system to drain the peritoneal cavity may be important in the absorption of proteins and large molecular weight compounds (Mactier et al. 1987)
	2. Mechanism of action studies (e.g. intraperitoneal glucose tolerance test in diabetes) to assess PD outcome independent of incretin response in the gut.	
	3. To avoid stress of oral dosing in some behavioral pharmacology studies.	

(continued)

Table 2.3 (continued)

Route of administration	Typical reasons for choice	Comments
IC/ICV/IT (IC: intracerebral; ICV: intracerebroventricular; IT: intrathecal)	1. To deliver high local concentrations of drug directly to CNS or specific tissues in the brain, to demonstrate/understand mechanism of action (pain pharmacology, neurodegenerative, and behavioral pharmacology).	Solubility could be a limiting factor, especially for compounds that will require high doses in order to saturate P-gp efflux.
	2. To overcome P-gp efflux that inhibits compounds crossing the blood–brain barrier from plasma.	
PO—oral gavage	1. Most common route. Applied across target-to-hit, hit-to-lead, and lead-optimization phases.	Gastric pH variability (in dog) can result in significant PK variability for basic compounds. Can be overcome by formulating the compound as in situ salts or in acidic media with sufficient buffer capacity.
	2. Essential route to bridge to efficacious exposures in humans for orally administered drugs.	
PO—in feed dosing (Mu et al. 2006), or in water	Convenience, sustained plasma concentration, large doses, combination therapies, or to induce disease states through drugs.	Taste factors and feeding patterns can affect intake. Food wafers have been reported to be useful (Ferguson and Boctor 2009). Modeling approaches may be used to guide amount incorporated into the water or feed.

regarding the duration of target engagement needed to elicit the desired pharmacologic response. As a result, formulations used in this phase of work are designed to provide maximum and/or sustained exposure and often use nonstandard routes of delivery to help further explore these concepts. Table 2.3 lists the various routes of administration used in in vivo studies and considerations for selecting one over another. Table 2.4 summarizes examples of vehicles and typical dose volumes for the preclinical species commonly used in pharmacology studies. In general, for projects that are focused on delivering clinical candidates for oral administration, pharmacology studies utilizing nonoral routes should be considered nonstandard and the exposure data obtained should not be used in making assessments of the potential for oral absorption.

The approach to identify formulations for these studies utilizes the strategy previously described, taking into account intended goal of the study, the need for single vs. chronic dosing, safety of formulation excipients, and any vehicle or dosing effects on the pharmacologic response or clinical parameters secondary to the pharmacologic response. The Lhasa Vitic Nexus database contains an exhaustive and continuously updated repository of information on excipient safety in preclinical species. Given that chronic dosing is often required in animal

Table 2.4 Summary of vehicles example and typical dose volumes for the preclinical species commonly used in pharmacology studies

Route	Species and dose volume (mL/kg)	Examples of vehicles	General comments on formulation
IV	Rat: 1 mL/kg (≤5 mL/kg recommended max) Mouse: ≤5 mL/kg Dog: 1 mL/kg (≤2.5 mL/kg may be acceptable if needed)	1. Deionized water with pH adjustment for solubility (acceptable pH range 2–8), or 25 mM buffers, pH 2 or pH 8. 2. ≤20 % cyclodextrin (e.g., SBEβCD, HPβCD), in 25 mM pH 2 or pH 8 buffer, and/or with pH adjustment in the range of 2–8. 3. Cosolvent based (e.g., 10 % DMA, 15 % EtOH, 30 % propylene glycol in pH 2 or pH 8 buffer).	Minimum volumes and lowest acceptable dose must be used to avoid formulation failure. Risk of precipitation upon injection must be tested using a precipitation screen.
SC	Mouse: 10 mL/kg Rat: 1 mL/ recommended (≤10 mL/kg may be acceptable) Dog: 1 mL/kg recommended (2 mL/kg may be acceptable)	Normal saline with small amounts of suspending and/or wetting agents (e.g., 10 % Cremophor EL, 1 % Hydroxypropylcellulose, 0.085 % Polyoxyl-50-stearate). Cosolvent based formulations may be used for osmotic pump infusions (e.g., 1:1 PEG400: DMSO).	Vehicles must be tested for irritation potential, especially for repeat dose studies. Osmotic pump formulations must use only excipients compatible with pump components. Formulations must be tested for delivery and precipitation potential.
PO	Mouse, rat: 10 mL/kg Dog: 5 mL/kg Monkey: 3 mL/kg	Solution: •DI water with pH adjustment for solubility (acceptable pH range 2–8), or 25 mM buffers, pH 2 or pH 8. •≤20 % Captisol or HPBCD with pH adjusted to 2 or 8. 0.5 M phosphoric acid, pH 2. Suspensions: •1 % hydroxyethylcellulose, 0.25 % polysorbate 80, 0.05 % Antifoam in DI water. •10 % acacia, 0.05 % antifoam in DI water.	*Examples of vehicle effects*: Cremophor EL, Triton X-100, Polysorbate 80, Solutol HS15, PEG400 have been known to alter plasma lipoproteins, resulting in significant interference with metabolic disorder studies. The mild anti-inflammatory effect of acacia has been known to impact arthritis models when used in suspension vehicles.
IP	Mouse: 20 mL/kg Rat: 10 mL/kg	Solution or suspension formulations	Solubility of the compound can be a limiting factor for absorption when dosed as a suspension.
IC/IV/IT	Rat: ≤10 μL Mouse: ≤5 μL	Normal saline, phosphate buffered saline, artificial CSF	

pharmacology models, this data is useful in selecting formulations that have sufficient safety for the proposed study duration. Several reviews on preclinical formulation and related topics describe examples (ten Tije et al. 2003; Gad et al. 2006; Neervannan 2006; Li and Zhao 2007; Maas et al. 2007; Pole 2008; Shah and Agnihotri 2011) of vehicle effects that interfere with pharmacology studies. Acacia (a commonly used suspending agent) interferes with pain and inflammation models (Lilly internal experience and (Dafallah and al-Mustafa 1996)), and therefore should generally be avoided for these types of studies. Figures 2.5 and 2.6 illustrate vehicle effects from Lilly internal experience. Figure 2.5 shows inflammation clinical scores upon dosing 10 % acacia and 1 % hydroxyethylcellulose as aqueous dispersions at a dose of 4 mL/kg. These are commonly used suspension vehicles, but as seen from the figure, the 10 % acacia vehicle has a positive anti-inflammatory effect that precludes the use of this vehicle in this pharmacology model. Figure 2.6 shows the effect of some standard formulation excipients on insulin release in the Oral Glucose Tolerance Test in mouse. As indicated in the figure, PEG400 has a statistically significant negative effect on insulin release that makes it unacceptable for use in this model. Another example of vehicle effects found in the literature is that of Cremophor EL (an emulsifier used in parenteral formulations), which affects plasma lipoproteins when used at concentrations greater than 0.4 mg/mL (Woodburn and Kessel 1994). These examples illustrate the importance of checking for vehicle effects either through experimentation or literature examples prior to making decisions for pharmacology studies.

2.4.1 Osmotic Pumps

The preceding paragraphs describe aspects of formulations and various routes of administration in animal models. However, overcoming poor exposures resulting from rapid clearance poses a unique challenge in pharmacology studies. Furthermore, there may be a need to maintain sustained plasma concentrations over several hours to gain useful mechanistic insights in some pharmacology models. Typical examples are kinase inhibitors where sustained target engagement is essential to block the signaling pathways responsible for cancer cell growth. This requirement, coupled with rapid clearance, is a fairly typical challenge faced during the target validation phase of many discovery projects. The simplest way to address clearance issues is to dose the compound multiple times a day. However, this is not always practical. For example, the increased handling of animals for administering multiple daily doses can cause stress that could confound the PD response in neuroscience studies (Gartner et al. 1980). Gastric-retained gel formulations have been used with some success to modulate pharmacokinetic profiles of rapidly cleared drugs (Foster et al. 2013) but in vivo performance is somewhat difficult to predict based on in vitro assessment.

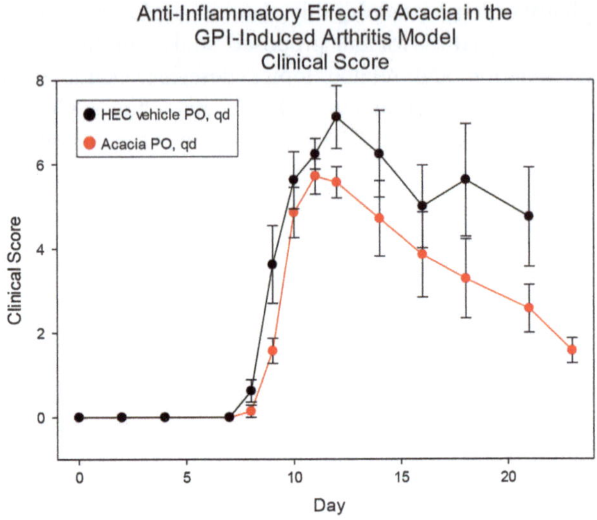

Fig. 2.5 Illustration of the effect of acacia on GPI-induced arthritis model in mouse

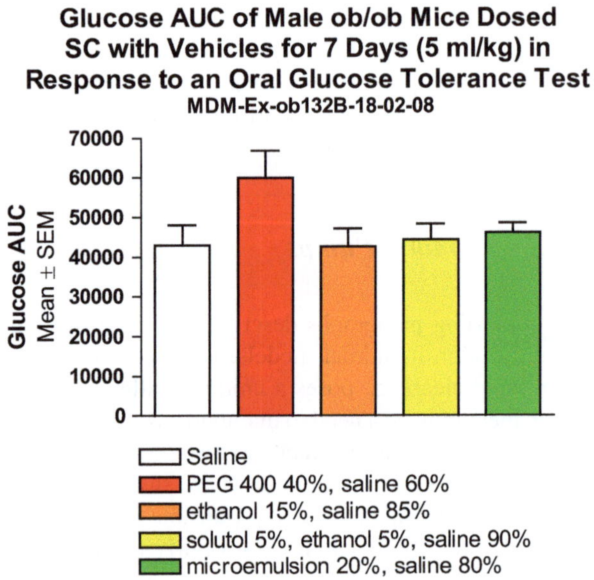

Fig. 2.6 Illustration of the effect of PEG400 on insulin response in the mouse OGTT model

One of the more widely used approaches to maintain sustained plasma profiles for extended durations of time is through infusion of the drug through parenteral routes. This is commonly achieved through the use of surgically implanted osmotic pumps. Osmotic pumps deliver at a constant rate and can be used to maintain a nearly constant plasma concentration and thus continuous target engagement for up to 2 weeks. In contrast to conventional dosing options that result in large peak to trough ratios, osmotic pumps also offer the secondary benefit of minimizing the total dose by eliminating the portion of the AUC in the plasma concentration profiles that is above the threshold for activity/efficacy. An excellent example of this was reported by Kumar et al. who reported on comparative in vivo efficacy responses upon dosing orally and via osmotic pumps implanted subcutaneously. Efficacy was assessed as a reduction in tumor volume and decrease in hemoglobin in the biopsy tissue (the latter is a measure of antiangiogenesis). As seen from their work, the doses required for efficacy were significantly lower when the drug was delivered via the SC osmotic pump as compared with oral (Kumar et al. 2007). These pumps are available in a wide range of capacities and delivery rates to suit the various preclinical animal models that are used in pharmacology studies (Alzet). Newer pump models such as those from Iprecio (Iprecio) are programmable for variable flow rate if needed, and re-fillable, thus enabling larger doses and/or longer duration studies. Formulation technologies that provide extended release are described in Chap. 3.

The first step in developing a formulation for osmotic pump studies is to select an appropriate pump model based on the animal species being used and the duration of the study. Based on the capacity of the pump, the volumetric delivery rate specific to the selected model, and the desired plasma concentration, the required formulation concentration is then estimated using the following equation with appropriate unit conversions:

$$\text{formulation concentration} = \frac{\text{steady state plasma concentration} \times \text{clearance}}{\text{pump delivery rate}}$$

In this equation, the steady state plasma concentration is the desired concentration that the pharmacologist intends to investigate and the clearance value is either estimated in silico or obtained from a previous pharmacokinetic study in the same species. It is important to note that while this equation is more frequently used to estimate steady state concentrations following administration by intravenous infusion, it can still serve as a simple method for approximating concentrations derived from continuous subcutaneous infusion as well. In doing so, an assumption must be made that the bioavailability following subcutaneous administration is 100 % relative to an intravenous dose.

Unlike subcutaneous bolus injections that can be formulated as suspensions in isotonic vehicles, only solution-based formulations are acceptable for osmotic pump delivery. However, given the small volumes delivered through the pump, high concentrations of nonaqueous solvents may be used, as long as they are water miscible and used in amounts that are compatible with the pump components. Alzet infusion pumps are known to be compatible with a wide variety of different types of media, and in the absence of available solubility data for the test compound in these media, extensive screening may be required to identify the optimal formulation. In practice, however, the screening and selection of a formulation may be done more efficiently and with less compound by evaluating an abbreviated list of solvent systems, generally categorized by the amount of organics present, and thus the overall aggressiveness of the formulation. For example, a typical set of solvent systems would likely include at least one aqueous-based system as well as a 1:1 mixture of an organic (i.e., DMSO) with water, and a very aggressive formulation consisting entirely of organics. Based on the results of this initial screening, further formulation optimization may be applied as needed. This approach allows for conservation of material which is often very limited at this stage of development. The formulation thus developed is then tested for precipitation potential during delivery. This is of critical importance since precipitation can result in clogging of the pump resulting in a complete failure of the study. This may be accomplished by simply filling the pumps with the proposed formulation(s) and incubating them in normal saline or a blood surrogate buffer and then monitoring the appearance of compound in the media as a function of time. Additional studies may also be performed to test the chemical stability of the compound in the selected formulation and compatibility with pump components if needed (Gullapalli et al. 2012).

2.4.2 *In-feed Dosing*

Sustained plasma exposures for pharmacology studies can also be achieved using in-feed dosing options. This approach is based on the fact that rodents eat at frequent intervals and their feeding patterns through the light and dark phases of the day are well understood. This information, along with the desired plasma concentration and clearance data, makes it possible to calculate the amounts of compounds to be incorporated into their feed. Both solid and liquid diets may be used, and combinations of multiple compounds may be dosed simultaneously with the feed as desired. Formulation of the active compound(s) with the feeds may be done in-house, or through labs that offer these services (Research Diets). As rodents eat approximately the equivalent of one tenth their body weight of food every day, the fraction of active drug in the feed is fairly small (e.g., about 0.2 % for a dose of 200 mg/kg).

In-feed dosing offers several advantages over continuous infusion pumps. It eliminates the need for solubilization of compounds in small volumes of formulation solvent and the associated risk of precipitation. It minimizes handling of the animals for implanting the pumps (and the wound healing process that follows) and allows for significantly longer term dosing. Lastly, with compounds that cause injection site irritation, or for pain and inflammation projects that want to avoid the injury caused by pump implantation, this is the preferred option for achieving sustained plasma concentrations of test compounds. One example of the application of in-feed dosing is with sitagliptin, a DPP-4 inhibitor that has a short half-life of 1–2 h in mice. Mu et al. (2009) were able to demonstrate chronic glycemic control over 10 weeks, with in-feed dosing of 280 mg/kg of this compound in mice (equivalent to 0.3 % w/w of the mouse diet). Important considerations in using this type of formulation include variability in exposure due to eating, binding to food, assurance of homogeneity of dosage form, and stability.

2.5 Pharmacokinetic Studies

Greater emphasis is increasingly being placed on early in vivo characterization and evaluation of key compound pharmacokinetic properties in order to select molecules that possess that greatest likelihood of long-term clinical success. After initial screening through batteries of in vitro biochemical and physicochemical assays, promising compounds are typically evaluated in a single-dose pharmacokinetic (PK) study, usually in a rodent species such as the rat. These studies are designed to include both an intravenous arm, as well as a second arm that approximates the intended route of administration in man, usually oral. The primary goal of these studies is to filter compounds with poor ADME characteristics, as well as to begin to develop a more mechanistic understanding of these properties in order to influence the SAR toward design of better molecules.

One of the more challenging aspects of conducting a pharmacokinetic study is the identification of a suitable vehicle to be used to solubilize the compound for use in the intravenous arm of the study. In the early discovery setting, compounds selected for testing often possess suboptimal physical properties (i.e., low solubility, high log *P*). In addition, the relatively large number of compounds selected for in vivo testing, as well as the need for very rapid data turnaround, presents significant challenges in screening and evaluating potential IV vehicles. A number of general approaches for the identification of a suitable IV formulation have been published previously. These approaches typically involve a very methodical screening of a variety of different options until a suitable solution is found. Lee and coworkers (2003) proposed a decision tree for use in early discovery that allows for selection of a suitable formulation using observations of experiments in which various pH and cosolvent concentrations are tested, based on the underlying physicochemical properties of the molecules in question. Similar approaches have been utilized across the pharmaceutical industry. In practice, however, these types of approaches are often impractical in the discovery setting, due to limitations on material for analysis and testing, large numbers of compounds under consideration, and challenging time constraints. In keeping with the formulation strategy described previously, the formulation scientist must be able to make formulation decisions by eliminating as many options as possible based solely on the properties of the molecules in question (i.e., p*K*a, log *P*, MW, melting point) and then rely on very limited in vitro screening to narrow the list of potential vehicle options. Initial attempts at solubilization often rely on the use of a complexing agent, such as a cyclodextrin in either an acidic or basic aqueous buffer solution, depending on the ionization properties of the molecule in question. Compounds that are not amenable to solubilization in these vehicles are then evaluated in more aggressive vehicles containing increasing amounts and types of organic cosolvents. It is important to note that these more aggressive vehicles, while well tolerated in single-dose studies, are not likely to be compatible with repeat dosing due to toxicity following chronic administration of the excipients. While solubilization is the primary goal of these studies, it is also important to evaluate the potential for precipitation of the compounds from the solubilizing vehicle. Several publications have described the use of in vitro precipitation screens to study precipitation from injectable formulations (Sheth 2011). Several in vitro screening methods are described, one utilizing static dilution into aqueous media and the other evaluating dynamic injection. These screens were used to assess the potential of in vivo precipitation of nine injectable formulations, which were selected from marketed products as well as several from internal Merck research programs. Good correlation was observed between results from the static and dynamic models. In addition, the in vitro data was found to correlate with instances of precipitation that were noted during preclinical testing in animals as well as clinical testing.

In addition to developing injectable formulations for individual compounds, the need to rapidly evaluate pharmacokinetic properties for larger numbers of compounds has led to the use of different approaches which have been proposed to improve the throughput and efficiency of these studies. One of the most promising techniques is cassette dosing, or N-in-1-dosing, which involves simultaneously administering a set of compounds in a common vehicle, as opposed to discrete dosing in which a single compound is administered (Nagilla et al. 2011). This approach allows for a reduction in the number of both studies and animals required, as well as the generation of fewer samples for bioanalytical analysis. Several independent analyses have demonstrated that PK parameters obtained through cassette dosing are comparable to those derived from dosing compounds (He et al. 2008; Nagilla et al. 2011) discretely. Despite these advantages, careful consideration must be given to the experimental design when choosing to conduct in vivo studies using cassette dosing. The identification and selection of a common vehicle for a series of structurally unique compounds can be challenging. When possible, compounds should be grouped together by structural class in order to take advantage of common chemical features. Compounds within a given class may behave similarly in terms of the mechanism and extent of solubilization. Combining compounds with significantly different functionality (i.e., mixing acidic and basic compounds) should generally be avoided, as these differences will likely make the identification of a suitable common vehicle very difficult if not impossible.

In addition to intravenous delivery, the administration of an oral dosage form is also included as a second arm in a typical pharmacokinetic study. Some minimal threshold for oral bioavailability (i.e., 20–30 %) is then used as filtering criteria in an effort to identify and de-prioritize compounds or structural classes that possess undesired absorption risks. This approach to selecting compounds to advance is problematic for a number of reasons. A recent report comparing measured animal (mouse, rat, dog, primate) and human bioavailabilities of 184 compounds extracted from the literature showed a very poor correlation (Musther et al. 2013). Additionally, the solid state properties of molecules are not typically controlled or even fully characterized in the discovery phase. As a result, the solubility and resulting bioavailability often decrease when subsequent lots of material become available which are more crystalline. Moreover, the doses selected for pharmacokinetic studies are often not reflective of what will be used in first in human trials. At lower doses, the absorption of compounds is often rapid and complete, even for compounds with low solubility. However, at higher doses, solubility and/or permeability limitations will begin to negatively impact absorption, resulting in oral bioavailabilities significantly less than the original target values. In general, to properly use this approach, several important aspects should be considered beyond simply using this threshold as a means to filter compounds. Developing a deeper understanding of the root cause of low bioavailability provides important feedback to the discovery team so that additional hypotheses can be proposed and tested in an attempt to resolve these issues. From the standpoint of the formulation scientist, it is critical to understand what impact formulation may have on bioavailability in order to correctly identify absorption risks and apply enabling formulation strategies when appropriate. Experimental approaches to diagnosing the cause of limited absorption represent the topic of Chap. 4, but are also treated below.

Bioavailability (%F) is defined as the product of absorption and metabolism and is represented by the following equation:

$$F = \mathrm{Fa} \times \mathrm{Fg} \times \mathrm{Fh}$$

where Fa = fraction absorbed, Fg = fraction escaping gut metabolism, and Fh = fraction escaping hepatic metabolism. Consider two hypothetical compounds, A and B (Fig. 2.7). Upon oral administration, a significant amount of Compound A remains unabsorbed, while Compound B is almost completely absorbed. However, first-pass extraction by either gut and/or liver metabolism is relatively minimal for Compound A while Compound B is significantly metabolized. As a result, both Compounds A and B would be found to have similar relative oral bioavailabilities. However, it is clear from this simple example that the underlying absorption and metabolism properties of the two molecules are quite different. Further analysis and additional experimental data may be needed to fully elucidate these differences. For Compound A, improvements in solubility and/or permeability or the use of enabled formulations should be targeted as a means to increase Fa, while for Compound B, additional SAR effort would be required to reduce metabolism of subsequent compounds. In cases where bioavailability is determined to be limited by Fa, it is necessary to distinguish between solubility and permeability limited absorption. In the discovery phase, this is often relatively straightforward due to the availability of in vitro systems designed for this purpose (Caco-2, MDCK, etc.). Once solubility has been identified as the primary issue, there is a need to further interrogate whether this is related to poor dissolution or low solubility. For compounds that are dissolution rate limited, micronization, either by milling the neat API or by probe sonication of a suspension, will result in enhanced bioavailability. The impact of particle size reduction on absorption may also be assessed using

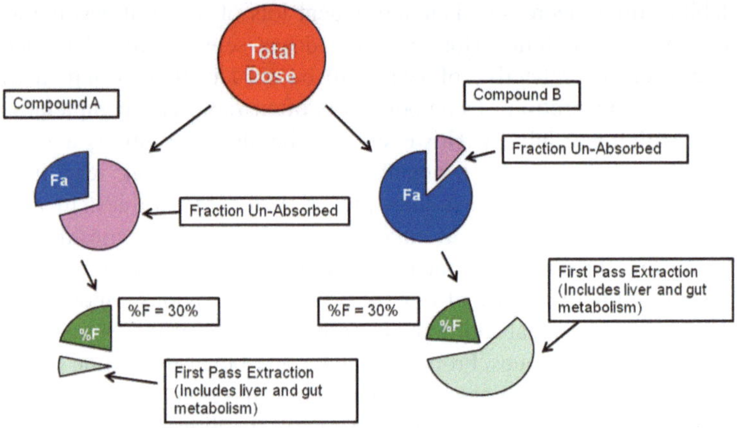

Fig. 2.7 Comparison of oral bioavailabilities of two hypothetical compounds A and B

computational approaches, such as the microscopic mass balance model described by Oh et al. (1993). In addition, commercially available software such as GastroPlus from Simulations Plus, LLC may be used to simulate the effects of changing particle size on absorption. Further reduction in particle size may be achieved through the production of nanosuspension formulations. If micronization alone is found to be ineffective in improving the dissolution properties of a compound, the addition of a surfactant such as polysorbate 80 may be added to improve the wetting properties of the material. In practice, particle size reduction is typically combined with the use of low levels of a surfactant in the formulation at the outset. When this approach still leads to lower than desired exposures, the use of solution-based formulations is then employed. This approach can range from the very simple, such as pH adjustment, to the use of complexing agents and cosolvents, and even to the development of stabilized amorphous formulation such as solid dispersions. When using these types of formulations for pharmacokinetic studies however, it is important to keep in mind the original goals of the study and to carefully evaluate the impact of the dosage form on the interpretation of the resulting data. A brief listing of general considerations for solutions and suspensions is presented in Table 2.5. For example, in cases where the type and/or extent of metabolism is being investigated, it may be desirable to utilize a solution formulation as a way to eliminate any impact of the solid state properties of the molecule. However, when the aim is to develop an understanding of the absorption properties of a solid oral dosage form, and thus gain insight into possible future development challenges, dosing of a suspension is preferred so as not to mask any absorption risks due to poor physicochemical properties. The exposure obtained following administration of a suspension in a pharmacokinetic study also provides an early indication of the likelihood of achieving sufficient exposures in subsequent toxicology studies, in which case the development of an enabled formulation might

Table 2.5 General properties of solution and suspensions

Solution	Suspension
Drug is completely dissolved	Drug is suspended homogeneously as fine particles
Drug directly available for absorption	Dissolution is necessary before drug becomes available for absorption
Unless properly formulated, drug might precipitate upon dosing	Rate of dissolution is a function of particle size and solubility
Solubility is a limiting factor for formulation; dose volume may be limiting factor due to excipient toxicity	Solubility is not a factor for formulation, but formulation viscosity may limit ability to use higher doses
Required for intravenous dosing; may be dosed orally to eliminate impact of solid state	Required when goal of study is to better understand oral bioavailability and absorption properties; solid state properties must be carefully considered

be required. It is important to note that if the data is to be used to draw conclusions about absorption, the solid state properties of the material used in the animal studies should be as representative as possible of the form that would be progressed in development. If the data is generated early in the project's lifecycle using material where solid state properties are either unknown or are found to be dramatically different than subsequent lots of material, exposure studies should be repeated with more representative material to ensure that the impact on absorption is characterized. Given this caveat, a reasonable argument can be made that in early discovery, it is not always appropriate to include an oral arm in a basic pharmacokinetic study, as the resulting data may not be relevant and may even at times result in absorption risks being over or underestimated.

Another key factor that must be considered in selecting a formulation is the potential for the formulation excipients to alter the pharmacokinetics of the test compound. The presence of suboptimal physical properties necessitates the use of vehicles containing organic cosolvents, cyclodextrins, and surfactants. A thorough review of the effects of common excipients on ADME properties was published by Buggins et al. (2007). Table 2.6 provides a summary of doses at which minimal effects are to be expected for a subset of the most commonly used excipients. For each route of administration, a maximum dose volume is specified at which the excipient's effect on the pharmacokinetics is expected to be minimal based on a relatively exhaustive search of the literature.

2.5.1 *Absorption Modeling of Animal Pharmacokinetic Data*

The underlying assumption in using preclinical species to conduct pharmacokinetic studies is that the results of these studies will have some relevance to absorption in

Table 2.6 Dose volumes of common pharmaceutical excipients at which minimal effects on pharmacokinetic properties are expected

Excipient	Recommended levels for in vivo studies
DMSO	IV and PO: Max 5 % DMSO with dose volume of 5 mL/kg (0.2 mL/kg DMSO)
EtOH	PO: Max 10 % with dose volume 10 mL/kg (1 mL/kg EtOH). Chronic dosing can influence PK due to effect on enzymes.
Propylene glycol	IV and PO: less than 3 mL/kg for pharmacology studies measuring plasma glucose levels.
PEG400	PO: Max 40 % PEG400 in formulation with 5 mL/kg dose volume (2 g/kg rats).
	IV: 40 % PEG400 at 1 mL/kg (0.4 mL/kg PEG400). Known inhibitor of drug efflux and also CYP3A, thus may enhance absorption of such compounds that are substrates.
HPβ cyclodextrin	PO: Max 20 % if 10 mL/kg (2 g/kg HPBCD).
	IV: Max 20 % if 2 mL/kg, 400 mg/kg HPBCD.
	Effect on distribution depends on protein binding, stability constant of complex.
SBEβ cyclodextrin	IV: Max 20 % if 4 mL/kg
Cremophor EL	IV and PO: Increased absorption due to inhibition of P-gp and CYP3A4, inhibits absorption by micellar entrapment.
PS 80	PO: Max 0.5 % if 10 mL/kg (0.05 g/kg PS 80).
Solutol HS 15	IV: Max 5 % with dose volume of 2 mL/kg.
	If compound is a P-gp substrate, may significantly alter PK after either IV or PO administration.
	Data suggest that Solutol or a component there-in is absorbed orally, PK of IV administered drug altered after PO administered Solutol HS 15.

humans. Chiou and Barve (1998) conducted studies to investigate the extrapolation of absorption experiments in rodents to humans. In this study, 64 compounds were selected from the literature where data existed for both species. Despite the fact that the compounds possessed a very broad range of physicochemical properties, including molecular weight (150–4,000), ionization state (acids, bases, neutral), and lipophilicity (log P −5 to +4), an excellent correlation was observed between absorption (Fa) in rats relative to humans ($r^2 = 0.975$). However, a very poor correlation was observed when a similar analysis was conducted to compare estimation of human bioavailability based on rat studies. This was attributed primarily to differences in expression of metabolizing enzymes in the intestines of the two species (Fg and Fh). Physiological differences between species must also be carefully considered both in the experimental design as well as in the interpretation of the data. For example, gastric pH has been shown to vary considerably in the dog, with values of fasting gastric pH reported to range from 1.8 to as high as 6.8 (Lui et al. 1986; Yamada and Haga 1990; Akimoto et al. 2000). In humans, however, gastric pH has consistently been shown to be less than 3, regardless of

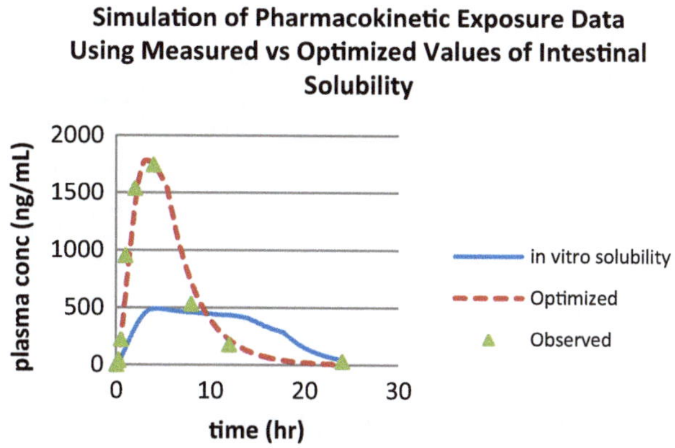

Fig. 2.8 Comparison of GastroPlus™ simulations using either in vitro solubility or optimized solubility to represent intestinal solubility of the test compound

the method used to measure it (Fancher et al. 2011). As a result of these differences, caution must be used in using the dog as a model of human absorption, especially for test compounds with ionization constants in the range of 5–8. To overcome this issue, the use of pentagastrin is common to control the canine stomach in a range that is more relevant to human fasting conditions, while the proton pump inhibitor (PPI) famotidine is recommended to simulate elevated stomach pH conditions.

Estimation of absorption risk in the discovery setting is a key activity that provides discovery scientists with a relatively straightforward method of selecting scaffolds that will ultimately achieve sufficient oral exposure to allow for testing of the clinical hypothesis. Once exposure data becomes available from rodent and/or canine pharmacokinetic studies, absorption modeling using commercial modeling packages such as GastroPlus™ may be used to form an initial assessment of absorption potential and risk. This approach requires a minimal set of measured parameters, which can be obtained from both in silico tools as well as experimentally. These parameters serve as inputs to the software, which when combined with prebuilt physiological variables, allow the user to evaluate an initial fit of the experimental plasma concentration data. In the event a good fit is not obtained, key parameters related to absorption (i.e., solubility, permeability, particle size, etc.) may then be optimized until an adequate fit is obtained. An example is highlighted in Fig. 2.8. Fitting of the observed exposure data was not possible using a value for intestinal solubility taken directly from an in vitro solubility experiment in fasted simulated intestinal fluid (FaSSIF). Optimization of the solubility parameter resulted in the estimation of an in vivo solubility that was significantly higher than the in vitro value. These optimized values may then be used to estimate absorption potential in humans, simply by applying a human physiology to the existing model (Fig. 2.9). This general method allows for rapid assessment of compounds and provides a means of either eliminating compounds with poor absorption potential or mitigating exposure limitations through the use of enabled formulations.

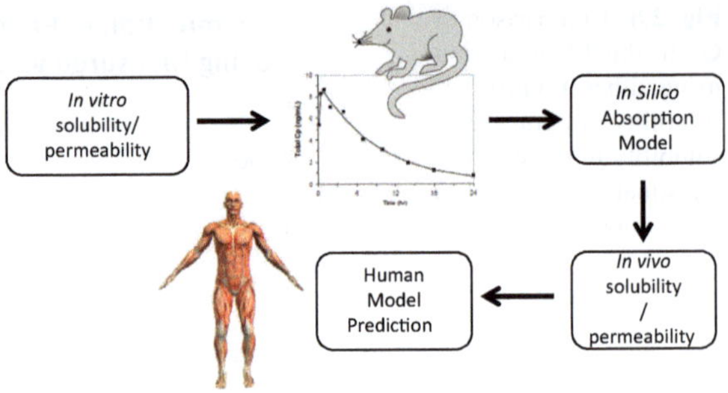

Fig. 2.9 Absorption modeling of in vivo pharmacokinetic data as a method for estimating human absorption risk

2.6 Toxicology Formulations

Toxicology or nonclinical safety assessment is a critical component of the discovery and clinical development of any pharmaceutical agent. There are many excellent reviews regarding toxicology testing in drug discovery and development and the reader is encouraged to consult these papers to gain a thorough understanding of toxicology studies necessary to support human dosing (Dorato and Buckley 2007; Buckley and Dorato 2009; Higgins et al. 2012). Regulatory guidance for conducting nonclinical safety studies have also been issued (M3(R2)). The safety term often used in the support of clinical trials is the Margin of Safety (MOS). The MOS relates to the No Observed Adverse Effect level (NOAEL—a dose that produces no relevant adverse effects) to the maximum effective dose and is displayed in Fig. 2.10. Dosing up to the maximum tolerated dose (MTD) to demonstrate target organ or dose limiting toxicity is a general expectation of regulatory authorities in support of clinical testing. In the absence of this, other equally appropriate dose-limiting criteria may be considered. These include a limit dose (1,000 or 2,000 mg/kg) that results in acceptable exposure margin relative to the clinical dose, a 50 fold exposure multiple, or an exposure limiting dose or maximum feasible dose (ICH M3(R2)).

The toxicology formulation must provide consistent plasma exposures with low variability and clear dose separation. This must be accomplished using excipients that have adequate safety data supporting the amounts used and the duration of the study. Conventional suspension formulations with particle size control, prepared in standard vehicles that are well understood and characterized in terms of safety, are preferred for oral toxicology studies and will ideally provide a linear increase in exposure with dose as depicted in Fig. 2.10. However, it is not uncommon for a compound to display solubility limited absorption in the toxicological dose range from a standard suspension vehicle, resulting in a plateau in the dose versus exposure relationship. If no dose limiting toxicity is observed, then avenues to enhance systemic exposure and/or justify a maximum feasible dose must be explored.

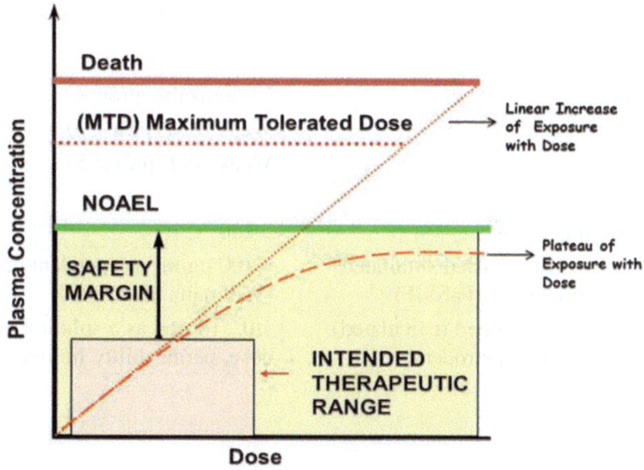

Fig. 2.10 Plasma concentration verse dose relationship in toxicology testing. Ideally, the formulation provides a linear increase in exposure with dose. Solubility limited molecules may exhibit a plateau of exposure with dose

Early alerts to the potential for exposure related issues at toxicological doses and a strategy for how and when to apply various formulation approaches are critical to avoid delays in the program timelines. Basic molecular properties that may be indicative of solubility limited absorption are listed in Table 2.7. The pharmacokinetic and absorption models previously described are particularly useful tools to understand absorption sensitivity (e.g., particle size, solubility, pH) and to identify early risk of saturation of exposure as dose is increased. It is not until actual studies are conducted however that dose dependency can be more fully understood. For example, in repeat dose-exposure response studies (e.g., toxicokinetics from a 4- or 14-day study), enzyme auto-induction may be observed as a confounding factor in hepatic clearance and may be mistaken for an absorption-limited phenomenon.

Once a plateau is observed in either predicted or observed exposure (in absence of a dose limiting toxicity), an appropriate amount of due diligence is needed to improve exposure. A strategy for toxicology studies is illustrated in the pyramid (Fig. 2.11). Two categories of options can be utilized; study design options involve the modification of dosing parameters (e.g., frequency and route of administration). Formulation options involve application of Alternative formulations that address the root cause of absorption issues. Considerations influencing the decision on which option to use include ease of implementation, cost, and/or logistics. A detailed review of the advantages and disadvantages of various options is provided in Table 2.8. Most pharmaceutical companies explore a number of options (formulation, dose frequency) to achieve higher exposure. The rationale and supporting data for the recommended strategy should be well documented to support regulatory submissions if needed.

The overall complexity of each method is represented by its relative height on the pyramid in Fig. 2.11. This ordering of formulation options is not meant to imply that each method must be tested in succession, starting from the bottom and working toward the top. Rather, key information as to the type and goal of the study, as well as physicochemical properties (e.g., lipid solubility, ionization) should be taken into account in order to determine the most appropriate level in

Table 2.7 Parameters suggesting solubility-limited absorption

Properties	Alerts
p*K*a	No ionizable groups
	Weak base (p*K*a < 4)
	Weak acid (p*K*a > 5)
Log *P*	>5
Melting point	>250 °C
Solubility in fasted simulated intestinal fluid (FaSSIF)	≤100 μg/mL (dependent on permeability and dose (Willmann et al. 2004)
Fraction absorbed (calculated) (rodent and nonrodent)	<0.2 (dosed as a solution or suspension at pharmacology dose, permeability limited absorption ruled out)

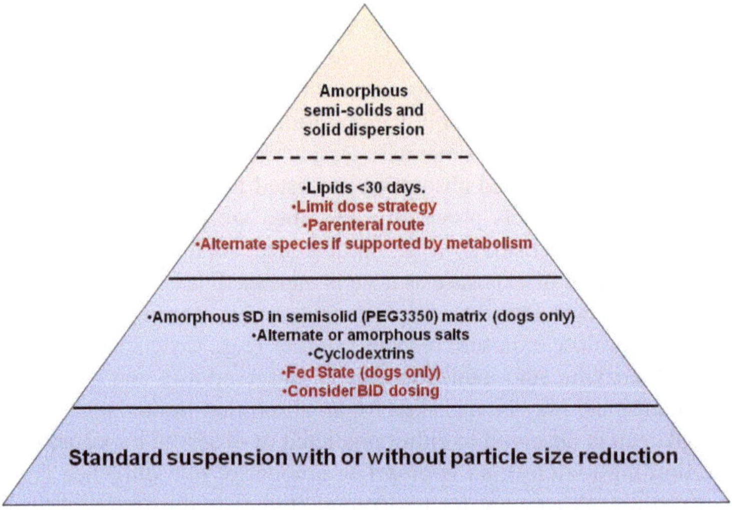

Fig. 2.11 The "pyramid" of dosing options or study design to increase exposure in toxicology studies

which to start. If one option fails, moving to the next level of complexity may be warranted in order to identify a suitable solution. This combined with in vitro dissolution tests such as the artificial stomach and duodenum and computational modeling can serve as valuable tools to choose the best formulation approach. The dissolution tools can be used to assess the rate and extent of dissolution, and also precipitation/supersaturation, properties of formulations under biorelevant conditions. Computational modeling tools can be used to identify the relative impact of absorption parameters (e.g., solubility, particle size, permeability, etc.) on exposure limitations (Bhattachar et al. 2011).

Solid dispersions are at the peak of the pyramid shown in Fig. 2.11. The use of solid dispersions has been very successful at Lilly for toxicology testing and this formulation option has provided high exposures for solubility limited molecules. An added advantage of solid dispersions is the clear "line of sight" to the clinical

Table 2.8 Summary of options to enhance toxicology exposure

Options	Advantages	Disadvantages
Study design		
b.i.d. dosing regimen	• Enhanced exposure Css (if appropriate $t_{½}$)	• Greater potential for stress-induced pathology and dosing accidents
		• Complicates kinetics and evaluation of exposure
	• Conventional formulation	• Increased compound requirements
		• May confound direct correlation to human QD dosing
Dose in fed state vs. fasted state (dog only)	• Conventional formulation	• Difficulty to control food intake
		• Pharmacological effect might affect food consumption during the study
		• Potential for variability in exposure
Limit dose strategy (e.g. 1,500 mg/kg)	• Conventional formulation	• High compound requirements
		• Potential safety issue if lack of premonitory signs for toxicity complicates clinical plan. Dose to exposure not MTD; requires real-time analysis of plasma and PK, and slower escalation
		• Acceptable MOS achieved, no MTD
		• Limit dose may not be deemed adequate by regulatory authority (e.g., certain divisions at FDA)
Increase dose volume[a]	• Conventional formulation	• Marginal improvement of AUC. May increase variability due to effects on GI motility or emesis
Change route of administration to define toxicity (e.g. IV infusion)	• Potentially higher systemic exposure with lower doses	• Potentially altered metabolite profile which may alter MTD or observed effects
		• Enabling vehicles still required for low-solubility compounds
		• Complicates long-term toxicology study designs as multiple routes may need to be tested
		• Route differs from clinical route
Change rodent or nonrodent species	• Enhance exposure	• Potentially altered metabolic profile compared to humans
Formulation		
Different salt form	• Enhance exposure	• Bridging studies between salt forms may be necessary

(continued)

Table 2.8 (continued)

Options	Advantages	Disadvantages
Toxicology vs. first in human studies		• Depending on the impurity profile of salt, may extend FHD timelines. New salts (other than those approved by FDA) may require additional long-term data
		• If amorphous, potential to crystallize in suspension
Solubilizing vehicles (lipid/cosolvent/surfactant)	• Can be prepared in toxicology and contract labs.	• Lack of experience with prolonged dosing
		• Limited dose volume and dose
		• Effect on pharmacokinetic parameters
		• Stability in presence of oxidizing excipients
		• Emesis, particularly in dogs
		• In vivo precipitation
		• May require a placebo arm
Solid dispersions	• Generally involves GRAS excipients	• API stability required to support long-term toxicology studies
	• High potential of significantly increasing exposure when designed appropriately for highly crystalline compounds	• Maximum loading dose limited (approximately 25 %)
		• Residual solvents
		• Requires sufficient characterization to assure chemical and physical stability of the solid form
		• Physical stability in suspension sufficient for dosing
Nanoparticles	• Use of conventional excipients	• May not enhance exposure if solubility limited.
		• Concentration of surfactant to stabilize nanosuspensions should be acceptable for long-term toxicology studies
		• Particulates from bead or equipment
		• Increase in particle size due to Ostwald ripening, hold time limits may be required

[a]Increasing dose volume may lead to physiological changes such as reflux, fasting, and alterations in gastrointestinal transit. Dose volumes higher than the recommended amount may be considered under the guidance of veterinary resources

dosage form since it can be used for both animal and human testing. A drawback is the additional resources and API is required for polymer screening, analytical testing (physical and chemical), and the use of larger scale specialized equipment (e.g., spray drying). However, the current state in understanding amorphous systems has improved in recent years and the development of automation and small-scale laboratory equipment to perform polymer screening/solid dispersion manufacture is making this an easier dosage form to execute (Friesen et al. 2008; Nagapudi and Jona 2008). Small-scale manufacture of amorphous dispersions is thoroughly discussed in Chap. 3.

As evident from Fig. 2.11 and Table 2.8, the use of nanoparticles, SMEDDS, and cosolvents in toxicology studies may be somewhat limited due to the following reasons.

Nanoparticles—The use of nanoparticles to enhance bioavailability at moderate doses is well recognized and there are commercial products that utilize this technology. At toxicological doses where absorption is solubility limited, these approaches are less successful based on Lilly experience. There are limited examples of the use of nanoparticles at high toxicological doses (Kesisoglou and Mitra 2012). The use of computational modeling tools to predict the exposure increase that can be achieved with this approach is somewhat limited at this time, as the in vivo absorption parameters of nanosuspensions are not very well understood.

SMEDDS (Self Microemulsifying Drug Delivery Systems) and Cosolvents—The amounts of surfactants and cosolvents necessary to achieve the solubility enhancements typically required in toxicology studies are generally poorly tolerated due to local gastrointestinal effects, and consequent effects on electrolytes and body weight over time. Thus, their use here has been limited to short-term exploratory studies.

Cosolvents—Formulations based on cosolvents may be used in small amounts for shorter term studies that generally do not exceed 4 days (Lilly internal experience) and thus these excipients are not considered to be preferred options. However, they are fairly simple to prepare and their use must always be based on a good understanding of the risk of precipitation upon administration.

Several decision trees for toxicology formulation development to conserve resource have been proposed in the literature (Higgins et al. 2012). Acceptable toxicology vehicles have also been previously published (Brewster et al. 2007). Additionally, a key resource can be an in-house database that archives details on different toxicology studies. This database could contain information on the API characteristics, formulation approach, exposure enhancement attained, and adverse effects reported in the animal model. Additional resource is the Vitic Excipient Database—Lhasa LTD as mentioned earlier.

As toxicology studies use large amounts of compound, the formulation and design options used in these studies have significant implications on the amounts of compound required, and this in turn can impact cost and timelines. Some enabled formulations may need slightly longer time and larger amounts of material to be made available for formulation and process development work. Studies using enabled formulations may help reduce the administered doses by improving the fraction absorbed, but depending on the type of formulation, there might be processing or handling losses incurred that need appropriate planning. Therefore, regardless of the formulation or study design options used, effective collaboration between toxicologists, formulators, and chemists is essential for planning and successful execution of toxicology studies.

2.7 Formulation Considerations for Alternate Drug Delivery

Alternate formulations or delivery routes are widely utilized as a means to modify the process of administration or in vivo release profile of pharmaceutical agents. Efforts to develop these formulations are often initiated late in the development process as part of a lifecycle management strategy (Chien and Ho 2011). In more recent times, however, successful drug delivery strategies have actively assessed alternate drug delivery systems in parallel with molecule selection, when molecule attributes may still be influenced, to help ensure the evolution of developable systems that meets the patients' and product lifecycle needs.

Alternate drug delivery may be particularly important for patient populations with elevated needs around cognition, behavior, or dexterity, including therapeutic areas such as Alzheimer's (Muramatsu et al. 2010), Parkinson's (Wright and Waters 2013) chronic and acute pain, and epilepsy (Anderson and Saneto 2012). Important therapeutic benefits and advantages may include a more favorable efficacy profile and/or alleviation of side effects. Examples include rivastigmine transdermal system, a cholinesterase inhibitor indicated for dementia of the Alzheimer's type and dementia associated with Parkinson's disease. The patch formulation provides an improved gastrointestinal side effect profile compared with oral administration (Exelonpatch.com) and caregiver convenience and preference for use (Bernabei, Rossini et al. 2012). Rotigotine, a dopamine agonist indicated for Parkinson's disease, is only available by transdermal patch form as a means to provide continuous delivery over 24 h. Continuous rather than pulsatile delivery is believed to more closely mimic physiological dopaminergic stimulation (Waters 2013). Intranasal sumatriptan and intranasal fentanyl provide more rapid onset for acute migraine and cancer breakthrough pain, respectively, with time to onset of 10–15 min (Dietrich and Gums 2012).

Alternate drug delivery may also play a role in the delivery of pharmaceutical doses that could not be administered effectively or safely through conventional routes. Sublingual tablets and sprays for nitroglycerin avoid extensive first-pass metabolism and provide rapid onset for treatment of angina pectoris. Similarly, nonoral administration of testosterone avoids the high first-pass metabolism and hepatotoxicity following conventional oral delivery and allows therapeutic exposures to be achieved (Pfeil and Dobs 2008). Marketed testosterone delivery systems include topical, transdermal patch, buccal, and IM depot (prodrug).

The decision to purse alternate drug delivery can occur at any stage of the drug discovery, development, or commercialization lifecycle. Early awareness by teams

Table 2.9 Preferred physicochemical properties for various routes of delivery.

Criteria	Intranasal	Pulmonary	Transdermal	Sublingual/buccal
MW	<1,000	<10,000	<500 (<350 preferred)	<500
Log *P*	1–4	−1 to 2	2–4	2–4
p*K*a	4–9	4–9	Unionized	4–9
pH range	4–7	3–7	4–7	3–8
Volume	50–150 μL	<200 μL	5–10 μL/cm^2	<500 μL

of potential benefits of alternate drug delivery and guidance around parameters required for various routes of administration can help to shape robust development strategies. This generally begins with cross-functional team discussion of the target product profile and potential therapeutic opportunities. Key questions center on:

1. Therapeutic benefits for the patient.
2. Efficacy and adverse effect profile and relationship to $C_{\max}$ and/or AUC.
3. EC_{50} or minimum concentration required to exert a therapeutic effect.

Once potential opportunities are identified, an initial assessment of feasibility can often be made with minor adjustments to the computational and formulation screens described previously. The physicochemical properties generally preferred for common routes of delivery have been reviewed (Mathias and Hussain 2010) and key parameters are summarized in Table 2.9. Further details may be found in recent reviews for transdermal (Neely et al. 2009; Paudel et al. 2010; Watkinson 2013), intranasal (Chapman et al. 2013), and sublingual/buccal delivery (Zhang et al. 2002; Goswami et al. 2013; Lam et al. 2013). Strict limitations around human efficacy dose (Table 2.10) for nonoral routes of administration are due primarily to permeation limitations or dose volume constraints at these sites. Consequently, good estimates of human efficacy dose and understanding of pharmacokinetic parameters as described for ADME studies are critical.

Equally important is ready access to formulations that allow exploration of pharmacokinetic and pharmacodynamic response in in vivo studies. Compared with oral administration, the range of vehicle options is more limited due in part to the high exposure to the excipients at the application site. Table 2.11 lists examples of vehicle systems employed for screening studies by various routes of administration. These relatively simple formulations may serve as a baseline for future formulation optimization studies. When possible, it is helpful to select excipients and concentrations with a history of prior use in humans so as to not over-enable exposures or generate adverse local effects. Good resources for this information include the FDA Inactive Ingredient Guide (FDA Inactive Ingredient Guide), major compendia (e.g., USP/NF, JP, PhEur), excipient suppliers, and external databases such as Lhasa (Lhasa Vitic Nexus database).

Alternate drug delivery may add significant value to a candidate if therapeutic advantages are realized for the patient. Early assessments of potential opportunities and technical feasibility in the discovery phase may serve to provide realistic expectations prior to investment in more costly development activities.

Table 2.10 Maximum human doses and potential therapeutic benefits of different dosage forms and routes of delivery

Dose form with assumptions around total unit size or wt	Preferred dose (mg/dose)	Rapid onset	Sustained plasma exposure	Decrease dose frequency	Minimize first-pass metabolism
Oral					
Conventional tablet, 150–450 mg	<100	–	–	–	–
Sustained release (matrix or multiparticulate)	<50	–	√	√	–
Orodispersible tablet	<25	√	–	–	–
Orodispersible film strip, 150 mg	<25	√	–	–	–
Soft gelatin capsule, #0, 0.68 mL	<100	√	–	–	–
Fine granules (5 g sprinkles, 20 % active)	<1,000	–	–	–	–
Sublingual					
Tablet	<10	√	–	–	√
Spray (0.5 mL)	<10	√	–	–	√
Transdermal					
Passive patch (delivered dose)	<10	–	√	√	√
Gel (delivered dose, 5 g applied product)	<10	–	√	√	√
Injectable					
IM depot	<2 mL	–	√	√	√
Subcutaneous	<1 mL	–	√	–	√

Table 2.11 Standard vehicles for in vivo evaluation of alternate routes of delivery.

Route	Species	Volume	Formulation examples	References
Transdermal	Rat, monkey, minipig, and ex vivo studies	Typically 5–25 μL/cm^2 for pharmacology studies, up to 10 % of BSA.	(a) Hydroalcoholic gel: EtOH or IPA 60–85 %, water 15–40 %, propylene glycol 0–20 %, hydroxypropylcellulose 1–2 % to increase viscosity	Lee et al. (2010), Lehman and Raney (2012)
			(b) Ex vivo: PEG400 45 %/PBS 55 % pH 6.4	
Intranasal	Rat, dog	Rat 10–20 μL, Dog 100–150 μL	Aqueous buffers pH 4–9 or saline	Sutton et al. (1993), Blagg et al. (2007)
Sublingual/buccal	Dog	Dog up to ~1 mL	(a) Aqueous buffers pH 4–9	Gayrard et al. (2013)
			(b) Ethanol 40 % in water	

2.8 Summary

A basic tenet of any in vivo study depends on reliable delivery of the drug to the target site of action and is profoundly influenced by the formulation. Formulations can impact drug release, absorption, metabolism, and the PK profile. The pharmaceutical scientist at the discovery–development interface is best qualified and ideally positioned to recognize the unique formulation needs of discovery teams and provide the necessary support. The need for this support has steadily increased in recent years due to the growing sophistication of the discovery engine and the shift toward a chemistry space characterized by lower solubility, greater lipophilicity, and thus greater challenges with in vivo release and absorption. Delivery of these drugs by "traditional" means where the compound is dosed as a simple formulation in an aqueous medium is increasingly not an option for eliciting the desired pharmacodynamic response or toxicological exposure. Active engagement of the pharmaceutical scientist and the utilization of appropriate formulations for in vivo studies, while maintaining a clear line of sight to commercial development are essential to the success of any discovery program.

References

Akimoto M, Nagahata N et al (2000) Gastric pH profiles of beagle dogs and their use as an alternative to human testing. Eur J Pharm Biopharm 49(2):99–102

Anderson GD, Saneto RP (2012) Current oral and non-oral routes of antiepileptic drug delivery. Adv Drug Deliv Rev 64(10):911–918

Balazs O (2011) Preclinical formulation in early drug research. In: Tihanyi K, Vastag M (eds) Solubility, delivery and ADME problems of drugs and drug candidates. Budapest, Bentham, p 22

Bernabei R, Rossini PM et al (2012) Compliance and caregiver satisfaction in Alzheimer's disease: results from the AXEPT study. Dement Geriatr Cogn Dis Extra 2(1):418–432

Bhattachar SN, Perkins EJ et al (2011) Effect of gastric pH on the pharmacokinetics of a BCS class II compound in dogs: utilization of an artificial stomach and duodenum dissolution model and GastroPlus, simulations to predict absorption. J Pharm Sci 100(11):4756–4765

Bikiaris DN (2011) Solid dispersions, part I: recent evolutions and future opportunities in manufacturing methods for dissolution rate enhancement of poorly water-soluble drugs. Expert Opin Drug Deliv 8(11):1501–1519

Bittner BG, Bravo RC, Walter I, Huwyler J (2005) Impact of oral administration of the surface-active excipient Solutol HS 15 on the pharmacokinetics of intravenously administered colchicine. Lett Drug Des Discov 2(3):3

Blagg J, Allerton CM et al (2007) Design and synthesis of a functionally selective D3 agonist and its in vivo delivery via the intranasal route. Bioorg Med Chem Lett 17(24):6691–6696

Bravo González RC, Huwyler J, Boess F, Walter I, Bittner B (2004) In vitro investigation on the impact of the surface-active excipients Cremophor EL, Tween 80 and Solutol HS 15 on the metabolism of midazolam. Biopharm Drug Dispos 25(1):13

Brewster M, Claire M, Noppe M, Lampo A, Loftsson T (2007) The use of solubilizing excipients and approaches to generate toxicology vehicles for contemporary drug pipelines. In: Augustijns P, Brewster M (eds) Solvent systems and their selection in pharamceutics and biopharmaceutics. Springer, New York, pp 221–256
Buckley LA, Dorato MA (2009) High dose selection in general toxicity studies for drug development: a pharmaceutical industry perspective. Regul Toxicol Pharmacol 54(3):301–307
Buggins TR, Dickinson PA et al (2007) The effects of pharmaceutical excipients on drug disposition. Adv Drug Deliv Rev 59(15):1482–1503
Carino SR, Sperry DC et al (2006) Relative bioavailability estimation of carbamazepine crystal forms using an artificial stomach-duodenum model. J Pharm Sci 95(1):116–125
Carrier RL, Miller LA et al (2007) The utility of cyclodextrins for enhancing oral bioavailability. J Control Release 123(2):78–99
Chapman CD, Frey WH 2nd et al (2013) Intranasal treatment of central nervous system dysfunction in humans. Pharm Res 30(10):2475–2484
Charman SA, Perry CS et al (2006) Alteration of the intravenous pharmacokinetics of a synthetic ozonide antimalarial in the presence of a modified cyclodextrin. J Pharm Sci 95(2):256–267
Chaubal MV (2004) Application of drug delivery technologies in lead candidate selection and optimization. Drug Discov Today 9(14):603–609
Chen XQ, Gudmundsson OS et al (2012) Application of lipid-based formulations in drug discovery. J Med Chem 55(18):7945–7956
Chien JY, Ho RJ (2011) Drug delivery trends in clinical trials and translational medicine: evaluation of pharmacokinetic properties in special populations. J Pharm Sci 100(1):53–58
Chiou WL, Barve A (1998) Linear correlation of the fraction of oral dose absorbed of 64 drugs between humans and rats. Pharm Res 15(11):1792–1795
Dafallah AA, al-Mustafa Z (1996) Investigation of the anti-inflammatory activity of Acacia nilotica and Hibiscus sabdariffa. Am J Chin Med 24(3–4):263–269
Dietrich E, Gums JG (2012) Intranasal fentanyl spray: a novel dosage form for the treatment of breakthrough cancer pain. Ann Pharmacother 46(10):1382–1391
Dimitrov NV, Meyer-Leece C et al (1996) Plasma alpha-tocopherol concentrations after supplementation with water- and fat-soluble vitamin E. Am J Clin Nutr 64(3):329–335
Dorato MA, Buckley LA (2007) Toxicology testing in drug discovery and development. Curr Protoc Toxicol Chapter 19: Unit 19 11
Dudley RE, Constantinides P (2011) Oral testosterone ester formulations and methods of treating testosterone deficiency comprising same. WIPO, Clarus Therapeutics, Inc
Fancher RM, Zhang H et al (2011) Development of a canine model to enable the preclinical assessment of pH-dependent absorption of test compounds. J Pharm Sci 100(7):2979–2988
Ferguson SA, Boctor SY (2009) Use of food wafers for multiple daily oral treatments in young rats. J Am Assoc Lab Anim Sci 48(3):292–295
Foster KA, Sun H et al (2013) Utility of gastric-retained alginate gels to modulate pharmacokinetic profiles in rats. J Pharm Sci 102(8):2440–2449
Friesen DT, Shanker R et al (2008) Hydroxypropyl methylcellulose acetate succinate-based spray-dried dispersions: an overview. Mol Pharm 5(6):1003–1019
Gad SC, Cassidy CD et al (2006) Nonclinical vehicle use in studies by multiple routes in multiple species. Int J Toxicol 25(6):499–521
Gao P, Guyton ME et al (2004) Enhanced oral bioavailability of a poorly water soluble drug PNU-91325 by supersaturatable formulations. Drug Dev Ind Pharm 30(2):221–229
Gao Y, Carr RA et al (2010) A pH-dilution method for estimation of biorelevant drug solubility along the gastrointestinal tract: application to physiologically based pharmacokinetic modeling. Mol Pharm 7(5):1516–1526
Gartner K, Buttner D et al (1980) Stress response of rats to handling and experimental procedures. Lab Anim 14(3):267–274
Gayrard V, Lacroix MZ et al (2013) High bioavailability of bisphenol A from sublingual exposure. Environ Health Perspect 121(8):951–956

Gleeson MP, Hersey A et al (2011) Probing the links between in vitro potency, ADMET and physicochemical parameters. Nat Rev Drug Discov 10(3):197–208
Goswami T, Kokate A et al (2013) In silico model of drug permeability across sublingual mucosa. Arch Oral Biol 58(5):545–551
Gullapalli R, Wong A et al (2012) Development of ALZET(R) osmotic pump compatible solvent compositions to solubilize poorly soluble compounds for preclinical studies. Drug Deliv 19 (5):239–246
Hageman MJ (2006) Solubility, solubilization and dissolution in drug delivery during lead optimization. In: Borchardt RT, Kerns EH, Hageman MJ, Thakker DR, Stevens JL (eds) Optimizing the drug-like properties of leads in drug discovery, vol VI, Biotechnology: pharmaceutical aspects. Springer, New York, pp 99–103
He K, Qian M et al (2008) N-in-1 dosing pharmacokinetics in drug discovery: experience, theoretical and practical considerations. J Pharm Sci 97(7):2568–2580
Higgins J, Cartwright ME et al (2012) Progressing preclinical drug candidates: strategies on preclinical safety studies and the quest for adequate exposure. Drug Discov Today 17 (15–16):828–836
Hopkins AL, Groom CR (2002) The druggable genome. Nat Rev Drug Discov 1:727–730
Jacquemin E, Hermeziu B et al (2009) Bioavailability of oral vitamin E formulations in adult volunteers and children with chronic cholestasis or cystic fibrosis. J Clin Pharm Ther 34 (5):515–522
Kesisoglou F, Mitra A (2012) Crystalline nanosuspensions as potential toxicology and clinical oral formulations for BCS II/IV compounds. AAPS J 14(4):677–687
Kesisoglou F, Panmai S et al (2007) Nanosizing—oral formulation development and biopharmaceutical evaluation. Adv Drug Deliv Rev 59(7):631–644
Kumar R, Knick VB et al (2007) Pharmacokinetic-pharmacodynamic correlation from mouse to human with pazopanib, a multikinase angiogenesis inhibitor with potent antitumor and antiangiogenic activity. Mol Cancer Ther 6(7):2012–2021
Kurkov SV, Loftsson T (2013) Cyclodextrins. Int J Pharm 453(1):167–180
Kurkov SV, Madden DE et al (2012) The effect of parenterally administered cyclodextrins on the pharmacokinetics of coadministered drugs. J Pharm Sci 101(12):4402–4408
Kwong E, Higgins J et al (2011) Strategies for bringing drug delivery tools into discovery. Int J Pharm 412(1–2):1–7
Lam JK, Xu Y et al (2013) Oral transmucosal drug delivery for pediatric use. Adv Drug Deliv Rev 73C:50–62
Larger PJ, Breda M et al (2005) Ion-suppression effects in liquid chromatography-tandem mass spectrometry due to a formulation agent, a case study in drug discovery bioanalysis. J Pharm Biomed Anal 39(1–2):206–216
Lee YC, Zocharski PD et al (2003) An intravenous formulation decision tree for discovery compound formulation development. Int J Pharm 253(1–2):111–119
Lee PH, Conradi R et al (2010) Development of an in silico model for human skin permeation based on a Franz cell skin permeability assay. Bioorg Med Chem Lett 20(1):69–73
Lehman PA, Raney SG (2012) In vitro percutaneous absorption of ketoprofen and testosterone: comparison of pluronic lecithin organogel vs. pentravan cream. Int J Pharm Compd 16 (3):248–252
Li P, Zhao L (2007) Developing early formulations: practice and perspective. Int J Pharm 341 (1–2):1–19
Li S, Pollock-Dove C et al (2012) Enhanced bioavailability of a poorly water-soluble weakly basic compound using a combination approach of solubilization agents and precipitation inhibitors: a case study. Mol Pharm 9(5):1100–1108
Loftsson T, Brewster ME (2010) Pharmaceutical applications of cyclodextrins: basic science and product development. J Pharm Pharmacol 62(11):1607–1621
Loftsson T, Brewster ME (2011) Pharmaceutical applications of cyclodextrins: effects on drug permeation through biological membranes. J Pharm Pharmacol 63(9):1119–1135

Long G, Works J (2013). Innovation in the biopharmaceutical pipeline: a multidimensional view. www.pharma.org
Lorenz W, Reimann HJ et al (1977) Histamine release in dogs by Cremophor E1 and its derivatives: oxethylated oleic acid is the most effective constituent. Agents Actions 7(1):63–67
Lorenz W, Schmal A et al (1982) Histamine release and hypotensive reactions in dogs by solubilizing agents and fatty acids: analysis of various components in cremophor El and development of a compound with reduced toxicity. Agents Actions 12(1–2):64–80
Lui CY, Amidon GL et al (1986) Comparison of gastrointestinal pH in dogs and humans: implications on the use of the beagle dog as a model for oral absorption in humans. J Pharm Sci 75(3):271–274
Lukas G, Brindle SD et al (1971) The route of absorption of intraperitoneally administered compounds. J Pharmacol Exp Ther 178(3):562–564
Luke DR, Tomaszewski K et al (2010) Review of the basic and clinical pharmacology of sulfobutylether-beta-cyclodextrin (SBECD). J Pharm Sci 99(8):3291–3301
M3(R2) Guidance on nonclinical safety studies for the conduct of human clinical trials and marketing authorization for pharmaceuticals. I. C. o. Harmonization
Maas J, Kamm W et al (2007) An integrated early formulation strategy—from hit evaluation to preclinical candidate profiling. Eur J Pharm Biopharm 66(1):1–10
Mactier RA, Khanna R et al (1987) Role of peritoneal cavity lymphatic absorption in peritoneal dialysis. Kidney Int 32(2):165–172
Mathias NR, Hussain MA (2010) Non-invasive systemic drug delivery: developability considerations for alternate routes of administration. J Pharm Sci 99(1):1–20
Merisko-Liversidge EM, Liversidge GG (2008) Drug nanoparticles: formulating poorly water-soluble compounds. Toxicol Pathol 36(1):43–48
Miller JM, Dahan A (2012) Predicting the solubility-permeability interplay when using cyclodextrins in solubility-enabling formulations: model validation. Int J Pharm 430(1–2):388–391
Morphy R (2010) Selectively nonselective kinase inhibition: striking the right balance. J Med Chem 53(4):1413–1437
Mu J, Woods J et al (2006) Chronic inhibition of dipeptidyl peptidase-4 with a sitagliptin analog preserves pancreatic beta-cell mass and function in a rodent model of type 2 diabetes. Diabetes 55(6):1695–1704
Mu J, Petrov A et al (2009) Inhibition of DPP-4 with sitagliptin improves glycemic control and restores islet cell mass and function in a rodent model of type 2 diabetes. Eur J Pharmacol 623 (1–3):148–154
Muramatsu RS, Litzinger MH et al (2010) Alternative formulations, delivery methods, and administration options for psychotropic medications in elderly patients with behavioral and psychological symptoms of dementia. Am J Geriatr Pharmacother 8(2):98–114
Musther H, Olivares-Morales A et al (2013) Animal versus human oral drug bioavailability: do they correlate? Eur J Pharm Sci 57:280–291
Nagapudi KJ, Jona J (2008) Amorphous active pharmaceutical ingredients in preclinical studies: preparation, characterization, and formulation. Curr Bioactive Compd 4(4):213–224
Nagilla R, Nord M et al (2011) Cassette dosing for pharmacokinetic screening in drug discovery: comparison of clearance, volume of distribution, half-life, mean residence time, and oral bioavailability obtained by cassette and discrete dosing in rats. J Pharm Sci 100(9):3862–3874
Neely BJ, Madihally SV et al (2009) Nonlinear quantitative structure-property relationship modeling of skin permeation coefficient. J Pharm Sci 98(11):4069–4084
Neervannan S (2006) Preclinical formulations for discovery and toxicology: physicochemical challenges. Expert Opin Drug Metab Toxicol 2(5):715–731
Oh DM, Curl RL et al (1993) Estimating the fraction dose absorbed from suspensions of poorly soluble compounds in humans: a mathematical model. Pharm Res 10(2):264–270
Pammolli F, Magazzini L et al (2011) The productivity crisis in pharmaceutical R&D. Nat Rev Drug Discov 10(6):428–438

Paudel KS, Milewski M et al (2010) Challenges and opportunities in dermal/transdermal delivery. Ther Deliv 1(1):109–131
Pfeil E, Dobs AS (2008) Current and future testosterone delivery systems for treatment of the hypogonadal male. Expert Opin Drug Deliv 5(4):471–481
Pole DL (2008) Physical and biological considerations for the use of nonaqueous solvents in oral bioavailability enhancement. J Pharm Sci 97(3):1071–1088
Porter CJ, Pouton CW et al (2008) Enhancing intestinal drug solubilisation using lipid-based delivery systems. Adv Drug Deliv Rev 60(6):673–691
Pouton CW, Porter CJ (2008) Formulation of lipid-based delivery systems for oral administration: materials, methods and strategies. Adv Drug Deliv Rev 60(6):625–637
Qian F, Huang J et al (2010) Drug-polymer solubility and miscibility: stability consideration and practical challenges in amorphous solid dispersion development. J Pharm Sci 99(7):2941–2947
Rabinow BE (2004) Nanosuspensions in drug delivery. Nat Rev Drug Discov 3(9):785–796
Rao VM, Stella VJ (2003) When can cyclodextrins be considered for solubilization purposes? J Pharm Sci 92(5):927–932
Reed KW, Yalkowsky SH (1985) Lysis of human red blood cells in the presence of various cosolvents. J Parenter Sci Technol 39(2):64–69
Roth BL, Sheffler DJ et al (2004) Magic shotguns versus magic bullets: selectively non-selective drugs for mood disorders and schizophrenia. Nat Rev Drug Discov 3(4):353–359
Saxena V, Panicucci R et al (2009) Developability assessment in pharmaceutical industry: an integrated group approach for selecting developable candidates. J Pharm Sci 98(6):1962–1979
Scannell JW, Blanckley A et al (2012) Diagnosing the decline in pharmaceutical R&D efficiency. Nat Rev Drug Discov 11(3):191–200
Shah AK, Agnihotri SA (2011) Recent advances and novel strategies in pre-clinical formulation development: an overview. J Control Release 156(3):281–296
Sheth AR (2011) In vitro screening methods to assess the potential of in vivo precipitation of injectable formulations upon intravenous administration. PDA J Pharm Sci Technol 65 (1):71–80
Shiotani K, Uehata K et al (1995) Differential effects of sulfate and sulfobutyl ether of beta-cyclodextrin on erythrocyte membranes in vitro. Pharm Res 12(1):78–84
Six K, Verreck G et al (2004) Increased physical stability and improved dissolution properties of itraconazole, a class II drug, by solid dispersions that combine fast- and slow-dissolving polymers. J Pharm Sci 93(1):124–131
Stella VJ, He Q (2008) Cyclodextrins. Toxicol Pathol 36(1):30–42
Stella VJ, Rao VM et al (1999) Mechanisms of drug release from cyclodextrin complexes. Adv Drug Deliv Rev 36(1):3–16
Strickley RG (2008) Formulation in drug discovery (chapter 24). Annu Rep Med Chem 43:419–451
Sutton SC, Engle K et al (1993) Intranasal delivery of the bisphosphonate alendronate in the rat and dog. Pharm Res 10(6):924–926
ten Tije AJ, Verweij J et al (2003) Pharmacological effects of formulation vehicles: implications for cancer chemotherapy. Clin Pharmacokinet 42(7):665–685
Tong WQ, Whitesell G (1998) In situ salt screening—a useful technique for discovery support and preformulation studies. Pharm Dev Technol 3(2):215–223
Traber MG, Kayden HJ et al (1986) Absorption of water-miscible forms of vitamin E in a patient with cholestasis and in thoracic duct-cannulated rats. Am J Clin Nutr 44(6):914–923
Turner PV, Brabb T et al (2011) Administration of substances to laboratory animals: routes of administration and factors to consider. J Am Assoc Lab Anim Sci 50(5):600–613
Van Eerdenbrugh B, Van den Mooter G et al (2008) Top-down production of drug nanocrystals: nanosuspension stabilization, miniaturization and transformation into solid products. Int J Pharm 364(1):64–75
Vandecruys R, Peeters J et al (2007) Use of a screening method to determine excipients which optimize the extent and stability of supersaturated drug solutions and application of this system to solid formulation design. Int J Pharm 342(1–2):168–175

Vasconcelos T, Sarmento B, Costa P (2007) Solid dispersions as strategy to improve oral bioavailability of poor water soluble drugs. Drug Discov Today 12(23–24):8

Venkatesh S, Lipper RA (2000) Role of the development scientist in compound lead selection and optimization. J Pharm Sci 89(2):145–154

Verreck G, Six K et al (2003) Characterization of solid dispersions of itraconazole and hydroxypropylmethylcellulose prepared by melt extrusion—part I. Int J Pharm 251 (1–2):165–174

Vitic Nexus. Lhasa Limited, Leeds, UK. http://www.lhasalimited.org/products/vitic-nexus.html

Vo CL, Park C et al (2013) Current trends and future perspectives of solid dispersions containing poorly water-soluble drugs. Eur J Pharm Biopharm 85(3):799–813

Warren DB, Benameur H et al (2010) Using polymeric precipitation inhibitors to improve the absorption of poorly water-soluble drugs: a mechanistic basis for utility. J Drug Target 18 (10):704–731

Waters C (2013) The development of the rotigotine transdermal patch: a historical perspective. Neurol Clin 31(3 Suppl):S37–S50

Watkinson AC (2013) A commentary on transdermal drug delivery systems in clinical trials. J Pharm Sci 102(9):3082–3088

Weiss RB, Donehower RC et al (1990) Hypersensitivity reactions from taxol. J Clin Oncol 8 (7):1263–1268

Williams HD, Trevaskis NL et al (2013) Strategies to address low drug solubility in discovery and development. Pharmacol Rev 65(1):315–499

Willmann S, Schmitt W et al (2004) A physiological model for the estimation of the fraction dose absorbed in humans. J Med Chem 47(16):4022–4031

Woodburn K, Kessel D (1994) The alteration of plasma lipoproteins by cremophor EL. J Photochem Photobiol B 22(3):197–201

Wright BA, Waters CH (2013) Continuous dopaminergic delivery to minimize motor complications in Parkinson's disease. Expert Rev Neurother 13(6):719–729

Xu S, Dai WG (2013) Drug precipitation inhibitors in supersaturable formulations. Int J Pharm 453 (1):36–43

Yalkowsky SH (1999) Solubility and solubilization in aqueous media. American Chemical Society, New York

Yamada I, Haga K (1990) Measurement of gastric pH during digestion of a solid meal in dogs. Chem Pharm Bull (Tokyo) 38(6):1755–1756

Yin AY, Htun M et al (2012) Reexamination of pharmacokinetics of oral testosterone undecanoate in hypogonadal men with a new self-emulsifying formulation. J Androl 33(2):190–201

Zhang H, Zhang J et al (2002) Oral mucosal drug delivery: clinical pharmacokinetics and therapeutic applications. Clin Pharmacokinet 41(9):661–680

Chapter 3
Enabling Discovery Through Leveraging and Miniaturizing Pharmaceutical Principles and Processes

Roy J. Haskell, Kimberly A. Foster, Ching Kim Tye, and Michael Morgen

3.1 Discovery Versus Development

The focus, goals, and work environments within discovery and development are quite different from each other and it is useful to describe those differences to understand why the tactics and strategies employed in each may appear divergent. Table 3.1 contrasts the two by focusing on how the nature of the problems and, hence, the solutions appropriate for each differ.

The basic substrate for work in both areas is the drug substance itself. It is discovery's charge to use the material available to find the molecules that will become candidates, and it is development's to bring those candidates into commercial reality. Thus, while there may be only one molecule selected for development, thousands or even tens of thousands of molecules must be synthesized and evaluated in discovery before that molecule is identified. Thus, activities are allocated across numerous molecules in discovery, whereas only one molecule receives the focus and resources applied in development. The experimental tools, resources, and timelines available in discovery make it inevitable that many molecules supplied to the pharmaceutical scientist are not well characterized and frequently have undesirable physical presentations (e.g., they are gums or oils). In contrast, the need to understand the drivers of properties such as polymorphism, powder flow, reproducibility, and stability requires that the active compounds used by development be well characterized and homogeneous. In addition, the quantity of material available

R.J. Haskell (✉)
Discovery Pharmaceutics, Bristol-Myers Squibb, 5 Research Parkway, Wallingford, CT 06492, USA
e-mail: Roy.Haskell@bms.com

K.A. Foster • C.K. Tye
Discovery Pharmaceutics, Bristol-Myers Squibb, Lawrenceville, NJ, USA

M. Morgen
Bend Research, Bend, OR, USA

A.C. Templeton et al. (eds.), *Discovering and Developing Molecules with Optimal Drug-Like Properties*, AAPS Advances in the Pharmaceutical Sciences Series 15,
DOI 10.1007/978-1-4939-1399-2_3

Table 3.1 Comparing the discovery and development environments

Parameter	Issue	Discovery	Development
Experimental substrate	Elements	Many compounds	Many formulations
	Sample definition	Poor	Good
	Quantities	1 mg to grams	100's mg - 1 kg
Execution	Motivation	Biological model	Clinical results
	Timelines	Hours/days/weeks	Weeks/months/years
	Disciplines	Highly integrated	Communities of practice
Limitations	Adulteration	Model integrity	Safety
	Compliance	<GLP	>GLP, GMP
Environment	Results	Affect compounds	Affect ideas
	Design	Extrapolation/ interpolation	QbD, DoE, PCA, etc.
	Major risk	False negative	False positive
	Molecule	Wide range	Candidate selected
In vivo	Dosing	Open ended	Well defined
	Species	Many	Few

GLP good laboratory practice, *GMP* good manufacturing practice, *QbD* quality by design, *DoE* design of experiments, *PCA* principal component analysis

for identification of a formulation in discovery is frequently small (e.g., 1 mg), whereas tens to thousands of grams of active compound are devoted to the analogous task in development.

Large communities of practice centering on pharmaceutical science exist in development, whereas the discovery scientist's results and influence are immediately integrated into multidisciplinary teams. Indeed, discussing the details of physical chemistry could actually be a disservice in discovery teams because it may impede understanding for those with significantly different technical backgrounds.

The determinant of progress in later stage R&D is in vivo results, achieved over years of human trials at preselected doses. By comparison, the primary drivers in the discovery space are the success and enablement of multispecies pharmacological models that employ variable/high doses. The results of these models are used to direct chemical synthesis in rapid-cycling iterations. Regulatory concerns do not affect day-to-day activities in discovery to the same extent as they do in development, but collection of artifact-free biological data from compounds with poor absorption, distribution, metabolism, and excretion (ADME) properties is no less of a challenge.

There is little discovery resource available to study the pharmaceutical properties of compounds that are inadequate in some area of performance (e.g., toxicology or efficacy). As a result, follow-up experiments often cannot be conducted no matter how useful they might be in understanding such properties, since the compound is never resynthesized. This results in interpolation or extrapolation from incomplete data sets, which were originally designed and collected for other purposes, as opposed to extraction from specific, comfortably undersaturated experimental designs.

The most significant difference between the two environments is, of course, the fact that in most stages of discovery the compound that could eventually command enormous development resources has not even been identified. While mitigation of undesirable properties is the aim of post-candidate selection activities, enabling discovery to focus on strategies that find, or better yet design, compounds requiring minimal investment in pharmaceutical science resources is a very cost-effective strategy.

Pharmaceutical sciences should play a major role in discovery, however. Rapid-cycling, multidisciplinary activities that study structurally diverse, scarce, intractable compounds using assays sensitive to vehicle effects necessitate the best approaches the discipline has to offer. While scientists must avail themselves of the drug delivery tools widely available in the development community, these tools must be adaptable to address the practical requirements of discovery. For instance, a highly effective delivery method that requires several hundred milligrams of crystalline solid and a week's lead time will find little practical application in the discovery setting.

Although the development tool or technology may not be directly applicable, the mechanism behind the method may be very useful in discovery. For example, amorphous dispersions are now an accepted approach to enhance oral exposure by inducing supersaturation and/or forming colloids that serve as sources of free drug. However, the question arises as to how to make dispersions in a manner compatible with the discovery environment described above. Solving this problem requires creativity that recognizes which aspects of a given development technology can be relaxed and which must be preserved to allow application to discovery. Using this approach, issues otherwise perceived as limitations may become irrelevant or actually become assets. An amorphous dispersion, for example, does not require the powder properties that enable good flow or dispersibility from a tablet if it is employed as an aqueous suspension prepared and manipulated with laboratory equipment. Similarly, an excipient that induces gel depots does not have to possess a long-term safety profile if it is to be used in a preclinical model that runs only a few weeks. The excipient does, however, need to be demonstrably innocuous with respect to the desired in vivo endpoint.

As noted above, there are a myriad of delivery technologies described in the literature, many with in vivo data collected from species common in discovery (e.g., rodents). In most cases, however, these studies are retroactive in the sense that they were first developed for clinical application and then scaled down to the compound-sparing smaller species to enable screening. This scenario benefits from the advantage of large compound supplies, significant practical experience, and existing understanding of in vivo behavior when tackling drug delivery problems. The discovery scientist has few of these advantages. Hence, developing an approach for species commonly used in discovery with development resources is not the same as developing an approach for discovery species when working solely in discovery.

This chapter seeks to show how a wide range of technologies and approaches generally considered to be the exclusive domain of the manufacturing floor can be applied to solve problems encountered in the early stages of research and development. Discovery applications of some types of formulations (e.g., nanosuspensions and amorphous dispersions) are described in great detail, whereas others (e.g., bead coating) are discussed more generally. This imbalance is not meant to suggest that some technologies are better suited than others, but rather reflects the different levels of maturity of the technologies as research-enabling tools. Indeed, the reader is encouraged to take ideas from the following discussion and consider how they might be developed and employed using this paradigm.

3.1.1 Enabling Exposure in a Discovery Setting

The drug discovery effort can be broadly divided into the following stage gates: target identification, target validation, hit-to-lead, lead evaluation, and lead optimization, with the ultimate goal being identification of a compound with properties suitable for clinical testing (Hughes et al. 2011). After the target identification stage, discovery teams focus on identifying a compound that can be used to conduct key pharmacology studies to validate the target. Significant work is performed to understand aspects of the target, such as what concentration is necessary for efficacy and whether the chemical template or modulation of the target has gross toxicological liabilities. Early assessment of the efficacy and safety of a new target often depends upon achieving adequate in vivo exposure to generate reliable results (Chiang et al. 2010). At this stage, teams often rely on lead or tool compounds that are potent enough to help answer key questions but have liabilities that would preclude further development. These compounds allow the discovery team to build knowledge of viable paths forward while continuing to improve the properties of newly synthesized compounds in the lead optimization stage. Leads in early discovery often exhibit undesirable pharmacokinetic (PK) attributes and physicochemical properties such as high clearance rates or poor solubility, making in vivo assessment difficult due to low exposure (Chiang et al. 2010). For example, a given combination of these properties may defeat efforts to maintain a given plasma concentration for a specified duration to achieve the desired pharmacological response (Thomas et al. 2006). The problem is exacerbated when teams need to determine the toxicity of an unoptimized compound in a study that often requires $\geq$10-fold exposure over the efficacious concentration.

Discovery formulators are tasked with designing drug delivery strategies that enable adequate exposure (magnitude and duration) in preclinical studies to help answer questions regarding efficacy and safety. This can be a daunting challenge for compounds with physicochemical and/or PK liabilities. Indeed, one publication recently cited that one of the main limitations to validating a target in vivo is the dosing regimen or formulation used (Chen and Du 2007). It is imperative that discovery formulators use exposure-enabling technologies to help advance programs in early preclinical studies.

3.1.2 Solution Dosing

Solution dosing is covered at length in Chap. 2, but a brief description of how it fits into the toolbox of delivery strategies is provided here. Early in the discovery process, solution dosing is employed to try to achieve the maximum exposure by eliminating the dissolution step in the drug absorption process. This is especially advantageous because the solid form of the compound has typically not been fully characterized in the discovery stage, perhaps apart from knowing whether it is amorphous or crystalline. Synthetic routes may change from lot to lot during scale-up and can result in multiple solid forms, which may affect exposure. Solution dosing—i.e., completely dissolving the compound—helps eliminate the variability associated with the solid form of the compound.

Solution dosing may involve solubilizing the compound with a cosolvent or complexation agent and/or adjusting the pH for ionizable compounds (Thomas et al. 2006). The solution formulations may also be prepared with the addition of precipitation inhibitors because the compound is likely to be more soluble in the dosing vehicle than in the gastrointestinal (GI) tract, where absorption occurs (Dai and Xu 2012; Dai 2010).

3.1.3 Suspension Dosing

Due to the undesirable properties of many of the compounds in discovery, teams may need to resort to suspension dosing if they are unable to solubilize the compound in a vehicle acceptable for in vivo studies. Suspension dosing may be the only option if the pharmacology model is sensitive to vehicles based on organic cosolvents. In these cases, discovery formulators are often restricted to aqueous-based suspension vehicles containing limited surfactant (e.g., 0.5 % methylcellulose/0.1 % Tween 80).

3.1.4 Tandem Dosing

One strategy often used in discovery to improve exposure is to increase the dosing frequency. While this may increase exposure, this strategy often requires a large amount of drug, which may not be available. In response, researchers recently developed a novel dosing paradigm known as "tandem dosing," in which multiple oral doses are administered in a short time to simulate an oral infusion (Chiang et al. 2010). This concept is based on using gastric residence time to define the dosing interval that enables the drug to be absorbed as efficiently as possible. Sequential oral doses are administered in such a way that the GI tract can accommodate a new dose as soon as the previous dose has passed the absorption window. In one published study, two different poorly soluble compounds were evaluated using this approach (Chiang et al. 2010). Both molecules had previously been dosed in rats at up to 1,000 mg/kg by quaque die (QD) administration of a standard suspension, but the area under the time-concentration curve (AUC) and maximum drug concentration (C_{max}) values had plateaued due to solubility limitations. The compounds were also dosed at 200 mg/kg as suspensions dosed ter in die (TID) at 2.5-h intervals in rats, which consumed only 600 mg/kg of material. Both the C_{max} and AUC values increased approximately twofold using the tandem dosing strategy compared with the exposure achieved with the original protocol (Chiang et al. 2010). In a follow-up study, the authors looked at optimizing the dosing interval (i.e., 1, 1.5, or 2.5 h) and found that the time interval that provided the best exposure depended on the dose (Chiang et al. 2011). In general, the higher the dose, the more beneficial the increased time between dosing. One advantage of tandem dosing is that it does not require additional formulation development resource—a simple suspension formulation was used in the published studies. In addition, higher exposure (i.e., higher C_{max} and AUC values) was achieved with this approach using less total drug. One potential disadvantage of this technique, however, is that animals must be handled several times in a short time frame. While this may not be a problem for some preclinical studies such as toxicology studies, many pharmacology models—particularly those with behavioral readouts—may be adversely affected by increased handling. Even when tandem dosing is used, the increased exposure potential may not be sufficient to achieve the desired in vivo concentrations, so more resource-intensive formulation technologies may be needed.

3.2 Nanosuspensions and Their Use

When the dimensions of material drop below the micron scale, two very different phenomena are observed. The first represents a linear extrapolation of behavior observed at macroscopic scales. The simplest example of this type of relationship is dissolution rate. Since this property is directly related to the solvent-accessible surface area, if large particles dissolve at a given rate, then small particles dissolve faster and nanoparticles dissolve faster still. The second phenomenon, generally not encountered until characteristic material dimensions are less than 100 nm, is behavior qualitatively different from that of conventional experience. This can manifest itself in the way matter interacts with biological systems—for example, the transformation of glass into a toxic substance because of the endocytotic mechanisms available solely to nanoparticulate forms of the material

(Stark 2011). Alternatively, the difference can be a discontinuous one in apparent conflict with everyday experience—for example, the several-hundred-degree drop in melting point (T_m) of nanogold compared to that of the corresponding bulk material (Roduner 2006).

While nanometer-sized particles represent a significant opportunity with respect to novel therapies that modulate biodistribution (Bertrand and Leroux 2012; Caron et al. 2012), this section describes the use of nanosuspensions as tools to identify and develop conventional molecular therapies based upon traditional medicinal chemistries. This discussion will be focused on the first type of behavior (i.e., continuous extrapolation) since leveraging the second requires that the intervention itself directly employs a form of nanotechnology—a topic beyond the scope of this chapter.

There is still much to be gained by employing nanoparticles' properties as a simple extrapolation from those of larger particles, however. For example, there is a finite amount of time available for dissolution as a solid passes through the GI tract or travels from the site of injection to the nearest reticuloendothelial organ. The poor physicochemical properties of many discovery compounds and/or samples may limit or prevent their bioavailability due to slow dissolution. Since, as noted above, nanosuspensions enable dissolution rates unattainable with larger materials (Kesisoglou and Amitava 2012; Quinn et al. 2012; Deschamps et al. 2009), they enable collection of useful biological data for such compounds. This use represents perhaps one of the most frequent applications of nanosuspensions.

As noted above, solution formulations also reduce concerns related to dissolution and solubility. However, in many instances, the requirement to identify a biocompatible solvent before a compound can be intravenously administered is a serious impediment to assessment of its ADME, pharmacological, and other properties. Though the particles in conventional microsuspensions are too large to easily pass through capillary beds—their use would likely result in embolism (Douglas et al. 1971)—reducing their size below 1 μm eliminates this concern and thus relaxes the need for solubilization in parenteral formulations.

Even when appropriate solvents can be identified, it may not be possible to find agents that retain the solubilized state when the solutions encounter aqueous environments (i.e., supersaturation), particularly when low potency or the need for hazard identification drives up dosing concentrations. In these cases, it may be better to employ a nanosuspension directly, which ensures that a material of known characteristics is present at the critical point of use instead of a solution that produces a poorly defined precipitate with unpredictable performance. When precipitation cannot be avoided, it may still be possible to choose additives that promote the formation of submicron solids. This can be accomplished restricting growth in one of the two ways: (1) by coating the particle surface or (2) by depleting the surrounding solution of compound via extensive nucleation (Ozaki et al. 2012). Such a presentation, employed either in vivo or in vitro, actually represents the in situ formation of a nanosuspension.

Table 3.2 Types of nanoparticles

Core	Shell	Surface	Type of nanosuspension
Crystalline drug	Crystalline drug	Stabilizer	Nanocrystal
Amorphous drug	Amorphous drug	Stabilizer	Amorphous drug nanoparticles
Oil and drug	NA	Stabilizer	Emulsion
Aqueous and drug	NA	Stabilizer and drug	Liposome
Stabilizer and drug	NA	Stabilizer and drug	Micelle
Lipid and drug	Lipid and drug	Stabilizer	Solid lipid nanoparticles
Drug and polymer	Drug and polymer	Stabilizer	Polymeric nanoparticle
Drug and polymer	Polymer	Stabilizer	Core-shell polymeric nanoparticle

NA not applicable

3.2.1 Types of Nanosuspensions and Their Production

While nanoparticle structure is closely related to the means of its production, the two issues should be addressed separately since a given structure can be achieved by more than one method. One flexible way of describing a nanoparticle is to consider it as being composed of an inner core, an encapsulating shell, and an outer surface, the last of which represents the interface between the particle and the surrounding medium. The distinction among the different possible nanostructures is then derived from the composition of these three components, as shown in Table 3.2. Note that this scheme does not preclude having more than one component made from the same material.

Nanosuspensions can be produced using the two general approaches shown in Table 3.3: (1) "size reduction" or top-down and (2) "size growth" or bottom-up. In the top-down approach, the size of particles is reduced to the point where they are small enough to take on the requisite properties. This is achieved by introducing large amounts of mechanical energy into a suspension via media milling, sonication, or high-pressure homogenization. The particles are broken up into smaller and smaller pieces until the energy represented by the total surface area of the particles is equivalent to that being supplied (Hennart et al. 2012). The efficiency with which mechanical energy can be delivered, regardless of scale, limits the minimum diameter accessible using this approach to 100–200 nm. The primary consideration for top-down methods is introducing energy into a stable system in a way that does not lead to destabilization.

In the bottom-up method, nanomaterials are produced by building them up from their individual constituents using one of a number of techniques (Chan and Kwok 2011). The simplest example of this approach is arrested precipitation. Since these approaches start with the size scale of individual molecules and grow from there, one of their advantages is that they can more easily be used to produce

Table 3.3 Means of making nanosuspensions

Approach	Method	Distinguishing features
Size reduction (top down)	Attrition	Grinding action of media
	Sonication	Cavitation generated by acoustic waves
	High-pressure homogenization	Shear and cavitation generated by turbulent flow
Size growth (bottom up)	Solvent displacement	Compound driven out of solution as particles by solvent/anti-solvent mixing
	Emulsion-diffusion	Emulsion/microemulsion droplets used as template for particle formation

nanomaterials with dimensions of less than 100 nm. However, this approach also presents the problem of producing an unstable (i.e., growing) system and then having to bring it rapidly to a steady state (Schwarzer and Peukert 2002). This problem is further complicated because the counterintuitive behavior of matter at these size scales (described above) immediately comes into play as growth occurs.

Comparing the two nanoparticle-production approaches, the top-down strategy is generally the most useful. This is not because it is perceived as universally applicable to discovery. Indeed, some drugs are available only as a solution or a glass-like film, immediately eliminating the possibility of the top-down approach for such compounds. Even if there is solid available, it might be in a form unsuitable to size reduction (e.g., it may desolvate in aqueous media to form an unworkable mass or exist as an unstructured material more likely to thermally decompose upon the input of energy than fracture into smaller size). In addition, unacceptable polydispersity can arise from solids that have adequate properties initially, but lose them in the continual presence of water and energy.

Instead, the preference for top-down method is based on the observation that bottom-up methods are even more likely to encounter difficulty. By necessity, such techniques involve a significant amount of dilution. So the problem is immediately encountered of maintaining colloidal stability while bringing the concentration up to the tens of milligrams per milliliter that might be necessary to pharmacologically assess low potency and/or high clearance compounds. A related issue is the frequent need to remove any solvents that are employed, again without compromising the suspension. The requirement that a compound demonstrate both very high solubility in some solvents and very low solubility in miscible solvents eliminates many compounds from consideration. Use of strong, water-miscible solvents such as dimethyl sulfoxide, dimethyl acetamide, dimethyl isosorbide, and glycofurol might make this less of a hurdle, but such solvents are also difficult to remove from the final product, exchanging one problem for another.

Finally, the most significant issue for bottom-up methods is that of control and reproducibility. As stated above, for matter to go from molecular dimensions to a stable colloid, it must by necessity pass through a size range where unexpected phenomena are observed. This in turn leads to a significant element of unpredictability in the process, making choices of excipients and processing parameters unique for each compound and/or scale rather than being broadly applicable across chemical space, as required by the higher throughput of discovery. Similarly, numerous studies have demonstrated the dependence of successful nanoparticle generation on subtle aspects of mixing (He et al. 2013; Thorat and Dalvi 2012), though carefully engineered mixing systems have been introduced to address such problems (Chiou et al. 2008).

3.2.2 *Analysis of Nanosuspensions*

3.2.2.1 Approach

By definition, the first characteristic to consider when discussing nanosuspensions is nanoparticle dimensions. As noted above, the size of the individual particles will determine both the performance and disposition in vitro and/or in vivo. Therefore, an accurate measure of nanoparticle dimensions will help in evaluating the suitability of various formulations for a specific task and also assist in interpreting the data collected using a given formulation. Such measurements can be made using a number of methods, which can be classified in the three ways shown in Table 3.4. The first are ensemble methods, which rapidly transform the superposition of all the signals generated by the sample's individual components into a size distribution. In the second category are the precise, yet slow, counting methods in which individual nanoparticles are counted and binned by dimension, producing a histogram that represents the sample's size population. In the last category are the elution methods in which a nanoparticle population is subjected to a size-related perturbation (e.g., centrifugal force), which physically separates the sample into components. Since sorption, in addition to size effects, can play a role in particle separation, these methods must be used with care to avoid artifacts.

Nanoparticle size is not the sole characteristic worth measuring, however. The magnitude and polarity of charge associated with a nanoparticle will play a significant role in determining its behavior and performance (Caron et al. 2012). Significant physical stability is afforded via charge-charge repulsion resulting from magnitudes in excess of 30 mV, with negative charges being more likely to avoid sticking to surfaces (Wu et al. 2011). A positive polarity can be used to enhance binding to cell membranes, which are negatively charged. There are several means of quantifying charge, but for the purposes of understanding the colloidal stability of nanoparticles, the most useful is zeta potential, which is the electric potential between the bulk solution and the interfacial slip plane, which separates the stationary layer of fluid diffusing with the particle from the fluid of the surrounding media (Wu et al. 2011).

The first consideration when choosing an analytical technique is the underlying reason for taking the measurement in the first place. Often, the analyst is trying to interpret data derived from use of the material in an experiment or to determine if the material has the characteristics needed for a specific purpose. For example, a

Table 3.4 Methods for characterizing nanosuspensions

Classification	Method	Range (nm)	Key elements	Simple	Sample sparing	Wide dynamic range	Comment	Reference
Ensemble	Fraunhofer diffraction	500 to >100,000	Diffraction of light by suspension produces pattern that is the Fourier complement of system microstructure			X	Imprecise at low end of size range	Keck and Muller (2008)
	Dynamic light scattering	3–3,000	Brownian motion measured by quantifying intensity fluctuations in scattered light		X		Intolerant of large particles	Villari and Micali (2008)
	Mie scattering	5 to >10,000	Angular dependence of scattered light intensity as a direct consequence of particle size			X	Requires real and imaginary refractive index	Villari and Micali (2008)
	Turbidity	50–5,000	Complement of light scattering, wavelength dependence of transmitted light	X	X		Quantitation requires real and imaginary refractive index	Oshchepkov and Sinyuk (1998)
Counting	Bright-field optical microscopy	>500	Individual particles imaged via the refraction of focused light	X	X		Statistical, limited low-end range	Murphy (2001)
	Dark-field optical microscopy	>10	Light scattered by individual particles detected at focal plane	X	X	X	Statistical	Murphy (2001)
	Electron microscopy	1–100	Short de Broglie wavelength of electrons decreases size detection limit		X		Statistical, limited high-end range	Murphy (2001)
	Single-particle optical sensing	>500	Particles passing singly through a laser generate signal in proportion to their size		X		Useful for mass balance	Driscoll et al. (2001)

(continued)

Table 3.4 (continued)

Classification	Method	Range (nm)	Key elements	Simple	Sample sparing	Wide dynamic range	Comment	Reference
Elution	Filtration	20–10,000	Particles separated based on ability to pass through membrane pores	X	X	X	Subject to sorption	
Charge	Electrophoretic light scattering	<50,000	Magnitude and phase of light scattered from particles in an oscillating electrical field		X	X	Standard method	Doane et al. (2012)
	Capillary zone electrophoresis	<50,000	Relative electrostatic migration of particles against electrophoretic flow		X	X	Also used for protein analysis	Lopez-Lorente et al. (2012)

nanosuspension is useful in evaluating the dependence of oral exposure on dissolution rate only if its particle size is known, it is monodisperse, and it is stable in vivo. Another consideration, particularly for discovery, is the amount of sample required for a given analysis. Typically, the preparation of nanosuspensions represents expenditure of a scarce active compound to answer a question, so little material is left over for characterization.

3.2.2.2 Methods

While the use of nanosuspensions suggests the need for sophisticated laboratory equipment, the truth is that the discovery-scale implementation makes analyses relatively straightforward. In the current application, microscopy is a very useful method of nanosuspension characterization (Murphy 2001). Nanoparticle size, aggregation, stability, and, if polarizing optics are employed, crystallinity can be assessed immediately with only 1 μL of sample. The inability of typical light microscopes to resolve particles less than approximately 0.7 μm renders the conventional technique best suited to characterization of starting materials and samples early in processing, since nanosuspensions themselves will appear only as a shimmering haze. Dark-field illumination, which reveals the presence of particles through light scattering rather than producing a direct image, is a useful means of probing below diffraction limits. This method also allows assessment of particle numbers and the detection of the presence/absence of Brownian motion—both practical, albeit indirect, indicators of nanosizing. However, use of such small volumes makes obtaining a statistically accurate sampling of the suspension highly unlikely. Analysts also need to be aware that cover slips tend to induce aggregation.

Measuring the amount of diluted suspension passing through syringe filters with a range of pore sizes is another straightforward, sample-preserving means of determining size. The most serious concern is that sorption will be misinterpreted as the intended sieving process. The influence of this artifact can be accommodated, but not eliminated, by ensuring that all of the filters have the same composition.

Another convenient, accessible method of characterization is wavelength-dependent turbidity analysis (Oshchepkov and Sinyuk 1998; Gulari et al. 1987). Light-scattering theory indicates that turbidity transitions from being relatively independent of wavelength at particle sizes above 2–3 μm to being skewed such that short wavelengths generate increasingly more signal as the size drops (Kerker 1969). Using ultraviolet-visible spectroscopy, measurements can be as simple as noting the ratio of transmittance at a wavelength of 400 nm to that at 800 nm. This ratio will increase as nanosizing continues and will stabilize as size stops changing, which makes it a reliable indicator of process status. More detailed information can be obtained by comparing the turbidity spectrum of the sample against those of previously measured standards and reconstructing the former as a linear combination of the latter (Haskell et al. 1998).

The need to obtain and interpret in vitro or in vivo results usually requires quantitative assessment of size rather than an inferential measurement. Dynamic light scattering (DLS), also known as photon correlation spectroscopy (PCS) or quasielastic light scattering (QELS), is frequently employed. The intensity variation of light scattered from particles in Brownian motion is quantified to produce a signal from which a corresponding diffusion coefficient can be derived. Applying the Stokes-Einstein relationship between diffusion and size in an environment of known temperature and viscosity completes the measurement (Villari and Micali 2008). Large particles passing through the scattering volume generate changes in intensity unrelated to size, which limits measurements to less than 3 μm. Hence, such particles must be eliminated from the sample before analysis. The need to avoid the artifacts induced by multiple scattering requires high dilution, which keeps sample consumption low, though some compounds are sufficiently soluble to dissolve under such conditions. While instrumentation for this measurement is moderately complex and may not exist in a discovery-oriented pharmaceutical science laboratory, the method is commonly employed to assess protein aggregation (Yadav et al. 2013), so it is likely available nearby.

Light diffraction, a method by which the angular-, wavelength-, and polarization-dependent scattering pattern is deconstructed to calculate a particle size distribution, can also be employed, although there are several limitations (Keck and Muller 2008). First, sample requirements can be as high as tens to hundreds of microliters, which may be more volume than is available in the entire formulation. Second, the real and imaginary refractive indexes may not be known. This limitation allows light passing through the particles to be misinterpreted as light derived from nanoparticles not actually present in the sample. Third, the necessary instrumentation is relatively scarce in discovery labs. Despite these limitations, the method is valuable, particularly in its ability to measure sizes ranging from 50 nm to 400 μm simultaneously. Thus, the presence of large particles does not present a problem the way it does for DLS.

Single-particle optical sensing is a particularly useful technique, which has been previously applied to parenteral emulsions (Driscoll et al. 2001). Particles of the suspension are introduced into a flow system, where they pass individually through a laser beam and produce a scattering signal in proportion to their size. Counts are binned into particle size ranges, producing a number histogram that is converted into a volume distribution, assuming sphericity. The area under this curve is an absolute measurement representing the total volume of material that passed through the sensor. This value can be normalized to the amount of material that is calculated to have been present and reported as a volume detected per unit mass of sample introduced. Because the size detection limit of this technique is approximately 0.7 μm, the entire distribution is observed in a microsuspension but is mostly undetected for a nanosuspension. As a result, the number of particles drops precipitously as processing continues. This selectivity renders the method reasonably useful for monitoring small changes in the large-size tail of a size distribution dominated by nanoparticles, providing information on the status of milling and stability and indicating aggregates (Tolla and Boldridge 2010).

3.2.3 Generating Top-Down Nanosuspensions at the Discovery Scale

Typically, large media mills are used to make top-down nanoparticles. These large mills (e.g., Dynomills and Netzsch mills) agitate a slurry of beads and suspension using rapidly rotating mixing elements. Such equipment are unsuitable for use in discovery for several reasons. First, these mills require a large amount of compound, ranging from thousands of liters of nanosuspension down to a few hundred milliliters. Even the smallest laboratory-based systems require 5–10 mL—far more volume than is available in most cases. Second, the equipment generates a large amount of frictional heat, which is an unavoidable by-product of a process based on grinding surfaces together. The chemical stability of discovery compounds in warm, aqueous environments is either poor or unknown, and the unstructured nature of the solid tends to encourage the production of an unusable putty in such conditions. Third, the time required to assemble, clean, and reassemble conventional equipment is prohibitive in an environment where tens of compounds may be formulated as nanosuspensions in a week. While the 150- to 200-nm nanoparticles produced by top-down approaches will enhance dissolution rate, they are not sufficiently small to enhance solubility (Van Eerdenbrugh et al. 2010).

Capturing the essential mechanism of such systems—i.e., a well-mixed slurry of beads, suspending vehicle, and drug—can also be accomplished using miniaturized equipment (Van Eerdenbrugh et al. 2009) or simple laboratory equipment at little cost. One approach is to employ glass vials, 0.5-μm glass beads, and a stir plate/stir bar in a configuration that can easily achieve stirring rates of 3,000 rpm (Haskell 2004). Because this approach uses commonly available laboratory equipment, it affords non-pharmaceutics laboratories to access the advantages of nanosuspensions (Nophsker and Haskell 2010). The most anticipated problem, decoupling of the magnets, is addressed by using higher strength magnetic stir bars, now commonly available from commercial sources, or by modifying the stir plate. By using equipment such as high-speed drills fitted with strong ceramic magnets, stirring speeds up to 10,000 rpm can easily be achieved, enabling use of highly viscous systems (e.g., thixotropic systems or lipids) as the continuous phase. The volume of beads is near 80 % of that of the total slurry and success is most generally encountered if the vessel is chosen so that the slurry fills approximately 50 % of the volume available. Maintaining thorough mixing is clearly an important concern, so the stir bar must be large enough to produce a vortex. Using vial combinations in the 0.5- to 20-mL range, this approach enables the production of nanosuspensions with active compound quantities ranging from 500 μg up to 2 g—the range required in the discovery environment.

The bead-bead interaction in this approach is of lower energy than that attained with large media mills. This is an important consideration for a process in which energy input is a key factor in determining the rate and extent of size reduction (Knieke et al. 2011). Original reports of this technique demonstrated that, at the laboratory scale, time can be substituted for aggressive agitation, since even

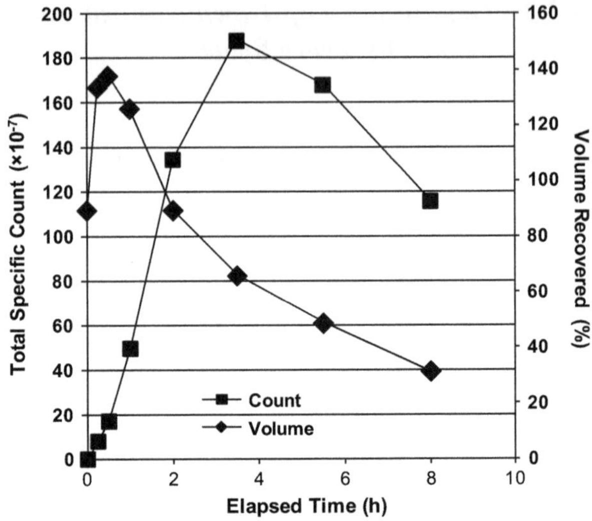

Fig. 3.1 Single-particle optical sensing data collected from suspension during size reduction. The volume of material detected (*diamond*) starts high and then drops as more and more material achieves a size below the detection limit. The number of particles counted (*square*) starts low, increases by necessity when large particles are broken up into detectable smaller particles, and then decreases as those that are newly produced become too small to be observed

low-energy roller mills were sufficient to produce particles as small as 180 nm (Liversidge et al. 1992). Figure 3.1 shows the time course of a process that produces a nanosuspension in a process taking place overnight. The plot, prepared using single-particle optical sensing, shows the number of particles counted per mass of suspension and the total volume corresponding to those same particles. The particle count rises with time, consistent with the larger observable particles breaking up into smaller observable particles, and then falls as the particles break up further, producing particle sizes below the detection limit of the sensor. Monitoring the volume leads to a first-order decrease in signal, as predicted by comminution models (Tangsathikulchai 2003).

After the suspension is produced, it is removed from the bed using transfer pipettes with openings smaller than that of the beads. Even though the surface tension is low due to the presence of surfactants, capillary action limits recovery to approximately 75 %. Milling at higher concentrations than is actually required allows the beads to be repeatedly rinsed with fresh vehicle, achieving recoveries in excess of 95 %. In general, milling efficiency is better at higher concentrations (Kapur et al. 1996), so this is a practical strategy in most cases. If other requirements eliminate dilution as an option, alternate approaches can be employed. Touching the vessel to a vibrating surface simultaneously reduces capillary effects and leverages the shear-thinning nature of the suspension, improving nonrinsing recoveries to 90 % or more. The problem of handling small-scale production (i.e., <0.75 mL) and/or the desire for even higher recoveries can be solved by poking holes in the bottom of the milling vessel, effectively converting it into a sieve. Gentle centrifuging of this with a collection vessel completely eliminates capillary effects and transfer losses, improving recoveries to more than 98 %. Obviously, the milling must be done in a plastic vial for this technique to be successful.

Polymers, such as polyvinylpyrrolidone (PVP), hydroxypropyl cellulose (HPC), carboxy methyl cellulose (CMC), and hydroxypropyl methylcellulose (HPMC, also known as hypromellose)—particularly those with lower molecular weights (Choi et al. 2008)—effectively achieve suspensions with the physical stability needed to obtain submicron particle dimensions during manufacture and maintain them before use.

Polymers—which achieve stability via steric mechanisms, osmotic pressure, and resistance to Ostwald ripening—are even more effective if used with charged surfactants such as sodium lauryl sulfate, dioctyl sulfosuccinate, and cetyl trimethylammonium bromide. These charged surfactants impart additional stabilization through electrostatic repulsion. Charged surfactants are less desirable for parenteral applications due to their hemolytic properties, which can in turn perturb pharmacological or toxicological studies, so poloxamers and polysorbates can be used in their place. Such surfactants avoid the formation of destabilizing lipophilic salts produced when basic compounds stabilized with sulfonic acids encounter reduced pH (Donoso et al. 2012).

While the mechanisms by which these excipients maintain stability are well understood (Verwey and Overbeek 1948), choosing which combinations to employ is still an empirical undertaking (Ghosh et al. 2011; Juhnke et al. 2010). As a result, formulation optimization is best carried out via screening—an unavoidably time-intensive activity. However, the need for as little as 0.5 mg per attempt makes it possible to try several conditions while consuming only a few milligrams of compound. Physicochemical stability must be maintained only long enough to perform the experiment for which the formulation was made in the first place, considerably relaxing the complexity of the stabilization problem. This is one of the primary reasons nanosuspensions of this type are so easily employed in discovery. Since the physiochemical properties are clustered for a given program's chemical space, it is generally not necessary to screen each compound.

One of the problems unique to this implementation of media milling is the tendency to generate foam because of the ease with which air is entrained into the slurry from the vessel headspace—something that is less of a consideration with large-scale equipment. Not merely an inconvenience, foam significantly reduces the efficacy of the process by dissipating much of the input energy. Fortunately, this is easily avoided via the use of small amounts (e.g., 0.2 ppm) of a silicone oil-based defoaming agent. Excessive use of these materials will destabilize the system, because the excess polymer will coat the particles and create a hydrophobic surface. That said, additional aliquots may be needed throughout the milling process because the effectiveness of the defoaming agent drops over time (Denkov et al. 2000).

Some combinations of available drug, suspension volume and concentration, bead quantity and diameter, and vial size require use of a custom stir bar to ensure adequate mixing. Custom stir bars also improve resistance to decoupling, as can happen when highly viscous slurries are milled using a high-strength magnetic system. This resistance is better described by the ability of a magnetic field to induce torque in the coupled object, quantified by units of oersted, than by the value of the vector field itself (in units of gauss). Typical stir bars use AlNiCo alloys with a resistance of ~1,000 oersteds. Substantially more torque is tolerated by SmCo magnets (~15,000 oersteds), but sometimes the even higher resistance to decoupling afforded by NdFeB magnets (up to 40,000 oersteds) is required. While the former are commercially available as cores for stir bars, the latter are not. Magnetic stir bars with any core can be easily made by inserting magnetic disks into chemically resistant tubing and then sealing the ends (Haskell 2003).

The major concerns working at the laboratory versus production scale are compared in Table 3.5, which makes it clear that generating nanosuspensions as described above renders their use more accessible in discovery than at the production scale for a variety of reasons. The two areas of concern specific to the small scale—i.e., avoiding foaming and maximizing recovery for discovery applications—are addressed through the use of antifoaming agents and creative use of laboratory equipment, respectively. An additional enabling feature of the laboratory-scale application is the relative ease of using radioactively-labeled compounds since the equipment employed is easily decontaminated or disposed of, if necessary. A similar argument applies to the formulation of compounds with occupational exposure band (OEB) ratings of four and higher.

Parenteral applications are also much more accessible when making and using nanosuspensions as discovery tools as opposed to a clinical drug product. Reasonable care alone leads to a high degree of sanitation adequate for many purposes, and sterile formulations can be produced using aseptic techniques without difficulty because of the simplicity of equipment. However, an easier approach is manufacture of the suspension at a higher concentration than necessary, followed by sterile filtration, assay of the filtrate for drug content, and appropriate dilution with additional sterile-filtered vehicle. Because of the lower energy process used in laboratory-scale media milling, filtration losses of compound can be higher than those encountered in a manufacturing environment, which typically employs higher energies that are more effective at reducing particle diameters below 0.22 μm. Thus, it is advisable to start with a suspension concentration that is as high as possible. As discussed above, higher concentrations actually enhance the effectiveness of the milling process. Use of sterile filtration does not obviate the need for clean excipients and careful handing since endotoxin contamination can be the result of carelessness in this regard.

3.3 Solid Amorphous Dispersions (SADs)

While crystalline nanosuspensions increase the surface area, which is often useful for addressing compounds that are dissolution rate limited, SADs are useful for addressing compounds whose absorption is limited by either dissolution rate or solubility since SADs exhibit both high thermodynamic activity and rapid dissolution of the amorphous drug. Dissolution can be especially rapid for spray-dried

Table 3.5 Comparison of nanosuspension preparation issues at the laboratory scale versus the production scale

Category	Issue	Concern at production scale	Concern at laboratory scale
Manufacturing	Equipment complexity and cost	High	Low
	Energy dissipation	High	Low
	Efficiency of energy use	High	Low
	Attrition of milling media	High	Moderate
	Poor recovery	Low	High
	Poor mixing	Moderate	Low
	Foaming	Low	High
Properties	Colloidal instability	High	High
	Characterization	High	Moderate
	Long-term physical stability	High	Low
	Bioburden	High	Moderate

dispersions (SDDs) made at small scale for discovery applications, as constitutive particles tend to be small in size (~1 μm) and therefore have high surface areas.

SADs can be formed by a variety of processes, including combining heated drug and excipient to produce hot melt extrusions (HMEs) (Crowley et al. 2007); evaporation that generates SDDs or rotoevaporated dispersions (Vehring 2008; Dobry et al. 2009); and precipitation from solvent-anti-solvents (Shah et al. 2013). SDDs have several advantages for discovery applications, including the following.

- Suitability for small-scale manufacture—Spray-drying is readily amenable to small-scale manufacture (tens of mg active). Some other processes for forming SADs, such as HMEs, have traditionally been less amenable to such small scale due to equipment limitations.
- Quick turnaround time—SDD formulations can often be produced at the rate of about one per hour.
- Broad applicability—Because they are kinetically stabilized using rapid drying, drug miscibility with the matrix material is not required, allowing compounds having diverse physicochemical properties to be formulated with great flexibility with respect to loading and excipient choice. Drug loading and physical stability can be balanced against one another for a particular discovery application, depending on the required dose and study requirements. In addition, a range of organic solvents can be used to assure mutual solubility of drug and matrix material in the spray solvent. Figure 3.2 indicates the breadth of applicability of SDDs to compounds with varying physicochemical properties. The plot shows the T_m divided by glass-transition temperature (T_g) versus the calculated octanol/water partition coefficient (CLogP) for 147 compounds that have been formulated at small scale as SDDs at Bend Research. More than 1,000 compounds have been formulated as SDDs for early preclinical evaluation.

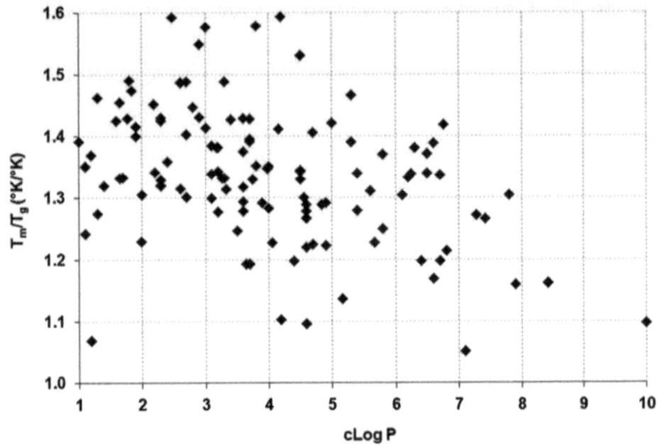

Fig. 3.2 T_m divided by glass-transition temperature (T_g) versus the calculated octanol/water partition coefficient (CLogP) for 147 compounds that have been formulated at small scale as SDDs at Bend Research

- Suitability for commercial progression—SDDs are scalable and they are used in products. They also typically use nontoxic, precedented excipients that can be used at high doses for preclinical toxicology testing of active compounds. The safety of the excipients and scalability of the process enable formulations assessed in discovery to transition rapidly into development.
- Flexibility with respect to drug form—The solid form of the active compound (e.g., amorphous or fully/partially crystalline) is not important, since the active compound will be dissolved in a solvent and sprayed into an amorphous powder. The approach is often amenable to compounds with unknown chemical stability, due to minimal exposure to elevated temperature, as opposed to HME, for example.

Due to the advantages of SDDs, they are the focus of this discussion of SADs. Dispersions formed in other ways share some of the characteristics described below.

3.3.1 SDD Manufacture

3.3.1.1 Materials

SDDs are typically manufactured using polymers with high T_gs to enhance the physical stability of the amorphous form. In addition, the polymers tend to be soluble in aqueous media at physiological pH. They are commonly water soluble or enteric, to ensure dissolution of the SDD in the GI tract. In addition, the dispersion polymer must have sufficient solubility in the organic spray solvents in which the active compound is also soluble. Examples of dispersion polymers include hydroxypropyl methyl cellulose acetate succinate (HPMCAS) (Curatolo et al. 2009; Friesen et al. 2008), HPMC (Raghavan et al. 2003), PVP (Lindfors et al. 2008), polyvinylpyrrolidone vinyl acetate copolymer (PVP-VA), and

Table 3.6 Key properties of select SDD polymers

Polymer (grade)	T_g (°C)		Water content at 25 °C/75 % RH (%)	Typical spray solvents	Ionizable groups
	Dry sample	Sample at 75 % RH			
HPMCAS (-M)	125	73	6	Acetone, methanol, tetrahydrofuran (THF), dichloromethane (DCM)	Acidic
HPMC (E3)	142	43	11	Methanol/water	None
PVP-VA (VA64)	106	17	21	Acetone, methanol	None
PVP (K-15)	116	−14	35	Methanol	None
Polymethacrylate (Eudragit L100)	191	97	12	Acetone, methanol, DCM	Acidic

RH relative humidity

polyacrylates. HPMCAS is often used because it is a particularly enabling polymer, due to its amphiphilic nature and associated ability to form drug/polymer colloids and to inhibit drug crystallization (Curatolo et al. 2009; Friesen et al. 2008). These drug-polymer colloids are often useful for resupplying free drug as it is absorbed from the GI tract (Friesen et al. 2008). Table 3.6 lists key properties of some select dispersion polymers.

For drugs that might be acid sensitive or that are not as prone to precipitation, particularly crystallization, neutral water-soluble polymers such as HPMC or PVP-VA are chosen. These neutral polymers are also appropriate if gastric, as opposed to intestinal, dissolution is important in vivo. In a discovery environment, several polymers and drug loadings can be screened for performance, depending on the quantity of active compound available. When the quantity of active compound is extremely limited, as is often the case, a single polymer and drug loading is chosen for formulation into an SDD with a high degree of success. A common selection is a conservative drug loading in the middle (-M) grade HPMCAS polymer. In situations where the quantity of active compound is limited, the map in Fig. 3.2 can be a useful guide for initial SDD formulation selection. The T_m/T_g ratio reflects the tendency of a drug to crystallize, with higher ratios representing a stronger crystallization tendency. As this ratio increases, it is often necessary to reduce the drug loading in the polymer and sometimes move toward a more hydrophobic polymer to mitigate crystallization in the solid state and in suspension. The ability to employ large dosing volumes preclinically and the lack of need for dosing forms other than suspensions allows low drug loadings to be useful and is an example of how some requirements can be relaxed in discovery. The CLogP value represents the lipophilicity of the active molecule. Compounds with higher CLogP values can have slower dissolution rates, particularly with more hydrophobic polymers, such as hydrophobic grades of HPMCAS. For compounds that are both highly lipophilic and have a tendency to crystallize, there is often a need to balance the need for rapid dissolution with the requirement for crystallization inhibition. This is sometimes accomplished through use of a moderately hydrophobic polymer, such as HPMCAS-M.

3.3.1.2 Process

Spray-drying can be carried out over a wide range of scales that is applicable to early development all the way through commercial manufacture. This offers the advantage of a progressible formulation that can be successfully transitioned from discovery into further development. Small-scale spray dryers are available in different designs from more than one source. We describe here a design used to manufacture SDDs at small scale, the Bend Research "mini" spray dryer, which has been optimized for rapid drying kinetics and good recovery at small scale (Curatolo et al. 2009).

In this spray dryer, the spray solution is fed via syringe to a two-fluid atomizing nozzle mounted at the top of a stainless-steel processing chamber. A syringe pump feeds the solution to the nozzle at a rate typically between 0.5 and 1.5 mL/min, while a mass flow controller meters nitrogen gas through the outside, annular nozzle opening. Heated nitrogen atomizes the feed solution into high-surface-area droplets and evaporates solvent for drying of the particles. The droplets dry rapidly and solidify as they are carried by the nitrogen downward toward the collection filter. In this application, the two-fluid nozzle is used to enable sufficient contact of the drying nitrogen for rapid drying kinetics, and to make small droplets amenable to drying in a short-residence-time chamber. Since the small spray-dried particles would be difficult to collect efficiently in cyclones that are used at larger scale due to the low drying-gas flow rate, they are collected on filter paper at the outlet of the chamber. The process parameters (e.g., flow rates, temperatures) are controlled with an electronic control system and can be optimized based on the SDD formulation or the spray solvent used. Batch sizes as small as ~100 mg of SDD are routinely collected at >90 % yield. Thus, at a low drug loading of 10 % in the SDD, which is often sufficient in discovery, only 10 mg of active compound is needed for each formulation. This allows efficient screening of multiple SDD formulations and/or compounds using minimal active compound, when both compound and time are in short supply.

Important considerations for spray-drying include the solvent choice. Typically, high-volatility organic solvents, sometimes mixed with small amounts of water, are chosen for processing because they are easy to remove from the SDD after spray-drying. It is important that both the active compound and polymer have sufficient solubility in the spray solvent to enable reasonable throughput. A solids concentration of approximately 2 % by weight is commonly used for small-scale spray-drying, although the aforementioned flexibility in discovery allows further reduction if the active compound or polymer has low solvent solubility. Solids should be sprayed at concentrations significantly lower (e.g., ~20 %) than their solubilities in the spray solvent to guard against precipitation of solids in the syringe, lines, or nozzle during processing.

After being sprayed, SDDs normally undergo secondary drying to remove residual solvent. At large scale, this step is performed using agitated dryers or tray dryers. For the small-scale process described above, the SDDs are often scraped from the filter paper into a vial or another small container and dried by vacuum desiccation. A nitrogen or airstream can be added to aid in solvent removal. The particle size of powders spray-dried at this scale are typically small and are best handled using a static wand to avoid significant loss of material.

3.3.2 *Characterization of SDDs at Small Scale*

3.3.2.1 Performance

A number of bulk-sparing performance tests have proven useful for characterizing the performance of SDDs early in discovery when active compound quantities are limited. Two such tests are the microcentrifuge dissolution test (Curatolo et al. 2009; Friesen et al. 2008) and the membrane permeation test (Vodak 2012; Babcock et al. 2009). The microcentrifuge test (MCT) is used to screen formulations by quantifying solubilized drug species that are important for oral absorption. In particular, it measures the time-dependent concentration of drug residing in small, neutrally buoyant species present in a biorelevant medium. Such species typically consist of freely dissolved drug, drug in bile-salt micelles, and drug/polymer colloids:

$$[D] = [D_{diss}] + [D_{mic}] + [D_{coll}] \tag{3.1}$$

where $[D]$, $[D_{diss}]$, $[D_{mic}]$, and $[D_{coll}]$ are the concentrations of total solubilized drug, dissolved drug, drug in bile-salt micelles, and drug in drug/polymer colloids, respectively. The drug/polymer colloids, which are typically present only when using amphiphilic polymers like HPMCAS, can be important for enhancing absorption because they are high-energy, high-surface-area species that can rapidly resupply freely dissolved drug and drug in bile-salt micelles (Friesen et al. 2008).

In the MCT, SDD is weighed into a microcentrifuge tube and a suitable biorelevant dissolution medium, such as a simulated gastric or intestinal fluid, is added to create a nonsink condition. The solution in the tube is vortexed briefly and at each time point the sample is centrifuged to pellet any undissolved SDD solids or dense, large-particle precipitate. An aliquot of supernatant is withdrawn, and the sum of the active compound in the species of Eq. (3.1) is quantified by high-performance liquid chromatography (HPLC). The test is performed with 1–2 mL of solution and often requires less than 1 mg of active compound to be run in duplicate, depending on the drug solubility in the medium. The dissolved drug concentrations provided by the amorphous drug form are often 5–10 times higher than the corresponding crystalline solubility (Friesen et al. 2008), so the test is typically run with drug concentrations that are approximately tenfold higher than

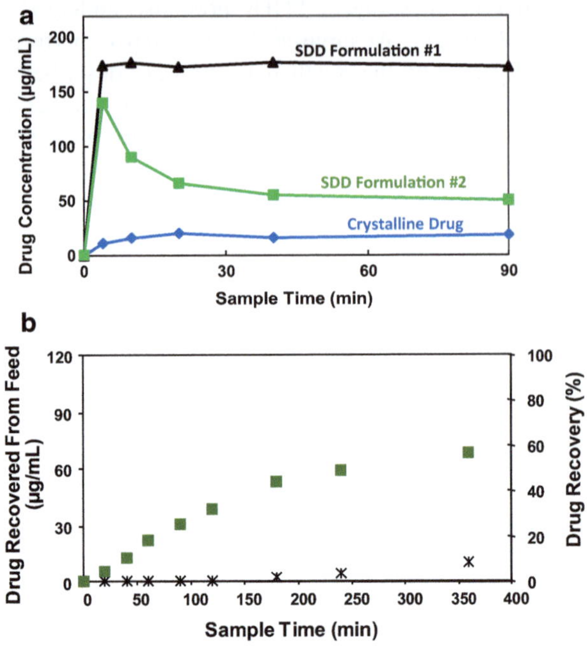

Fig. 3.3 Example microcentrifuge data for two SDDs compared to data for crystalline drug (**a**), and example membrane permeation data for an SDD compared with crystalline drug (**b**). See text for discussion

the latter value. The MCT can employ a single biorelevant medium, or can involve a transfer between media to mimic transit through the GI tract. The transfer test can be particularly valuable when evaluating ionizable active compounds with pH-dependent solubility.

The MCT is particularly useful for evaluating supersaturation levels and sustainment of those levels—i.e., crystallization inhibition—provided by SDDs. Figure 3.3a shows an example of microcentrifuge data for two SDDs compared to data for crystalline drug. As the figure shows, SDD #1, in particular, shows a substantially higher level of solubilized drug species than the crystalline drug due to the supersaturation provided by the amorphous drug form and/or drug/polymer colloids. SDD formulation #2, although it outperforms crystalline drug, is not optimum for the drug compound shown in this example.

The MCT can be combined with an ultracentrifugation step to further separate solubilized drug species. Ultracentrifugation typically pellets the less buoyant drug/polymer colloids, leaving freely dissolved drug and drug in bile-salt micelles in the supernatant. In conjunction with micro- and ultracentrifugation, nuclear magnetic resonance (NMR) analysis of the suspension can likewise provide further information on drug speciation by separating drug species based on mobility (Curatolo et al. 2009; Friesen et al. 2008). Drug in colloidal particles is generally not detected by solution NMR due to the particles' relatively long rotational time. Dynamic light scattering can be used to confirm that colloids have been pelleted from the supernatant of ultracentrifuged samples, with a large reduction in size/scattering intensity indicating a successful separation. Finally, the pellet from the MCT can be inspected by polarized light microscopy or analyzed by powder X-ray diffraction (PXRD) to determine if drug concentrations have decreased over the course of the test due to precipitation of crystalline or amorphous drug. In addition, the pellet from the MCT can be redispersed into fresh biorelevant medium in a "redissolution" test to assess whether the precipitate can quickly source high levels of solubilized drug. A number of these drug-speciation tests can be conducted using a sample from the same microcentrifuge test (ultracentrifugation, dynamic light scattering, PXRD, and polarized light microscopy) or using very little additional SDD (in the case of NMR analysis), thus conserving valuable active compound.

The MCT is useful not only in assessing the performance of the as-manufactured SDD but is also a valuable tool for assessing the physical stability of an SDD suspension in the dosing vehicle. Suspension stability can be particularly important for longer term toxicology studies where frequent dosing over many days is necessary. Suspension stability of the SDD versus crystallization or particle aggregation is important because it facilitates dosing by allowing preparation of a large volume of suspension for use over several days or even weeks, so that separate suspensions do not have to be prepared one or more times per day. For suspension stability studies, instead of adding solid SDD powder to the biorelevant dissolution media, a concentrated suspension of SDD can be held for various periods of time before dilution into the dissolution media and assessment of dissolution performance. Typical suspending agents include HPMC and HPMCAS in buffered media. Consistent performance in the test after extended hold times (e.g., days or weeks) suggests good physical stability. These data are especially useful for establishing acceptable formulation protocols for preclinical dosing, especially for multiple-dose studies in which it can be inconvenient to make new suspensions just before each dosing.

The membrane permeation test is another useful in vitro test of SDD performance and has been used for more than a decade to screen SDDs. The test measures the flux of drug across a semipermeable synthetic membrane into a secondary liquid phase that provides a sink for the drug. Like the MCT, the membrane test uses minimal active compound—typically, a few milligrams, depending on the solubility of drug in the aqueous feed media. As in the MCT, the SDD is suspended in the biorelevant medium such that the active compound is at several times its solubility. In the test, a porous (1-μm pore size) polypropylene membrane separates an aqueous donor solution from an organic permeate solution in which the drug is highly soluble. The ideal organic solvent has limited miscibility with the aqueous feed solution and a low vapor pressure to limit evaporation; a representative organic solution is 80:20 (w/w) decanol:decane. An SDD is dosed to the aqueous donor solution and the organic permeate solution is sampled over time. These samples are then analyzed by UV-VIS absorption with or without a chromatographic separation step, to determine the drug concentration in the organic permeate solution. Since it is believed that only freely dissolved drug is transported across the membrane, and since the flux is proportional to freely dissolved drug levels (Babcock et al. 2009), the test offers an indirect measure of the sustainment of free drug levels in the aqueous donor solution. This information, in conjunction with the sustainment/speciation data from the MCT, can be useful in ranking formulations or compounds. Figure 3.3b shows example membrane permeation data for an SDD compared with crystalline drug.

3.3.2.2 Characterization of the State of Matter

When formulating SDDs for discovery, it is useful to verify that a homogeneous drug/polymer dispersion has been formed, since phase separation, and particularly the presence of crystalline drug, can adversely affect the solubilization performance of the SDD by lowering drug activity, dissolution rate, and long-term physical stability (Friesen et al. 2008). Several bulk-sparing methods can be used to detect phase separation. These include PXRD, scanning-electron microscopy (SEM), and modulated differential scanning calorimetry (mDSC). Depending on the specifics of the instrument and sample, PXRD can be used to detect the presence of crystalline drug to a level of approximately 1 wt% of the SDD (Palermo et al. 2012). As such, PXRD is not as sensitive as some other techniques, but is nondestructive and consumes no active compound. SEM, which uses <1 mg of material, is often used as a highly sensitive, but nonquantitative, method of detecting drug crystallinity in SDD particles (crystals often form first on the surface of SDD particles due to the high mobility and the high-energy state of the drug at the particle surface).

mDSC is a valuable tool for assessing amorphous phase separation in dispersions. In particular, a single, sharp T_g is typically interpreted as an indicator that there is no large-scale ($> \sim$15-nm domains) phase separation of drug and polymer (Utracki 1990). The T_g of the SDD, both dry and at various RH values, is a valuable indicator of solid-state physical stability of the SDD when evaluating formulations or compounds. At the earliest stages of research, less attention is typically given to long-term physical stability when using the formulations to enable discovery. However, a good understanding of this property is required to assess SDD progression into development, as described in Chap. 1.

3.3.2.3 In Vivo Performance

For early preclinical in vivo studies, such as pharmacokinetic or toxicity studies in rodents, SDDs are typically prepared as suspensions and dosed by oral gavage (e.g., through a 16G needle). Well-dispersed and stable suspensions are typically made by adding a suspension vehicle dropwise to the SDD powder using a mortar and pestle to uniformly wet the particles. The suspension is made at a concentration that is high enough to deliver the dose in a reasonable dose volume (e.g., 5–10 mL/kg), but low enough that it is syringeable and has sufficient physical stability to achieve the dosing protocol. The maximum syringeable concentration of SDD in suspension depends on the choice of polymer and the drug, and sometimes on the drug loading in the polymer. For HPMCAS SDDs, suspensions with polymer concentrations up to 150–200 mg/mL are usually achievable.

3.4 Discovery Formulation Approaches for Modulating Oral Exposure

We have just focused on two mature techniques used in the discovery environment to help achieve desired exposure. As more is understood about the efficacy/toxicity of a given molecule, teams may need to find a way to achieve tunable exposure to further understanding of the molecule and target attributes. In addition, some therapeutic areas (e.g., diabetes and oncology) may need to modulate the PK profile to avoid tolerability issues associated with a high C_{max} (Reddy et al. 2013; Sampath Kumar et al. 2012; Eichenbaum et al. 2006). Modulating exposure of compounds through modified-release delivery approaches may be useful in discovery settings for a variety of reasons, including the following:

- Overcoming the PK deficiencies of tool compounds [e.g., excessive ratios of C_{max} to the minimum drug concentration (C_{min}), rapid clearance], so that teams can use them to validate targets and answer other questions in preclinical animal models: This allows the use of nonoptimized molecules and enables programs to proceed without the need to wait for the "right" compound to be discovered, synthesized, and characterized.
- Manipulating the exposure profiles of high-clearance compounds, which allows pharmacological parameters such as occupancy to be studied more easily.
- Reducing the need for repeated dosing, decreasing in-life resource burden, and simultaneously minimizing the risks that repeat dosing and the associated handling stress will introduce artifacts into animal studies.

Drawbacks and challenges include the following:

- The likely need to increase dose (i.e., consume scarce material) to compensate for the reduced absorption rate that many approaches use to generate sustained release.
- The potential that sustained release will result in increased intersubject variability in exposure.
- The need to reconcile the technology employed to achieve sustained release with the physiological dimensions and characteristics of preclinical species: For example, the needed volume for a formulation in which excipient constitutes 90 % of the dosed material may be too large to be administered.
- The tendency of modified-release technologies to be highly compound dependent, which could render a technology unusable in discovery where dozens of molecules may require controlled-release delivery.

Modified-release drug delivery technologies have seen limited application in the discovery space (Eichenbaum et al. 2006), with the possible exception of their use to identify if a given compound has suitable properties for controlled release as a development candidate (Thombre 2005). Modified-release technology has primarily been applied for post-Phase 2 clinical candidates or for lifecycle management applications (Eichenbaum et al. 2006). While lengthy late-stage preclinical animal studies may be conducted with compounds formulated in food to enable sustained exposure (Strickley 2013), this may not be an option early in discovery programs when active compound is limited. In the early discovery setting, teams have historically relied on subcutaneously implanted mini-pumps to achieve sustained delivery of compounds (as is described below in this chapter). These pumps can provide a constant plasma concentration for in vivo studies, but their utility is subject to solubility limitations and the pumps require surgical implantation, which may interfere with pharmacology results. Discovery formulators are beginning to pursue alternative technologies that modulate exposure in preclinical settings. This topic has been increasing in popularity in the last few years to the point where exposure modulation in discovery was the subject of a recent mini-symposium at the 2013 AAPS National Meeting (Fraunhofer et al. 2013).

On a numbers basis, most preclinical studies are conducted in rodents because they are the primary species used in the pharmacological models, and because their use entails much less resource (e.g., husbandry, compound) than larger species do. Since many formulation technologies are designed for use in humans, their use in rodents represents a significant challenge because of the differences in physiology (Kararli 1995). Many of the formulations used for modulation of exposure are solid oral dosage forms, which may not be easily administered to rodents. There have been several publications in which modified-release dosage forms have been successfully dosed in rats. However, these are typically studies conducted with mature/development compounds to evaluate a novel formulation strategy for potential commercial application. This research is thus afforded extensive background, formulation development, and analytical characterization resources not available in the discovery environment. For example, alginate microspheres of diclofenac sodium have been formulated to protect the stomach from irritation and tested in rats (Fenyvesi et al. 2009). In this study, 19 different formulations were prepared and optimized by modifying pH, polymer ratio within the microspheres, gel-hardening time, and concentration of calcium chelator. Other formulation development studies have been conducted in rats in which various microspheres were prepared and coated with enteric polymers to enable site-specific delivery to the GI tract (Gan et al. 2013; Lo et al. 2013; Ferrari et al. 2013; Ahmadi et al. 2011). In addition, microspheres designed to be retained in the GI tract by giving off CO_2 in vivo have been tested in rats to enable sustained delivery of ginger extract to the stomach in an ulcer prevention model (Singh and Kaur 2011; Choi et al. 2002). While these formulation strategies may not be directly applicable to a discovery program where 20 mg of material is available, the concepts behind these technologies can be scaled back to allow fit-for-purpose application in the discovery environment. These strategies are highlighted in the following section.

3.4.1 Enteric Coated Capsules

Enteric coating polymers can be used to protect a molecule that is labile in the acidic media of the stomach and can protect the stomach from materials that cause irritation. From a delivery perspective, compounds have been formulated with enteric polymers to enable delivery to the small/large intestine. Several researchers have published the results of studies on administering enteric coated formulations in rats using size 9 capsules (Reix et al. 2012; Saphier et al. 2010; Sonaje et al. 2010; Kremser et al. 2008). In one study, capsules were coated with Eudragit L 100-55, a polymer that was designed to achieve enteric properties as it dissolves at pH 5.5 in the duodenum. Imaging studies were conducted in rats using barium chloride to show that the enteric coated capsules remained intact in the stomach for 5 h, whereas uncoated capsules disintegrated in the stomach in less than 1 h (Reix et al. 2012). In addition, a pharmacokinetic study was conducted using acetaminophen (5 mg active plus 11 mg of lactose per capsule) dosed in either enteric coated or uncoated capsules (Reix et al. 2012). The coated capsules increased the time of maximum drug concentration (T_{max}) approximately 6-fold, C_{max} approximately 2.5-fold, and AUC approximately 2-fold compared with values for the uncoated capsules. In that study, a fluid-bed apparatus was used to administer a uniform coating on the capsule. This equipment is not widely available in the discovery realm, but other researchers have manually coated capsules to employ enteric coating technology on a small scale (Saphier et al. 2010; Sonaje et al. 2010; Kremser et al. 2008). While this approach may introduce some variability in polymer coating thickness, it may be acceptable in the discovery environment. One potential disadvantage of this technology is that the amount of material that can be dosed to rodents using capsules is limited. In addition, capsules may not be able to accommodate the wide range of doses frequently encountered in discovery. Thus, it would be advantageous to use a formulation technology that provided more flexibility with respect to dose range and could easily be administered via gavage tube.

3.4.2 In Situ Alginate Gels for Modified Release of Drug in Rodent Models

Polymer systems, such as alginate gels, which increase in size after swelling and retard passage through the pylorus, have been used to enable gastric retention (Hoffman et al. 2004; Curatolo and Lo 1995). Alginate is a linear, biodegradable copolymer of β-D-mannuronic acid and α-L-guluronic acid (George and Abraham 2006; Khotimchenko et al. 2001). It is soluble above pH 2.5 and in the lower pH range converts to insoluble alginic acid (Moustafine et al. 2005). Cross-linked gels result from interaction of alginate with metal ions and divalent cations such as calcium (Fenyvesi et al. 2009). These gels are physically stable in the acidic pH environment of the stomach and erode in the higher pH environment of the intestines (Ain et al. 2003). Over time, mechanical degradation of the gel occurs by the normal actions of the stomach, and fragments of the gel pass into the small intestine where erosion processes can come into play and promote drug release.

Alginate systems have previously been employed to test the concepts of sustained-release drug delivery in rodents (Satapathy et al. 2010; Rastogi et al. 2007; Kubo et al. 2003; Miyazaki et al. 2000; Katayama et al. 1999). Some authors have reported using the gels in preclinical species to compare release profiles with commercial sustained-release formulations (Kubo et al. 2003; Miyazaki et al. 2000). In studies evaluating the sustained release of paracetamol, alginate was formulated with calcium ions with the calcium chelator sodium citrate being added to prevent gel formation *ex vivo* (Kubo et al. 2003). Gel formation was accomplished in situ by the release of the complexed calcium ions in the low-pH environment of the stomach, resulting in cross-linking of the alginate matrix. Another published dosing strategy involved the separate oral administration of a solution of a sodium alginate immediately followed by a calcium salt in rats to evaluate sustained release of ampicillin in the treatment for *Helicobacter pylori* (Katayama et al. 1999).

Studies using the in situ gelling technique with the radio-opaque tracer barium sulfate have shown that alginate gels form in rats (Fig. 3.4) and are retained in the stomach for up to 8 h (Foster et al. 2012). Compounds can be formulated in the alginate matrix and release over time in vivo through a diffusion and erosion process that enables modulation of the PK curves versus a standard formulation. Modulation of exposure using in situ alginate gels compared with a standard formulation was recently shown for celecoxib, a poorly soluble compound, as well as for ibuprofen, a moderately soluble compound, when dosed in rats (Foster et al. 2013). In those studies, the final concentration (C_{last}) of both compounds was more than threefold higher than the concentrations obtained from suspension dosing, indicating that in situ gels may be an option for discovery teams that hope to stay above a desired concentration for a given duration. In contrast, no modulation in exposure was observed for metoprolol, a highly soluble compound, when dosed in the alginate formulation versus a control solution (Foster et al. 2013). Advantages of the in situ gel system include that it minimizes the resources needed for formulation development and accommodates a wide range of drug loadings. Both of these advantages make this system amenable to the discovery environment. Limitations of in situ gel systems include variability in the duration of gastric retention from animal to animal. Relying on in situ gel formation may not provide robust gels—that is, they may have insufficient cross-linking, compared with gel-based microspheres prepared *ex vivo*. In addition, the simple approach described here is not applicable to compounds that are highly soluble in the low pH environment of the stomach. Without additional improvements to the technology, such as coating of drug containing particles that are encapsulated in the gel, application of this approach can be limited to lower solubility compounds. This can be a constraint in a discovery program where multiple chemistry options are still being pursued.

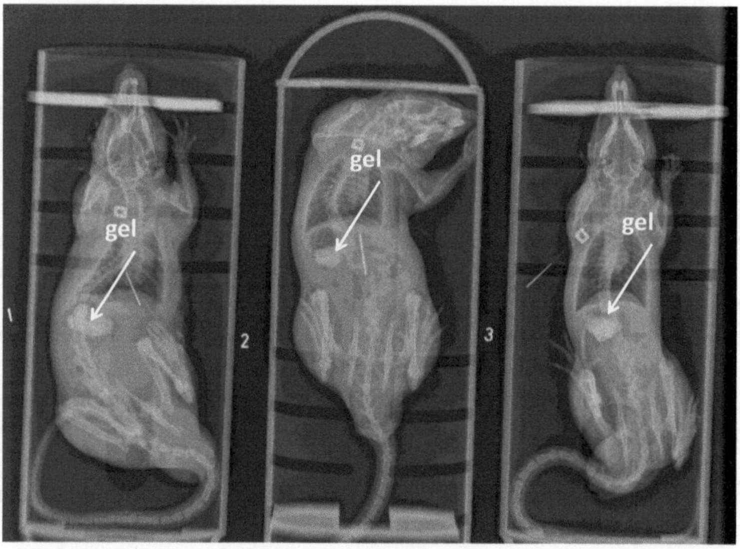

Fig. 3.4 Digital radiograph showing gastric retained alginate gel containing barium sulfate in rats (Foster et al. 2012)

3.4.3 Bioadhesive Liquid Crystals for Modified Release of Drug in Rodent Models

Mucoadhesive formulations are designed to adhere to the mucin at the absorption site in the GI tract and facilitate flux across the intestinal membrane. The concept behind this strategy is that the increased contact time of a compound in the intestine may facilitate increased absorption, particularly for poorly permeable compounds. This approach has been illustrated with pravastatin sodium, a highly soluble/poorly permeable compound, which was formulated as microparticles with the bioadhesive polymers ethylcellulose or Eudragit S100 (Garg and Pathak 2011). The flux of pravastatin with the optimized Eudragit S100 bioadhesive formulation was 1.6-fold higher than that of drug alone across goat ileum tissue. These results demonstrate that mucoadhesive formulations may have the potential to overcome poor permeability of tool compounds and thus improve bioavailability.

Liquid crystals are one formulation approach that has recently been explored for bioadhesive utility in rodents (Nguyen et al. 2010, 2011; Lee et al. 2009). Surfactant and polymer systems form supra-assemblies that can be used as drug delivery vehicles (Garg et al. 2007). These lyotropic systems include liquid crystalline aggregates, such as liposomes or cubosomes, which load, stabilize, and deliver active ingredients (Garg et al. 2007). These self-assembled structures can be formed by some amphiphilic lipids in excess water and have been identified as potential sustained-release drug delivery vehicles (Nguyen et al. 2011; Garg et al. 2007). In this strategy, the surfactant assembles into bilayers that are twisted into a three-dimensional honeycombed structure with bicontinuous domains of water and lipid (Garg et al. 2007). Drugs can be incorporated into the internal domains of these structures to enable diffusion-controlled release into the surrounding aqueous environment (Nguyen et al. 2011; Shah et al. 2001; Drummond and Fong 2000; Burrows et al. 1994). One advantage cubosomes have over liposomes is that they can accommodate water-soluble, lipid-soluble, and amphiphilic molecules (Garg et al. 2007). Phytantriol and glycerol monooleate/dioleate are commonly used for this type of delivery system. Phytantriol offers the advantage of lacking an ester bond, which improves stability in the gastric environment and resistance to digestion (Nguyen et al. 2010).

The use of cubosomes to achieve a sustained-release profile was described in a study of the poorly soluble compound cinnarizine (Nguyen et al. 2010). In this study, rats were dosed orally (7 mg/kg cinnarizine in 300 mg of lipid vehicle) with phytantriol or glyceryl monooleate (GMO) cubosomes and the results were compared with those for a standard aqueous suspension. The bioavailability for the phytantriol formulation was 41 % versus 19 % for the GMO formulation and 6 % for the standard aqueous suspension (Nguyen et al. 2010). The phytantriol formulation had a T_{max} of 33 h versus 5 h for the GMO formulation and 1 h for the suspension. The increased drug release of the GMO and phytantriol formulations are presumed to be due to the bioadhesion of the cubosomes versus the suspension. The greater than sixfold increase in release time of the phytantriol relative to GMO cubosome formulation was due to the previously discussed stability difference of GMO versus phytantriol in the gut (Nguyen et al. 2010). In fact, 56 % of the phytantriol formulation was retained in the stomach at 24 h. In contrast, only 66 % of the GMO formulation remained at 1 h, and none remained at 24 h (Nguyen et al. 2010). In a separate study comparing phytantriol cubosomes to an emulsion of cinnarizine, 7 mg/kg cinnarizine was prepared in 3 g vehicle (10 % w/v lipid, 1 % w/v Pluronic F127 in water for the cubosome formulation versus 40 mg of oleic acid in 0.2 % Tween 80 for the emulsion) and dosed in rats. The T_{max} of the cinnarizine increased from 1 h with the emulsion to 5 h with the phytantriol cubosome formulation (Nguyen et al. 2011). In that study, the gastric retention of phytantriol cubosomes provided sustained absorption of cinnarizine from 12 to 48 h (Nguyen et al. 2011).

One advantage of the cubosome approach is that formulations can be prepared with basic laboratory equipment, such as a probe sonicator. However, definitive characterization of the liquid crystalline phase requires small-angle X-ray scattering (SAXS) and cryo-transmission electron microscopy (TEM) capabilities (Garg et al. 2007), which are not typically available in the discovery environment. Therefore, formulation characterization of these systems may be limited.

A potential concern with the use of phytantriol is that published data illustrate that phytantriol can be absorbed from the intestine (Nguyen et al. 2010) even though the material historically has been used as a topical agent (McLain 2007). While an understanding of the pharmacokinetics and toxicology of oral ingestion of phytantriol needs to be further studied if it is to be used clinically, the lack of such information does not limit its use in discovery. Based on literature reports, the oral median lethal dose (LD_{50}) of phytantriol is more than 5,000 mg/kg for mice and rats (DSM 2011; McLain 2007; Roche 1999). In addition, reports showed that phytantriol is not cytotoxic in cultured human lymphocytes, not phototoxic, and not mutagenic in Ames tests, with or without metabolic activation (Nguyen et al. 2010; McLain 2007). Another potential concern for this formulation strategy is that gastric retention of the formulation may impact food intake of the study animals; however, normal turnover of the enterocytes may help alleviate these concerns. In the cinnarizine studies described above, the gastric retention of the cubosome formulations appeared to have no impact on the food intake of the rats (Nguyen et al. 2011).

Thus far, the approaches discussed in this section have been demonstrated in an acute setting. While a given therapeutic area may have acute pharmacology/toxicology models at their disposal, multi-day studies will ultimately be needed to demonstrate the utility of these formulation strategies in a chronic setting. Thus, the formulation strategy must allow repeated dosing using a regimen that ensures reliable, predictable exposure. This means that formulation strategy must not be adversely affected by the presence of food.

While oral delivery is typically the preferred dosing route, the application of in situ gels and liquid crystal technology is not limited to oral delivery (Guo et al. 2010). Indeed, subcutaneous dosing may be required to enable desired exposure or test compounds by the intended route of administration. The next section focuses on modulating exposure via parenteral administration.

3.5 Discovery Formulation Approaches for Modulating Exposure via Parenteral Routes

3.5.1 Parenteral Routes for Discovery

Drugs administered via any route other than oral or rectal routes are considered to be parenteral. Parenteral drug delivery can be via injection (intravenous, intramuscular, subcutaneous, or intraperitoneal), transdermal delivery, or inhalation. Injections, especially via the intravenous and intraperitoneal routes, are the most commonly used in discovery research for evaluating PK because they bypass the GI tract and directly access systemic circulation. For compounds with inadequate oral exposure or compounds with high variability, parenteral routes offer the best alternative for achieving the required systemic exposure. Also, while aseptic practice and formulation sterility are major burdens in the clinical setting, the use of injectables in the discovery setting, especially in rodents, has minimal aseptic requirements. This makes parenteral route practical and extremely important in drug discovery research.

3.5.2 Parenteral Modified-Release Drug Delivery Technologies

As described earlier, most of the drug delivery technologies employed in the development setting are not applicable in the discovery environment due to the amount and type of resources required. Today, limited off-the-shelf modified-release technologies are available to discovery formulators for parenteral delivery. The most common technology is the implantable Alzet® mini osmotic infusion pump, which is manufactured by Durect Corporation (previously Alza Corporation). There are also Infu-Disk™ infusion pumps, based on electrochemical Prime Mover™ technology, made by Med-e-Cell. Recently, a more advanced type of peristaltic infusion pump, the iPrecio® programmable pump developed by PrimeTech Corporation, was introduced. Other technologies have been used, but those are mainly employed by individual groups with specific knowledge for a particular technology, and are not widely available.

As drug hunting evolves, the ability to modulate a compound's PK early in the discovery process has become increasingly important, which means that modified-release drug delivery is no longer just a product lifecycle management tool. The ability to modify compound release by switching the drug delivery device on and off or purposefully changing the intensity or duration of exposure will enable PK/pharmacodynamic (PD) studies. Furthermore, since PD studies are typically longer in duration and require chronic administration, modified-release drug delivery platform technologies will enable them to be conducted more effectively and productively with nonoptimized compounds. Preclinical formulations are important in supporting these early in vivo studies and are the critical link between key pharmacology, PK, and toxicology studies (Kwong et al. 2011). The ability to modulate the compound PK allows the discovery team to study targets more effectively, obtain meaningful data earlier, and better understand toxicological responses, helping to bring the best candidate forward.

Therefore, discovery formulation scientists are increasingly interested in modified-release delivery technologies that can be applied in the discovery stage. The approach has included adapting technologies from development, miniaturizing equipment, and leveraging advances in electronics that enable "smart" drug delivery devices. Some of these technologies are described below.

3.5.3 Mini Osmotic Infusion Pumps

Osmotic pumps were conceived in the 1970s (Theeuwes and Yum 1976) and the Alzet® mini osmotic infusion pumps have been the most popular, dependable, and, until recently, the only off-the-shelf controlled-release drug delivery tool available for discovery scientists for the last three decades. The mini osmotic infusion pumps are single-use, miniature, implantable devices that release drug in a controlled, predictable manner. Cited in 13,500 published articles, they have been used in mice, rats, and a variety of laboratory animals, to deliver a broad range of agents, including small molecules, antibodies, cytokines, growth factors, hormones, peptides, and proteins. These references have been compiled and categorized by the agent types they have been used to deliver. This reference list, which is searchable and updated periodically, can be found on the manufacturer's website (www.alzet.com).

The pumps are powered by the osmotic pressure difference between a compartment within the pump, called the salt sleeve, and the body fluid of the animal in to which the pump is implanted. The high osmolality of the salt sleeve causes water be drawn into the pump through a semipermeable membrane that forms the outer surface of the pump. As the water enters the salt sleeve, it compresses the flexible reservoir, displacing the test solution from the pump at a controlled, predetermined rate in a zero-order fashion. Since the delivery rate is controlled by osmotic pressure and the water permeability of the pump's outer membrane, the physical and chemical properties of the drug molecule have no bearing on its release rate. The fact that the device is insensitive to physicochemical properties of the active compound makes this technology an extremely useful device in the discovery environment, where a large amount and a broad range of investigational molecules are studied quickly. A detailed technical explanation of the operation of osmotic pumps has been previously published (Theeuwes and Yum 1976).

Alzet® mini osmotic pumps consist of 0.5- to 6.5-mL capsules that are surgically implanted. They can deliver compounds at continuous, controlled rates from 0.1 to 10 μL/h for 1 day to 6 weeks without external connections. Most commonly, they are implanted subcutaneously or in the peritoneal cavity. They also can easily be attached to a catheter and used to infuse a blood vessel, a target organ, or a target tissue, such as the brain.

For targeted delivery to the brain (which is needed because many compounds do not cross the blood-brain barrier adequately), the pumps are used with special commercial kits that allow infusion of compounds directly into the brain. This setup has been used in many neuroscience studies to infuse active compounds locally to the central nervous system. These compounds have included growth factors (Paliouras et al. 2012), small interfering ribonucleic acid (siRNA) (Wang et al. 2011; Kim and Kang 2011), and psychoactive drugs (Abrams et al. 2008).

The mini osmotic pumps have been used with suitable catheters to achieve targeted delivery to many other organs. These include direct infusion into the kidneys of mice (Villar et al. 2013), the uterus of a sheep (Dorniak et al. 2012), the inner ears of guinea pigs (Toyota et al. 2012), and the tumors inside the pancreas of mice (Casazza et al. 2012). The use of these pumps in toxicology and in preclinical experiments is discussed extensively elsewhere (Perkins et al. 2000).

The mini osmotic pumps are compatible with organic and inorganic solvents commonly used in preclinical formulations, such as acids, bases, cyclodextrins, dimethyl sulfoxide (DMSO) concentrations up to 50 %, polyethylene glycol, and others. The minipumps can also be used with solvents that are incompatible with the pump's reservoir material by attaching an external reservoir to the pump, the contents of which are displaced by the action of the pump. The external reservoir can be made by coiling polyethylene tubing, a technique known as Lynch coil (Lynch et al. 1980). Detailed information is available on the manufacturer's website (http://www.alzet.com/products/guide_to_use/catheter_use.html).

Although the osmotic minipumps have numerous attributes that made them the gold standard for preclinical parenteral controlled-release delivery for many years, they have some drawbacks that hinder their broader application. Firstly, these include limited, non-refillable reservoir volumes (100–200 μL for small pumps and 2 mL for larger pumps) that make administration of high doses to small rodents challenging. Replacement pumps can be implanted, but doing so requires a second surgical procedure. Secondly, the potential exists for precipitation or plugging of the minipump at release site. Typically, drug is formulated in organic solvents as a concentrate in the reservoir. When the formulation is released from the pump, it is diluted with the surrounding body fluid and precipitation sometimes occurs within the cannula, which results in the blockage of the pump. Thirdly, implantation of the pump requires specialized surgical procedure, which may induce inflammation and thus could interfere with the animal model. Because of these drawbacks, discovery formulation scientists are continually seeking new PK-modulating tools that can offer higher precision, broader flexibility, and easier execution.

3.5.4 Mini Peristaltic Infusion Pumps

A new type of mini infusion pump based on peristaltic action, the iPRECIO® infusion pump, was invented by the PrimeTech Corporation. Available since 2007, these pumps offer an alternative to the mini osmotic pumps for use in small laboratory animals. These electromechanical pumps are implanted subcutaneously, and allow the drug dosage to be programmed to infuse at specified time points and durations. They are self-contained intelligent devices consisting of a microcontroller with associated memory, embedded software, inputs/outputs, a micropulse stepping motor, crystal oscillators, electronic circuitry, and a battery power source. Once programmed, implanted, and activated, the infusion proceeds unattended following the user-specified protocol (Tan et al. 2011; Abe et al. 2009; Tan and Tsuru 2009). The animals are unrestrained and can move freely in their cages.

In comparison to an osmotic version, this new pump has three attractive features: it is (1) more precise, (2) refillable, and (3) programmable for complex release patterns. However, it has a significant higher up-front cost. In 2014, each disposable pump costs about $250 USD and required a software investment of $1,600 USD. The first commercial version of the peristaltic pump weighs about 9 g and is too big to be used with mice, but is suitable for use with rats and animals that are larger than 230 g. The pump is programmable and can delay the start of infusion to allow a surgical recovery period. During this time, only saline is infused to keep the catheter open. The pump can be programmed for up to 10 variable flow rates of 1–30 μL/h with ±5 % accuracy. The reservoir of the pump can be refilled percutaneously via a refill port with a resealable septum. The duration of total infusion is limited by the battery, which can be used to deliver a total infusion volume of 4.5–6.5 mL (Tan and Tsuru 2009). The pump can also be attached to a catheter and thus used to deliver formulation to a specific part of the body.

While the technology is still relatively new, reports of successful application of the pumps have been published. In one study, the pumps were used to infuse serotonin, a compound known to cause a decrease in blood pressure, into Sprague-Dawley rats. The flow rate started at 5 μg/kg/min for 48 h and was increased stepwise by 10 μg/kg/min to achieve a final rate of 45 μg/kg/min (Davis et al. 2010). Delivery intervals were followed by an interval in which the pumps were stopped for 48 h. The ability to program the device to start, stop, and deliver different doses at different time points allowed the researchers to clearly see the effect of different doses of serotonin on blood pressure as the doses were gradually increased. This study demonstrated that a smaller rate of serotonin is required to elicit a hypotensive effect in rats compared with previous data established using osmotic pumps. In a different study, the pumps were shown to reproducibly inject the minimum programmable volume of 100 nL of bicuculline methiodide over 6 min into the hypothalamus in the brain of conscious rats to produce cardiovascular and locomotor responses (Zaretsky et al. 2012).

Recently a smaller pump, which is suitable for subcutaneous implantation in mice, has become available. This device weighs 3.3 g and has a built-in elastic reservoir. In comparison to a larger system, it allows wireless reprogramming after implantation, which enables researchers to change the flow rate and duration of dose administration during the experiment. This degree of flexibility is highly desired in discovery, where the doses required to elicit responses from pharmacological models are frequently poorly defined. According to the manufacturer, this new pump can be used in a variety of research areas, including cardiovascular, neuroscience, oncology, and pharmacology, although reports of its use in these areas have not been published.

3.5.5 Multilayered Modified-Release Microspheres

Multilayered modified-release microspheres are a mature technology in the pharmaceutical industry. However, even the smallest commercially available bead coater is too big for practical use in discovery. For example, the Micro Flo-coater® by Freund-Vector or the MP-MICRO™ fluid-bed coater by GEA Niro, can typically be operated with particles no smaller than 300–500 μm in diameter, a spray bed size of minimally 20 g, and at least a few grams of active compound. These minimum requirements, especially the gram scale of active compound, make the fluid-bed coating technology unsuitable for drug discovery setting. The minimum particle size, approximately 300–500 μm, is also too large for injectable application.

To extend the application of multilayered microspheres into discovery research, a miniature fluid-bed coater was built that can operate at 1-g spray-bed size and requires only about 100 mg of active compound (Morgen et al. 2013; Tye et al. 2013). Multilayered microspheres made with this so-called micro coater were successfully used to demonstrate modified release of active compounds in rats (Tye et al. 2014).

The micro coater developed was based on the Wurster coating process. The engineering challenges of equipment miniaturization, including overcoming the electrostatic charges experienced by the microspheres, were addressed by unique coater design, careful equipment-surface conditioning, and appropriate temperature and humidity control. The modified-release microspheres were monodisperse multilayer particles approximately 100 μm in diameter, consisting of a glass bead core, a drug layer, and a second modified-release polymer layer. Glass beads approximately 60 μm diameter were chosen as the core, because they resulted in coated beads of an acceptable size for parenteral injections for rodent species, and they offer the right combination of particle size and density to allow fluidization in the miniature coater.

The drug layer was applied by spraying a solution mixture of a drug and a binding agent (HPC or HPMC) in an organic solvent. The modified-release polymer layer was applied by spraying a solution mixture of water-insoluble polymer (ethylcellulose) and water-soluble polymer (HPMC). The water-insoluble polymer acts as the barrier for dissolution, whereas the water-soluble polymer dissolves to leave tortuous water passages that modulate drug diffusion.

Multilayered microspheres were successfully manufactured using four model compounds—phenytoin, rapamycin, metoprolol, and atazanavir—in some cases using less than 200 mg of active compound per batch. In in vitro dissolution tests using sink conditions, the drug release rates from the microspheres were related to the solubility of the drug. Metoprolol, with a high aqueous solubility of 169 mg/mL, was found to be completely released from the microspheres within 2 h; phenytoin, rapamycin, and atazanavir, which have low aqueous solubility (<40 μg/mL), showed close to zero-order release over several days.

The release profiles of these multilayer microspheres do not show the undesired high initial "burst" commonly seen in monolithic systems, because the release rate is controlled by the diffusion of compounds across the outer modified-release polymer layer, which maintains a barrier of constant dimension. To modulate drug release rates, the thickness of the modified-release polymer layer of the microspheres was varied. Microspheres were manufactured with various coating thicknesses. Thicker coatings were found to be correlated with slower drug releases, as expected from a diffusion-controlled modified-release system.

In an in vivo study, multilayered microspheres containing atazanavir were dosed in rats and successful modification of drug PK was observed (Tye et al. 2014). In a 24-h study, one dose of the atazanavir modified-release microspheres was injected intraperitoneally and it sustained plasma drug concentration at a close to constant level for 24 h. These results were in contrast to intraperitoneal injections of atazanavir solution, which resulted in a PK profile with only 1 hr half-life in rats.

In a follow-up 5-day study, the single dose of atazanavir microspheres provided plasma concentration that gradually decreased throughout the duration of the study. While constant plasma concentration was not obtained, the pharmacokinetic profile was clearly extended relative to the solution control. The atazanavir microspheres were also given to rats via subcutaneous injections, and were found to continuously release drug throughout the whole period in an 8-day study (Tye et al. 2014). The plasma level peaked at 7 h and then gradually decreased. Even though zero-order release was not achieved, the multi-day sustained-release profile achieved was significant, as the plasma concentration from the control solution dropped to undetectable levels by 24 h. The subcutaneous route appeared to provide slower and prolonged release compared with the intraperitoneal route, probably due to the physiological differences between the tissues that affect drug dissolution and diffusion.

Another potential use of the microspheres is via the oral route. The microspheres could be coated with an enteric polymer using a similar spray-coating process, to produce beads that protect drug molecules from the acidic conditions of the stomach. Furthermore, since the microspheres are not exposed to high temperatures in the spray-coating process due to evaporative cooling, use of this technology Could be expanded beyond small molecules to peptides and proteins.

3.5.6 Microneedles

The mechanical use of micron-scale needles (microneedles) provides a minimally invasive drug delivery route in which a microneedle array is placed on the skin to enhance skin permeability. Microneedles were introduced about two decades ago, when they were prepared as a mechanical tool to pierce the stratum corneum to create drug delivery channels without stimulating underlying pain nerves (Henry et al. 1998). Painless application was the most attractive feature of this approach. Since then, the field of microneedles has expanded quickly, and they have been investigated for a wide range of transdermal applications, including those in the pharmaceutical and cosmetic industries. For pharmaceuticals, microneedles have been used to deliver drugs into the intradermal space. They have also been investigated as a means of delivering peptides, proteins, oligonucleotides, and deoxyribonucleic acid (DNA) vaccines; reducing blood glucose levels by insulin; and inducing immune responses from foreign proteins (Hussein et al. 2013). The potential advantages of microneedles for drug and vaccine delivery, as well as the challenges associated with commercialization of such technology, are reviewed elsewhere (Mikszta et al. 2006).

Microneedles have been made of different types of materials, including silicon, metals, and biodegradable polymers. The various methods of microneedle fabrication and drug incorporation into the microneedles for different types of drug delivery have been previously summarized and critically reviewed (Donnelly et al. 2010). More recently, there is a comprehensive review about microneedle design strategies (e.g., solids, hollow center, coatings, water soluble, and hydrogels) and their related limitations (Tuan-Mahmood et al. 2013). A review for the construction and application of microneedles for peptides and proteins was recently published (Chandrasekhar et al. 2013), and there is also a review that describes newer polymeric microdevices, including microneedle arrays with micropumps, that are in early stages of research and development (Ochoa et al. 2012). Hollow microneedles with a fine bore can be integrated with a drug reservoir for continuous drug delivery. With the use of a single hollow glass microneedle inserted into human cadaver skin, Martanto et al. (2006) studied the relationship between infusion rate and other parameters, such as insertion depth, retraction distance, infusion pressure, needle-tip opening size and geometry, and the presence of hyaluronidase. Strategies were demonstrated to increase the flow rate using well-designed infusion protocols and needle geometries.

While most microneedle research is focused on application for use in humans, this technology has the right attributes to be converted for use in drug discovery for preclinical animals. The micron size of the needles is directly applicable to small animals and the different lengths of microneedles allow selection of the appropriate piercing depth for the target of interest. Microneedle is a delivery technology that can potentially be used from discovery, on animals, to development, on human. Furthermore, since discovery technologies do not require regulatory data packages and long-term stability data, the application of microneedles in discovery will be accompanied by significantly less regulatory burden than in development and can progress much faster. Considering the significant advancement of nanotechnology in recent years, and the fact that microneedles have been successful in delivering macromolecules, there is a lot of interest in the microneedle delivery technology for application in drug discovery.

3.5.7 Utility of Parenteral Controlled Release

As drug-discovery processes evolve, new targets come onboard, and chemistry space of interest expands to include oligonucleotides, peptides, proteins, stem cells, and more. In addition, the pharmaceutical industry today is facing huge macro environmental pressure to significantly reduce their total time and cost for bringing new drugs to the market. Therefore, currently there is a compelling needs and interest exist to develop better PK-modulating drug delivery tools for drug discovery to effectively facilitate target validation research and mechanistic understanding of new targets to speed up drug discovery process. In the area of parenteral delivery, it is foreseen that these new tools will be created by adapting or miniaturizing existing tools from development, and by taking advantage of the recent advancement in new micro- and nanotechnologies in combination with electronic intelligence to create specialized micro-intelligent devices for preclinical animals.

3.6 Summary

There are many drug delivery challenges encountered when enabling the discovery of new compounds. Well-developed techniques used in development such as spray-dried dispersions, nanosuspensions, and sustained-release formulations can be used to address these needs if they are sufficiently scaled down and appropriately modified. In many cases, however, their application to discovery is facilitated by the reduced requirements for stability, robustness, and scale-up. As demonstrated in this chapter, the key to success in this area is to identify the core characteristics of a technology and then creatively adapt them to the tasks at hand.

References

Abe C, Tashiro T, Tanaka K, Ogihara R, Morita H (2009) A novel type of implantable and programmable infusion pump for small laboratory animals. J Pharmacol Toxicol 59(1):7–12

Abrams DJ, Zheng L, Choo KS, Yang JJ, Wei W, Anchordoquy TJ, Zawia NH, Stevens KE (2008) An initial animal proof-of-concept study for central administration of clozapine to schizophrenia patients. Schizophr Res 100(1–3):86–96

Ahmadi F, Varshosaz J et al (2011) Preparation and in vitro/ in vivo evaluation of dextran matrix tablets of bedesonide in experimental ulcerative colitis in rats. Drug Deliv 18:122–130. doi:10.3109/10717544.2010.520352

Ain Q, Sharma S et al (2003) Alginate-based oral drug delivery system for tuberculosis: pharmacokinetics and therapy effects. J Antimicrob Chemother 51:931–938

Babcock C, Friesen T et al (2009) Method and device for evaluation of pharmaceutical compositions. US Patent 7611630 B2, 3 Nov 2009

Bertrand N, Leroux J (2012) The journey of a drug-carrier in the body: an anatomo-physiological perspective. J Control Release 161:152–163

Burrows R, Collett J et al (1994) The release of drugs from monoglyceride-water liquid crystalline phases. Int J Pharm 111:283–293

Caron W, Song G, Kumar P et al (2012) Interpatient pharmacokinetic and pharmacodynamic variability of carrier-mediated anticancer agents. Clin Pharmacol Ther 91:802–812

Casazza A, Kigel B, Maione F, Capparuccia L, Kessler O, Giraudo E, Mazzone M, Neufeld G, Tamagnone L (2012) Tumour growth inhibition and anti-metastatic activity of a mutated furin-resistant Semaphorin 3E isoform. EMBO Mol Med 4(3):234–250

Chan H, Kwok P (2011) Production methods for nanodrug particles using the bottom-up approach. Adv Drug Deliv Rev 63:406–416

Chandrasekhar S, Iyer LK, Panchal JP, Topp EM, Cannon JB, Ranade VV (2013) Microarrays and microneedle arrays for delivery of peptides, proteins, vaccines and other applications. Expert Opin Drug Deliv 10(8):1155–1170

Chen P, Du H (2007) Target validation: a door to drug discovery. Drug Discov Ther 1(1):23–29

Chiang P, South S et al (2010) Utilizing a novel tandem oral dosing strategy to enhance exposure of low-solubility drug candidates in a preclinical setting. J Pharm Sci 99(7):3132–3140

Chiang P, South S et al (2011) The impact of dosing interval in a novel tandem oral dosing strategy: Enhancing the exposure of low solubility drug candidates in a preclinical setting. J Drug Deliv 2011:9 pp. doi:10.1155/2011/528284

Chiou C, Chan H, Prud'homme R et al (2008) Evaluation on the use of confined liquid impinging jets for the synthesis of nanodrug particles. Drug Dev Ind Pharm 34:59–64

Choi B, Park H et al (2002) Preparation of alginate beads for floating drug delivery systems: effects of CO_2 gas-forming agents. Int J Pharm 239:81–91

Choi J, Park C, Lee J (2008) Effect of polymer molecular weight on nanocomminution of poorly soluble drug. Drug Deliv 15:347–353

Crowley M, Zhang F et al (2007) Pharmaceutical applications of hot-melt extrusion: Part 1. Drug Dev Ind Pharm 33:909–9267

Curatolo W, Lo J (1995) Gastric retention systems for controlled drug release. Canadian Patent 1,335,351C, 25 Apr 1995

Curatolo W, Nightingale J et al (2009) Utility of hydroxypropylmethylcellulose acetate succinate (HPMCAS) for initiation and maintenance of drug supersaturation in the GI milieu. Pharm Res 26:1419–1431

Dai W (2010) In vitro methods to assess drug precipitation. Int J Pharm 393:1–16

Dai W, Xu S (2012) In vitro high-throughput drug precipitation methods for oral formulations. Am Pharm Rev posted online June 25 2012

Davis P, Fink GD, Seitz B, Watts SW (2010) Serotonin infusion via the iPrecio® micro infusion pump results in repeated reductions in blood pressure in the normotensive Sprague Dawley rat. FASEB J (Meeting Abstract Supplement) lb551. http://www.fasebj.org/cgi/content/meeting_abstract/24/1_MeetingAbstracts/lb551

Denkov N, Marinova K, Christova C et al (2000) Mechanisms of action of mixed solid-liquid antifoams: 3. Exhaustion and reactivation. Langmuir 16:2515–2528

Deschamps B, Musaji N, Gillespie J (2009) Food effect on the bioavailability of two distinct formulations of megestrol acetate oral suspension. Int J Nanomedicine 4:185–192

Doane T, Chuang C, Hill R et al (2012) Nanoparticle zeta potentials. Acc Chem Res 45:317–326

Dobry D, Settell D et al (2009) A model-based methodology for spray-drying process development. J Pharm Innov 4:133–142

Donnelly RF, Raj Singh TR, Woolfson AD (2010) Microneedle-based drug delivery systems: microfabrication, drug delivery, and safety. Drug Deliv 17(4):187–207

Donoso M, Haskell R, Schartman R (2012) Surfactant choice and the physical stability of nanosuspensions as a function of pH. Int J Pharm 439:1–7

Dorniak P, Bazer FW, Wu G, Spencer TE (2012) Conceptus-derived prostaglandins regulate endometrial function in sheep. Biol Reprod 87(1):9

Douglas F, Fafilmout K, Patt N (1971) Foreign particle embolism in drug addicts: respiratory pathophysiology. Ann Intern Med 75:865–872

Driscoll D, Etzler F, Barber T et al (2001) Physicochemical assessments of parenteral lipid emulsions: light obscuration versus laser diffraction. Int J Pharm 219:21–37

Drummond C, Fong C (2000) Surfactant self assembly objects as novel drug delivery vehicles. Curr Opin Colloid Interface Sci 4:449–456

DSM (2011) Phytantriol; MSDS No. 0429082. DSM Nutritional Products, Parsippany, NJ, 22 November 2011

Eichenbaum G, Pollock-Dove C et al (2006) Preclinical assessment of the feasibility of applying controlled release oral drug delivery to a lead series of typical antipsychotics. J Pharm Sci 95:883–895. doi:10.1002/jps.20550

Fenyvesi Z, Auner A et al (2009) Preparation of pH-sensitive beads for NSAD using three-component gel system. J Pharm Sci 98(11):4285–4295

Ferrari PC, Souza F et al (2013) Development and in vitro evaluation of coated pellets containing chitosan to potential colonic drug delivery. Carbohyd Polym 91:244–252

Foster K, Morgen M et al (2012) Utility of in situ sodium alginate/karaya gum gels to facilitate gastric retention in rodents. Int J Pharm 434:406–412. doi:10.1016/j.ijpharm.2012.06.009

Foster K, Sun H et al (2013) Utility of gastric retained alginate gels to modulate pharmacokinetic profiles in rats. J Pharm Sci 102(8):2440–2449. doi:10.1002/jps

Fraunhofer W, Marra M et al (2013) Sustained release applications in preclinical settings. Paper presented at the American Association of Pharmaceutical Scientists Annual Meeting and Exposition, San Antonio, TX, 13 Nov 2013

Friesen D, Shanker R et al (2008) Hydroxypropyl methylcellulose acetate succinate-based spray-dried dispersions: an overview. Mol Pharm 5:1003–1019

Gan L, Gao Y et al (2013) Novel pH sensitive lipid-polymer composite microspheres of 10-hydroxycamptothecin exhibiting colon specific biodistribution and reduced systemic absorption. J Pharm Sci 102:1752–1759. doi:10.1002/jps

Garg Y, Pathak K (2011) Design and in vitro performance evaluation of purified microparticles of pravastatin sodium for intestinal delivery. AAPS PharmSciTech 12:673–682

Garg G, Saraf S et al (2007) Cubosomes: an overview. Biol Pharm Bull 30:350–353

George M, Abraham T (2006) Polyionic hydrocolloids for the intestinal delivery of protein drugs: alginate and chitosan—a review. J Control Release 114:1–14

Ghosh I, Bose S, Vippagunta R et al (2011) Nanosuspension for improving the bioavailability of a poorly soluble drug and screening of stabilizing agents to inhibit crystal growth. Int J Pharm 409:260–268

Gulari E, Bazzi G, Gluari E et al (1987) Latex particle size distributions from multiwavelength turbidity spectra. Part Part Syst Charact 4:96–100

Guo C, Wang J et al (2010) Lyotropic liquid crystal systems in drug delivery. Drug Discov Today 15:1032–1040

Haskell R (2003) High strength magnetic stir bar. Patent Application WO2003049845, 19 Jun 2003

Haskell R (2004) Laboratory scale milling process. US Patent 6,814,319, 9 Nov 2004

Haskell R, Shifflet J, Elzinga P (1998) Particle-sizing technologies for submicron emulsions. In: Benita S (ed) Submicron emulsions in drug targeting and delivery, vol 9, Drug targeting and delivery. Harwood Academic Publishers, Amsterdam, pp 21–98

He W, Lu Y, Qi J et al (2013) Food proteins as novel nanosuspension stabilizers for poorly water-soluble drugs. Int J Pharm 441:269–278

Hennart S, van Hee P, Drouet V et al (2012) Characterization and modeling of a submicron milling process limited by agglomeration phenomena. Chem Eng Sci 71:484–495

Henry S, McAllister DV, Allen MG, Prausnitz MR (1998) Microfabricated microneedles: a novel approach to transdermal drug delivery. J Pharm Sci 87(8):922–925

Hoffman A, Stepensky D et al (2004) Pharmacokinetic and pharmacodynamic aspects of gastroretentive dosage forms. Int J Pharm 277:141–153

Hughes J, Rees S et al (2011) Principles of early drug discovery. Br J Pharmacol 162:1239–1249

Hussein MZ, Meihua JT, Fakurazi S, Ithnin H (2013) The evolutionary development in drug discovery and delivery. J Drug Deliv Sci Technol 23(3):195–205

Juhnke M, Berghausen J, Timpe C (2010) Accelerated formulation development for nanomilled active pharmaceutical ingredients using a screening approach. Chem Eng Technol 33:1412–1418

Kapur P, Healy T, Scales P et al (1996) Role of dispersants in kinetics and energetics of stirred ball mill grinding. Int J Miner Process 47:141–152

Kararli T (1995) Comparison of the gastrointestinal anatomy, physiology, and biochemistry of humans and commonly used laboratory animals. Biopharm Drug Dispos 16:351–380

Katayama H, Nishimura T et al (1999) Sustained release liquid preparation using sodium alginate for eradication of *Helicobacter pyroli*. Biol Pharm Bull 22:55–60

Keck C, Muller R (2008) Size analysis of submicron particles by laser diffractometry—90% of the published measurements are false. Int J Pharm 355:150–163

Kerker M (1969) The scattering of light and other electromagnetic radiation. Academic, New York

Kesisoglou F, Amitava M (2012) Crystalline nanosuspensions as potential toxicology and clinical oral formulations for BCS II/IV compounds. AAPS J 4:677–687

Khotimchenko Y, Koralev V et al (2001) Physical-chemical properties, physiological activity, and usage of alginates, the polysaccharides of brown algae. Russ J Mar Biol 27(S1):533–564

Kim JE, Kang TC (2011) The P2X7 receptor-pannexin-1 complex decreases muscarinic acetylcholine receptor-mediated seizure susceptibility in mice. J Clin Invest 121(5):2037–2047

Knieke C, Romeis S, Peukert W (2011) Influence of process parameters on breakage kinetics and grinding limit at the nanoscale. AIChE J 57:1751–1758

Kremser C, Albrecht K et al (2008) In vivo determination of the time and location of mucoadhesive drug delivery systems disintegration in the GI tract. Magn Reson Imaging 26:638–643

Kubo W, Miyazaki S et al (2003) Oral sustained delivery of paracetamol from *in situ* gelling gellan and sodium alginate formulations. Int J Pharm 258:55–64

Kwong E, Higgins J, Templeton AC (2011) Strategies for bringing drug delivery tools into discovery. Int J Pharm 412(1–2):1–7

Lee K, Nguyen T et al (2009) Nanostructure of liquid crystalline matrix determines in vitro sustained release and in vivo oral absorption kinetics for hydrophilic model drugs. Int J Pharm 365:190–199

Lindfors L, Forssén S et al (2008) Nucleation and crystal growth in supersaturated solutions of a model drug. J Colloid Interface Sci 325:404–413

Liversidge G, Cundy K, Bishop J et al (1992) Surface modified drug nanoparticles. US Patent 5,145,684, 8 Sep 1992

Lo Y, Hsu C et al (2013) pH-and thermo-sensitive pluronic/ poly(acrylic acid) in situ hydrogels for sustained release of an anticancer drug. J Drug Target 21:54–66

Lopez-Lorente A, Simonet B, Valcarcel M (2012) Analysis of nanoparticles based on electrophoretic separations. In: Damia B, Farre M (eds) Comprehensive analytical chemistry, vol 59. Elsevier, Amsterdam, pp 33–89

Lynch HJ, Rivest RW, Wurtman RJ (1980) Artificial induction of melatonin rhythms by programmed microinfusion. Neuroendocrinology 31(2):106–111

Martanto W, Moore JS, Kashlan O, Kamath R, Wang PM, O'Neal JM, Prausnitz MR (2006) Microinfusion using hollow microneedles. Pharm Res 23(1):104–113

McLain V (2007) Final report on the safety assessment of phytantriol. Int J Toxicol 26(Suppl 1):107–114. doi:10.1080/10915810601163947

Mikszta JA, Haider MI, Pettis RJ (2006) Microneedles for drug and vaccine delivery: when will the dream become a reality. In: Wille J (ed) Skin delivery systems. Transdermals, dermatologicals, and cosmetic actives. Blackwell Press, Ames, IA, pp 309–325

Miyazaki S, Kubo W et al (2000) Oral sustained delivery of theophylline using *in-situ* gelation of sodium alginate. J Control Release 67:275–280

Morgen M, Tye CK, LaChapelle E, Markovich M, Murri B, Millard D, Konagurthu S, Haskell RJ, Su CC, Hageman MJ (2013) Small scale manufacture of modified release coated beads for preclinical drug discovery research. Poster presentation at the American Association of Pharmaceutical Scientists annual meeting and exposition, San Antonio, TX, 10–14 Nov 2013

Moustafine R, Kemenova V et al (2005) Characteristics of interpolyelectrolyte complexes of Eudragit E 100 with sodium alginate. Int J Pharm 294:113–130

Murphy D (2001) Fundamentals of light microscopy and electronic imaging. Wiley, New York

Nguyen T, Hanley T et al (2010) Phytantriol and glyceryl monooleate cubic liquid crystalline phases as sustained release oral drug delivery systems for poorly water-soluble drugs II. In-vivo evaluation. J Pharm Pharmacol 62:856–865

Nguyen T, Hanley T et al (2011) Nanostructured liquid crystalline particles provide long duration sustained-release effect for a poorly water soluble drug after oral administration. J Control Release 153:180–186

Nophsker M, Haskell R (2010) Shaken not stirred: preparation of submicron suspensions in the discovery setting. Paper presented at the AAPS Northeast Regional Discussion Group, Rocky Hill, CT, 23 April 2010

Ochoa M, Mousoulis C, Ziaie B (2012) Polymeric microdevices for transdermal and subcutaneous drug delivery. Adv Drug Deliv Rev 64(14):1603–1616

Oshchepkov S, Sinyuk A (1998) Optical sizing of ultrafine metallic particles: retrieval of particle size distribution from spectral extinction measurements. J Colloid Interface Sci 208:137–146

Ozaki S, Minamisono T, Yamashita T et al (2012) Supersaturation—nucleation behavior of poorly soluble drugs and its impact on the oral absorption of drugs in thermodynamically high-energy forms. J Pharm Sci 101:214–222

Palermo R, Short S et al (2012) Determination of figures of merit for near-infrared, Raman and powder x-ray diffraction by net analyte signal analysis for a compacted amorphous dispersion with spiked crystallinity. J Pharm Innov 7:56–68

Paliouras GN, Hamilton LK, Aumont A, Joppé SE, Barnab́-Heider F, Fernandes KJL (2012) Mammalian target of rapamycin signaling is a key regulator of the transit-amplifying progenitor pool in the adult and aging forebrain. J Neurosci 32(43):15012–15026

Perkins L, Peer C, Murphy-Hackley P (2000) The use of mini-osmotic pumps in continuous infusion studies. In: Healing G, Smith D (eds) Handbook of preclinical continuous intravenous infusion. Taylor and Francis, London

Quinn K, Gullapalli R, Merisko-Liversidge E et al (2012) A formulation strategy for gamma secretase inhibitor ELND006, a BCS class II compound: development of a nanosuspension formulation with improved oral bioavailability and reduced food effects in dogs. J Pharm Sci 101:1462–1474

Raghavan S, Schuesse K et al (2003) Formation and stabilization of triclosan colloidal suspension using supersaturated systems. Int J Pharm 261:153–158

Rastogi R, Sultana Y et al (2007) Alginate microspheres of isoniazid for oral sustained drug delivery. Int J Pharm 334:71–77

Reddy J, Nagaraja T et al (2013) Design and characterization of sustained release matrix tablets of glimepiride by using synthetic and natural polymers. Int J Drug Discov Herb Res 3:573–578

Reix N, Guhmann P et al (2012) Duodenum-site specific delivery: in vivo assessment of a pharmaceutically developed enteric-coated capsule for broad applicability in rat studies. Int J Pharm 422:338–340. doi:10.1016/j.ijpharm.2011.10.017, e-published 13 October 2011

Roche (1999) Safety data sheet for phytantriol. http://euroweb.roche.com/msds/english/0429082.pdf. Accessed Feb 2003

Roduner E (2006) Size matters: why nanomaterials are different. Chem Soc Rev 35:583–592

Sampath Kumar K, Bhowmilk D et al (2012) Sustained release drug delivery system potential. Pharm J 1:46–56

Saphier S, Rosner A et al (2010) Gastro intestinal tracking and gastric emptying of solid dosage forms in rats using X-ray imaging. Int J Pharm 388:190–195

Satapathy T, Panda P et al (2010) Evaluation of anti-GERD activity of gastro retentive drug delivery system of itopride hydrochloride. Artif Cells Blood Substit Immobil Biotechnol 38:200–207

Schwarzer H, Peukert W (2002) Experimental investigation into the influence of mixing on nanoparticle precipitation. Chem Eng Technol 6:657–661

Shah J, Sadhale Y et al (2001) Cubic phase gels as drug delivery systems. Adv Drug Deliv Rev 47:229–250

Shah N, Iver R et al (2013) Improved human bioavailability of vemurafenib, a practically insoluble drug, using an amorphous polymer-stabilized solid dispersion prepared by a solvent-controlled coprecipitation process. J Pharm Sci 102:967–981

Singh P, Kaur I (2011) Development and evaluation of a gastro-retentive delivery system for improved antiulcer activity of ginger extract (*Zingiber officinale*). J Drug Target 19:741–751

Sonaje K, Chen Y et al (2010) Enteric-coated capsules filled with freeze-dried chitosan/poly (γ-glutamic acid) nanoparticles for oral insulin delivery. Biomaterials 31:3384–3394

Stark W (2011) Nanoparticles in biological systems. Angew Chem Int Ed 50:1242–1258

Strickley R (2013) Drug-in-food as a preclinical efficacy formulation. Paper presented at the American Association of Pharmaceutical Scientists Annual Meeting and Exposition, San Antonio, TX, 13 November 2013

Tan T, Tsuru Y (2009) Infusion pumps for small laboratory animals. ALN Magazine, vol May/-June. http://www.alnmag.com/article/infusion-pumps-small-laboratory-animals

Tan T, Watts SW, Davis RP (2011) Drug delivery: enabling technology for drug discovery and development. iPRECIO® Micro infusion pump: programmable, refillable, and implantable. Front Pharmacol 2:44

Tangsathikulchai C (2003) Effects of slurry concentrations and powder filling on the net mill power of a laboratory ball mill. Powder Technol 137:131–138

Theeuwes F, Yum SI (1976) Principles of the design and operation of generic osmotic pumps for the delivery of semisolid or liquid drug formulations. Ann Biomed Eng 4(4):343–353

Thomas V, Bhattachar S et al (2006) The road map to oral bioavailability: an industrial perspective. Expert Opin Drug Metab Toxicol 2:591–608

Thombre A (2005) Assessment of the feasibility of oral controlled release in an exploratory development setting. Drug Discov Today 10:1159–1166

Thorat A, Dalvi S (2012) Liquid antisolvent precipitation and stabilization of nanoparticles of poorly water soluble drugs in aqueous suspensions: recent developments and future perspective. Chem Eng J 181–182:1–34

Tolla B, Boldridge D (2010) Distortion of single-particle optical sensing (SPOS) particle count by sub-countable particles. Part Part Syst Charact 27:21–31

Toyota H, Shimogori H, Sugahara K, Yamashita H (2012) A novel treatment for vestibular disorder with FGLM-NH2 plus SSSR. Neurosci Lett 526(2):128–132

Tuan-Mahmood TM, McCrudden MTC, Torrisi BM, McAlister E, Garland MJ, Singh TRR, Donnelly RF (2013) Microneedles for intradermal and transdermal drug delivery. Eur J Pharm Sci 50(5):623–637

Tye CK, Su CC, Haskell RJ, Morgen M, Markovich M, LaChapelle E, Murri B, Millard D, Konagurthu S, Hageman MJ (2013) Manufacture of modified release coated-beads for a preclinical drug discovery research. American Association of Pharmaceutical Scientists Northeast Regional Discussion Group annual conference, Rocky Hill, CT, 19 Apr 2013

Tye CK, Su CC, Haskell RJ, Morgen M, Shaffer M, Holenarsipur VK, Sinha J, Rao IR, Palanisamy K, Italia JL, Desai S, Nigam A, Mandlekar S, Desikan S, Hageman MJ (2014) Designer modified release microspheres for preclinical drug discovery research – in vivo performance. American Association of Pharmaceutical Scientists annual meeting and exposition, San Diego, CA, 2–6 Nov 2014

Utracki L (1990) Polymer alloys and blends: thermodynamics and rheology. Carl Hanser Verlag, Munich, p 93

Van Eerdenbrugh B, Stuyven B, Froyen L et al (2009) Downscaling drug nanosuspension production: processing aspects and physicochemical characterization. AAPS PharmSciTech 10:44–53

Van Eerdenbrugh B, Vermant J, Martens J et al (2010) Solubility increases associated with crystalline drug. Mol Pharm 7:1858

Vehring R (2008) Pharmaceutical engineering via spray drying. Pharm Res 25:999–1022

Verwey E, Overbeek J (1948) Theory of the stability of lyophobic colloids. The interaction of sol particles having an electric double layer. Elsevier, New York

Villar VAM, Jones JE, Armando I, Asico LD, Escano CS Jr, Lee H, Wang X, Yang Y, Pascua-Crusan AM, Palmes-Saloma CP, Felder RA, Jose PA (2013) Sorting nexin 1 loss results in D5 dopamine receptor dysfunction in human renal proximal tubule cells and hypertension in mice. J Biol Chem 288(1):152–163

Villari V, Micali N (2008) Light scattering as spectroscopic tool for the study of disperse systems useful in pharmaceutical sciences. J Pharm Sci 97:1703–1730

Vodak D (2012) In vitro tests for reliable selection of active amorphous form and solid dosage form: Methods for quantifying aqueous species and rate and extent of dissolution. Paper presented at predicting in vivo oral drug absorption using bio-relevant in vitro testing: current approaches and future directions symposium, American Association of Pharmaceutical Scientists Annual Meeting and Exposition, Chicago, IL, 14–18 Oct 2012

Wang Y, Guo SZ, Bonen A, Li RC, Kheirandish-Gozal L, Zhang SXL, Brittian KR, Gozal D (2011) Monocarboxylate transporter 2 and stroke severity in a rodent model of sleep apnea. J Neurosci 31(28):10241–10248

Wu L, Zhang J, Wanatabe W (2011) Physical and chemical stability of drug nanoparticles. Adv Drug Deliv Rev 63:456–469

Yadav S, Liu J, Scherer T et al (2013) Assessment and significance of protein–protein interactions during development of protein biopharmaceuticals. Biophys Rev 5:121–136

Zaretsky DV, Zaretskaia MV, Durant PJ, Rusyniak DE (2012) The use of microinfusion pump to perform intrahypothalamic injections in conscious rats. Neuroscience annual meeting, New Orleans, USA, 13–17 Oct 2012

Chapter 4
Diagnosing Biopharmaceutical Limitations

Susan M. Jenkins and Dawn D. Parker

4.1 Introduction

The drug discovery process involves several iterative processes of increasing potency, efficacy, and ADME properties while avoiding off-target and systemic toxicological effects. This chapter will focus on the part of the discovery process where the compounds of interest are functional from a biological and chemical perspective but are sub-optimal due to poor oral exposure.

Often, issues related to poor exposure are addressed by screening compounds in an in vivo model until the targeted exposure is reached and one compound is selected. This approach may work if the cause of the low exposure is known and a rational approach has been designed to address the issue. However, if the reasons for poor exposure cannot be identified readily, it may be beneficial to investigate representative compounds that are not intended for development to better understand the issues and determine the best path forward.

Oral bioavailability($\%F$) is a result of the following relationship:

$$\%F = 100 * \text{Fa} * \text{Fg} * \text{Fh} * \text{Fother}$$

where

Fa = fraction of compound that is absorbed
Fg = the fraction that passes through the gut
Fh = the fraction passing through the liver

S.M. Jenkins (✉)
Metabolism and Pharmacokinetics, Bristol-Myers Squibb, 5 Research Parkway, Wallingford, CT 06492, USA
e-mail: Susan.Jenkins@bms.com

D.D. Parker
Discovery Pharmaceutics, Bristol-Myers Squibb, Wallingford, CT, USA

A.C. Templeton et al. (eds.), *Discovering and Developing Molecules with Optimal Drug-Like Properties*, AAPS Advances in the Pharmaceutical Sciences Series 15,
DOI 10.1007/978-1-4939-1399-2_4

Fother = the fraction passing through other clearance pathways, such as the fraction passing through the renal system (Fr) or the fraction passing through the lungs (Fl) (Hyun-Jong 2013).

Once it is understood where the problems occur, it becomes easier to address the specific issues (Ballard et al. 2012). This chapter addresses these issues individually to improve understanding of causes and effects and to develop plans to solve the problems, if possible. For the purposes of this discussion, we have assumed that the major route of elimination occurs through the gut or liver. However, other elimination pathways (such as renal) should always be evaluated and taken into account.

The Biopharmaceutics Classification System (BCS) and Biopharmaceutics Drug Disposition Classification System (BDDCS) have been introduced to help scientists understand the underlying issues affecting compound progression (Benet 2010). The systems are complimentary, each classifying compounds into four categories based on solubility and permeability characteristics. The BDDCS expands on the BCS by including metabolism and transporter-related properties. Therefore, Fh can be further defined by the hepatic fraction due to metabolism (Fh,m) and the hepatic fraction due to transporter processes (Fh,t). The same can be described by the fraction passing through the gut (Fg,m and Fg,t). An excellent review has been written for both classification systems (Benet 2013). These systems can be broken down into a decision tree as follows:

Each of the biopharmaceutics classes has limitations related to the class it is in. For compounds with high permeability, metabolism may be a concern because of the compounds' unimpeded access to the enzymes involved with metabolism. Also, high-permeability compounds may be reabsorbed from drug excretion routes and thus can be eliminated only through metabolism (Benet 2013). Low-permeability compounds may have absorption issues, due to reduced ability to penetrate the intestinal lumen. Transporter issues may be common to all classes except the high-permeability/high-solubility Class 1 compounds. Although it is not known why Class 1 compounds are not susceptible to transporter involvement in the intestine and liver, it may be that the permeable nature of these compounds overcomes transporter effects or they are just not transporter substrates by their nature (Benet 2010).

This review discusses the decision tree in Fig. 4.1 in terms of each component so that the cause of the biopharmaceutical limitation(s) may be identified and approached in a logical fashion.

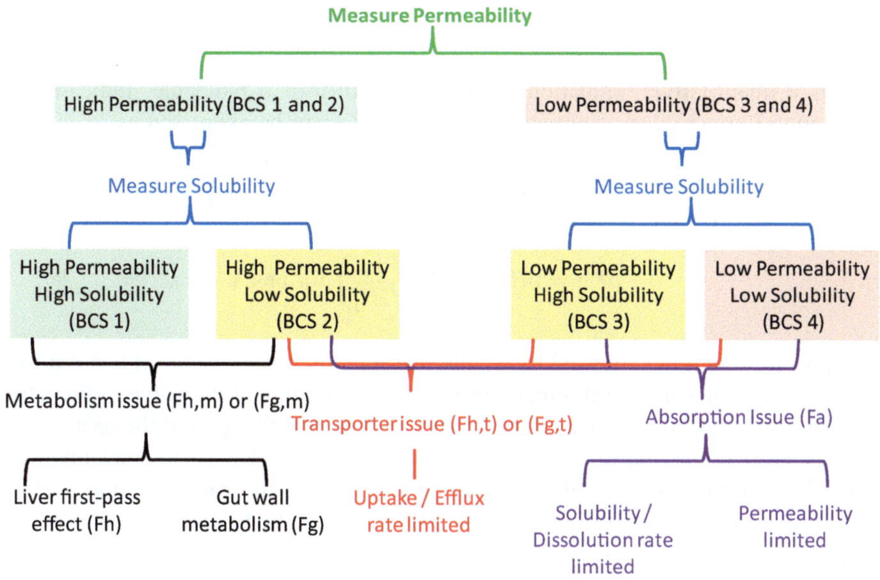

Fig. 4.1 Decision tree for determination of metabolic, transporter, and physicochemical limitations to absorption

4.2 Permeability and Solubility

Early in the drug discovery process, several parameters should be evaluated for each compound of interest. Many drug candidates are halted due to pharmacokinetic-related issues; many others are stopped because they have physicochemical properties that render them unsuitable for development. Since these delays can be costly if encountered at a later stage in development, in vitro predictions must be made in the discovery environment where speed and throughput are high (Avdeef 2001). Parameters relevant to in vivo performance, such as the distribution coefficient (log *D*), partition coefficient (log *P*), logarithmic acid dissociation constant (p*K*a), polar surface area (PSA), and molecular weight (MW) can be readily calculated. These calculations can be used as a basis for determining a compound's ability to be progressed using Lipinski's "rule of 5." In the discovery setting, the "rule of 5" has shown that poor absorption or permeation is more likely when there are more than 5 H-bond donors, 10 H-bond acceptors, the molecular weight (MW) is greater than 500 and the calculated log *P* (*c* Log *P*) is greater than 5 (Lipinski et al. 2001). Physical measurement of these parameters is often performed to better define the physical and chemical properties of compounds in a given chemical space. Several reviews (Meanwell 2011; Gleeson et al. 2011; Keldenich 2009; Takano et al. 2012) have been written that discuss the principles and approaches for physiochemical profiling and determination of the tools needed to develop oral drug products.

4.2.1 pKa

p*K*a is a quantitative evaluation of the ionizability of a molecule, indicating if subgroup(s) on a molecule are acidic or basic. p*K*a can be expressed as:

$$pKa = pH - \log\left(\frac{[A-]}{[HA]}\right) \text{ for acids and}$$

$$pKb = pH - \log\left(\frac{[B+]}{[BOH]}\right) \text{ for bases}$$

The p*K*a is sometimes referred to as "the dissociation constant" but it is not a true constant; it is dependent on temperature, ionic strength, and the solvent dielectric constant. In most cases, however, these variables do not change and the p*K*a can be calculated at physiologically relevant conditions. Several experimental methods exist for determining the p*K*a of a compound, with the most common being potentiometric titration (Reijenga et al. 2013). Lipophilicity and p*K*a can also be calculated using computation methods (calculated parameters have a c to distinguish the difference, such as cLogP). However, the reliability of these methods is dependent on starting assumptions and the choice in parameters used for the calculation. It still may be some time before computational methods as accurate as actual measurements (Reijenga et al. 2013).

The p*K*a of a compound has a direct impact on its solubility and permeability. Acidic compounds may have a difficult time solubilizing in the low-pH environment of the stomach, whereas a basic compound may precipitate while traversing the duodenum, jejunum, or ileum, where neutral conditions prevail. This situation is further complicated by the pH variability in the gastrointestinal (GI) tract, depending on the species and whether that species is in the fasted or fed state. In the human fasted state, the stomach pH varies from 1.1 to 2.1 and in the fed state from 4.3 to 5.4 (Kararli 1995). Gastric pH in fasted dogs is higher than in humans ranging from pH 2 to 7, with a small but significant decrease in gastric pH with feeding (Mahar et al. 2012; Lui et al. 1986; Sagawa et al. 2009). The p*K*a of a molecule can readily be chemically modified to adjust this balance. It often happens that several structural modifications are attempted to address p*K*a issues while retaining potency and other desirable characteristics of the compound class. However, when structural modifications cannot improve permeability, then increasing solubility may be the only way to improve absorption (Curatolo 1998). Section 4.5.2 further explores effects of food and pH on compound absorption.

4.2.2 Lipophilicity (log P/log D)

Lipophilicity is a fundamental physicochemical property that plays a key role in the ADME of a compound (Waterhouse 2003). It is defined as a compound's ability to partition into a nonpolar lipid environment versus that of an aqueous environment.

Lipophilicity is typically measured to assess a compound's ability to penetrate membranes, which can include the GI tract or the blood–brain barrier. Log *P* and log *D* are the measurements used in the pharmaceutical industry to describe lipophilicity. These measurements are the ratio of equilibrium concentrations in immiscible solvents, most commonly octanol/water. Log *P* is defined as the logarithm of the ratio of the concentration of the unionized compound in octanol to that of water, whereas log *D* is the ratio of the sum of the concentration of all forms of the compound (ionized plus unionized) in each of the two phases. Log *P* and log *D* differ in that only the unionized species are measured for log *P* and all species (ionized and unionized) are measured for log *D*. Since log *D* is pH-dependent, the pH at which log *D* is measured must be specified. Log *D* is commonly measured at pH 6.5 to estimate oral intestinal absorption and pH 7.4 to simulate the pH of blood for IV administration. For unionizable compounds, log $P = \log D$ at any pH. To obtain an accurate log *P*, the pH is adjusted to 2 units above or below the p*K*a, depending on whether the compound is an acid or base.

Lipophilicity is a main factor in determining the PK properties of drug candidates (Leucuta 2014). Lipinski's "rule of 5" suggests that poor absorption or permeability is more likely when log *P* is greater than 5 (Lipinski et al. 2001). This is due to the fact that compounds with higher log *P* values are more nonpolar and typically have poorer aqueous solubility. It is also true that compounds with low log *P* values (<1) are more polar and have poorer lipid-membrane permeability (Curatolo 1998; van de Waterbeemd et al. 2001).

4.2.3 Solubility

Solubility is the quantitation of the degree of mixing that occurs when a substance (solute) is placed in contact with another (solvent) and allowed to reach equilibrium (Yalkowsky 1999). Solubility is an important physicochemical property that is measured in the discovery setting because it can be used to predict absorption and oral bioavailability in accordance with the BCS (Amidon et al. 1995). Solubility has a direct effect on absorption, since only compounds presented in solution form can be absorbed in the intestine. Therefore, it is important to dose (1) solutions that remain solutions in vivo or (2) suspensions containing solids that are amenable to dissolution within the transit time of the GI tract. The speed at which a solid dissolves into a solvent system is the dissolution rate. The rate of dissolution is often expressed as the Noyes–Whitney equation:

$$\frac{\mathrm{d}m}{\mathrm{d}t} = A\frac{D}{h}(\mathrm{Cs} - C(t))$$

where:

$m =$ mass of dissolved material
$t =$ time
$A =$ surface area of the interface between the dissolving substance and the solvent

D = diffusion coefficient

h = thickness of the boundary layer of the solvent at the surface of the dissolving substance

Cs = concentration of the solid (equilibrium solubility)

$C(t)$ = concentration of dissolved solute over time

Solubility measurements can be made thermodynamically or kinetically. Thermodynamic solubility is the concentration of compound in a saturated solution when excess solid is present. In this case, the solution and the solid are in equilibrium. A typical method of determining thermodynamic solubility is from solid material by a shake-flask test, which is an accurate but time-consuming method (Box et al. 2006). Alternatively, kinetic solubility measurements are made from a predissolved solution of the compound, which is added to an aqueous solvent system. The solubility is determined at the time when the compound begins to precipitate from solution. These methods answer two different questions. The thermodynamic method asks "To what extent does the compound dissolve?" whereas the kinetic method asks "To what extent does the compound precipitate?" (Saal and Petereit 2012).

Solubility and dissolution are dependent on many factors, including the structural attributes of the compound (e.g., molecular weight, lipophilicity, p*K*a); the solvent being used (e.g., pH, composition) and the in vitro study conditions (e.g., solvent, equilibration time, temperature, separation technique of supernatant from solid). The solid form of the compound can also affect solubility and dissolution, as it may exist in many different forms, amorphous, crystalline, polymorphs, or different particle sizes. Amorphous solids do not have a crystal lattice structure, therefore dissolution is achieved faster than a crystalline compound (Fig. 4.2). Their lack of order means amorphous solids are all different from one another, leading to variable solubility values and may be more hygroscopic, less pure, and/or less stable than their crystalline counterparts (Yu 2001).

Crystalline solids are defined unit cells with repeated regularity in three dimensions (Fig. 4.2). Crystalline solids of the same form provide reproducibility in equilibrium solubility, stability, particle size, hygroscopicity, and shape. Crystalline solids also provide a consistent dissolution rate, albeit slower than that of amorphous solids. One exception to this reproducibility occurs when polymorphs (different forms of the same crystalline compound) are present. Polymorphs often have different characteristics one from another.

Fig. 4.2 Schematic representation of a molecule (m) in amorphous (**a**), crystalline containing several polymorphs (**b**) and crystalline (**c**) states (After Yu 2001)

Due to the differences that can exist in the solid state, various lots of the same compound are often tested in assays in the discovery setting. Solid-state characterizations are often performed to distinguish one lot from another.

Particle size can also affect solubility and dissolution. By reducing particle size, the surface area of the solid is increased. As noted in the Noyes–Whitney equation, this increase allows for faster dissolution and more compounds in solution in a shorter period of time. This can be critical when performing in vivo suspension dosing or when solid precipitates from a supersaturated solution (Sect. 4.5.3).

Low-solubility compounds can be difficult to formulate for in vivo dosing. Solution formulations involve more than simply dissolving compound into a solvent system. The solution needs to traverse the GI tract without the compound precipitating. This can be achieved through the use of an excipient that aids in solubilization and precipitation inhibition. Another way to improve the solubility of a compound in solution is to produce a supersaturated solution, which is a kinetically unstable state in which the concentration exceeds the thermodynamic solubility (Gao and Shi 2012). This unstable state causes precipitation of solid from solution due to the increased concentration or can occur if the compound is in this state for an extended period of time. The solid that precipitates may have variable particle sizes or be a different form of material than what was originally dissolved. No matter what size or form precipitates, if the solid does not re-dissolve in the GI tract, then low oral bioavailability can be expected.

4.2.4 Biorelevant Solubility

Solubility can be measured in many different solvents or solvent systems. Most commonly referred to is the aqueous solubility. However, solubility values determined in water or buffer systems at a given pH are not usually predictive of solubilities in the GI tract. This may lead to underestimates of oral bioavailability, especially for lipophilic or poorly soluble compounds (Dressman et al. 2007). Solubility determined as a function of pH provides information about where in the GI tract the compound may be soluble and gives rise to pH-dependent exposure. These systems do not represent the conditions of the GI tract, which is much more complicated than pH alone, as GI luminal contents can affect solubility. Therefore, the use of simulated gastric and simulated intestinal fluids (SGF and SIF, respectively) more closely mimics the in vivo environment. In the pharmaceutical industry, instant SIFs are commonly used due to their predictive results and ease of use. These powders contain bile acids and can be made into either fasted-state or fed-state simulated intestinal fluids (Boni et al. 2009). The solubility values generated in such solvents provide solubility values that are closer to those encountered in vivo, but may also be used to aid in prediction of food effect (i.e., in in silico modeling, as discussed in Sects. 4.6) (Lentz 2008). In vivo work would be needed to confirm a food effect (Sect. 4.5.2), but this in vitro work could be the first indication that further work needs to be performed.

Overall, determination of solubility is a useful tool in comparing compounds and aiding in assessing in vivo performance.

4.2.5 *Chemical Stability*

In vitro chemical stability measurements are typically done in response to an in vivo issue being addressed or at the request of the chemist who may notice instability during synthesis. These measurements are performed at physiologically relevant pH and temperature and take into account pH ranges throughout the GI tract. These data can provide insight into potential chemical instability in the GI tract since pH may play a significant role in degradation pathways. For example, if a compound is hydrolyzed under acidic conditions, then it may be degraded in the stomach, leading to low bioavailability since the parent compound was never introduced to the site of absorption. Solution-state chemical stability is also monitored in dosing solutions to ensure a compound is stable through the duration of dosing, whether an acute or chronic dosing schedule.

4.2.6 *Parallel Artificial Membrane Permeability Assay (PAMPA)*

Permeability is often measured using PAMPA, which determines the permeability of a substance through an artificial lipid membrane. The system simulates the passive permeability through the intestinal membrane; compounds are usually tested using pH 5.5 and 7.4 buffers to cover the pH range in the intestinal lumen. Differences in permeability at the two pH levels may indicate potential for pH-limited absorption. A permeability coefficient (Pc) is calculated based on the rate of passage of a test compound across the lipid membrane and can be expressed as nanomoles of compound traversed across the membrane per second (nM/s). High intrinsic permeability is predicted for compounds with a Pc >100 nM/s (>75 % fraction absorbed) and low permeability is predicted for compounds with a Pc < 20 nM/s (<15 % fraction absorbed). Difficulties with interpretation of this assay can arise if there is poor recovery (less than 30 %) due to binding to the apparatus or membrane or with compounds that are undetected due to solubility or chromatography/detection issues. An extensive review of the PAMPA and other permeability assays is covered in this review (Balimane 2012).

4.2.7 *Caco-2*

A more comprehensive permeability assay is the human colon adenocarcinoma (Caco-2) bi-directional permeability assay. This assay has lower throughput compared to the PAMPA assay, but is physiologically more relevant. Caco-2 cells possess morphological and functional qualities similar to enterocytes which serve as a barrier to the absorption of drugs that are administered orally. The Caco-2 cells undergo differentiation in culture to form cell monolayers with well-established tight junctions that function as intestinal epithelial cells. The cells are seeded on a collagen-coated transwell plate and are incubated for 2–4 weeks to allow the cells to differentiate. Compound is either added to the apical (A) or basolateral (B) side of the membrane and after an incubation time of approximately 2 h, samples are collected from both sides and analyzed. The permeability coefficient (Pc) is calculated as a rate of passage of test compound across the Caco-2 monolayer in nanomoles per second. In the absence of transporters, the efflux ratio Pc (B → A)/ Pc (A → B) should equal 1. As with the PAMPA assay, compounds with a Pc greater than 100 nM/s are considered highly permeable with prediction of high in vivo absorption, whereas compounds with a Pc of less than 40 nM/s are predicted to be poorly absorbed. Also like the PAMPA assay, poor recovery (less than 70 %) from the apparatus could cause significant variation in permeability values and may give erroneous values. Poor recovery is often associated with compounds that have high lipophilicity, poor solubility, or nonspecific binding of the compound to the device (Thomas et al. 2008). The presence of bovine serum albumin (BSA) in the basolateral assay buffer at levels up to 4 % can improve permeability and recovery by mimicking the physiological levels of human albumin within the capillary lumen (Press 2011). Although Caco-2 cells are not suitable to predict intestinal absorption when transporter-mediated routes and/or first-pass metabolism is involved, they can be cloned to express multiple efflux and uptake transporters, which may assist in discerning if transporter involvement is a factor in absorption (Sect. 4.4).

4.2.8 The Role and Interpretation of In Vitro Measurements

In vitro measurements are important in the discovery setting as they provide useful data and can be performed quickly and with minimal material requirements in most cases. Compounds may be tested many times in these assays as more lots of material are made, providing more pure or different forms of the material. The data provide insight into compound behavior in the in vivo environment, albeit not a perfect mimic. Therefore, in vitro measurements such as solubility are not typically used as decisional data for a compound, but can be informative on a comparative basis to other compounds or chemical series.

4.3 Metabolism

Quite often, compounds have excellent physicochemical properties and complete absorption from the intestinal lumen but still suffer from low oral exposure due to metabolism and resulting high first-pass effects. Highly permeable, lipophilic compounds are subject to metabolic processes, especially Cytochrome P450 oxidative metabolism, glucuronidation, and other Phase 1 and Phase 2 metabolism.

Metabolic processes continue to be extensively researched, due to the importance to the drug discovery and development process. Metabolite identification (Met ID) is an important aspect in drug discovery, not only to determine the “soft spot” on the molecule for SAR purposes but also to determine if circulating metabolites have importance from a biological or toxicological perspective. Metabolism may also be species-selective. Often there is considerable variability in metabolic stability in the preclinical species that could make a difference between further progression and abandonment of a compound. Several reviews have been written that specifically address CYP450 and other metabolism issues in a drug discovery setting (Yengi et al. 2007; Bohnert 2009; Emoto et al. 2010; White 2011).

4.3.1 In Vitro Studies

The use of liver microsomes and hepatocytes as a measure of the rate of metabolic drug oxidation is being extensively used in drug discovery. Incubation of the compound of interest with liver microsomes in the presence of nicotinamide adenine dinucleotide phosphate (NADPH) (or an NADPH-generating system) and determination of the percent of parent remaining over time is a relatively quick way of gaining valuable information regarding the capacity for oxidative metabolic turnover. Hepatocyte studies are lower-throughput compared to microsomes, but can provide a better estimate for metabolic turnover (Camenisch and Umehara 2012), especially if the compound is metabolized by Phase 2 metabolism, such as glucuronidation or sulfation. Several papers have been written regarding the prediction of in vivo metabolite clearance of compound based on microsomal or hepatocyte studies. These predictions are often based on the well stirred and parallel tube models of drug metabolism and incorporate the protein binding (both plasma and microsomal protein binding) in their calculation (Obach 1996; Iwatsubo et al. 1997; Austin et al. 2002; Bell and Wang 2012; Sohlenius-Sternbeck et al. 2012). These calculations are valuable for compounds that are primarily cleared by hepatic metabolism (Fh,m) and do not take extra-hepatic metabolism and other clearance pathways into account. By increasing the stability of compounds using in vitro studies, bioavailability can be greatly enhanced in vivo, if the reason for the poor exposure was due to metabolism.

4.3.2 Intravenous and Oral Administration

Intravenous (IV) and oral (PO) pharmacokinetic studies in different animal species have traditionally been used to select lead candidates. Several important parameters, such as plasma clearance (CL), volume of distribution (Vss), half-life (T½), and oral bioavailability (*%F*) can be obtained in this manner. Because of the value of this type of study, several approaches such as cassette dosing, pooling samples, screening PK, and other processes have been proposed to increase compound throughput while sparing animal resources (Nagilla et al. 2011). Once CL information is obtained, it is useful to determine the extraction ratio (ER) as shown by the following equation, where Q is the hepatic blood flow of the species under investigation.

$$\mathrm{ER} = \frac{\mathrm{CL}}{Q}$$

Clearance can then be classified as low (ER < 0.3), moderate (ER between 0.3 and 0.7), and high (ER > 0.7). High clearance can be due to several processes, including transporter involvement, tissue partitioning or metabolism. Met ID and correlation of in vivo with in vitro (metabolic stability) findings can assist with determining the role of metabolism to high clearance. Correlation of in vitro metabolic stability with in vivo results also appears to be valuable in screening and prioritizing compounds with poor metabolic stability (Lau et al. 2002). Oral bioavailability is determined by the following equation:

$$\%F = 100 \times \frac{\mathrm{AUCn\ PO}}{\mathrm{AUCn\ IV}}$$

where AUCn PO is the area under the curve following oral administration normalized to dose and AUCn IV is the area under the curve following intravenous administration normalized to dose. Since bioavailability is based on a ratio of PO to IV exposure, it is possible to have relatively high bioavailability but if the compound has high clearance, oral exposure can still be low on an absolute basis. Conversely, oral exposure may be high but if the compound has low clearance, the resulting bioavailability may be low. It is often useful to calculate the fraction absorbed (f) once bioavailability is known (Nomeir et al. 2009) based on the following equation:

$$f = \frac{\%F/100}{1 - \mathrm{ER}}$$

This term reflects both Fa and Fg as described in Sect. 4.1. If the fraction absorbed is less than 1, incomplete absorption of compound can be assumed. If the fraction absorbed is higher than 1, a saturable (nonlinear) clearance process may be responsible. In either case, the reason for the inconsistency should be examined to better understand compound liabilities. PK studies listed below (Sect. 4.3.3–4.3.6) assist in addressing the potential issues in a specific manner.

4.3.3 Gut Metabolism

Determination of gut metabolism (Fg,m) as a major clearance pathway is not as easy to ascertain as hepatic metabolism (Fh,m). Although intestinal microsomes may be useful in obtaining relative contribution of gut metabolism, this method is not as predictive to in vivo as liver microsomes or hepatocytes (Shen et al. 1997; Galetin et al. 2010). In general, the intestinal enzymes that are involved with gut metabolism have a much lower contribution to the first-pass effect compared to hepatic involvement (Thummel 2002). Therefore, any compound with extensive gut metabolism generally also has a high first-pass systemic clearance. However, determination of gut metabolism may be important for compounds with high intestinal first-pass extraction. Several studies are described in the following sections that are able to separate intestinal from hepatic first-pass effects. The use of specific inhibitors (Sect. 4.3.5) and the portal vein cannulation study (Sect. 4.3.7) have been used successfully to determine the extent of intestinal extraction.

4.3.4 Bile-Duct Cannulation

In vivo PK screens in rodents (especially rat) are the most direct method of determining compounds with high clearance, but high clearance does not necessarily point to metabolism-related issues; transporters may also be involved (Sect. 4.4). A bile-duct cannulated (BDC) study may be needed to determine the primary route of elimination. In this study, the compound is administered by IV or PO route and urine, bile, and feces are collected and analyzed for parent and metabolites. A description of the surgical procedure involved with this study has been described (Landskroner et al. 2011). If quantitative recovery of compound and metabolites is possible through the use of radiolabel or other means, the percentage of the administered dose that is excreted via the bile and urine can be used to directly calculate the extent of oral absorption. In a rat study with a selective opioid agonist, recovery of parent and metabolites in urine and feces from intact rats was compared with recovery in bile, urine, and feces of BDC animals. Since fecal recovery was 90.4 % in intact female rats and 46.6 % in BDC animals, it was concluded that a minimum of 51 % of administered dose was absorbed in rats (Guo 2011).

Table 4.1 describes the potential information that can be obtained from a BDC study. For example, following IV or PO administration of a compound, where parent is present as the major component in the bile, the results point to hepatic transporter involvement (Fh,t) or passive permeability into the bile cannicula as exposure-limiting mechanisms. However, if parent is only present as a minor component in the bile with metabolites being the major drug-related components, then hepatic (or gut) metabolism is apparent (Fh,m $\pm$ Fg,m). For example, in a rat BDC study with a novel dopamine agonist, only 30 % of the oral dose was recovered in the bile, with no parent compound detected. When combined with

Table 4.1 Potential outcomes from a bile-duct cannulated study

Route	Bile	Urine	Feces
IV	Primarily parent = Fh,t Primarily metabolites = Fh,m	Primarily parent = Fr,t Primarily metabolites = Fh, m ± Fother,m	Primarily parent = Fg,t Primarily metabolites = Fg, m ± Fother,m
PO	Primarily parent = Fh,t Primarily metabolites = Fh,m ± Fg,m	Primarily parent = Fr,t Primarily metabolites = Fh, m ± Fother,m	Primarily parent = Fg, t ± Fa Primarily metabolites = Fg, m ± Fother,m

other studies, it was concluded that the low bioavailability of this compound was a consequence of poor absorption as well as extensive first-pass metabolism (Bollard 2000). If the compound is administered IV and parent is observed in the feces in a bile-duct cannulated study, it is assumed that an active gut efflux mechanism is responsible (Fg,t), since the fraction excreted in the bile is already accounted for. Following oral administration, parent compound present in the feces may be due to lack of absorption (Fa) or efflux transporter involvement (Fg,t). An example of this was described using a radiolabeled NK2 antagonist in BDC rats. Following oral administration, 87 % of the dose was recovered in the feces, consistent with poor oral absorption. Following intravenous administration, 17 % of the dose was recovered in the feces, suggesting that direct secretion of radioactivity across the gut wall occurred and thus is partially responsible for the low exposure (Beaumont et al. 2000). Metabolites present in the feces following oral administration indicate the potential for gut metabolism, although there is also the potential for systemic metabolism and direct transport into the gut.

Altogether, a BDC study can greatly assist in the determination of absorption, metabolism, and elimination aspects for the compound under investigation. Radiolabeled material may not be available early in the drug discovery process, but a qualitative view may be all that is needed to differentiate the primary pathway for elimination using unlabeled compound. However, careful interpretation of the results is still necessary, due to the interplay of several processes at work in the in vivo system combined with the potential lack of quantitative information. By understanding the elimination pathway through a BDC study, a hypothesis can be established for improving oral exposure and the appropriate studies can be designed with follow-up compounds to address the issue.

4.3.5 Metabolism Inhibition

Inhibitors may be used in vitro (Khojasteh et al. 2011) and in vivo to probe involvement of specific CYP isoforms in the metabolism of test compounds. Administration of an appropriate inhibitor, such as ketoconazole (Ward et al. 2004) (a CYP3A4 inhibitor) or sulfaphenazole (CYP2C9 inhibitor) (Baldwin 1995) have been used to determine the extent of CYP involvement in the metabolism and first-pass effects of xenobiotics. 1-Aminobenzotriazole (ABT) has also been used in vivo (Strelevitz et al. 2006; Landskroner et al. 2011) as a nonspecific mechanism-based CYP450 inhibitor to distinguish P450-dependent first-pass effects from other factors such as solubility, permeability, and active efflux. Antipyrine, a probe for measuring efficiency of hepatic oxidative metabolism, was evaluated in the rat, dog and monkey in the presence of ABT. A 93 % inhibition of antipyrine clearance was observed following a 2 h oral pretreatment of ABT at 100 mg/kg in the rat and 20 mg/kg in dog and monkey (Balani et al. 2002), indicating the effectiveness of CYP inhibition using this method. However, these studies need to be interpreted carefully. For example, ketoconazole may also inhibit p-glycoprotein (an efflux transporter) (Zhang 1998) and UDP-glucuronosyltransferase (glucuronidation) (Takeda 2006), resulting in a more complex interpretation of the results.

4.3.6 Duodenal Administration

Duodenal administration should be considered in special cases, especially if there is a potential for acidic degradation or if there is reason to suspect that precipitation of the compound is occurring in the stomach. For example, duodenal administration of idarubicin as a solid lipid nanoparticle in rats resulted in prolonged release of compound, improving bioavailability compared with solution (Zara et al. 2002). The authors concluded that this prolonged release approach may be useful in reducing the toxicity of idarubicin. Duodenal administration in the rat was used to demonstrate an increase in bioavailability of an in-situ gelling formulation that may provide a new approach to improved bioavailability for BCS Class 3 drugs (Dai et al. 2013). An ion-pair strategy was used to improve intestinal absorption and oral bioavailability of amifostine, a prodrug with low lipophilicity and charge at physiological pH (Samiei et al. 2013). Using duodenal dosing to bypass drug degradation in the stomach, these authors demonstrated a 20- to 30-fold increase in amifostine bioavailability. Due to the invasive nature of this approach, duodenal administration is most often done in a rodent, such as mouse or rat, but duodenal administration can also be done in the dog with surgically ported animals (Sinko et al. 1997; Lee et al. 2001). Although duodenal administration is not feasible as an eventual clinical approach, it can be used to determine the formulation strategy for improved bioavailability.

4.3.7 Portal-Vein Cannulation

A useful in vivo method to separately assess the first-pass effect of a compound in the intestine and liver is the portal-vein cannulation study. A detailed description of

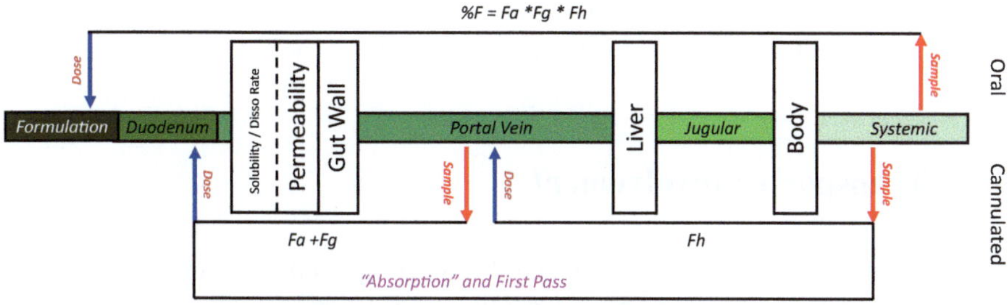

Fig. 4.3 The portal vein cannulation study (courtesy R. Haskell)

the surgical procedure for this study has been published (Landskroner et al. 2011). In this design, compound is dosed by oral (or duodenal) gavage and blood is collected from the portal vein to obtain Fg*Fa and systemic (jugular, carotid) veins to obtain Fg*Fa*Fh. These results are compared with results obtained when the compound is administered in the portal vein and samples are collected systemically to obtain Fh (Fig. 4.3). Using this method, it was determined that the low bioavailability of a thromboxane A2 antagonist was due to extensive gut and hepatic first-pass clearance rather than solubility or permeability issues (Murakami et al. 2003). Another example of this method was used to determine the intestinal and hepatic first-pass effect of metoprolol in the rat. Following IV, intraportal and intraduodenal administration of metoprolol at 1 and 2 mg/kg, the resulting intestinal and hepatic first-pass extraction ratio was approximately 0.45 and 0.60 (respectively) indicating considerable contribution of intestinal first-pass extraction to the low bioavailability of metoprolol in rats (Yoon et al. 2011). Evaluation of the metabolites present in the samples may also assist in determining if extensive metabolism is a factor in the clearance process. This procedure is most often conducted in rats, but other species have been evaluated using this approach.

4.3.8 Biopharmaceutical Considerations

From a biopharmaceutical perspective, it is difficult to increase oral exposure in compounds where metabolism is a major component. Intra-nasal (Ali 2012), buccal (Garg 2011), and transdermal (Perumal 2013) approaches have been used as alternatives to other more common drug administration (PO, IV, IM, SC, etc.) to avoid gut and hepatic first-pass effects with success. Another alternative is the use of prodrugs that are capable of being metabolized and released as parent compound, thereby avoiding first-pass effects. An example of this is an amide prodrug of gemcitabine that avoided the extensive first-pass metabolism that occurs following administration of gemcitabine (Wickremsinhe et al. 2013). However, these approaches may not have an appropriate preclinical species for evaluation, leading to extended development times and potentially more risk than a "traditional" approach. Determination of the metabolic "soft spots" on a molecule and rapid evaluation of metabolic stability are often better approaches in this case.

4.4 Transporter Involvement

Until recently, it was not well understood how compounds that are metabolically stable and have acceptable biopharmaceutical properties can still have high clearance or low oral bioavailability. For years, transporter involvement had been suspected, since what was observed in vivo could not be explained by simple permeability, solubility, or metabolic processes. Great strides have been made in the last 10–15 years in the understanding of transporter involvement in drug disposition. An International Transporter Consortium has been established which explored clinically important transporters involved with drug absorption and distribution and provided recommendations to help guide in vitro and in vivo transporter studies necessary for filing new drug applications (Giacomini et al. 2010).

Transporters involved with drug disposition can be roughly divided into two groups: uptake transporters or efflux transporters. Uptake transporters have been demonstrated to improve drug absorption through the GI tract but can also increase blood clearance by active uptake into hepatocytes. These transporters include members of the organic anion transporting polypeptide (OATP) family, peptide transporters (PEPT), sodium/bile acid transporters (ASBT), and monocarboxylic acid transporters (MCT). Efflux transporters can restrict drug absorption by pumping drugs out of intestinal epithelial cells back into the intestinal lumen. They include multidrug resistance protein 2 (MRP2), breast cancer resistance protein (BCRP), and p-glycoprotein (P-gp, also known as multi-drug resistance transporter 1 (MDR1)). Many other transporters have been described and characterized but this chapter will only cover the most relevant transporters involved with drug disposition.

4.4.1 Assessing Transporter Activity In Vitro

One of the most prominent transporters responsible for the efflux of compounds across biological membranes is P-gp. Also known as MDR1 and ABCB1, it has an important role in limiting the entry of various drugs into the central nervous system and has a large role in intestinal absorption and biliary (or renal) excretion of drugs. A useful in vitro tool to determine P-gp involvement is the Caco-2 system overexpressing the P-gp enzyme. Since P-gp is localized in the apical membrane in polarized cell monolayers, a high efflux ratio of basal-to-apical ($B \rightarrow A$) compared with apical-to-basal ($A \rightarrow B$) indicates a potentially significant role for P-gp involvement in absorption. Conversely, if the efflux ratio is significantly lower than 1, the potential for an active uptake process should be investigated. However, a high efflux ratio in vitro does not always translate to poor oral absorption. If a compound has high permeability and solubility, the rate of absorption may exceed efflux, depending on the Km and Vmax of the transporter and the physical properties of the molecule. Therefore, the involvement of P-gp is more apparent in compounds that have a poor apparent permeability coefficient (Papp). Another efflux transporter expressed in intestine and liver is BCRP (ABCG2). BCRP was found to be an important factor in the clearance of pitavastin in vitro by using Madin-Darby canine kidney II (MDCK II) cells expressing human canalicular efflux transporters and in vivo using BCRP1 knockout mice (Hirano et al. 2005).

Hepatic uptake processes play an important role in the hepatic clearance of drugs. An important transporter in this regard is OATP, which is responsible for the disposition of many drugs such as the HMG-CoA reductase inhibitors (statins) and angiotensin-converting enzyme (ACE) inhibitors. A review of the OATP family of transporters and their impact on pharmacokinetic properties of pharmaceutically relevant compounds has been written (Kalliokoski and Niemi 2009). Two approaches have been described to evaluate uptake using hepatocytes. One approach is the media-loss assay (Soars et al. 2007), which evaluates the loss of parent compound from the incubation medium into hepatocytes. This assay is used to predict the in vivo clearance of drugs with significant hepatic uptake. A direct hepatocyte uptake assay has also been established that can determine the uptake clearance of test substrates by passive diffusion or transporters (Watanabe et al. 2010).

4.4.2 Assessing Transporter Activity In Vivo

For compounds that have low in vivo clearance and unaccountably low oral bioavailability that cannot be explained by first-pass metabolism, the role of P-gp and other efflux transporters should be investigated, especially if Caco-2 studies indicate the potential for transporter involvement. Conversely, if higher than expected in vivo clearance is observed, the role of OATP and other uptake transporters should be considered. Although mammalian P-gps display a high level of primary structure similarity and homology across several species (Bosch and Croop 1998), there are differences in substrate selectivity that have been observed in vitro (Suzuyama et al. 2007) and in vivo (Xia et al. 2006; Nishimura et al. 2008). Therefore, relating results from preclinical species must always be considered in light of eventual human studies. With that in mind, there are several available in vivo methods to ascertain the effect of transporters on oral bioavailability.

One method is to determine pharmacokinetic parameters in knockout animals. In this way, involvement of specific transporters may be evaluated. A Balb/c, MDR1a, or MDR1a/b gene-deficient (knock-out) (Chen et al. 2003) mouse has been used as a powerful tool to evaluate the role of P-gp in vivo, using an FVB mouse (control wild-type strain for MDR1a mouse) as a comparator. Caution should be used, however, in interpreting results from these studies. For example, although MDR1a-deficient mice lack P-gp at both the apical membrane of intestinal

Table 4.2 Inhibitors commonly used to determine transporter involvement

Inhibitor	Transporter	Reference
Elacridar	P-gp and BCRP	Darby et al. (2011)
Ko143	BCRP	Allen et al. (2002)
Rifampicin	OATP	Imaoka (2013)
Cyclosporin A	OATP, BSEP	Shitara (2011)
Probenecid	OAT	Feng et al. (2013)
Cimetidine	OCT	Feng et al. (2013)

epithelial cells and the biliary canicular membrane of hepatocytes (Schinkel et al. 1995), up-regulation of the MDR1b gene has been observed when MDR1a is disrupted. Additionally, by limiting intracellular substrate availability, P-gp may influence CYP3A-mediated metabolism in MDR1a and MDR1a/b-deficient mice (Lan et al. 2000) and must be considered when interpreting the data from the MDR1a knock-out mouse studies.

Another way to evaluate transporter involvement is co-dosing an inhibitor when administering the test compound. Common transporter inhibitors used both in vitro and in vivo as tools to elucidate pharmacokinetic parameters and understand absorption mechanisms are shown in Table 4.2.

Due to potential drug–drug interactions and high clearance that is often observed with compounds that are substrates for hepatic uptake, SAR studies are often conducted to help eliminate compounds that display transporter involvement, thereby improving pharmacokinetic properties. Unfortunately, conducting hepatocyte uptake assays for multiple compounds can be difficult, due to the low throughput and labor-intensive nature of this assay. In this case, a PK screen often assists in rapidly disqualifying compounds that have high clearance and corresponding high liver exposure. Compound is administered IV in the test species (usually rat) and plasma is sampled up to 5 h post-dose, with livers removed and analyzed for exposure. In this way, clearance and liver to plasma ratios can be rapidly assessed to find compounds with more favorable PK properties.

4.4.3 Biopharmaceutical Considerations

Formulation strategies have been evaluated to enhance membrane permeability and inhibit efflux transporters (P-gp). Agents such as Tween 80 (polyoxyethylene(20)-sorbitanemonooleate), Pluronic P85 (polyoxyethylene-polyoxypropylene copolymer), Cremophor EL (polyoxyethyleneglycoltriticinoleate), and TPGS (D-α-Tocopheryl polyethylene glycol 1000 succinate) (Akhtar et al. 2011; Guo et al. 2013) can modulate drug absorption by multiple mechanisms including inhibition of intestinal P-gp. Other approaches that have been described in Sect. 4.3.6 may also be used to avoid first-pass effects due to transporters. Improvement in permeability or solubility (Sect. 4.5) so that efflux transporters may be saturated or bypassed (by increasing passive permeability, for example) may also be used. This includes the synthesis of prodrugs that are more efficiently absorbed and are cleaved once in systemic circulation, but the opportunity for hepatic transporter involvement may still be an issue.

4.5 Absorption

As compounds are evaluated further, it becomes necessary to achieve multiples of exposure necessary for toxicology studies. Although there is no firm criteria in the multiples of efficacious exposure necessary to move a compound forward, due to the nature of the compound and the safety required for the indication (FDA 2010), a rough exposure multiple that is often used is 30 times the expected human efficacious AUC for initial toxicology studies and 30 times the expected human efficacious Cmax for preliminary cardiovascular safety studies. Achieving high exposure for toxicology studies can be a daunting prospect for a biopharmaceutical scientist, as the choice for safe, tolerable vehicles is diminished. Figure 4.4 shows possible AUC exposure increases when dose is increased for a representative preclinical species. The desired effect is for dose-proportionality (squares); however, this is often not the case. Permeability-limiting exposure is often observed as slowly increasing AUC with increasing dose, but at a less than dose proportional level (diamonds). Solubility-limiting AUC increases proportionally until the solubility limit has been reached; as dose is increased beyond the solubility limit, exposure does not increase further (triangles). When a clearance process becomes saturated, such as P-gp efflux or CYP450 metabolism, a greater than dose proportional increases is observed with increasing dose (circles). This is a general rule of thumb and other process, such as transporter involvement, may be involved that could change the conclusion. For example, what appears to be solubility-limiting exposure may be uptake-limiting for a poorly permeable compound. A thorough

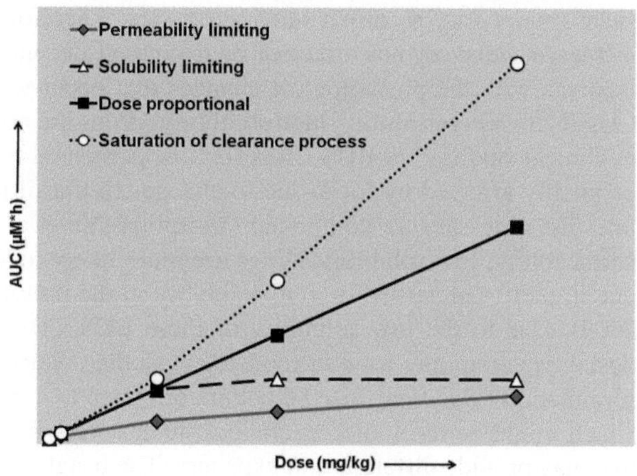

Fig. 4.4 Assessing limitations for achieving exposure in dose-escalation studies

understanding of the physical chemical properties of the molecule and the use of applicable studies (such as a transporter inhibitor in this example) may be necessary to determine the cause of nonlinearity and to establish the best path forward for the molecule or the compound series.

4.5.1 *In Vivo Studies*

As discussed earlier, BDC or PVC studies are able to determine the fraction of a compound that is absorbed (Sect. 4.3). Another way to determine fraction absorbed is if the compound has a renal elimination component. Determination of the percent of compound-related material eliminated in the urine following IV administration compared to oral administration provides a rough estimate of the fraction absorbed. An example of this was the relative urinary recoveries of radioactivity following IV and PO administration of an NK2 antagonist in the rat. A recovery of 19.7 % was observed in the urine following IV administration and 1.7 % following oral administration indicating the compound is approximately 9 % absorbed in the rat (Beaumont et al. 2000). However, the molecular weight of a compound is known to be an important factor affecting the renal elimination of certain compounds. The molecular weight threshold appears to be approximately 325 ± 50 for biliary excretion of polar compounds in the rat (Hughes et al. 1973). Compounds with a high molecular weight will be primarily eliminated by biliary processes, with very little renal elimination.

4.5.2 *Food Effect Studies*

Food can dramatically alter the absorption of a drug through a variety of mechanisms. Food may affect drug solubility, increase GI motility and bile flow, delay gastric emptying, or affect other processes. Therefore, drug absorption may be increased, delayed, not affected, or decreased depending on how the compound interacts with the physiological changes that occur with food consumption. BCS Class I (high permeability, high solubility) drugs are not affected by fed conditions in clinical studies, but BCS Class III (low permeability, high solubility) drugs can be greatly affected by food, due to changes in transit time through regions of the intestine necessary for compound absorption (Yu et al. 2004). BCS Class II (high permeability, low solubility) drugs are more likely to show a positive food effect, due in part to increased in vivo solubility in the presence of food (Wu and Benet 2005). Due to the low solubility of these BCS Class II compounds, suspension dosing is commonly used in food-effect studies. Since suspensions are solid-state formulations, they may also be better correlated to a traditional tablet formulation, which might be used later in clinical studies, versus using a solution formulation that may provide different oral exposure. The beagle dog has been used as a model to examine the effect of food on preclinical compound bioavailability (Lentz et al. 2007). In this model, a high-fat (FDA) meal in coordination with pentagastrin pretreatment was predictive of the human food effect for a validation set of compounds. Pentagastrin administered at 6 μg/kg via IM injection 15–30 min prior to dosing lowered the gastric pH in the dog to mimic the basal gastric pH in the human (Zhou et al. 2005). Famotidine (Pepsid) can also be used to increase gastric pH by oral administration 3 h before dosing the test compound (Zhou et al. 2005). When applied to fasted animals, these treatments are also helpful in determining the pH-dependent effects on oral drug absorption and for screening formulations to overcome pH dependency on oral bioavailability. The rat has also been shown to be a useful model to examine pH-dependent solubility on oral exposure using pentagastrin and famotidine as gastric pH modulators (Lubach et al. 2013).

4.5.3 Formulations to Diagnose Limited Exposure

When compounds have good IV pharmacokinetic properties but low oral bioavailability, formulation may play a key role in achieving exposure needed for compound advancement. The use of formulations to improve oral exposure is dependent on the factor that is limiting absorption. Solubilization, dissolution rate, and permeability are the most common rate-limiting steps to drug absorption. While permeability is generally not affected by formulation, the following properties can be influenced by formulation technology:

- Solubility—determined by compound properties
- Supersaturation—delivering solution over the solubility limit of the compound
- Dissolution rate—influenced by particle size

While traditional dosage forms, such as solutions and tablets, are preferred for development, many poorly soluble compounds are not amenable to these dosage forms. Fortunately, recent advances in formulation technologies have allowed many of these compounds to be advanced to development (Chap. 3).

4.5.3.1 Effect of Precipitating vs. Nonprecipitating Solution

Due to the poor aqueous solubility of many compounds, in vivo precipitation from solution is a common problem, which leads to low oral bioavailability. Low solubility compounds may precipitate in vivo when they come in contact with GI fluids, blood, or other aqueous body fluids (Schroeder and DeLuca 1974). Compounds may precipitate in vivo before absorption due to loss of the supersaturated state, thereby changing the compounds precipitation kinetic profile and affecting drug absorption and bioavailability (Dai et al. 2007). The form(s) and particle size (s) of the in vivo precipitate are unknown, leading to many questions about what was *actually* dosed. To assess the potential for a solution formulation to precipitate in vivo, in vitro precipitation studies at physiologically relevant pHs and in SIFs have become a standard practice in the many discovery environments. Solubility-enhancing solutions, supersaturated solutions, and high concentration solutions are of the most concern when dosing in vivo due to their likelihood of precipitation. The in vitro precipitation screening of these formulations can be performed in a 96-well plate format, which enables many solutions to be evaluated at once with a rapid turnaround time. The precipitation can be evaluated by spectroscopy using a UV plate reader, which evaluates light scattering in a given sample that changes when precipitation occurs. The light scattering is affected by particle size, concentration, and wavelength making spectroscopy a more informative tool than visual observations alone (Morrison and Nophsker 2009). If the in vitro screening is performed post-dosing on the dosing solution administered to animals, the precipitation data obtained can be correlated back to in vivo results, since formulation performance is often the first place to look when oral bioavailability is low or an unanticipated result is observed in vivo. If the in vitro screening is performed prior to in vivo dosing and precipitation is observed, a change in the solution formulation to resist precipitation (i.e., by adding a precipitation inhibitor) can be made and reevaluation in the in vitro precipitation assay should be performed.

Solutions that are precipitation resistant have a significant impact on the amount of oral exposure obtained from solution formulation dosing compared to nonprecipitation resistant solutions. In Fig. 4.5, two solutions of a compound were dosed to rats, one a solution known to precipitate and the other a nonprecipitating solution. In this case, there was approximately a fivefold increase in exposure when comparing the two solutions. This shows that not all solutions are created equal; dissolution of a compound into a solution formulation does not ensure good exposure. Dosing solutions must be precipitation-resistant to ensure the compound can be absorbed in the GI tract, as only solubilized compound can be absorbed.

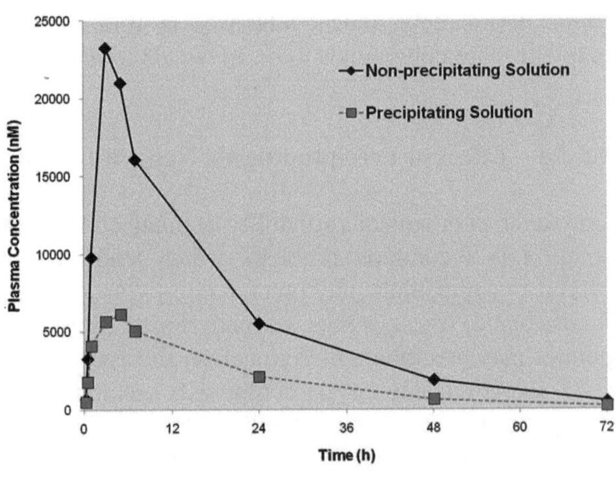

Fig. 4.5 Comparing rat oral exposures from precipitating vs. nonprecipitating solutions

4.5.3.2 Effect of Solid on Dissolution

In late-stage development, tablet dosing is the most common formulation utilized due to its straightforward nature and prior knowledge of many pharmaceutical scientists in developing such formulations. Due to this, there is a need to evaluate solid dosage forms in discovery at a relatively early stage. Suspension dosing is often used as a stand-in for tablet formulations. In the discovery setting, suspensions can be easily and quickly prepared compared to the time it takes to develop a well behaved tablet for in vivo dosing.

When dosing a solid, an additional step, dissolution of the compound, is added to the process of absorption that is not present from solution dosing. As discussed in the previous section, a compound can only be absorbed from the solution state. Therefore, when a solid formulation is dosed, the compound must dissolve before it can be absorbed. The time it takes for a compound to dissolve can vary and is affected by different physicochemical properties, including the crystallinity of the material and the form being dosed (i.e., free or salt form).

When low solubility compounds are dosed in vivo and a plateau or loss of exposure is observed with respect to increasing dose, solubility-limited exposure may be the problem. This is a common issue observed with low solubility compounds. Solubility-enhancing formulations or supersaturated solutions aid in solubilizing low solubility compounds but may have a concentration limit where they are either insoluble or begin to precipitate. Target concentrations of solution dosing are study-dependent due to the doses required. For example, oral PK studies at lower doses show high exposure through a solubilized formulation, while high doses required for toxicology studies may not provide any further increase in exposure. Solubility-limited exposure can be diagnosed by dosing a suspension of a low solubility compound in vivo and comparing the exposure obtained to a solution formulation (Parker et al. 2013). Due to the extra step of dissolution, the suspension may provide little compound in solution, making it unavailable for absorption (Fig. 4.6).

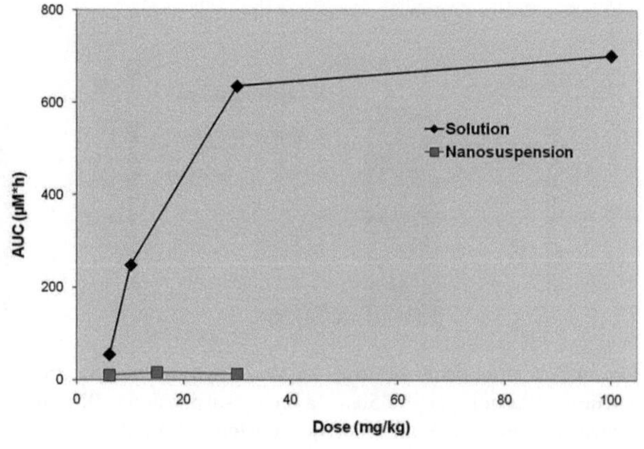

Fig. 4.6 Solubility-limited oral exposure graph comparing solution and nanosuspension dosing (Parker et al. 2013)

4.5.3.3 Effect of Surface Area on Dissolution

The particle size, or the surface area of a solid, can greatly affect the dissolution of the compound when dosing a suspension formulation. When good oral exposure is dependent on its dissolution properties, it is defined as dissolution-rate-limited exposure. This can most effectively be probed by dosing suspensions of different particle sizes and evaluating the exposure achieved. By dosing a suspension with reduced particle size, the surface area of the suspension is increased. This increase allows for faster dissolution and more compounds in solution in the GI tract, which can readily be absorbed. In Fig. 4.7, two compounds with low aqueous solubility were dosed in rats or dogs as microsuspensions or nanosuspensions. The oral bioavailability was improved in each of these cases with the nanosuspensions, indicating dissolution rate was limiting exposure from microsuspensions.

Another way to address exposure limitations in vivo is to look to stabilized amorphous systems. Spray–dried dispersions (SDDs) are one of the technologies that can be utilized. SDDs are amorphous systems that are stabilized within a polymer matrix. Their primary objectives are fourfold (Friesen et al. 2008):

1. Enhance oral absorption by sustaining supersaturated concentrations in vivo
2. Provide physical stability by avoiding crystallization or phase separation of the amorphous material
3. Provide a way to reproducibly manufacture material
4. Is applicable to structurally diverse molecules with a wide range of physico-chemical properties

The applicability of this technology for a given compound can be assessed with a few simple in vitro measurements, including log *P*, melting point, and glass transition temperature. The in vitro feasibility assessment of this alternate technology can be done quickly, which is highly valued in the discovery setting.

SDDs improve the fraction of drug absorbed by increasing dissolution rate, increasing the dissolved drug concentration (solubility), and sustaining supersaturated concentrations in vivo. Insoluble compounds can benefit from SDD technology to enhance oral bioavailability, especially when traditional formulation

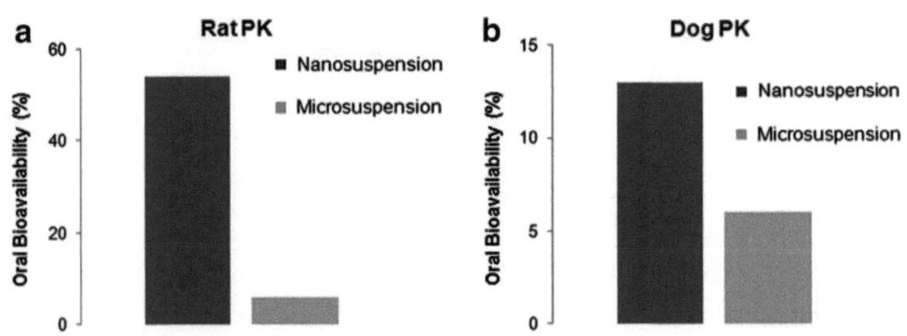

Fig. 4.7 Comparison of oral bioavailability achieved from nanosuspensions vs. microsuspensions. Particle size d50s: (**a**) nanosuspension = 0.5 μm, microsuspension = 45 μm; (**b**) nanosuspension = 0.3 μm, microsuspension = 1.5 μm

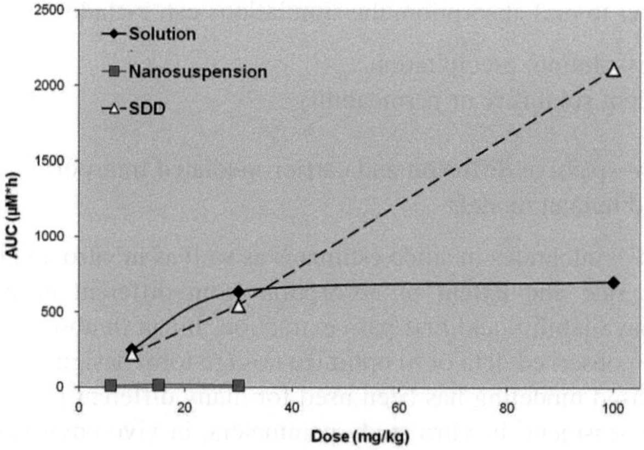

Fig. 4.8 Rat oral exposures comparing different formulations, where solubility-limited absorption is overcome by SDD formulation technology (Parker et al. 2013)

strategies fail. In Fig. 4.8, rat PO dosing of an SDD was conducted for a nonionizable compound that had solubility limitations (precipitation likely at higher doses of a supersaturating solution), as well as poor exposure from a nanosuspension. The SDD formulation provided dose proportional increases in exposure, thereby allowing a path forward for higher dosing, a necessity for compound development. This exemplifies an example of solubility-limited absorption being overcome by formulation technology, in this case an SDD (Parker et al. 2013). The topic of amorphous dispersions in general and SDDs in particular is covered in more detail in Chap. 3.

4.6 In Silico Modeling

In silico modeling provides specific information that offers insight into a drug's in vivo behavior that would otherwise be unknown. A simulation software that is commonly used in the pharmaceutical industry is GastroPlus™ (by Simulations Plus, Inc.). This specific model incorporates the proprietary Advanced Compartmental Absorption and Transit Model (ACAT™), which accounts for all drug mass present in vivo in all possible states. To understand the limitations to oral exposure, the drug's presence in the solid, dissolved, and absorbed states, are the most relevant. It can model oral absorption, pharmacokinetics, and pharmacodynamics.

With respect to oral absorption, the simulations can include:

- Release, dissolution, precipitation
- pH-dependent solubility or permeability
- Transit
- Absorption—passive diffusion and carrier-mediated transport
- Animal and human models

GastroPlus™ integrates in silico estimates as well as in vitro and in vivo data. It simulates the rate and extent of absorption from different dosage forms and calculates bioavailability and first-pass extraction. It can fit absorption, PK, or PD data to explain observed data or to optimize dosage form design. This quantitative, mechanism-based modeling has been used for many different purposes, including formulation assessment, in vitro study parameters, in vivo environment effect on absorption, formulation development, and predicting human PK. Formulation assessments have been done comparing dosage forms aiding in the selection or improvements needed to formulate a compound to obtain adequate oral exposure. It has also been used from an in vitro perspective, where modeling aided in the selection of dissolution media in setting in vitro bioequivalence dissolution standards for weak acids (Tsume et al. 2012). The in vivo environment is quite complex but modeling has been utilized to predict the magnitude of gastric pH changes on absorption relative to the dog's stomach pH, which can vary from pH 2 to 7. The modeling suggested improved bioavailability could be achieved through modulation of the gastric pH (Bhattachar et al. 2011). In formulation development, modeling has been used to guide form selection and particle size properties, steer dosage form design, assess dosage form performance, and influence clinical study design (Mathias and Crison 2012). Human PK from intestinally metabolized compounds was predicted using modeling and scaling from human liver microsome data (Heikkinen et al. 2012).

Another software system used in the Pharmaceutical industry is Simcyp (Certara). Simcyp has been organized as a cooperative venture with a consortium of pharmaceutical companies. The objective of this consortium is to develop a user-friendly program that could be used as a framework for integrating in vitro data to predict in vivo pharmacokinetics–pharmacodynamics in virtual patient populations (Rostami-Hodjegan and Tucker 2007). Simcyp contains numerous databases that contain human physiological genetic and epidemiological information based on existing information. By integrating these databases with in vitro and other data from a specific compound, the program allows for prediction of PK behavior in a "real world" population. Simcyp has been used for modeling of clearance, drug–drug interactions, transporter involvement, food effect, and assessment of population variability as part of its comprehensive package (Dong 2008). A combination of two or more physiologically based modeling approaches has also been used to optimize predictions. For example, GastroPlus was used to study and model the absorption characteristics in the rat. The results were incorporated into Simcyp to assess the human oral PK profile and drug–drug interaction potential of a BCS Class II compound when co-administered with a potent CYP3A4 inhibitor (Sinha et al. 2012).

Modeling is a useful tool in both the discovery and development settings. It can aid in prediction and explanation of in vivo results as well as aid in decision making for both animal and human models. As the models used in these programs continue to be refined and made more user-friendly, acceptance of in-silico modeling in the pharmaceutical industry continues to grow and expand.

4.7 Summary and Conclusions

Understanding the cause of poor exposure and what can be done to improve the molecule or the formulation while still maintaining the relevant pharmacological effect is a key aspect for compound progression. In order to accomplish this, appropriate studies need to be conducted with a limited number of compounds to better understand where the problems are occurring. By using a decision tree (Fig. 4.1) to understand the source of the issue and addressing the potential issue directly (is the low oral exposure due to transporter effects?) the proper study could be applied. Several in vitro (Table 4.3) and in vivo (Table 4.4) studies have been discussed in this chapter to address these specific issues.

As the cost of developing pharmaceutical compounds continues to rise and the risk of failure great, there is an ever increasing need to learn as much as possible early in discovery from limited material and tight timelines. By starting with calculated physical chemical properties and expanding to relevant in vitro assays, the drug discovery process can focus on the compounds that have the highest potential for success. However, as compounds gain therapeutic potency, quite often the desirable properties, solubility and permeability, are compromised leading to compounds with poor exposure. Therefore, a careful balance must be maintained in compound selection; excluding compounds based on physicochemical properties and in vitro results alone could eliminate the most pharmacologically active agents. The best approach is the careful investigation of new compound classes in selected relevant pharmacokinetic and pharmaceutical studies and relating the results back to the higher throughput studies so that the most appropriate properties of the series being considered are investigated.

Table 4.3 Sections describing in vitro studies to address potential issues discussed in this chapter

Potential issue / Study to consider	Active transport	Passive permeability	Solubility	Dissolution rate	GI precipitation	Chemical stability	pH effects	Rapid clearance/first-pass elimination	Gut metabolism
Determination of log *D*/log *P*		4.2.2	4.2.2						
Solubility measurements			4.2.3 4.5.3	4.2.3 4.5.3	4.2.4	4.2.5	4.2.4		
PAMPA		4.2.6					4.2.6		
Caco-2	4.4.1	4.2.7							
Microsomal stability								4.3.1	4.3.3
Hepatocyte stability								4.3.1	
Hepatocyte uptake	4.4.1								

Table 4.4 Sections describing in vivo studies to address potential issues discussed in this chapter

Potential issue / Study to consider	Active transport	Passive permeability	Solubility	Dissolution rate	GI precipitation	Food effects	pH effects	Rapid clearance/first-pass elimination	Gut metabolism
Solution formulation		4.5.3	4.5.3		4.5.3				
Precipitation-inhibiting solution	4.4.3 4.5.3	4.4.3 4.5.3	4.2.3 4.4.3		4.5.3				
Suspension formulation			4.2.3	4.5.3		4.5.2			
Knockout animals	4.4.2								
Fed versus fasted						4.5.2			
Pentagastrin/famotadine						4.5.2	4.5.2		
Dose escalation	4.5	4.5	4.5					4.5	
Intravenous administration	4.3.2 4.4.2							4.3.2	
Bile duct cannulation								4.3.4	
Portal vein administration/sampling								4.3.7	4.3.7
CYP450 or transporter inhibition	4.4.2								4.3.5
Duodenal administration					4.3.6		4.3.6		

References

Akhtar N, Ahad A, Khar RK, Jaggi M, Aqil M, Iqbal Z, Ahmad FJ, Talegaonkar S (2011) The emerging role of P-glycoprotein inhibitors in drug delivery: a patent review. Expert Opin Ther Pat 21(4):561–576

Ali A (2012) A novel intranasal drug delivery system: review. Novel Sci Int Pharm Sci 1(8):550

Allen JD, van Loevezijn A, Lakhai JM, van der Valk M, van Tellingen O, Reid G, Schellens JH, Koomen GJ, Schinkel AH (2002) Potent and specific inhibition of the breast cancer resistance protein multidrug transporter in vitro and in mouse intestine by a novel analogue of fumitremorgin C. Mol Cancer Ther 1(6):417–425

Amidon GL, Lennernas H, Shah VP, Crison JR (1995) A theoretical basis for a biopharmaceutic drug classification: the correlation of in vitro drug product dissolution and in vivo bioavailability. Pharm Res 12(3):413–420

Austin RP, Barton P, Cockroft SL, Wenlock MC, Riley RJ (2002) The influence of nonspecific microsomal binding on apparent intrinsic clearance, and its prediction from physicochemical properties. Drug Metab Dispos 30(12):1497–1503

Avdeef A (2001) Physicochemical profiling (solubility, permeability and charge state). Curr Top Med Chem 1(4):277–351

Balani SK, Zhu T, Yang TJ, Liu Z, He B, Lee FW (2002) Effective dosing regimen of 1-aminobenzotriazole for inhibition of antipyrine clearance in rats, dogs, and monkeys. Drug Metab Dispos 30(10):1059–1062

Baldwin SJ (1995) Ketoconazole and sulphaphenazole as the respective selective inhibitors of P4503A and 2C9. Xenobiotica 25(3):261–270

Balimane PV (2012) Permeability and transporter models in drug discovery and development. Wiley, Hoboken, NJ, pp 161–168

Ballard P, Brassil P, Bui KH, Dolgos H, Petersson C, Tunek A, Webborn PJ (2012) The right compound in the right assay at the right time: an integrated discovery DMPK strategy. Drug Metab Rev 44(3):224–252

Beaumont K, Harper A, Smith DA, Bennett J (2000) The role of P-glycoprotein in determining the oral absorption and clearance of the NK2 antagonist, UK-224,671. Eur J Pharm Sci 12(1):41–50

Bell LC, Wang J (2012) Probe ADME and test hypotheses: a PATH beyond clearance in vitro-in vivo correlations in early drug discovery. Expert Opin Drug Metab Toxicol 8(9):1131–1155

Benet LZ (2010) Predicting drug disposition via application of a biopharmaceutics drug disposition classification system. Basic Clin Pharmacol Toxicol 106(3):162–167

Benet LZ (2013) The role of BCS (biopharmaceutics classification system) and BDDCS (biopharmaceutics drug disposition classification system) in drug development. J Pharm Sci 102(1):34–42

Bhattachar SN, Perkins EJ, Tan JS, Burns LJ (2011) Effect of gastric pH on the pharmacokinetics of a BCS class II compound in dogs: utilization of an artificial stomach and duodenum dissolution model and GastroPlus, simulations to predict absorption. J Pharm Sci 100 (11):4756–4765

Bohnert T (2009) The role of drug metabolism in drug discovery. Wiley, Hoboken, NJ, pp 91–176

Bollard M (2000) Pharmacokinetics and disposition of the novel dopamine agonist Z-7760 in rat after intravenous and oral administration. Xenobiotica 30(10):983–991

Boni JE, Brickl RS, Dressman J, Pfefferle ML (2009) Instant FaSSIF and FeSSIF—biorelevance meets practicality. Dissolution Technol 16(3):41–45

Bosch I, Croop JM (1998) P-glycoprotein structure and evolutionary homologies. Cytotechnology 27(1–3):1–30

Box KJ, Volgyi G, Baka E, Stuart M, Takacs-Novak K, Comer JE (2006) Equilibrium versus kinetic measurements of aqueous solubility, and the ability of compounds to supersaturate in solution–a validation study. J Pharm Sci 95(6):1298–1307

Camenisch G, Umehara K (2012) Predicting human hepatic clearance from in vitro drug metabolism and transport data: a scientific and pharmaceutical perspective for assessing drug-drug interactions. Biopharm Drug Dispos 33(4):179–194

Chen C, Liu X, Smith BJ (2003) Utility of Mdr1-gene deficient mice in assessing the impact of P-glycoprotein on pharmacokinetics and pharmacodynamics in drug discovery and development. Curr Drug Metab 4(4):272–291

Curatolo W (1998) Physical chemical properties of drug candidates in the discovery and exploratory development settings. PSTT 1(9):387–393

Dai W-G, Dong LC, Song Y (2013) Enhanced bioavailability of poorly absorbed hydrophilic compounds through drug complex/in situ gelling formulation. Int J Pharm 457(1):63–70

Dai WG, Dong LC, Shi X, Nguyen J, Evans J, Xu Y, Creasey AA (2007) Evaluation of drug precipitation of solubility-enhancing liquid formulations using milligram quantities of a new molecular entity (NME). J Pharm Sci 96(11):2957–2969

Darby RA, Callaghan R, McMahon RM (2011) P-glycoprotein inhibition: the past, the present and the future. Curr Drug Metab 12(8):722–731

Dong JQ (2008) Applications of computer-aided pharmacokinetic and pharmacodynamic methods from drug discovery through registration. Curr Comput-Aided Drug Design 4(1):54

Dressman JB, Vertzoni M, Goumas K, Reppas C (2007) Estimating drug solubility in the gastrointestinal tract. Adv Drug Deliv Rev 59(7):591–602

Emoto C, Murayama N, Rostami-Hodjegan A, Yamazaki H (2010) Methodologies for investigating drug metabolism at the early drug discovery stage: prediction of hepatic drug clearance and P450 contribution. Curr Drug Metab 11(8):678–685

FDA (2010) Guidance for industry M3(R2) nonclinical safety studies for the conduct of human clinical trials and marketing authorization for pharmaceuticals

Feng B, Hurst S, Lu Y, Varma MV, Rotter CJ, El-Kattan A, Lockwood P, Corrigan B (2013) Quantitative prediction of renal transporter-mediated clinical drug-drug interactions. Mol Pharm 10(11):4207–4215

Friesen DT, Shanker R, Crew M, Smithey DT, Curatolo WJ, Nightingale JA (2008) Hydroxypropyl methylcellulose acetate succinate-based spray-dried dispersions: an overview. Mol Pharm 5(6):1003–1019

Galetin A, Gertz M, Houston JB (2010) Contribution of intestinal cytochrome p450-mediated metabolism to drug-drug inhibition and induction interactions. Drug Metab Pharmacokinet 25 (1):28–47

Gao P, Shi Y (2012) Characterization of supersaturable formulations for improved absorption of poorly soluble drugs. AAPS J 14(4):703–713

Garg S (2011) Buccal adhesive drug delivery system: safer delivery of biotherapeutics. J Drug Delivery Therapeutics 1(2):35

Giacomini KM, Huang SM, Tweedie DJ, Benet LZ, Brouwer KL, Chu X, Dahlin A, Evers R, Fischer V, Hillgren KM, Hoffmaster KA, Ishikawa T, Keppler D, Kim RB, Lee CA, Niemi M, Polli JW, Sugiyama Y, Swaan PW, Ware JA, Wright SH, Yee SW, Zamek-Gliszczynski MJ, Zhang L (2010) Membrane transporters in drug development. Nat Rev Drug Discov 9(3):215–236

Gleeson MP, Hersey A, Montanari D, Overington J (2011) Probing the links between in vitro potency, ADMET and physicochemical parameters. Nat Rev Drug Discov 10(3):197–208

Guo J (2011) In vitro and in vivo metabolism of a selective -opioid receptor. Drug Metab Dispos 39(10):1883–1894

Guo Y, Luo J, Tan S, Otieno BO, Zhang Z (2013) The applications of vitamin E TPGS in drug delivery. Eur J Pharm Sci 49(2):175–186

Heikkinen AT, Baneyx G, Caruso A, Parrott N (2012) Application of PBPK modeling to predict human intestinal metabolism of CYP3A substrates—an evaluation and case study using GastroPlus. Eur J Pharm Sci 47(2):375–386

Hirano M, Maeda K, Matsushima S, Nozaki Y, Kusuhara H, Sugiyama Y (2005) Involvement of BCRP (ABCG2) in the biliary excretion of pitavastatin. Mol Pharmacol 68(3):800–807

Hughes RD, Millburn P, Williams RT (1973) Molecular weight as a factor in the excretion of monoquaternary ammonium cations in the bile of the rat, rabbit and guinea pig. Biochem J 136 (4):967–978

Hyun-Jong C (2013) In vitro-in vivo extrapolation (IVIVE) for predicting human intestinal absorption and first-pass elimination of drugs: principles and applications. Drug Dev Ind Pharm

Imaoka T (2013) Integrated approach of in vivo and in vitro evaluation of the involvement of hepatic uptake organic anion transporters in the drug disposition in rats using rifampicin as an inhibitor. Drug Metab Dispos 41(7):1442–1449

Iwatsubo T, Hirota N, Ooie T, Suzuki H, Shimada N, Chiba K, Ishizaki T, Green CE, Tyson CA, Sugiyama Y (1997) Prediction of in vivo drug metabolism in the human liver from in vitro metabolism data. Pharmacol Ther 73(2):147–171

Kalliokoski A, Niemi M (2009) Impact of OATP transporters on pharmacokinetics. Br J Pharmacol 158(3):693–705

Kararli TT (1995) Comparison of the gastrointestinal anatomy, physiology, and biochemistry of humans and commonly used laboratory animals. Biopharm Drug Dispos 16(5):351–380

Keldenich J (2009) Measurement and prediction of oral absorption. Chem Biodivers 6(11):2000–2013

Khojasteh SC, Prabhu S, Kenny JR, Halladay JS, Lu AY (2011) Chemical inhibitors of cytochrome P450 isoforms in human liver microsomes: a re-evaluation of P450 isoform selectivity. Eur J Drug Metab Pharmacokinet 36(1):1–16

Lan LB, Dalton JT, Schuetz EG (2000) Mdr1 limits CYP3A metabolism in vivo. Mol Pharmacol 58(4):863–869

Landskroner KA, Hess P, Treiber A (2011) Surgical and pharmacological animal models used in drug metabolism and pharmacokinetics. Xenobiotica 41(8):687–700

Lau YY, Krishna G, Yumibe NP, Grotz DE, Sapidou E, Norton L, Chu I, Chen C, Soares AD, Lin CC (2002) The use of in vitro metabolic stability for rapid selection of compounds in early discovery based on their expected hepatic extraction ratios. Pharm Res 19(11):1606–1610

Lee YH, Perry BA, Lee HS, Kunta JR, Sutyak JP, Sinko PJ (2001) Differentiation of gut and hepatic first-pass effect of drugs: 1. Studies of verapamil in ported dogs. Pharm Res 18 (12):1721–1728

Lentz KA (2008) Current methods for predicting human food effect. AAPS J 10(2):282–288

Lentz KA, Quitko M, Morgan DG, Grace JE Jr, Gleason C, Marathe PH (2007) Development and validation of a preclinical food effect model. J Pharm Sci 96(2):459–472

Leucuta SE (2014) Selecting oral bioavailability enhancing formulations during drug discovery and development. Expert Opin Drug Discov 9(2):139–150

Lipinski CA, Lombardo F, Dominy BW, Feeney PJ (2001) Experimental and computational approaches to estimate solubility and permeability in drug discovery and development settings. Adv Drug Deliv Rev 46(1–3):3–26

Lubach JW, Chen JZ, Hau J, Imperio J, Coraggio M, Liu L, Wong H (2013) Investigation of the Rat model for preclinical evaluation of pH-dependent oral absorption in humans. Mol Pharm 10(11):3997–4004

Lui CY, Amidon GL, Berardi RR, Fleisher D, Youngberg C, Dressman JB (1986) Comparison of gastrointestinal pH in dogs and humans: implications on the use of the beagle dog as a model for oral absorption in humans. J Pharm Sci 75(3):271–274

Mahar KM, Portelli S, Coatney R, Chen EP (2012) Gastric pH and gastric residence time in fasted and fed conscious beagle dogs using the Bravo pH system. J Pharm Sci 101(7):2439–2448

Mathias NR, Crison J (2012) The use of modeling tools to drive efficient oral product design. AAPS J 14(3):591–600

Meanwell NA (2011) Improving drug candidates by design: a focus on physicochemical properties as a means of improving compound disposition and safety. Chem Res Toxicol 24(9):1420–1456

Morrison J, Nophsker M (2009) Exploring and mapping "formulation space" for precipitation resistant vehicle development personal communication

Murakami T, Nakanishi M, Yoshimori T, Okamura N, Norikura R, Mizojiri K (2003) Separate assessment of intestinal and hepatic first-pass effects using a rat model with double cannulation of the portal and jugular veins. Drug Metab Pharmacokinet 18(4):252–260

Nagilla R, Nord M, McAtee JJ, Jolivette LJ (2011) Cassette dosing for pharmacokinetic screening in drug discovery: comparison of clearance, volume of distribution, half-life, mean residence time, and oral bioavailability obtained by cassette and discrete dosing in rats. J Pharm Sci 100 (9):3862–3874

Nishimura T, Kato Y, Amano N, Ono M, Kubo Y, Kimura Y, Fujita H, Tsuji A (2008) Species difference in intestinal absorption mechanism of etoposide and digoxin between cynomolgus monkey and rat. Pharm Res 25(11):2467–2476

Nomeir AA, Morrison R, Prelusky D, Korfmacher W, Broske L, Hesk D, McNamara P, Mei H (2009) Estimation of the extent of oral absorption in animals from oral and intravenous pharmacokinetic data in drug discovery. J Pharm Sci 98(11):4027–4038

Obach RS (1996) The importance of nonspecific binding in in vitro matrices, its impact on enzyme kinetic studies of drug metabolism reactions, and implications for in vitro-in vivo correlations. Drug Metab Dispos 24(10):1047–1049

Parker DD, Haskell RJ, Stefanski KJ (2013) Spray-dried dispersion performance. AAPS NERDG annual meeting, Rocky Hill, CT

Perumal O (2013) Turning theory into practice: the development of modern transdermal drug delivery systems and future trends. Skin Pharmacol Physiol 26(4–6):331–342

Press B (2011) Optimization of the Caco-2 permeability assay to screen drug compounds for intestinal absorption and efflux. Methods Mol Biol 763:139–154

Reijenga J, van Hoof A, van Loon A, Teunissen B (2013) Development of methods for the determination of pKa values. Anal Chem Insights 8:53–71

Rostami-Hodjegan A, Tucker GT (2007) Simulation and prediction of in vivo drug metabolism in human populations from in vitro data. Nat Rev Drug Discov 6(2):140–148

Saal C, Petereit AC (2012) Optimizing solubility: kinetic versus thermodynamic solubility temptations and risks. Eur J Pharm Sci 47(3):589–595

Sagawa K, Li F, Liese R, Sutton SC (2009) Fed and fasted gastric pH and gastric residence time in conscious beagle dogs. J Pharm Sci 98(7):2494–2500

Samiei N, Mangas-Sanjuan V, González-Álvarez I, Foroutan M, Shafaati A, Zarghi A, Bermejo M (2013) Ion-pair strategy for enabling amifostine oral absorption: Rat in situ and in vivo experiments. Eur J Pharm Sci 49(4):499–504

Schinkel AH, Mol CA, Wagenaar E, van Deemter L, Smit JJ, Borst P (1995) Multidrug resistance and the role of P-glycoprotein knockout mice. Eur J Cancer 31A(7–8):1295–1298

Schroeder HG, DeLuca PP (1974) A study on the in vitro precipitation of poorly soluble drugs from nonaqueous vehicles in human plasma. Bull Parenter Drug Assoc 28(1):1–14

Shen DD, Kunze KL, Thummel KE (1997) Enzyme-catalyzed processes of first-pass hepatic and intestinal drug extraction. Adv Drug Deliv Rev 27(2–3):99–127

Shitara Y (2011) Clinical importance of OATP1B1 and OATP1B3 in drug-drug interactions. Drug Metab Pharmacokinet 26(3):220–227

Sinha VK, Snoeys J, Osselaer NV, Peer AV, Mackie C, Heald D (2012) From preclinical to human—prediction of oral absorption and drug–drug interaction potential using physiologically based pharmacokinetic (PBPK) modeling approach in an industrial setting: a workflow by using case example. Biopharm Drug Dispos 33(2):111–121

Sinko PJ, Sutyak JP, Leesman GD, Hu P, Makhey VD, Yu H, Smith CL (1997) Oral absorption of anti-aids nucleoside analogues: 3. Regional absorption and in vivo permeability of 2′,3′-dideoxyinosine in an intestinal-vascular access port (IVAP) dog model. Biopharm Drug Dispos 18(8):697–710

Soars MG, Grime K, Sproston JL, Webborn PJ, Riley RJ (2007) Use of hepatocytes to assess the contribution of hepatic uptake to clearance in vivo. Drug Metab Dispos 35(6):859–865

Sohlenius-Sternbeck AK, Jones C, Ferguson D, Middleton BJ, Projean D, Floby E, Bylund J, Afzelius L (2012) Practical use of the regression offset approach for the prediction of in vivo intrinsic clearance from hepatocytes. Xenobiotica 42(9):841–853

Strelevitz TJ, Foti RS, Fisher MB (2006) In vivo use of the P450 inactivator 1-aminobenzotriazole in the rat: varied dosing route to elucidate gut and liver contributions to first-pass and systemic clearance. J Pharm Sci 95(6):1334–1341

Suzuyama N, Katoh M, Takeuchi T, Yoshitomi S, Higuchi T, Asashi S, Yokoi T (2007) Species differences of inhibitory effects on P-glycoprotein-mediated drug transport. J Pharm Sci 96 (6):1609–1618

Takano R, Kataoka M, Yamashita S (2012) Integrating drug permeability with dissolution profile to develop IVIVC. Biopharm Drug Dispos 33(7):354–365

Takeda S (2006) Inhibition of UDP-glucuronosyltransferase 2b7-catalyzed morphine glucuronidation by ketoconazole: dual mechanisms involving a novel noncompetitive mode. Drug Metab Dispos 34(8):1277–1282

Thomas S, Brightman F, Gill H, Lee S, Pufong B (2008) Simulation modelling of human intestinal absorption using Caco-2 permeability and kinetic solubility data for early drug discovery. J Pharm Sci 97(10):4557–4574

Thummel KE (2002) The role of the gut mucosa in metabolically based drug-drug interaction. Drugs Pharm Sci 116(drug-drug interactions):359

Tsume Y, Langguth P, Garcia-Arieta A, Amidon GL (2012) In silico prediction of drug dissolution and absorption with variation in intestinal pH for BCS class II weak acid drugs: ibuprofen and ketoprofen. Biopharm Drug Dispos 33(7):366–377

van de Waterbeemd H, Smith DA, Jones BC (2001) Lipophilicity in PK design: methyl, ethyl, futile. J Comput Aided Mol Des 15(3):273–286

Ward KW, Stelman GJ, Morgan JA, Zeigler KS, Azzarano LM, Kehler JR, McSurdy-Freed JE, Proksch JW, Smith BR (2004) Development of an in vivo preclinical screen model to estimate absorption and first-pass hepatic extraction of xenobiotics. II. Use of ketoconazole to identify P-glycoprotein/CYP3A-limited bioavailability in the monkey. Drug Metab Dispos 32(2):172–177

Watanabe T, Kusuhara H, Maeda K, Kanamaru H, Saito Y, Hu Z, Sugiyama Y (2010) Investigation of the rate-determining process in the hepatic elimination of HMG-CoA reductase inhibitors in rats and humans. Drug Metab Dispos 38(2):215–222

Waterhouse RN (2003) Determination of lipophilicity and its use as a predictor of blood-brain barrier penetration of molecular imaging agents. Mol Imaging Biol 5(6):376–389

White RE (2011) Review of drug metabolism. Drug Discov Dev 1:3

Wickremsinhe E, Bao J, Smith R, Burton R, Dow S, Perkins E (2013) Preclinical absorption, distribution, metabolism, and excretion of an oral amide prodrug of gemcitabine designed to deliver prolonged systemic exposure. Pharmaceutics 5(2):261–276

Wu CY, Benet LZ (2005) Predicting drug disposition via application of BCS: transport/absorption/elimination interplay and development of a biopharmaceutics drug disposition classification system. Pharm Res 22(1):11–23

Xia CQ, Xiao G, Liu N, Pimprale S, Fox L, Patten CJ, Crespi CL, Miwa G, Gan LS (2006) Comparison of species differences of P-glycoproteins in beagle dog, rhesus monkey, and human using Atpase activity assays. Mol Pharm 3(1):78–86

Yalkowsky S (1999) Solubility in aqueous media. Oxford University Press, New York

Yengi LG, Leung L, Kao J (2007) The evolving role of drug metabolism in drug discovery and development. Pharm Res 24(5):842–858

Yoon IS, Choi MK, Kim JS, Shim CK, Chung SJ, Kim DD (2011) Pharmacokinetics and first-pass elimination of metoprolol in rats: contribution of intestinal first-pass extraction to low bioavailability of metoprolol. Xenobiotica 41(3):243–251

Yu L (2001) Amorphous pharmaceutical solids: preparation, characterization and stabilization. Adv Drug Deliv Rev 48(1):27–42

Yu LX, Straughn AB, Faustino PJ, Yang Y, Parekh A, Ciavarella AB, Asafu-Adjaye E, Mehta MU, Conner DP, Lesko LJ, Hussain AS (2004) The effect of food on the relative bioavailability of rapidly dissolving immediate-release solid oral products containing highly soluble drugs. Mol Pharm 1(5):357–362

Zara GP, Bargoni A, Cavalli R, Fundaro A, Vighetto D, Gasco MR (2002) Pharmacokinetics and tissue distribution of idarubicin-loaded solid lipid nanoparticles after duodenal administration to rats. J Pharm Sci 91(5):1324–1333

Zhang Y (1998) Effects of ketoconazole on the intestinal metabolism, transport and oral bioavailability of K02, a novel vinylsulfone peptidomimetic cysteine protease inhibitor and a P450 3A, P-glycoprotein dual substrate, in male Sprague-Dawley rats. J Pharmacol Exp Ther 287(1):246

Zhou R, Moench P, Heran C, Lu X, Mathias N, Faria TN, Wall DA, Hussain MA, Smith RL, Sun D (2005) pH-dependent dissolution in vitro and absorption in vivo of weakly basic drugs: development of a canine model. Pharm Res 22(2):188–192

Chapter 5
The Importance of Molecular Design Principles in Delivering High Quality Pharmaceutical Candidates

Thomas E. Prisinzano

5.1 Introduction

The successfully delivery of a high quality pharmaceutical candidate requires the combined efforts of many different scientific disciplines toward a common goal. Among these disciplines are molecular biology, behavioral pharmacology, and medicinal chemistry. Historically, medicinal chemistry has played an important role in the drug development process. This is based on its foundation in organic chemistry or the art and science of constructing molecules. The discipline of medicinal chemistry is similar in this regard that it places an emphasis on constructing molecules. However, it has the added art and science of knowing which molecules are worth making for medicinal purposes. In general, similar molecules exert similar biological activities. However, each biological target has its own idiosyncrasies and challenges that provide exceptions to this otherwise widely held belief. To the untrained organic chemist, all molecules look like drugs. To a medicinal chemist, it is clear from experience that most molecules are useless medicinally.

Drug discovery has become a lengthy and expensive process. The time required to deliver a new drug generally exceeds ten years and is estimated to be more than $1 billion in the United States. Therefore, time is expensive and Competition is increasingly difficult. Historically, the role of the medicinal chemist has been to repair deficiencies in the molecular properties through the synthesis and biological testing of analogs. In addition, the medicinal chemist has been tasked to provide backup candidates to increase the likelihood of success. Presently, the role of the medicinal chemist is evolving and now often encompasses the synthesis of probe molecules to further the understanding of a given biological target. Medicinal

T.E. Prisinzano (✉)
Department of Medicinal Chemistry, University of Kansas School of Pharmacy, 1251 Wescoe Hall Dr., 4070 Malott Hall, Lawrence, KS 66045, USA
e-mail: prisinza@ku.edu

A.C. Templeton et al. (eds.), *Discovering and Developing Molecules with Optimal Drug-Like Properties*, AAPS Advances in the Pharmaceutical Sciences Series 15,
DOI 10.1007/978-1-4939-1399-2_5

chemists now routinely prepare fluorescent analogs and other labeled derivatives to assist screening campaigns as well as further elucidate the biology associated with a given target.

Hits and leads rarely have sufficient properties for direct clinical use or for further pharmaceutical development. The molecular transformations needed to improve a given hit or lead are dependent on the particular deficiency in the hit or lead. Unfortunately, there is no single approach that works in all cases. Consequently, one must be familiar with general approaches and their proper employment. While somewhat daunting, it does provide the opportunity for creativity and individuality. It is not uncommon for multiple solutions to the same deficiency from different medicinal chemists.

5.2 Current Thinking on High Throughput Screening

High throughput screening of large chemical libraries of compounds is a proven way to identify novel chemical entities that target a biological system of interest. This technology is being used by pharmaceutical industry and a growing number of academic screening centers for drug discovery campaigns. While these types of campaigns are often successful in identifying agents for a given molecular target, many of the components of screening libraries have poor drug-like properties. This often hinders further development of the hit or lead through medicinal chemistry. A seasoned medicinal chemist is able to identify the tractability of a high throughput screening hit upon quick examination of the chemical structure. Key features that will be ascertained are potential metabolic liabilities, reactive groups, or functionality that may be unstable under physiological or assay conditions. Furthermore, the potential synthetic route to analogs will be envisioned. This may also play in the decision making process for potential prosecution.

The characteristics of a high throughput screening library have evolved as the technology has become more mature. Initially, libraries were constructed based on diversity. The greater the diversity of the library, the more likely one was to find a novel hit for a given biological target. The thought behind this type of library construction was that this strategy would maximize diversity in chemical space and any functional deficiencies in the members of the library could be addressed through medicinal chemistry optimization. Typical components of this type of library are flat aromatic compounds that can be prepared using well-established palladium coupling methods or through other simple chemical steps. However, enthusiasm for this type of library construction has cooled because it has been noted that it is often very difficult to optimize the pharmaceutical properties of certain types of molecules present in a diversity focused library.

A more recent type of library has emerged where drug-like properties are favored. In this type of library construction, libraries are constructed where drug-like properties are favored. The idea behind this type of library is that drug-like properties are the most important factor and that other key parameters such as potency and selectivity can be retrofit later through medicinal optimization.

Generally, this type of library is thought to be productive due to its attention on drug design principles.

One final type of library design is that where computational chemistry plays an important role. In this type of library design, it is assumed that rational design principles can be calculated and applied to library construction. This type of library construction assumes that potency, selectivity, and drug-like properties can be added later through a medicinal chemistry optimization campaign. Enthusiasm for this type of high throughput screening library continues to grow.

5.3 Pharmaceutical Candidate vs. Biological Probe

In addition to its clear role in pharmaceutical development, medicinal chemistry, may be used further to provide probe compounds to further interrogate biology (Bunnage et al. 2013). The same techniques used to develop pharmaceutical candidates can be used to enhance biological probes to assist in target validation. However, there are many differences in the criteria for what one would consider developing a drug compared to a useful biological probe.

As can be seen in Fig. 5.1, developing a drug is far more complex than a biological probe. One can see that for a successful drug one needs a mixture of potency, efficacy, selectivity, and toxicity. Additional pharmaceutical development concerns relate to stability and formulation. In developing a drug for market, one must also consider the financial aspects of the drug. Thus, it is clear that one must juggle a number of factors and make a successful drug. Unfortunately, each one of these factors has a tendency to be codependent on one another. This makes solving any potential issues difficult as when one solves one problem or shortcoming, another deficiency forms based on the previous solution. In this regard, it is very similar to trying to solve a Rubik's Cube®.

The criteria for a biological probe are considerably less stringent than that for a drug. There are some key overlaps including potency and selectivity. Other key considerations directly relevant to the pharmaceutical scientist are aqueous

Drug = ***a[potency] + b[efficacy] + c[selectivity]***+d[toxicity] + e[absorption] + f[distribution] + g[metabolism] + h[excretion] + i[protein binding] + j{acute toxicity] + k[chronic toxicity] + l[mutagenicity] + m[stability] + n[accessibility] + o[cost] + p[patentability] + q[clinical efficacy] + r[solubility] + s[taste] + t[formulability] + u[idiosyncratic problem]

Probe =
cellular potency < 1 μM
biochemical potency < 100 nM
selectivity > 100-fold
aqueous solubility
active or passive transport
reversible or covalent mechanism

Fig. 5.1 The difference between a drug and chemical probe

solubility and membrane permeability. A key difference between a drug and probe is the acceptability of a covalent mechanism of action. Generally, one prefers a reversible mechanism of action to a covalent mechanism of action due to the reduced potential toxic side effects through protein modification. This is not as great a concern for cancer therapeutics where causing cellular death is generally viewed as a positive biological outcome. However, this widely held belief is being reexamined as covalent drugs are being explored for the treatment of a number of other conditions (Liu et al. 2013).

As suggested earlier the development of a chemical probe or high quality pharmaceutical candidate relies heavily on the intuition and knowledge of a skilled medicinal chemist. Key to the development will be the utilization of the molecular design principles needed to favorably alter the characteristics of a given molecule.

5.4 Molecular Design Principles

The essence of medicinal chemistry is to know the relationship between structure and molecular properties so that a molecule may be designed to fit a particular need. This section deals with the common design principles used to solve molecular deficiencies. While there is a growing number of computational metrics that can used to help guide optimization (Lipinski et al. 2001; Meanwell 2011), it is often not clear what changes one could or should make to alter the properties of a given molecule. Often even if one knows the computed values necessary to remove a given deficiency, it is difficult to find a known arrangement of atoms to satisfy the calculated metrics. However, there is no doubt that these types of metrics are useful in the optimization process and will have an increasing level of influence in the design process (Hann and Keseru 2012).

There are many examples of structural modifications that have been used successfully to alter the molecular properties of pharmaceutical candidates and chemical probes. Unfortunately, there are too many to efficiently tabulate here. However, experience is an excellent teacher and what has worked previously for one series of molecules is likely to work for a new series with the same type of deficiency. While this may not always work, it can certainly help serve as a starting point for an optimization campaign through medicinal chemistry. For these structural modifications to be implemented optimally, one should first have identified the pharmacophore and auxophore for a given molecule.

The pharmacophore is the minimum structural unit of a molecule that interacts with its associated molecular target and retains its pharmacological response. A complete molecular dissection will identify a given pharmacophore. Generally, the pharmacophore is usually much less active and selective than a complete pharmaceutical candidate. Alterations to the chemical structure of a pharmacophore lead to profound decreases in biological activity. The auxophore is the nonessential portion of a molecule that supports the pharmacophore and modulates the pharmacokinetic properties of a molecule. Generally, there is a significant range of modifications tolerated in this part of the molecule.

If the pharmacophore for a given hit or lead has not been identified, this should be done first. Preparing simple derivatives that alter the functional groups present in the molecule and observing how this affects the biological activity should achieve this quickly. For example, carboxyl groups can be esterified and alcohol or amino groups may be acetylated. These modifications have the added value of potentially identifying opportunities for future prodrug campaigns. With the biological results in hand, one can proceed to delete functional groups through synthesis. Generally, this sequence decreases potency and/or selectivity. However, this loss in biological activity may be overcome through the introduction of new functional groups.

As mentioned earlier, a previously described structural modification may or may not be effective in optimizing a new chemotype. However, there are several design principles that have been used extensively and have a high probability of success. These include (1) molecular dissection, (2) conformational constraint, (3) use of bioisosteres, (4) introduction of fluorine, and (5) modification of hydrogen bonding. In molecular dissection, surplus atoms are removed. At its extreme, this technique is useful in identifying the pharmacophore and auxophore. It is also useful in simplifying the synthetic route to a given series of molecules and avoiding potential problems with stereochemical construction (Schaeffer et al. 1971). Conformational constraint is the process where conformational flexibility is removed in a given ligand (Smissman et al. 1966). It can be viewed as the opposite of molecular dissection. The guiding principle is that if the molecule can be preorganized into a suitable conformation then the entropy of association with a given biological target will be reduced. However, this technique assumes that the atoms used to rigidify a molecule are not forbidden by space limitations at the biological target. In some cases, rigidification has been shown to greatly enhance potency and/or selectivity. Bioisosteres are atoms or groups of atoms that may be interchanged in a given molecule without a significant loss in potency. Proper bioisosterism allows simplification of synthesis and an opportunity to manipulate the ADMET properties of a given molecule. This technique is considered to be the most commonly used drug design technique (Hamada and Kiso 2012). Installation of a strategic fluorine atom very often has a beneficial effect on molecules (Purser et al. 2008). The reasons for this positive effect are not always clear. A fluorine atom is larger than a hydrogen atom and can lead to conformational preferences. However, a fluorine atom has significant polarity and this may be responsible for its beneficial effects. Finally, hydrogen bonding plays a significant role in many drug properties. The addition or deletion of a hydrogen donor or hydrogen bond acceptor can often have a profound effect on the pharmaceutical properties of a molecule. Intramolecular hydrogen bonds generally decrease water solubility, decrease cLogP, and increase uptake. However, there are complicating factors that include the ratio of intra and intermolecular hydrogen bonds and the strength of a given hydrogen bond. In the discussion below, a series of studies using the natural product salvinorin A will be used to highlight several important molecular design principles. These examples were selected based on their potential for wide applicability rather than their individual success.

5.5 Case Study: Salvinorin A as a Kappa Opioid Agonist

Salvinorin A is a neoclerodane diterpene and the major active component of the hallucinogenic mint plant *Salvia divinorum* Epling & Játiva (Lamiaceae). Since the finding that salvinorin A exerts its potent psychotropic actions through the activation of opioid receptors, the site of action of morphine and related analogs, there has been much interest in elucidating the underlying mechanisms behind its effects. These effects are particularly remarkable, because (1) salvinorin A is the first reported nonnitrogenous opioid receptor agonist, and (2) its effects are not mediated by 5-HT_{2A} receptors, the classical target of hallucinogens such as LSD and mescaline (Roth et al. 2002). Rigorous investigation into the structural features of salvinorin A responsible for opioid receptor affinity and selectivity has produced numerous receptor probes, affinity labels, and tools for evaluating the biological processes responsible for its observed psychological effects (Cunningham et al. 2011).

As a kappa opioid receptor agonist, salvinorin A has the potential to treat a number of CNS disorders including drug abuse and pain (Kivell and Prisinzano 2010). Given its unique structure relative to other opioid agonists, this provided a distinct opportunity for the identification of an opioid with reduced side effects relative to morphine. In addition, this unique structure also suggested that salvinorin A is interacting in a novel way at the kappa receptor compared to other known opioids.

Upon closer examination of salvinorin A, there are also a number of structural liabilities that potentially hinder its further development (Fig. 5.2). First, the natural product contains a number of ester moieties that are likely to undergo metabolism by plasma esterases. This signaled that the half-life was likely to be short. Second, salvinorin A has no ionizable groups. This suggested water solubility was likely to be poor similar to many high throughput screening hits. Finally, the natural product contained a furan ring. This heterocycle was easily identified as a metabolic liability due to its potential for forming reactive metabolites. Collectively, these liabilities were addressed through analog preparation using several different molecular design principles: (1) molecular dissection, (2) conformation constraint, (3) use of bioisosteres, (4) introduction of fluorine, and (5) modification of hydrogen bonding.

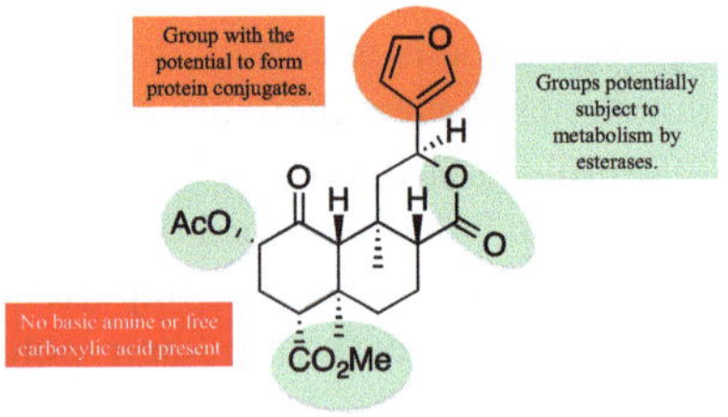

Fig. 5.2 Analysis of salvinorin A using medicinal chemistry design principles

5.5.1 Molecular Dissection

It is widely known in medicinal chemistry that furan rings should be avoided in drug development campaigns due to their potential for forming reactive metabolites (Dalvie et al. 2002). As a furan containing natural product, salvinorin A also possesses the potential for reactive metabolites. Previous studies showed that teucrin A, a structurally natural product found in germander (*Teucrium chamaedrys* L.; Lamiaceae), produced hepatotoxicity in humans likely resulting from the formation of a reactive metabolite by cytochrome P450 enzymes (Kouzi et al. 1994; Peterson 2013).

One way to mitigate this potent metabolic liability is to remove the furan completely. This was accomplished as described in Fig. 5.3. Treatment of salvinorin A with a mixture of $NaIO_4$ and $RuCl_3{\cdot}3H_2O$ gave the corresponding carboxylic acid. The conversion of the carboxylic acid to the corresponding selenoester was accomplished with selenophenol after activation of the carboxyl group with $PhOPOCl_2$. Finally, reaction of the selenoester with tributyltin hydride and AIBN afforded the corresponding desfuryl analog. On the basis of its structure, the desfuryl analog would be expected to have greatly reduced potential for reactive intermediates. However, this analog has greatly reduced activity at the kappa opioid receptor (Simpson et al. 2007). Therefore, other strategies need to be utilized to mitigate this deficiency and retain biological activity.

5.5.2 Conformational Constraint

One analog found to be considerably more potent than salvinorin A was EOM Sal B (Munro et al. 2008). It should be noted that the C-2 ether moiety is flexible and able to adopt a number of conformations when interacting at the kappa opioid receptor.

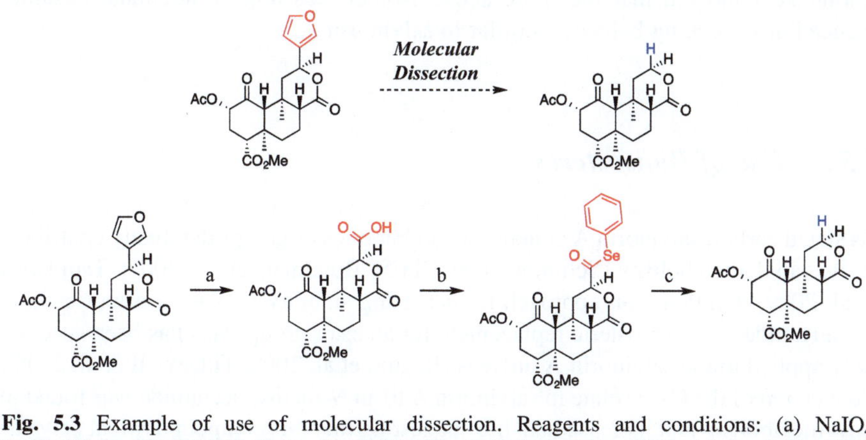

Fig. 5.3 Example of use of molecular dissection. Reagents and conditions: (a) $NaIO_4$, $RuCl_3{\cdot}3H_2O$, $CH_3CN/CCl_4/H_2O$; (b) $PhOPOCl_2$, C_6H_5SeH, NEt_3, THF; (c) Bu_3SnH, AIBN, Toluene

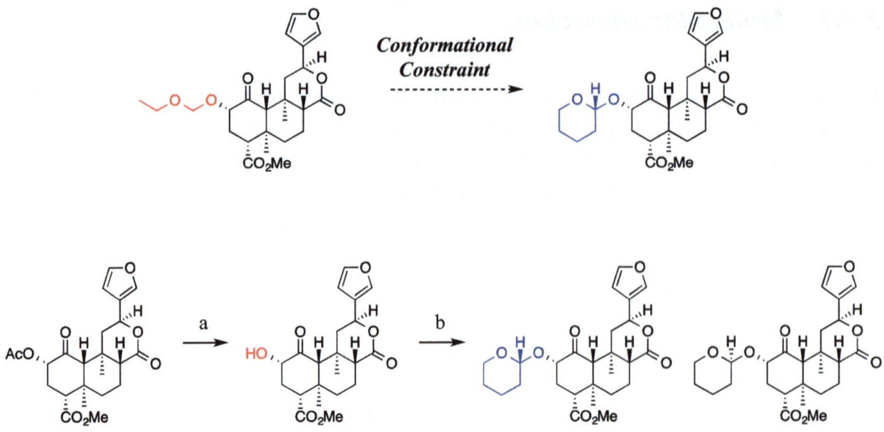

Fig. 5.4 Example of use of conformational constraint. Reagents and conditions: (a) Na_2CO_3, MeOH; (b) (i) Dihydropyran, PPTS, CH_2Cl_2; (ii) HPLC, CH_3CN/H_2O

One way to gain insight into this observation is to use conformation constraint (Smissman et al. 1966). In this technique, the flexible C-2 ether is constrained into a ring of varying sizes trying to identify the compound with optimal potency (Prevatt-Smith et al. 2011). When the ethoxymethyl ether is constrained into a 6-membered tetrahydropyran ring, a new stereocenter is formed. This then suggests that one of the two possible epimers is likely more active than the other. Both epimers were prepared as outlined in Fig. 5.4. The methanolysis of salvinorin A under basic conditions afforded the desacetyl derivative, salvinorin B. Treatment of salvinorin B with dihydropyran under acidic conditions gave the corresponding tetrahydropyran analogs. The absolute stereochemistry for each epimer was determined using a combination of spectroscopic methods and X-ray crystallography (Prevatt-Smith et al. 2011). Biological evaluation of both compounds revealed that a hydrogen in the β position or R configuration was preferential. Additional in vivo testing also showed that the more active isomer was able to attenuate cocaine-induced drug seeking behavior similar to salvinorin A.

5.5.3 *Use of Bioisosteres*

As stated earlier, salvinorin A contains a number of ester groups that likely contribute to its rapid metabolism (Schmidt et al. 2005; Butelman et al. 2007; Tsujikawa et al. 2009). On common approach to increasing stability to plasma esterases is to use an amide as a bioisosteric replacement for an ester group. This has been successfully applied to the salvinorin A nucleus (Beguin et al. 2008; Tidgewell et al. 2008). Conversion of the C-2 acetate in salvinorin A to an *N*-methyl acetamide was found to slow metabolism and increase half-life suggesting that this compound might serve as useful probe for kappa opioid receptors in animals and humans (Beguin et al. 2008).

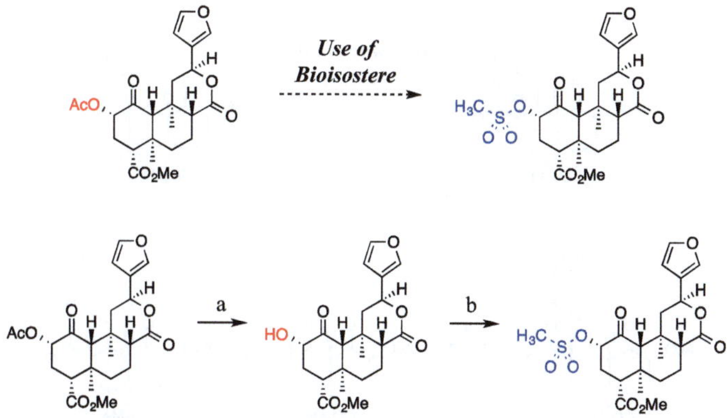

Fig. 5.5 Example of use of bioisostere. Reagents and conditions: (a) Na_2CO_3, MeOH; (b) CH_3SO_2Cl, NEt_3, CH_2Cl_2

Another option to potentially increase metabolic stability of an acetyl group is to convert it into a mesylate ester (Fig. 5.5). Methanolysis of salvinorin A under basic conditions afforded salvinorin B. Treatment of salvinorin B with methanesulfonyl chloride gave the corresponding C-2 mesylate group. This molecule possesses activity similar to salvinorin A and has been found to have an increased duration of antinociceptive activity in rodents (Harding et al. 2005; Simonson et al. 2014).

5.5.4 Introduction of Fluorine

The introduction of fluorinated functional groups to a given molecule often perturbs the biophysical properties and imparts unique characteristics that distinguish the fluorinated analog from the parent compound. This design strategy is widely employed in medicinal chemistry to improve the pharmacokinetic properties of a given molecule. However, the introduction of fluorine to molecules is not trivial and there are few synthetic methods available to do so. This is an area of increasing attention in organic chemistry and newer methods are becoming available (Ji et al. 2011; Nagib and MacMillan 2011; Fujiwara et al. 2012). In particular, the incorporation of a trifluoromethyl group can enhance potency by promoting electrostatic interactions with a biological target, improving cellular membrane permeability, and decreasing metabolism by cytochromes P450 enzymes by deactivating the aromatic ring (Wiley et al. 1979).

As mentioned earlier, the presence of a furan ring is a potential metabolic liability and could potentially hinder development of salvinorin A based biological probes or a pharmaceutical candidate. It was also clear that removal of the furan ring greatly decreases potency (Simpson et al. 2007). Another potential design strategy would be to incorporate a trifluoromethyl group into the furan ring of salvinorin A. This would have the benefit of keeping the furan present but would be

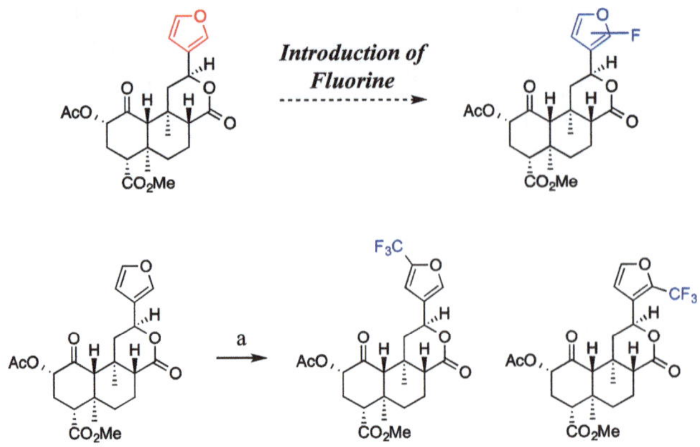

Fig. 5.6 Example of introduction of fluorine. Reagents and conditions: (a) K_2HPO_4, CF_3SO_2Cl, $RuCl_2(phen)H_2O$, CH_3CN, Light, 48 h

expected to render the furan ring less reactive toward oxidative metabolism by cytochromes P450 enzymes. As shown in Fig. 5.6, a trifluoromethyl group can be added to the furan ring of salvinorin A in a one step procedure using photocatalysis (Nagib and MacMillan 2011). This particular reaction gives a mixture of regioisomers but this is advantageous if one does not know the optimal location for the trifluoromethyl group. Unfortunately, initial biological testing suggests that the incorporation of a trifluoromethyl group to the furan ring of salvinorin A is not well tolerated.

5.5.5 *Modification of Hydrogen Bonding*

Electrostatic interactions generally help orient ligands to interact with their given biological target. These types of interactions are quite strong and have a range of 34–48 kJ/mol. Hydrogen bonds are also important and are relatively strong once dehydration and rehydration are taken into consideration. One should note that hydrogen bonds are strongly directional and are often mediated through chains of associated water molecules. The energy range for a hydrogen bond is from 5 to 14 kJ/mol. However, this may be considerably stronger if the hydrogen bond is charge assisted. Generally, introducing more polar functionality to a given molecule may increase aqueous solubility. One way to do this is to add functionality that is capable of hydrogen bonding. However, one must note that the number of hydrogen bonds in a molecule plays an important role in determining its oral bioavailability (Lipinski et al. 2001; Veber et al. 2002).

One way to alter the physical properties of a molecule is to modify an existing hydrogen bonding group present in the molecule. One example is the furanyl oxygen in salvinorin A. This particular group has been suggested to play a key role in the binding of salvinorin A to the kappa opioid receptor (Vardy et al. 2013).

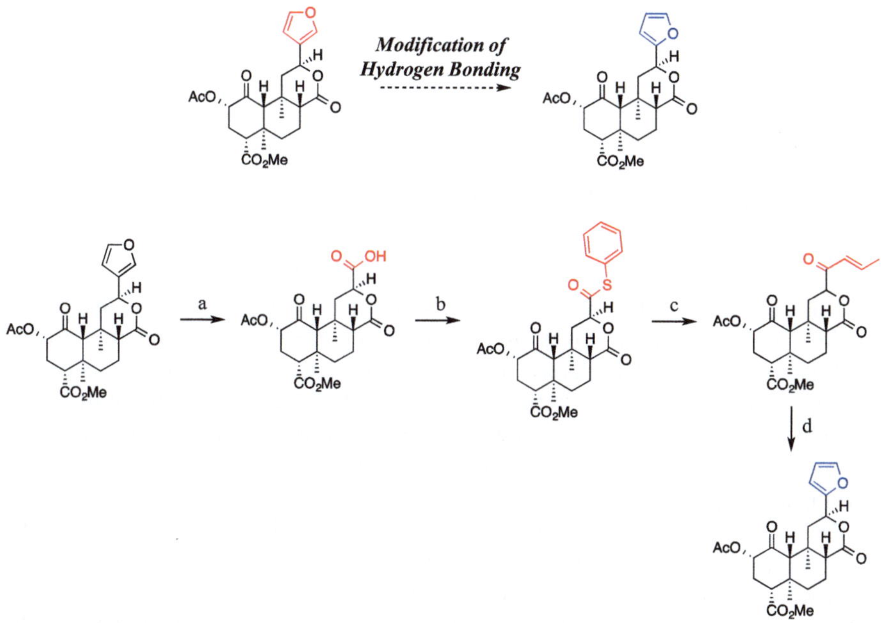

Fig. 5.7 Example of use of modification of hydrogen bonding. Reagents and conditions: (a) $NaIO_4$, $RuCl_3{\cdot}3H_2O$, $CH_3CN/CCl_4/H_2O$; (b) (i) CDMT, NMM, CH_2Cl_2; (ii) NMM, PhSH, CH_2Cl_2; (c) $RB(OH)_2$, CuTC, $Pd(PPh_3)_4$, THF; (d) Se_2O, C_6H_5Cl

One way to probe the importance of this interaction as well as alter the physical properties of salvinorin A is to manipulate this hydrogen bonding group. Ways that one could consider manipulating this hydrogen bonding group would be to add either an electron donating group or an electron withdrawing group to the furan ring of salvinorin A. This would be expected to potentially increase or decrease the hydrogen bonding potential of the furanyl oxygen. Another approach would be to change the location of the furanyl oxygen relative to the carbon backbone. This would be expected to either increase or decrease potency based on either helping or hurting the directionality of the critical hydrogen bond from the furan ring to the kappa opioid receptor.

The synthesis of the 2-furanyl analog of salvinorin A is shown in Fig. 5.7 (Lovell et al. 2012). Treatment of salvinorin A with a mixture of $NaIO_4$ and $RuCl_3{\cdot}3H_2O$ gave the corresponding carboxylic acid. The conversion of the carboxylic acid to the corresponding thioester was accomplished with thiophenol after activation of the carboxyl group with CDMT.

Using palladium catalysis and the appropriate boronic acid, the thioester was converted to the corresponding alkenyl ketone in good yield. Conversion of the alkenyl ketone to the 2-furan was accomplished by refluxing with SeO_2 in bromobenzene.

Biological testing showed that the 2-furanyl analog had similar activity at kappa opioid receptors compared to salvinorin A. Interestingly, the 2-furanyl analog had a melting point of 198–200 °C. This is considerably lower than that described for salvinorin A (Ortega et al. 1982). It also illustrates that the modification of a hydrogen bonding group can have a significant effect on the physical properties of a molecule.

5.6 Summary and Perspective

As described earlier, hits and leads rarely have sufficient properties for direct clinical use or for further pharmaceutical development. Unfortunately, there is no single approach that improves all molecular deficiencies in a pharmaceutical candidate in all cases. Ultimately, the delivery of high quality candidates is dependent upon the talents of a multidisciplinary team. One key component of that team is a skilled medicinal chemist whose creativity, intuition, and application of molecular design principles is critical to identify compounds that can be successfully developed.

The goal of preparing high quality pharmaceutical candidates quickly and cheaply has remained largely elusive likely due to the complexity of the endeavor. Even with the diversity of structures exhibiting affinity and activity at a growing number of biological targets, there is still ample opportunity for improvement. The components of many high throughput screening libraries are not ideal due to their lack of attention paid to important development parameters such as aqueous solubility. However, this is changing as chemical libraries are now being developed with greater attention being paid to drug-like properties. In addition, there is a growing influence of computational techniques to assist library development.

While molecular design principles have traditionally been used to develop high quality pharmaceutical candidates, they can also be used to improve chemical probes. This is a growing trend in medicinal chemistry as the role of a medicinal chemist continues to evolve. There are clear differences in the criteria needed for a drug versus that of a chemical probe. However, there is generally an overemphasis on compound potency in many drug seeking campaigns (Gleeson et al. 2011). In such campaigns, potency is driven to the exclusion of all other criteria needed for development. The general thinking is that if the potency is high enough the window for optimizing other properties is large and should be relatively trivial to do. However, in vitro potency does not always correlate with therapeutic dose (Gleeson et al. 2011). While this is no doubt that potency is an important characteristic, one must keep in mind it is not the only criteria needed to successfully advance a drug molecule or chemical probe. Attention must also be paid to solubility, permeability, and toxicity. Often the candidate that advances the farthest in a given series is not the most potent analog. To develop a high quality pharmaceutical candidate, one must take a balanced approach and keep in mind that success in one parameter likely means a deficiency in a different parameter. This holistic approach is also likely to benefit chemical probe development. While the criteria for a chemical probe are relaxed relative to a drug, potency should not be driven to the exclusion of other parameters. More soluble and permeable chemical probes are likely to be of greater utility and will increase the type of experiments that can be conducted with the probe.

The key to effective use of molecular design principles is to know the relationship between chemical structure and molecular properties. Once this relationship has been established, one may effectively design a molecule with enhanced properties. To maximize the effectiveness of molecular design principles, one must know both the pharmacophore and auxophore for a chemical series. Unfortunately even with this knowledge in hand, a given structural modification may or may not be effective in overcoming a molecular deficiency. While there are a growing number of computational metrics that can used to help guide optimization (Lipinski et al. 2001; Meanwell 2011), it is often not clear one should make to give the desired result. However, there are several design principles that have been used extensively and have a high probability of success. These include (1) molecular dissection, (2) conformational constraint, (3) bioisosteres, (4) introduction of fluorine, and (5) modification of hydrogen bonding. Successful implementation of these design principles relies on an understanding of both the strengths and weaknesses of a given design strategy. One must also keep in mind that there are no guarantees that a given design strategy will work in all cases. Experience and chemical intuition are key components of any optimization campaign. It is not uncommon for different medicinal chemists using the same design principle to come up with diverse but effective solutions.

In closing, the delivery of high quality pharmaceutical candidates is a daunting process. Multiple parameters must be juggled and optimizing any single one may be at the expense of another. Success requires judgment and the ability to find the best overall compromises quickly. Experience suggests that in some chemical series certain factors are trivial and can be neglected. It is therefore important to identify the most productive parameter to manipulate keeping in mind that this is not always potency. Successful implementation of the molecular design principles described earlier is predicated on a solid understanding of the influence chemical structure has on these parameters. With this knowledge in hand, one will be able to delivery high quality pharmaceutical candidates as well as chemical probes.

Acknowledgements This chapter is based in part on a short course in Contemporary Medicinal Chemistry given by Professors Lester A. Mitscher and Thomas E. Prisinzano provided by the International Organization for Chemical Sciences in Development (IOCD). The author wishes to thank the National Institute on Drug Abuse (DA018151) for financial support of the medicinal chemistry described in the case study. The content is the sole responsibility of the author and does not necessarily represent the official views of the National Institute on Drug Abuse, National Institutes of Health.

References

Beguin C, Potter DN, Dinieri JA, Munro TA, Richards MR, Paine TA, Berry L, Zhao Z, Roth BL, Xu W, Liu-Chen LY, Carlezon WA Jr, Cohen BM (2008) *N*-methylacetamide analog of salvinorin A: a highly potent and selective kappa-opioid receptor agonist with oral efficacy. J Pharmacol Exp Ther 324(1):188–195

Bunnage ME, Chekler EL, Jones LH (2013) Target validation using chemical probes. Nat Chem Biol 9(4):195–199

Butelman ER, Mandau M, Tidgewell K, Prisinzano TE, Yuferov V, Kreek MJ (2007) Effects of salvinorin A, a kappa-opioid hallucinogen, on a neuroendocrine biomarker assay in nonhuman primates with high kappa-receptor homology to humans. J Pharmacol Exp Ther 320 (1):300–306

Cunningham CW, Rothman RB, Prisinzano TE (2011) Neuropharmacology of the naturally occurring kappa-opioid hallucinogen salvinorin A. Pharmacol Rev 63(2):316–347

Dalvie DK, Kalgutkar AS, Khojasteh-Bakht SC, Obach RS, O'Donnell JP (2002) Biotransformation reactions of five-membered aromatic heterocyclic rings. Chem Res Toxicol 15 (3):269–299

Fujiwara Y, Dixon JA, Rodriguez RA, Baxter RD, Dixon DD, Collins MR, Blackmond DG, Baran PS (2012) A new reagent for direct difluoromethylation. J Am Chem Soc 134(3):1494–1497

Gleeson MP, Hersey A, Montanari D, Overington J (2011) Probing the links between in vitro potency, ADMET and physicochemical parameters. Nat Rev Drug Discov 10(3):197–208

Hamada Y, Kiso Y (2012) The application of bioisosteres in drug design for novel drug discovery: focusing on acid protease inhibitors. Expert Opin Drug Discov 7(10):903–922

Hann MM, Keseru GM (2012) Finding the sweet spot: the role of nature and nurture in medicinal chemistry. Nat Rev Drug Discov 11(5):355–365

Harding WW, Tidgewell K, Byrd N, Cobb H, Dersch CM, Butelman ER, Rothman RB, Prisinzano TE (2005) Neoclerodane diterpenes as a novel scaffold for mu opioid receptor ligands. J Med Chem 48(15):4765–4771

Ji Y, Brueckl T, Baxter RD, Fujiwara Y, Seiple IB, Su S, Blackmond DG, Baran PS (2011) Innate C-H trifluoromethylation of heterocycles. Proc Natl Acad Sci U S A 108(35):14411–14415

Kivell B, Prisinzano TE (2010) Kappa opioids and the modulation of pain. Psychopharmacology (Berl) 210(2):109–119

Kouzi SA, McMurtry RJ, Nelson SD (1994) Hepatotoxicity of germander (*Teucrium chamaedrys* L.) and one of its constituent neoclerodane diterpenes teucrin A in the mouse. Chem Res Toxicol 7(6):850–856

Lipinski CA, Lombardo F, Dominy BW, Feeney PJ (2001) Experimental and computational approaches to estimate solubility and permeability in drug discovery and development settings. Adv Drug Deliv Rev 46(1–3):3–26

Liu Q, Sabnis Y, Zhao Z, Zhang T, Buhrlage SJ, Jones LH, Gray NS (2013) Developing irreversible inhibitors of the protein kinase cysteinome. Chem Biol 20(2):146–159

Lovell KM, Vasiljevik T, Araya JJ, Lozama A, Prevatt-Smith KM, Day VW, Dersch CM, Rothman RB, Butelman ER, Kreek MJ, Prisinzano TE (2012) Semisynthetic neoclerodanes as kappa opioid receptor probes. Bioorg Med Chem 20(9):3100–3110

Meanwell NA (2011) Improving drug candidates by design: a focus on physicochemical properties as a means of improving compound disposition and safety. Chem Res Toxicol 24 (9):1420–1456

Munro TA, Duncan KK, Xu W, Wang Y, Liu-Chen LY, Carlezon WA Jr, Cohen BM, Beguin C (2008) Standard protecting groups create potent and selective kappa opioids: salvinorin B alkoxymethyl ethers. Bioorg Med Chem 16(3):1279–1286

Nagib DA, MacMillan DW (2011) Trifluoromethylation of arenes and heteroarenes by means of photoredox catalysis. Nature 480(7376):224–228

Ortega A, Blount JF, Manchand PS (1982) Salvinorin, a new trans-neoclerodane diterpene from salvia-divinorum (Labiatae). J Chem Soc Perkin Trans 1(10):2505–2508

Peterson LA (2013) Reactive metabolites in the biotransformation of molecules containing a furan ring. Chem Res Toxicol 26:6–25

Prevatt-Smith KM, Lovell KM, Simpson DS, Day VW, Douglas JT, Bosch P, Dersch CM, Rothman RB, Kivell B, Prisinzano TE (2011) Potential drug abuse therapeutics derived from the hallucinogenic natural product salvinorin A. Medchemcomm 2(12):1217–1222

Purser S, Moore PR, Swallow S, Gouverneur V (2008) Fluorine in medicinal chemistry. Chem Soc Rev 37(2):320–330

Roth BL, Baner K, Westkaemper R, Siebert D, Rice KC, Steinberg S, Ernsberger P, Rothman RB (2002) Salvinorin A: a potent naturally occurring nonnitrogenous kappa opioid selective agonist. Proc Natl Acad Sci U S A 99(18):11934–11939

Schaeffer HJ, Gurwara S, Vince R, Bittner S (1971) Novel substrate of adenosine deaminase. J Med Chem 14(4):367–369

Schmidt MD, Schmidt MS, Butelman ER, Harding WW, Tidgewell K, Murry DJ, Kreek MJ, Prisinzano TE (2005) Pharmacokinetics of the plant-derived kappa-opioid hallucinogen salvinorin A in nonhuman primates. Synapse 58(3):208–210

Simonson B, Morani AS, Ewald AW, Walker L, Kumar N, Simpson D, Miller JH, Prisinzano TE, Kivell BM (2014) Pharmacology and anti-addiction effects of the novel κ opioid receptor agonist Mesyl Sal B, a potent and long-acting analogue of salvinorin A. Br J Pharmacol doi: 10.1111/bph.12692

Simpson DS, Katavic PL, Lozama A, Harding WW, Parrish D, Deschamps JR, Dersch CM, Partilla JS, Rothman RB, Navarro H, Prisinzano TE (2007) Synthetic studies of neoclerodane diterpenes from Salvia divinorum: preparation and opioid receptor activity of salvinicin analogues. J Med Chem 50(15):3596–3603

Smissman EE, Nelson WL, LaPidus JB, Day JL (1966) Conformational aspects of acetylcholine receptor sites. The Isomeric 3-Trimethylammonium-2-acetoxy-trans-decalin Halides1 and the Isomeric α, β-Dimethylacetylcholine Halides2. J Med Chem 9(4):458–465

Tidgewell K, Groer CE, Harding WW, Lozama A, Schmidt M, Marquam A, Hiemstra J, Partilla JS, Dersch CM, Rothman RB, Bohn LM, Prisinzano TE (2008) Herkinorin analogues with differential beta-arrestin-2 interactions. J Med Chem 51(8):2421–2431

Tsujikawa K, Kuwayama K, Miyaguchi H, Kanamori T, Iwata YT, Inoue H (2009) In vitro stability and metabolism of salvinorin A in rat plasma. Xenobiotica 39(5):391–398

Vardy E, Mosier PD, Frankowski KJ, Wu H, Katritch V, Westkaemper RB, Aube J, Stevens RC, Roth BL (2013) Chemotype-selective modes of action of kappa-opioid receptor agonists. J Biol Chem 288(48):34470–34483

Veber DF, Johnson SR, Cheng HY, Smith BR, Ward KW, Kopple KD (2002) Molecular properties that influence the oral bioavailability of drug candidates. J Med Chem 45(12):2615–2623

Wiley RA, Hanzlik RP, Gillesse T (1979) Effect of substituents on in vitro metabolism and covalent binding of substituted bromobenzenes. Toxicol Appl Pharmacol 49(2):249–255

Part II
Predictive Approaches to Establishing, Understanding, and Communicating Risk in Early Development

Chapter 6
Predictive Approaches to Establishing, Understanding, and Communicating Risk with Emphasis on Early Development

Stephen R. Byrn

To understand risk, the solid-state chemistry of the API and drug product must be fully understood. ICHQ6A outlines the information required for full characterization of the drug substance and drug product. It also discusses stability studies and how those relate to understanding API properties and drug product properties. This guidance suggests that the scientist should determine what solid form is present in the drug substance (API) and drug product. It directs the manufacturer to "know what they have." Clearly, knowing what you have mitigates risk. Additionally, the ICHQ8 guidance on development and the ICHQ9 guidance on risk require a firm understanding of how the medicine was developed and any risks involved.

In the early stages of development, often the time and quantities of API are not available to allow full analysis of the solid-state chemistry of the compound (API). In these cases, determination of a partial or abbreviated solid-state chemistry landscape is in order. For example, knowledge of the API solid-state chemistry, as presented, with some simple screening of the solid-state chemistry space may be appropriate. This strategy is outlined in the second chapter in this section (McNevin and Higgins). In many cases, it is advantageous to use the most stable form for experiments at this early stage in order to reduce the risk of transformation during a toxicological or clinical trial. Then, a process of preparing this form at large enough scale to perform toxicology and ultimately human experiments must be developed. The solvents used in the crystallization experiments will usually provide methods to prepare these larger amounts of materials. As more API and time become available then additional screening can be performed.

Determination of the solid-state chemistry landscape is even more important when a poorly soluble drug is under development. In these cases, the solid form of the API and the solid form and formulation in the drug product determine apparent

S.R. Byrn (✉)
Department of Industrial and Physical Pharmacy, Purdue University, 575 Stadium Mall Drive, West Lafayette, IN 47906, USA
e-mail: sbyrn@purdue.edu

A.C. Templeton et al. (eds.), *Discovering and Developing Molecules with Optimal Drug-Like Properties*, AAPS Advances in the Pharmaceutical Sciences Series 15,
DOI 10.1007/978-1-4939-1399-2_6

solubility which in turn determines blood levels. That is, the formulation determines bioavailability and therapeutic response. Thus, it is important to physically characterize the API form and the formulations in order to reduce risk of product failure. Furthermore, the vast majority of medicines (drug products) are solids and those drug products that are not solids often start with solid APIs. In addition to solubility and bioavailability, the solid form may affect stability, flow, compression, hygroscopicity, and a number of other properties.

6.1 Learning Before Doing and Quality by Design to Reduce Risk

Learning before doing is a good way to reduce risk during development. Figure 6.1 shows the quality by design wheel used by the FDA. This wheel illustrates the importance of product design and process design. In effect, we discuss product design in this chapter. By using the optimum solid form, it is possible to design a product with the desired solubility. Once this has been achieved, the goal is to develop a formulation that optimizes other properties (e.g. flow) so that a viable manufacturing process is available. Additionally, a process that can reproducibly make the form and formulation must be achieved. As outlined above, often with proper attention to the solid form a quality product can be designed rapidly. In many respects, quality by design can reduce risk of product failures.

For early studies, a simple crystallization/amorphization process followed by filling a solid form in a capsule is often sufficient. This type of product can be manufactured from solid produced in a variety of ways including spray drying or crystallization followed by capsule filling using one of the commercially available automated capsule filling machines.

Instability, both chemical and physical, is another major risk in early development. Often it is possible to identify stability problems using a few simple experiments. Exposing the solid form to stress conditions and accelerated stability conditions can provide invaluable information on stability and risk of instability. Armed with this information, adjustments in the designed product can be made.

Further, it is important to examine the risk of failure due to the mechanical properties of the drug substance. If the drug substance cannot be filled into capsules, then it is necessary to prepare trial formulations. In this case, excipient compatibility must be determined to assess risk of failure due to drug–excipient interactions. Also, the formulation should be designed with the manufacturing method in mind. In some cases, a wet or dry granulation will be needed. Alternatively, softgels containing dissolved API are a viable alternative to improve solubility and bioavailability.

In all cases, the polymorph in the formulation must be known and controlled in order to reduce risk. In many instances, it should be possible to prepare drug product on small scale and assess its risk for instability and its risk for insolubility. In our laboratory, we have had good success carrying out dissolution studies on

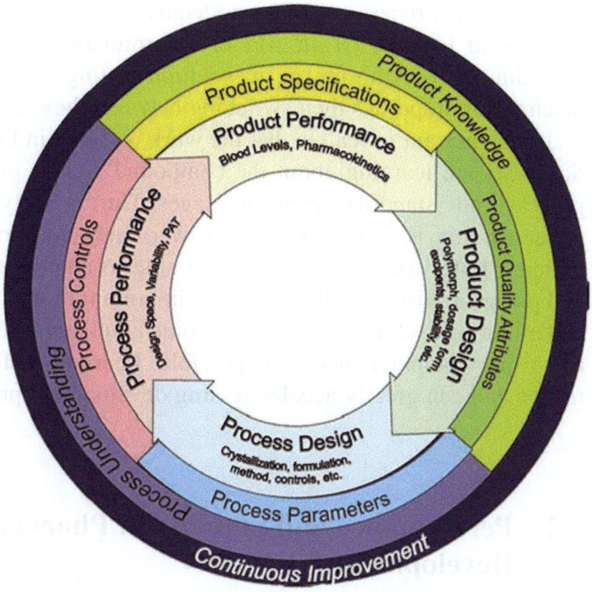

Fig. 6.1 Quality by design strategies

formulations filled in capsules in very early in development. In some cases, it is helpful to determine the pharmacokinetics of the early formulation in animals. Also, preliminary stability studies are always helpful in understanding the drug substance and drug product. All of the initial studies must be done with an eye to the final manufacturing process.

Other challenges faced during initial development are also apparent. Of paramount importance is the short timeline. Another potential problem is that a broad dose range is required. This is because the MTD (maximum tolerated dose) and MAD (maximum absorbable dose) are not known. These dose levels will be determined in animal studies and in the first in human clinical trials. Up until that time, solid-state chemistry and polymorphism studies as well as formulation design must be able to accommodate broad dose ranges. Additionally, usually minimal amounts of API are available and sometimes API lots with different attributes are obtained. The early formulation design and solid-state chemistry work must be able to accommodate these variations.

All of the work outlined in Fig. 6.1 must be carried out from a risk management perspective. The largest risk is that a successful formulation could not be reproduced. Thus, no matter how short the timeline the formulation developed must be reproducible. Other risks include unstable formulations with unacceptable impurity levels. In addition, it is important to carry out a formal risk assessment at each stage of development (Phase I/IND), Phase II, and Phase III. This risk assessment should be documented and communicated to all affected parties.

One strategy that can be quite helpful during the development process outlined in Fig. 6.1 is to utilize a sameness/equivalence protocol. A sameness/equivalence protocol is useful in comparing APIs and formulations made during the development process. A sameness/equivalence protocol includes the parameters outlined in Fig. 6.1. If an API or a formulation can be shown to be the same according to polymorphism, dissolution testing, biopharmaceutics, stability and mechanical properties then switching between these materials is clearly supported.

It is important to realize that the work outlined in Fig. 6.1 is a learning process. Scientists are learning about the compound and its properties. Like any learning process understanding occurs in stages. During early development, the level of understanding is just enough to answer early questions of clinical efficacy and toxicity. Later learning must also be able to address manufacturability, stability, and polymorphism. In all stages, it is important to maximize learning before doing and minimize risk. Learning before doing emphasizes learning on a laboratory scale and using computer models if possible. Learning before doing enhances performance and can greatly accelerate drug development programs.

6.2 Performance and Stability in Pharmaceutical Development

Physical characterization and solid-state chemical analysis is a critical aspect of evaluating the risk of failure as well as the performance and stability of compositions under development. The performance of a pharmaceutical composition is intimately related to its dissolution in a GI tract full of variability (e.g. variable pH). For this reason, it is important to determine solubility and whether a supersaturated solution is formed. If a supersaturated solution is formed, then it is important to determine if crystallization/precipitation occurs. Other studies important to evaluating performance include dissolution tests and tests in animals.

If crystallization occurs, then it is important to determine which form or forms has crystallized. The form that is produced will affect a number of pharmaceutical properties including solubility and particle size.

Further, solid forms can interconvert in the solid state upon storage, in dosage forms and during processing. Thus it is important to understand physical stability and these physical interconversions in order to assess risk of product changes. Of particular interest is conversion from a less stable form to a more stable form which would have lower solubility and perhaps lower bioavailability. Of particular concern is crystallization of amorphous forms, Oswald ripening of nanocrystalline products and conversion of polymorphs to more stable forms or hydrates.

Transitions from one solid phase to another can occur in the absence of solvent. The mechanisms and kinetics of such solid-state transitions can be very complex. For example, Kitaigorodskii et al. (1965) showed that a pin-prick can initiate the solid-state transformation of α-*p*-dichlorobenzene to β-*p*-dichlorobenzene within a single crystal. The transformation of α- to β-*p*-dichlorobenzene is delineated by the spread of the reaction front from the nucleation site through the crystal. Studies of the nucleation process showed that, in "perfect crystals" (i.e., relatively "defect-free" crystals) of either phase, it was often impossible to induce a phase transformation, and these crystals simply melted at the melting point of the respective initial phase. However, with less perfect crystals, it was often possible to induce the phase transformation by artificial creation of a defect using a pin.

Studies of repeated transformations, $\alpha \rightarrow \beta$ $\alpha \rightarrow \beta$, showed that when the transformation of a crystal began at one site, subsequent transformations always began at that same site, whether it was a $\alpha \rightarrow \beta$ or an $\beta \rightarrow \alpha$ transformation. Thus, the crystal seemed to "remember" its nucleation site. This memory effect is explained by the conservation of the nucleation site at crystal defects. In other words, repeated transformations do not destroy the nucleation site for this transformation.

The rates of the $\alpha \rightarrow \beta$ and $\beta \rightarrow \alpha$ polymorphic transformations were best studied by measuring the rate of advancement of the phase boundary or front under a microscope during heating. Each rate was determined by measuring the volume of the crystal transformed per unit time and was shown to be dependent on the location of the nucleus in the crystal, as well as the crystal shape. For example, if the nucleus is in the center of the crystal and the reaction expands symmetrically, after a fixed time, twice as much crystal will be reacted than if the reaction started on the end. Thus, the rate of volume transformation depends on noncrystallographic factors and is, therefore, not a true measure of the absolute rate. This observation probably explains (at least in part) why there is great variation in the measured rates of solid-state reactions when measurements depend on the volume transformed (amount transformed) and not the rate of front advancement. Because of these observations, the rates were measured by determining the rate of advancement of the front in crystals that nucleated at a single site. Even with these measurements, the rates of transformation of apparently similar crystals differed by a factor of 6. The rates were determined at different temperatures allowing calculation of the activation energy of the process. The activation energy for the $\alpha \rightarrow \beta$ reaction was 72.8 ± 10 kJ/mol, while that for the $\alpha \rightarrow \beta$ process was 71.6 ± 10 kJ/mol. Thus, these measurements indicate the α- and β-phases are within experimental error of each other in energy.

A related process is the thermally induced rearrangement of the *a* to *b* form of *p*-nitrophenol (Coppens and Schmidt 1965). In this reaction, needle-shaped single crystals rearrange with the phase boundary moving approximately perpendicular to the needle axis. Grinding or other input of mechanical energy induces the polymorphic transformation of chlorpropamide (Otsuka and Kaneniwa 1990), fostedil (Takahashi et al. 1985), chloramphenicol palmitate (Kaneniwa and Otsuka 1985), and several other drugs (Chan and Doelker 1985).

Studies of phenylbutazone illustrate several of the practical consequences of

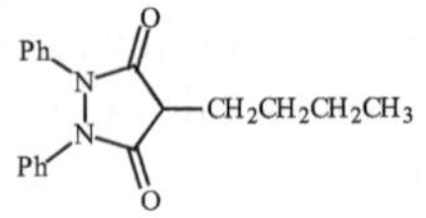

phenylbutazone

polymorphic interconversions. Ibrahim and co-workers (1977) prepared four batches of phenylbutazone by crystallization: Form I, from *tert*-butyl alcohol; Form II, from cyclohexane; Form III, from heptane; and Form IV, from 2-propanol–water. The X-ray powder patterns and infrared spectra of these four forms were different. However, because it is possible that some batches of crystals contained mixtures of forms, single-crystal X-ray diffraction studies of these forms are in order. At any rate, studies of the dissolution rates of these four forms show that the crystals from cyclohexane have the slowest dissolution rate and that the other three batches have nearly equal dissolution rates. The differences in dissolution rate were attributed to differences in surface area. However, the possibility that a solution-mediated phase transformation converted all three forms to the same form during the dissolution tests was not ruled out. All four batches of crystals were pressed into disks. This treatment caused Form III to transform to Form IV, and Forms I and II to transform to a fifth phase. Grinding produced the same changes. These studies indicate that high pressure can cause polymorphic transformations. This suggestion is consistent with other studies reported by Ibrahim and co-workers (1977). Matsuda et al. (1984) have extended these studies on the polymorphic interconversions of phenylbutazone by determining the effect of relative humidity and temperature on the rate of polymorphic transformation. The kinetic data for the transformations were fitted to the Avrami–Erof'ev model and first-order kinetic model. One of the transformations was affected by humidity, and the temperature dependence of this transformation rate was remarkable. Matsuda et al.

6.3 Moisture Uptake

Some crystalline solids take up water from the atmosphere and are termed hygroscopic solids in the literature. Water sorption by hygroscopic solids represents a substantial risk in development.

Unfortunately, there can be no clear definition of hygroscopic solids because hygroscopicity is a relative term. Hygroscopicity is determined by both a kinetic and a thermodynamic term and is a function of the atmospheric relative humidity. In high relative humidities, many solids are hygroscopic. In atmospheres of low humidity, only a few solids will be hygroscopic. Another factor influencing hygroscopicity is surface area and porosity. The larger the surface area of the solid, the more rapid the uptake of moisture. This is because solids with larger surface areas have more sites for adsorption of water molecules.

When solids that are not solvates contain large amounts of water, it has been hypothesized that water must be taken up into the solid by disordered or high-energy regions such as defects and amorphous sites. Such effects might be exaggerated by manufacturing processes that reduce particle size, such as micronization, milling, or related processes known to increase the number of high energy sites. Of course, some solids can take up so much water during these processes that they become damp or even liquefy at RH_0 (Zografi et al. 1991). This tendency is usually easily detected by microscopic observation.

The formation of crystal hydrates, of course, is another way for water to be incorporated into a solid. In these cases, the water molecules generally occupy a specific crystallographic site in the solid. This site can be determined by X-ray crystallography which unequivocally proves the existence and composition of the hydrate. However, many hydrates exist in which the water is located in tunnels within the crystal. The water can be located accurately only by determination of the crystal structure at low temperatures (if even then). In these cases, the water content may change rather easily with changes in relative humidity.

Plots of vapor pressure versus relative humidity are an excellent way to determine the nature of a solid with respect to water sorption. The different kinds of behavior that these plots may be expected to show include:

1. Virtually no water uptake.
2. Gradual water uptake, characteristic of an amorphous material or a nonstoichiometric hydrate (a hydrate without a simple ratio of water to host molecule).
3. "Stair-step" water uptake, characteristic of a stoichiometric hydrate.

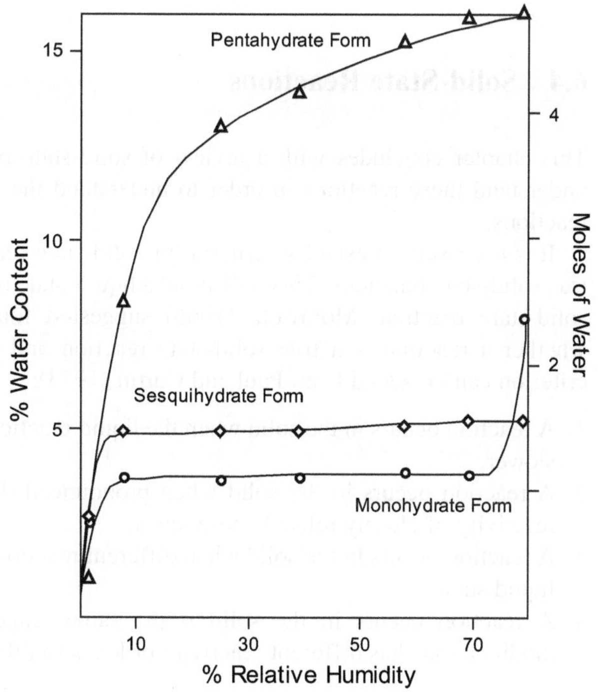

Fig. 6.2 Vapor pressure versus relative humidity diagrams for three hydrates of sodium cefazolin. The sesquihydrate and monohydrate behave normally and the "pentahydrate" is actually a nonstoichiometric hydrate (Osawa et al. 1988)

Figure 6.2 shows the behavior of two stoichiometric hydrates (a monohydrate and a sesquihydrate) as well as a nonstoichiometric hydrate of sodium cefazolin. In addition, amorphous sodium cefazolin also exists and takes up and loses water in a more or less gradual fashion as described below.

Sorption of water into amorphous solids or regions of a solid involves dispersion or dissolution of the water molecules within a solid. The more polar a solid, the greater the amount of water taken up. Obviously, in such systems the water content depends upon relative humidity.

In summary, Zografi et al. (1991) made the following recommendations with regard to water specifications:

1. A complete profile of relative humidity versus water content (weight) should be reported for all reference standards.
2. For amorphous solids, both T_g and W_g should be reported.
3. For deliquescent solids, the RH_0 and an appropriate warning on the label should be provided.
4. For stoichiometric hydrates, the water specifications should reflect the stoichiometry.
5. Attention should be paid to materials which do not form well-defined hydrates and can take up or lose water as the humidity is varied. Any structural changes that accompany changes in water content should be noted. (Some hydrates can lose water without changing crystal structure. This is due to the formation of an extremely stable crystal packing network by the host molecule.)

6.4 Solid-State Reactions

This chapter concludes with a review of solid-state reactions. It is important to understand these reactions in order to understand the risk of occurrence of these reactions.

It is necessary to establish criteria for solid-state reactions in order to focus on true solid-state reactions. This will avoid a liquid-state reaction being identified as a solid-state reaction. Morawetz (1966) suggested four criteria for determining whether a reaction is a true solid-state reaction and a fifth and very important criterion can be added from Paul and Curtin (1973):

1. A reaction occurs in the solid when the liquid reaction does not occur or is much slower.
2. A reaction occurs in the solid when pronounced differences are found in the reactivity of closely related compounds.
3. A reaction occurs in the solid when different reaction products are formed in the liquid state.
4. A reaction occurs in the solid if the same reagent in different crystalline modifications has different reactivity or leads to different reaction products.

5. A reaction occurs in the solid phase if it occurs at a temperature below the eutectic point of a mixture of the starting material and products.

Once it has been established that the reaction is occurring in the solid state, the reaction can be understood in terms of a four-step process (Paul and Curtin 1973):

1. Loosening of the molecules at the reaction site.
2. Molecular change.
3. Solid solution formation.
4. Separation of the product phase.

It is important to realize that many solid-state reactions of drugs involve drug degradations which have been studied mostly on the macroscopic level. In fact, few studies aimed at determining the molecular aspects of the solid-state chemistry of drugs have been published. These reactions are of interest because of a desire to prevent such degradation. Even for such common agents as vitamin D_2 and vitamin A, the structures of only a few of the solid-state degradation products have been published. Therefore, in many respects, the solid-state chemistry of drugs is synonymous with drug degradation.

An assessment of risk of degradation in the solid state is assisted by an analysis of the known types of solid-state reactions. Typically, these solid-state reactions fall into the following categories

- Solid-state oxidations.
- Solid-state hydrolyses including solid–gas hydrolyses (additions of water to solids).
- Solid-state decompositions reactions giving off a gas.
- Solid-state photochemical reactions.
- Solid-state thermal reactions.
- Solid–solid reactions.

Table 6.1 summarizes examples of these categories.

Table 6.1 Solid-state chemical reactions of drugs

Solid-state oxidations
Dialuric acid (Clay et al. 1982)
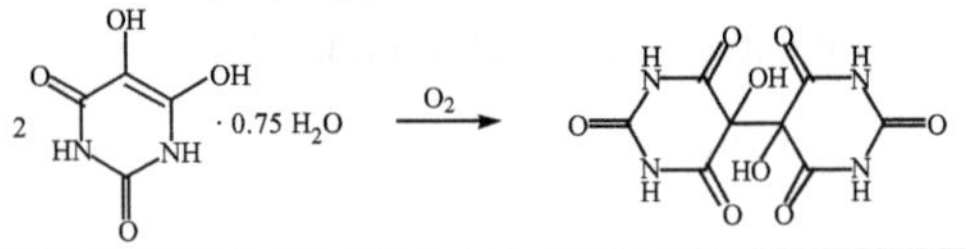
Dihydrophenylalanine (Ressler 1972)
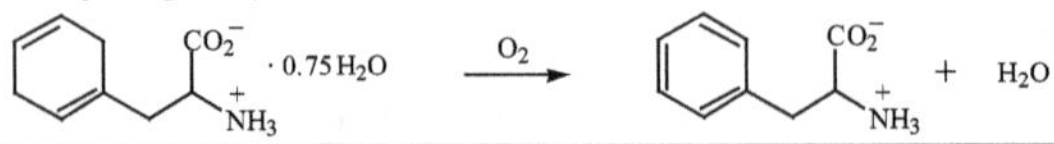
Phorbol esters (Schmidt and Hecker 1975)
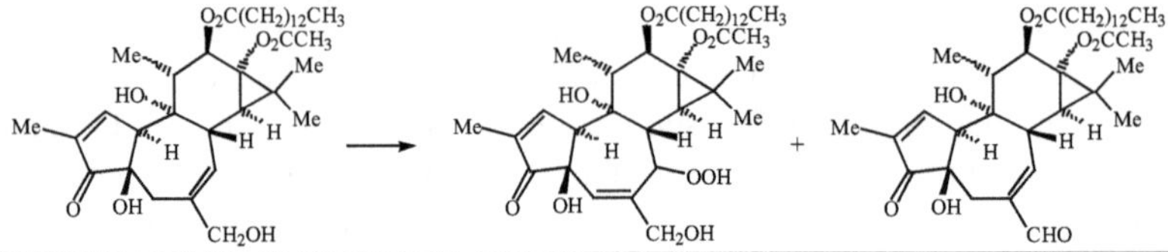
Vitamin C (Rubin et al. 1976)
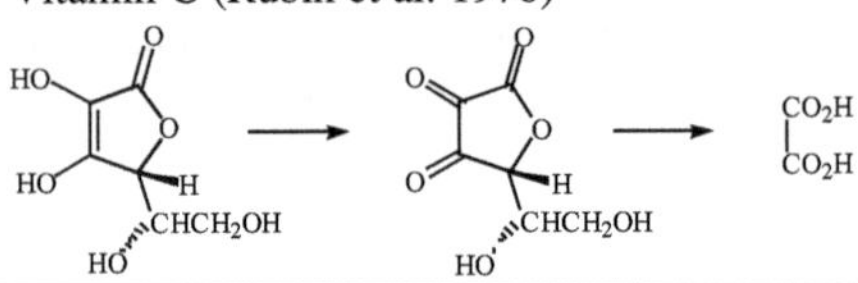
Hydrocortisone *tert*-butylacetate (Brenner et al. 1969)
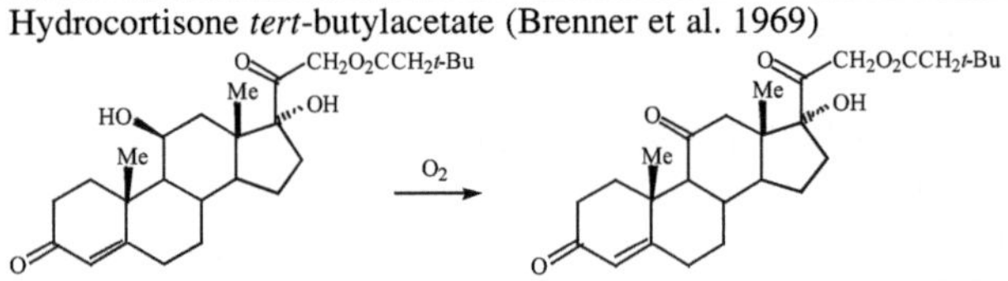
Vitamin D_2 (Kanzawa and Kotaku 1953)
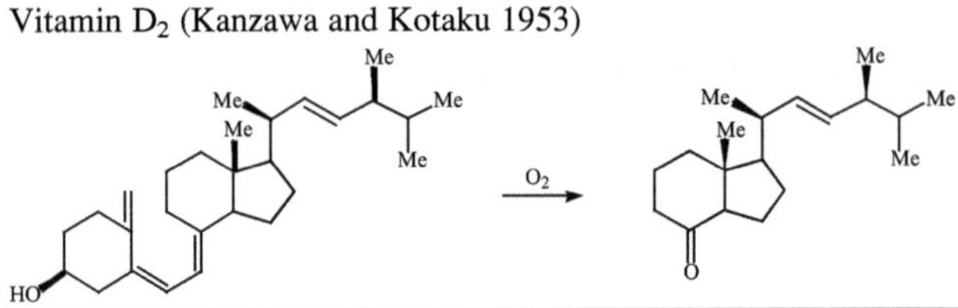
Additions of gases to solids—solid-state hydrolyses
Aspirin (Leeson and Mattocks 1958)
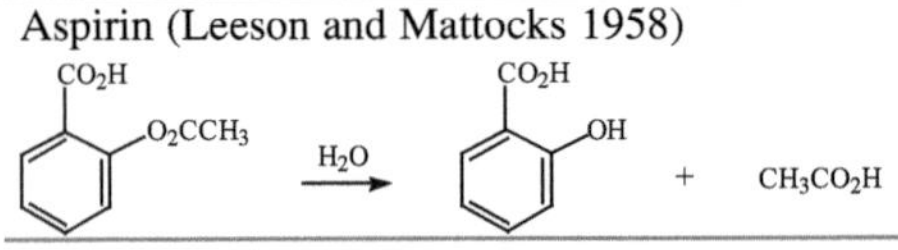

(continued)

Table 6.1 (continued)

Nitrazepam (Genton and Kesselring 1977)

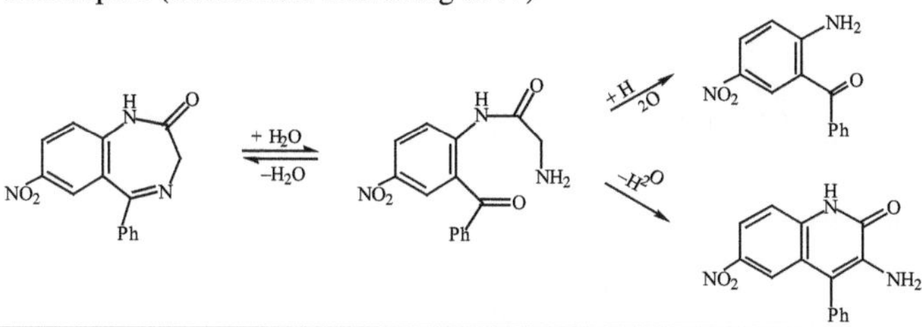

Solid-state decomposition reactions: A (solid) → B (solid) + C (gas)

Dehydration of tetracyclines (Simmons et al. 1966)

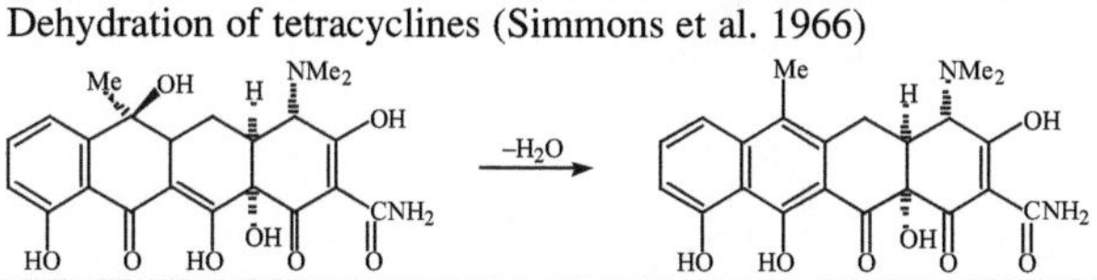

Dehydrochlorination of caffeine hydrochloride (Biedermann 1883)

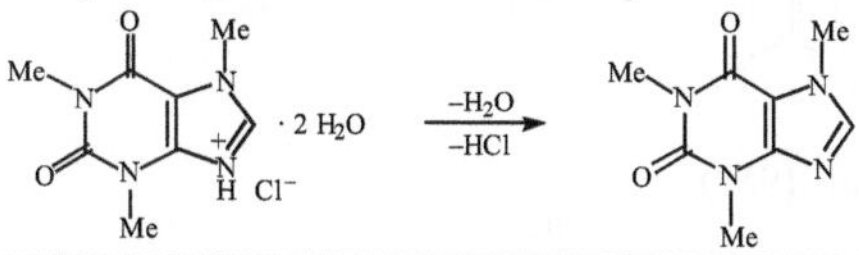

Decarboxylation of *p*-aminosalicylic acid (Lin et al. 1978)

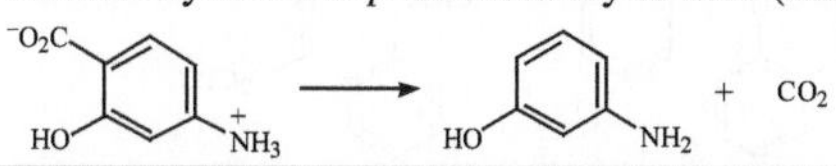

Solid-state decomposition reactions: A (solid) → B (solid) + C (gas)

Decarboxylation of moxalactam (Pikal and Dellerman 1989)

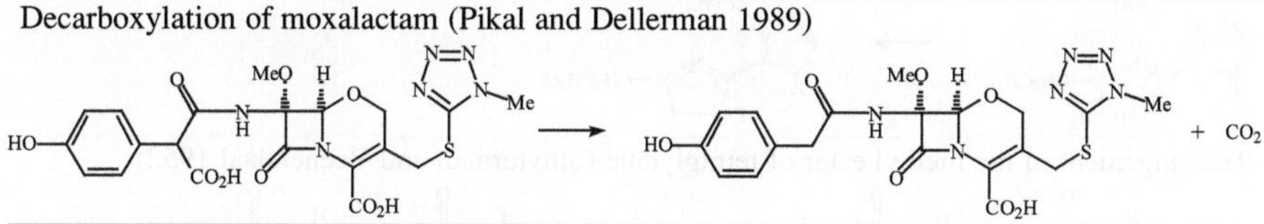

Dehydrochlorination of aminosalicylic acid hydrochlorides (Lin et al. 1978)

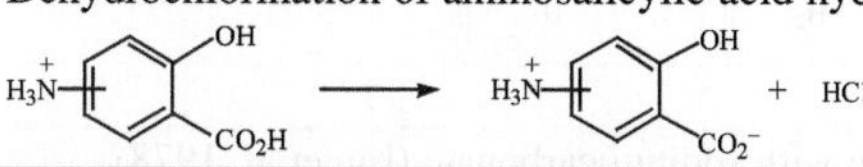

Solid-state photochemical reactions

Gibberellins (Adam and Voigt 1971)

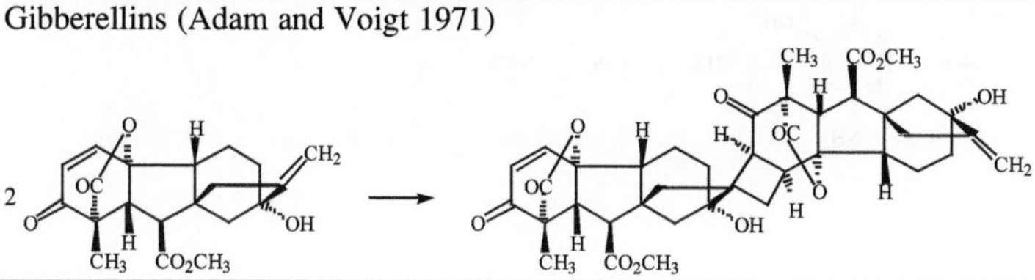

(continued)

Table 6.1 (continued)

Dieldrin (Benson 1971)

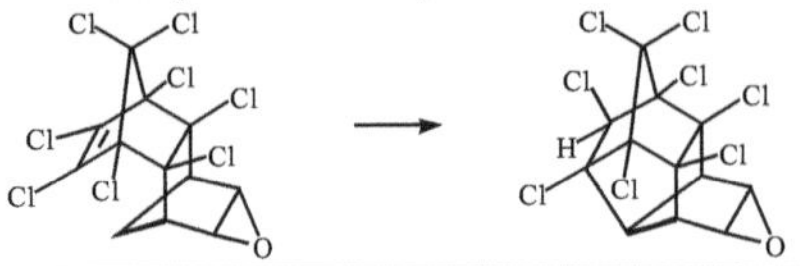

Nifedipine (Hayase et al. 1994)

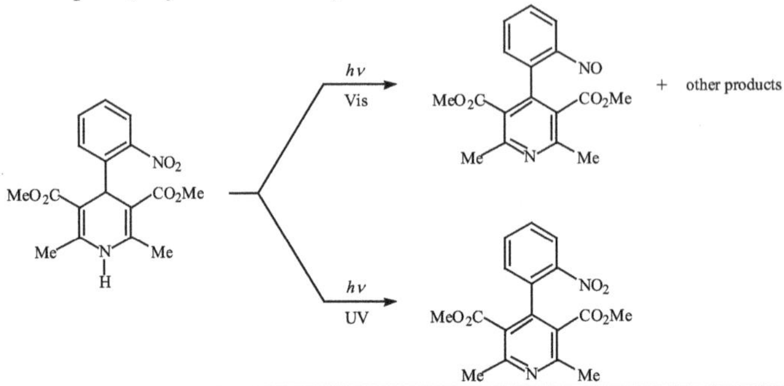

Solid-state thermal reactions

Rearrangement of aspirin anhydride (Garrett et al. 1959)

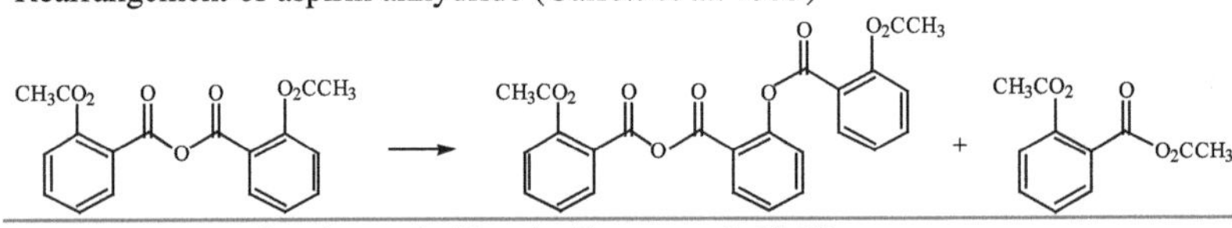

Rearrangement of a triazenoimidazole (James et al. 1969)

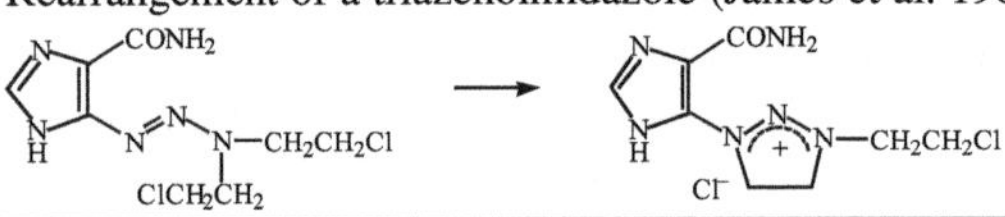

Rearrangement of the methyl ester of tetraglycine (Sluyterman and Veenendaal 1952)

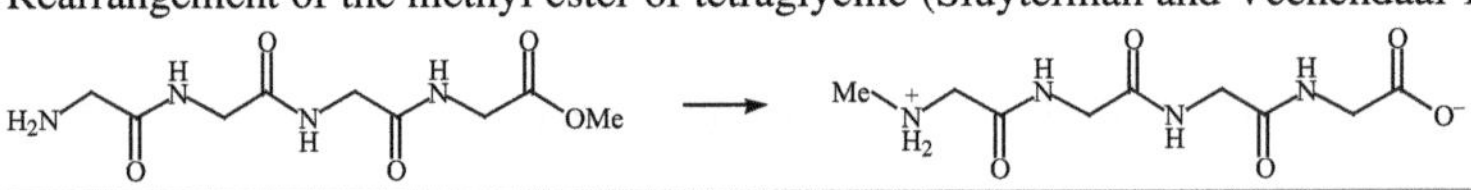

Solid–solid reactions

Reaction of *p*-aminosalicylic acid hydrochloride with sodium carbonate (Lin et al. 1978)

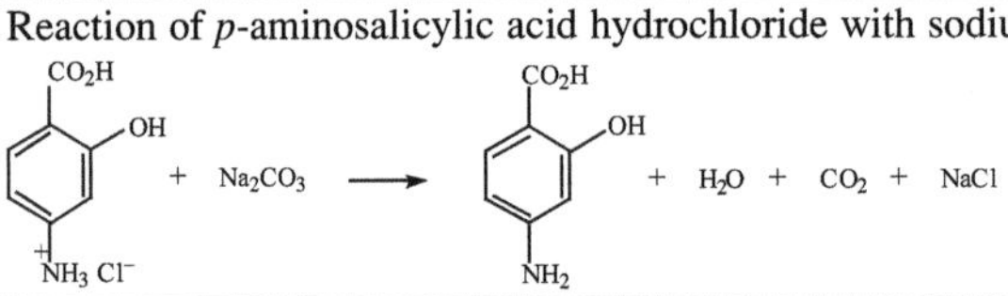

6.5 Summary

This chapter provided an overview of factors that are responsible for risk in early development. The FDA quality by design wheel is used as a guide and a focus is put on the idea of learning before doing. Next the idea of understanding the risk of instability, both chemical and physical, is addressed. In this section, several historical examples of physical transformations in the solid state are summarized. This chapter concludes by addressing the importance of understanding moisture uptake and the solid-state reactivity of the system.

References

Adam G, Voigt B (1971) Solid state photoaromatization of 3-keto-gibberellin A3. Tetrahedron Lett 4601–4604

Benson WR (1971) Photolysis of solid and dissolved dieldrin. J Agric Food Chem 19:66–72

Biedermann H (1883) Concerning caffeine and its salts. Arch Pharm 221:175–186

Brenner G, Roberts FE, Hoinowski A, Budavari J, Powell B, Hinkley D, Schoenewaldt E (1969) Effect of crystalline form on the air-oxidation of steroidal 11β-ols to 11-ones. Angew Chem Int Ed 8:975–976

Chan HK, Doelker E (1985) Polymorphic transformation of some drugs under compression. Drug Dev Ind Pharm 11:315–332

Clay RJ, Knevel AM, Byrn SR (1982) The desolvation and oxidation of crystals of dialuric acid monohydrate. J Pharm Sci 71:1289–1291

Coppens P, Schmidt GMJ (1965) The crystal structure of the metastable (β) modification of *p*-nitrophenol. Acta Cryst 18:654–663

Garrett ER, Schumann EL, Grostic MF (1959) Prediction of stability in pharmaceutical preparations. VI. Stability, products, and mechanism in the pyrolytic degradation of aspirin anhydrate. J Am Pharm Assoc Sci Ed 68:684–691

Genton D, Kesselring UW (1977) Effect of temperature and relative humidity on nitrazepam stability in solid state. J Pharm Sci 66:676–680

Hayase N, Itagaki Y-I, Ogawa S, Akutsu S, Inagaki S-I, Abiko Y (1994) Newly discovered photodegradation products of nifedipine in hospital prescriptions. J Pharm Sci 83:532–538

Ibrahim HG, Pisano F, Bruno A (1977) Polymorphism of phenylbutazone: properties and compressional behavior of crystals. J Pharm Sci 66:669–673

James RH, Sternglanz PD, Fulmer Shealy Y (1969) 5(or 4)-[3, 3-Bis(2-chloroethyl)-1-triazeno] imidazole-4(or 5)-carboxamide: a titrametric determination of its *v*-triazolinium transformation product and studies of its stability. J Pharm Sci 58:1193–1195

Kaneniwa N, Otsuka M (1985) Effects of grinding on the transformation of polymorphs of chloramphenicol palmitate. Chem Pharm Bull 33:1660–1668

Kanzawa T, Kotaku S (1953) Stability of vitamin D. J Pharm Soc Jpn 73:1357–1360

Kitaigorodskii AI, Mnyukh YV, Asadov YG (1965) Relations for single-crystal growth during polymorphic transformation. J Phys Chem Solids 26:463–472

Leeson LJ, Mattocks AM (1958) Decomposition of aspirin in the solid state. J Am Pharm Assoc Sci Ed 67:329–333

Lin C-T, Siew P-Y, Byrn SR (1978) Solid-state dehydrochlorination and decarboxylation reactions. I. Reactions of *p*-aminosalicylic acid hydrochloride and *p*-aminosalicylic acid, revised crystal structure of *p*-aminosalicylic acid. J Chem Soc Perkin Trans 2:957–962

Matsuda Y, Tatsumi E, Chiba E, Miwa Y (1984) Kinetic study of the polymorphic transformations of phenylbutazone. J Pharm Sci 73:1453–1460

Morawetz H (1966) Reactivity of organic crystals. Science 152:705–711

Osawa T, Kamat MS, DeLuca PP (1988) Hygroscopicity of cefazolin sodium: application to evaluate the crystallinity of freezer-dried products. Pharm Res 5:421–425

Otsuka M, Kaneniwa N (1990) Effect of grinding on the crystallinity and chemical stability in the solid state of cephalothin sodium. Int J Pharm 62:65–73

Paul IC, Curtin DY (1973) Thermally induced organic reactions in the solid state. Acc Chem Res 6:217–225

Pikal MJ, Dellerman KM (1989) Stability testing of pharmaceuticals by high-sensitivity isothermal calorimetry at 25 °C: cephalosporins in the solid and aqueous solution states. Int J Pharm 50:233–252

Ressler C (1972) Solid-state dehydrogenation of L-1,4-cyclohexadiene-1-alanine hydrate to L-phenylalanine. J Org Chem 37:2933–2936

Rubin SH, DeRitter E, Johnson JB (1976) Stability of vitamin C (ascorbic acid) in tablets. J Pharm Sci 65:963–968

Schmidt R, Hecker E (1975) Autooxidation of phorbol esters under normal storage conditions. Cancer Res 35:1375–1377

Simmons DL, Woo HSL, Koorengevel CM, Seers P (1966) Quantitative determination by thin-layer chromatography of anhydrous tetracycline in degraded tetracycline tablets. J Pharm Sci 55:1313–1315

Sluyterman LAÆ, Veenendaal HJ (1952) Reactions of polypeptide esters in the solid state. I. Migration of methyl groups. Recl Trav Chim Pays-Bas 71:137–152

Takahashi Y, Nakashima K, Ishihara T, Nakagawa H, Sugimoto I (1985) Polymorphism of fostedil: characterization and polymorphic change by mechanical treatments. Drug Dev Ind Pharm 11:1543–1563

Zografi G, Gary Hollenbeck R, Laughlin SM, Pikal MJ, Schwartz JP, Van Campen L (1991) Report of the advisory panel on moisture specifications. Pharmacopeial Forum 17:1459–1474

Chapter 7
Strategies and Methods for Drug Candidate Phase Optimization in Discovery Space

Michael McNevin and John Higgins

7.1 Introduction

The advantages of identifying optimal and robust solid-state phases of both marketed drugs and new chemical entities are well known to the pharmaceutical industry and were partially discussed in Chap. 1 in this section. An optimized active pharmaceutical ingredient (API) solid-state phase can provide a variety of positive attributes such as improved purity, solubility, physical and chemical stability, and formulation performance when compared to a suboptimal phase. Historically, phase optimization was pursued during preclinical development rather than in discovery, which mainly focused on structure–activity relationships. Considering the fact that many drug candidates fall short of commercialization due to toxicity, poor pharmacokinetics, or lack of efficacy (DiMasi et al. 2010) it made sense to delay spending resources on solid-state phase optimization until a candidate reached a certain milestone during clinical development (e.g., Ph II clinical trials). Over the past decade, however, scientific, risk, and business drivers have moved the pharmaceutical industry to more closely align drug discovery and drug development efforts (Palucki et al. 2010). Indeed, discovery and development scientists now often work in partnership to not only optimize the pharmacological properties of promising chemical scaffolds, but also identify an associated solid-state phase that has appropriate physicochemical characteristics which facilitate optimal performance in in vivo *studies* ranging from animal toxicology studies to human clinical trials and commercialization. Hence, medicinal chemistry efforts in the discovery lead optimization (LO) space now routinely focus on structure–activity

M. McNevin (✉)
Formulation Sciences, Merck Research Laboratories, Merck & Co., 556 Morris Avenue, S7-E2-2679, Summit, NJ 07901, USA
e-mail: Michael_McNevin@merck.com

J. Higgins
Discovery Pharmaceutical Sciences, Merck & Co., West Point, PA, USA

A.C. Templeton et al. (eds.), *Discovering and Developing Molecules with Optimal Drug-Like Properties*, AAPS Advances in the Pharmaceutical Sciences Series 15,
DOI 10.1007/978-1-4939-1399-2_7

Table 7.1 Variability of properties and attributes impacted by polymorphic phase

Differing properties and attributes of polymorphs	
Melting points	Hygroscopicity
Chemical stability	Density
Hardness	Dissolution rate
Morphology	Solubility

relationships (SAR) *and* structure–property relationship (SPR) studies as well (Lipinski et al. 1997; Kerns and Di 2008).

Through their diversity in chemical structure, intra- and intermolecular bonding, and conformational flexibility, drug molecules have the ability to order into crystalline materials with varying packing arrangements. Compounds with identical chemical composition but differences in their crystal lattice structure are known as polymorphs, which have been extensively described (Bernstein 2002 and Hilfiker 2006). Table 7.1 shows properties and attributes that can differ between polymorphs and which ultimately impact the processability and bioavailability of the drug. If a guest molecule is included in the packing arrangement through ionic or hydrogen bonding effects, a crystalline salt, co-crystal, or solvate is formed. These phases may also have benefits over polymorphs of a compound or its amorphous material. Informed selection between the observed phases can yield vastly different outcomes for storage requirements, processing operating space, formulation, and biopharmaceutical performance (Gibson 2009). For commercialized drugs, perhaps the most highly publicized example of the risks with polymorphism is seen with Norvir (Chemburkar 2000; Bauer et al. 2001; Morissette et al. 2003). Here, the emergence of a less soluble phase caused slowed dissolution of the marketed dosage form and compromised the oral bioavailability of the drug. Events such as these which may endanger the supply chain of a drug product provide an obvious rationale for optimizing and selecting robust development phases.

Crystalline forms are often preferred as they provide a reproducible means of purifying a compound with consistent physicochemical properties and are often more physically and chemically stable than the corresponding amorphous phase. The identification of a robust crystalline phase also is appreciated by regulatory agencies whose task it is to ensure that all is being done to ensure that a drug product performs as intended. Finally, the identification of various solid-state phases and their unpredictable properties results in additional intellectual property.

7.1.1 A Rationale for Drug Phase Screening/Optimization Before Development Entry

A prime example of perhaps the most worrisome of solid-state risks in the preclinical drug stage development is when an amorphous phase of a drug converts to a crystalline phase, with concomitant decrease in solubility and oral absorption. This can wreak havoc on a preclinical toxicology program (which usually requires high

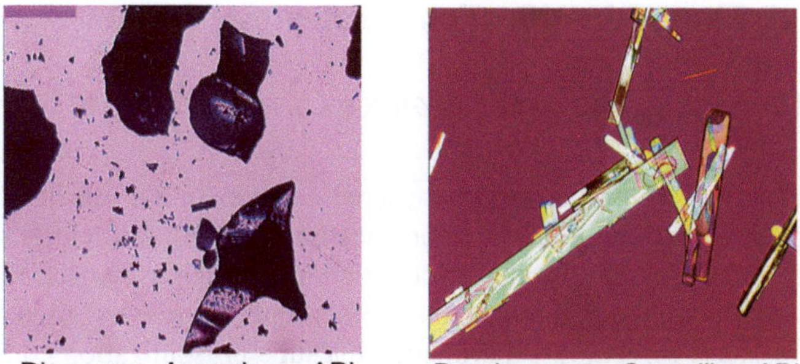

Toxicology Program: API phase vs. Exposure

API Phase	Solubility (mg/mL)	AUC (750 MPK, μM h)
Amorphous (Discovery)	>25	~200
Crystalline (Development)	<0.01	<30

Fig. 7.1 Impact of crystallization on aqueous solubility and oral absorption in male Wistar rats (Higgins 2009, Image reproduced with permission from *American Pharmaceutical Review*. Copyright 2009 CompareNetworks)

exposure), and such a situation is well documented in the literature (Palucki et al. 2010).

An example where a compound initially progressed as an amorphous solid and was later converted to a crystalline phase upon scale-up is shown in Fig. 7.1 (Higgins 2009). It is not surprising that the crystalline phase was significantly less soluble than the corresponding amorphous phase and led to considerably lower oral absorption (Hancock and Parks 2000). Fixing this drop in exposure caused significant delays to the preclinical toxicology program. If the crystalline phase had been identified earlier, the choice of candidate to advance to later development stages would likely have been different.

Hence, with a robust crystalline phase in hand, one can better understand solubility limits and associated oral absorption, which leads to acceptable and reproducible safety margins in preclinical toxicity studies. Similarly, early identification of an optimal phase minimizes the potential for multiple phase changes during clinical development, which in turn reduces resource expenditures on in vitro physicochemical studies and in vivo bridging biocomparison studies.

7.2 Early Phase Identification Strategies

7.2.1 *Material, Time, and Resource Considerations*

Consideration of the limitations of phase screening capabilities during discovery is crucial to determining the level of rigor one should invest in identifying a viable phase for development. Early phase identification strategies can be restricted by various factors, including compound availability and resource and time constraints.

Empirical phase screening experiments that tune crystallization conditions and processes are the widely accepted approach for discovering API crystalline phases and will be discussed at length throughout this chapter. There are various in silico methods for predicting crystallization and polymorphism described elsewhere (Ouvard and Price 2004; Neumann and Perrin 2005; Florence et al. 2006; Lancaster et al. 2006; Hulme and Price 2007). However, even with advances in predictive methods, there are no guarantees that a crystalline phase will be identified due to variables including, but not limited to, the presence of viable nucleation sites, impurities, and suitable growth conditions.

During the drug discovery stage, material constraints are often the major limiting factor for phase screening activities. Fortunately, it has been shown that even relatively small quantities of a compound can be utilized to identify viable crystalline phases. Furthermore, crystalline seeds from small-scale batches can be used to initiate larger scale crystallization.

Crystallization studies are often thought of as a time-intensive process that require hours to days. This is generally a misconception, since the experiments do not require dedicated hands-on time as the crystallization process proceeds. In fact, the time invested is mainly consumed in the design of the experiments to achieve a diverse set of conditions intended to yield the highest chance of success. Frequently, fateful experimental discards such as the material in the bottom of an NMR tube can result in successful crystallizations. Once the crystallization experiments are complete, additional resources are expended on the characterization and differentiation of any "hits" that are observed. Experiments as simple as visual polarized light microscopy (Fig. 7.2), Raman spectroscopy, or X-ray powder diffraction (XRPD) analysis may be employed and are discussed later in this chapter.

7.2.2 *Preferred Physicochemical Attributes of Drug Candidates*

Desirable properties of an optimal solid-state phase of a drug have been described in many places over the past 50 years (Byrn et al. 1999). Some typical physicochemical criteria are summarized in Table 7.2, along with some desirable properties

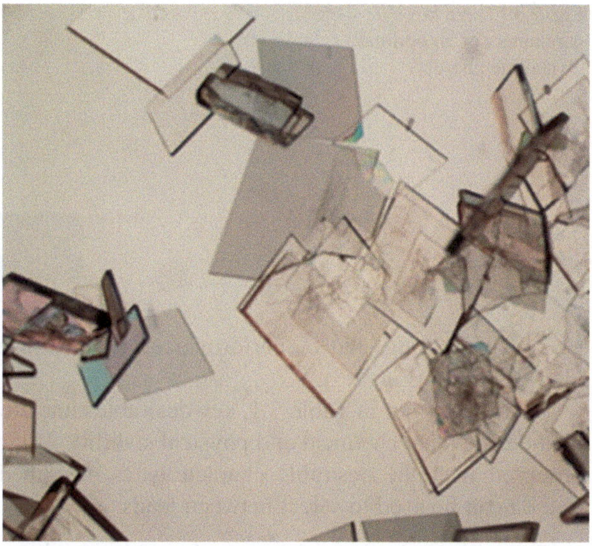

Fig. 7.2 Flake crystals as imaged via polarized light microscopy

Table 7.2 Desirable attributes of a robust crystalline phase

Compound attribute	Desired properties/specifications
Solid-state form	Crystalline solid; mp > 80 °C with difference > 25 °C above the Tg, single phase identifiable by X-ray and vibrational spectroscopy or other selective techniques and compound degradation (>1 % LCAP) not noted until > 20 °C above the mp
Physical stability	No physical change in phase above 40 °C and 75 % RH > 1 mo or below 20 % RH
Chemical stability	No impurity generated >0.5 % above 40 °C and 75 % RH > 1 mo
Hydration state	Preferentially anhydrous phase if stable from 20 to 75 % RH
Hygroscopicity	Non-hygroscopic or stepwise hygroscopicity within typical terrestrial range
Photostability	Photostable to 100,000 lux 24 h

and specifications. The attributes in Table 7.2 are discussed in various levels of detail throughout this chapter.

7.2.3 Differentiating Characteristics of Phases

Experimentalists are likely to spend much of their time differentiating between crystalline phases and attempting to define which is suitable for the proposed development strategy. An API phase suitable for a formulated drug product is dependent on the balance between a triangle of formulation stability upon storage,

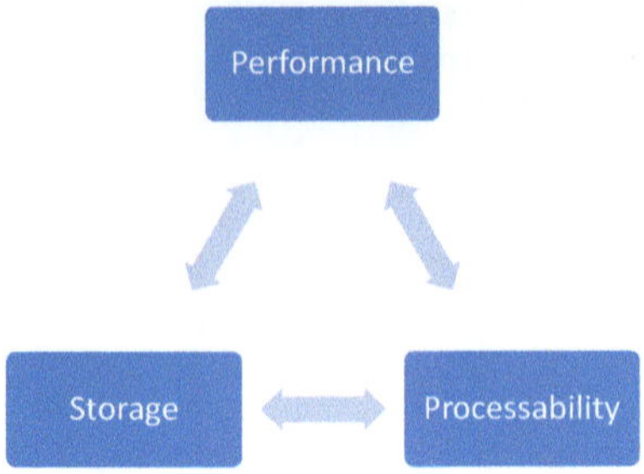

Fig. 7.3 Three key parameters of an optimal crystalline phase

ease of processing, and bioperformance in the dosage form to give reproducible bioavailability (Fig. 7.3).

As noted earlier in Table 7.1, key desirable characteristics of a crystalline phase include adequate chemical and physical stability. If multiple crystalline phases are observed with the desirable characteristics described above, the following key criteria can be used to select between leads:

- Crystalline salts formed with preferred counterions
- Degree of crystallinity; hydration/solvation state
- Solubility in processing solvents and biorelevant media
- Ease of processing (including use of preferred solvents) and yield

Typically, it is preferable that a single phase exists within both terrestrial temperature and humidity ranges. If other phases exist which straddle these limits, boundaries on processing and subsequent storage conditions may be warranted for both the API and drug product to ensure phase purity.

7.3 Crystalline Hit Generation, Characterization, and Evaluation of a Suitable Phase

7.3.1 Strategic Approaches to Phase Discovery and Evaluation

In the discovery stage, only a few hundred milligrams of API are generally available. Early batches tend to be less pure and it has been postulated that impurities can retard crystal formation (Meenan et al. 2002). Furthermore, the same impurities may not be generated by the final synthetic route used to make the drug. With these issues in mind, crystallization efforts are more likely to be successful if a high-purity batch can be isolated or purified. A "blueprint" as shown in Fig. 7.4 is useful for generating and selecting suitable phases in an optimally compound- and resource-sparing manner.

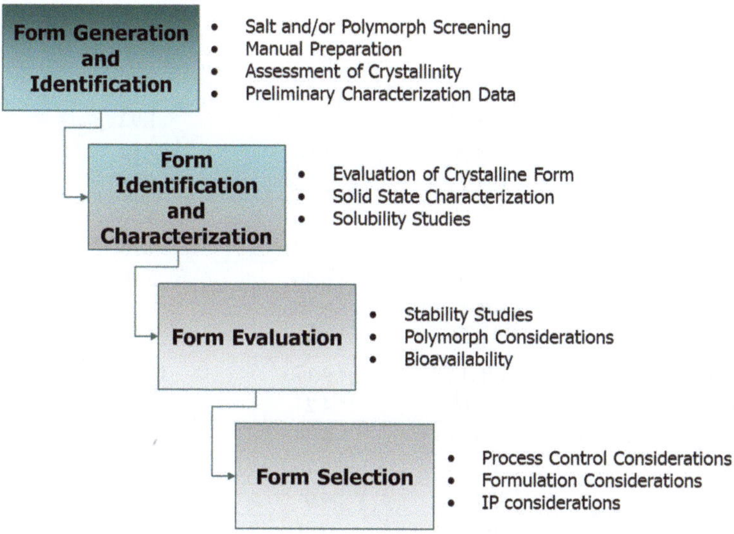

Fig. 7.4 A blueprint for generating and selecting suitable phases

7.3.2 Crystalline Hit Generation Methods for Salts and Co-crystals

If a compound has an ionizable functional group, screening potential counterions for salt formation may provide access to salts with improved physicochemical or biopharmaceutical properties (Berge et al. 1977; Bastin et al. 2000). Determination of the p*K*a of the free acid or base will guide the salt selection process and help determine the likelihood of disproportionation of the corresponding salts (Stephenson et al. 2011). Solubility and chemical stability throughout the physiological pH range, as well as physical stability, are essential for rapid definition of a viable developable salt phase (Nakanishi et al. 1998). Counterions for pharmaceutical salt formation have been preferentially classified via their GRAS status, toxicity, and tolerability. Tables 7.3a, 7.3b, and 7.3c (adapted from Stahl and Wermuth 2011) outline this classification for pharmaceutically acceptable acids to be utilized with free base compounds to form salts or co-crystals of the active pharmaceutical ingredient. Table 7.4 similarly outlines pharmaceutically acceptable bases with which to complex an acidic API. Historically, hydrochloride salts of basic compounds and sodium salts of acidic compounds are utilized in development with the greatest frequency (Gould 1986; Newman et al. 2008). Other commonly used acid salts include hydrobromide, sulfate, tartrate, maleate, methanesulfonate, phosphate, and citrate. Other occasionally used basic salts include calcium, potassium, magnesium, and meglumine.

There also has been increasing interest in the use of co-crystals (Bak et al. 2008). In contrast to salts, co-crystals can be defined as multicomponent assemblies held together by freely reversible, noncovalent, nonionic interactions. As with

Table 7.3a List of class 1 acid salt formers

Name	Mw	p*K*a1	p*K*a2	p*K*a3
Hydrochloric	36.46	−6		
Sulfuric	98.08	−3	1.92	
Phosphoric	98.00	1.96	7.12	12.32
*Glutamic (L)**	147.13	2.19	4.25	9.67
*Tartaric (+L)**	150.09	3.02	4.36	
Citric	192.12	3.13	4.76	6.40
*Malic (−L)**	134.09	3.46	5.10	
Gluconic (D)	196.16	3.76		
*Lactic (L)**	90.08	3.86		
Ascorbic (L)	176.12	4.17	11.57	
Succinic	118.09	4.21	5.64	
Adipic	146.14	4.44	5.44	
Acetic	60.05	4.76		
Stearic	284.49	4.9		
Carbonic	62.02	6.46	10.3	
Thiocyanic	59.09	*−1.33*		
Glycerophosphoric	172.07	1.47	6.19	
*Aspartic (L)**	133.10	1.88	3.65	9.60 B
Maleic	116.07	1.92	6.23	
Pyruvic	88.06	2.39		
Glucaric (D)	210.14	3.01	3.94	
Fumaric	116.07	3.03	4.38	
Galactaric	210.14	3.08	3.63	
Glucuronic (D)	194.14	3.18		
Lactiobionic	358.29	*3.2*		
Glucoheptanoic (D)	226.18	*3.3*		
Hippuric	179.17	*3.55*		
Aceturic	117.10	3.82		
Glycolic	76.05	3.82		
Glutaric	132.11	4.34	5.27	
Sebacic	202.25	4.59	5.59	
Butyric	88.11	4.82		
Capric	172.27	4.9		
Palmitic	256.43	4.9		
Lauric	200.32	5.3		

Chiral salt formers are in italic
Chiral salt formers with an asterisk (*) available as enantiomer
Class 1: Physiologically ubiquitous ions

counterions for API salts, a suitable "coformer" must be identified in order to form a co-crystal. The interaction is similar to that of solvation, with the caveat that the individual coformer instead is a solid when under standard conditions. However, the overall crystallization process for co-crystals may be similar to that of salts or

Table 7.3b List of class 2 acid salt formers

Name	Mw	p*K*a1	p*K*a2
Alginic	24000	*3.5*	
Benzoic	122.12	4.19	
Nicotinic	123.11	4.85	2.07
Propionic	74.08	4.87	
Caprylic	144.21	4.91	
Naphthalene-1,5-disulfonic	288.30	*−3.37*	*−2.64*
Di(*t*-butyl)naphthalenedisulfonic	400.51	*−3.05*	*−2.32*
Di(*t*-butyl)naphthalenesulfonic	320.45	*−2.48*	
Ethane-1,2-disulfonic	190.20	*−2.1*	*−1.5*
Ethanesulfonic	110.13	*−2.05*	
Cyclamic	179.24	*−2.01*	
Ethanesulfonic, 2-hydroxy-	126.13	*−1.66*	
Toluenesulfonic	172.20	−1.34	
Methanesulfonic	96.11	−1.2	
Dodecylsulfuric	266.40	*−0.09*	
Naphthalene-2-sulfonic	208.24	0.17	
Benzenesulfonic	158.18	0.7	
Oxalic	90.03	1.27	4.27
Pamoic	388.37	2.51	3.1
Glutaric, 2-oxo-	146.10	*2.7*	*4.5*
Napthoic, 1-hydroxy (2-)	188.18	*2.7*	*13.5*
Orotic	156.10	2.8	9.45
Malonic	104.06	2.83	5.70
Gentisic	154.12	2.93	
*Pyroglutamic (-L)**	129.11	3.32	
Benzoic 2-(4-hydroxybenzoyl)-	242.23	3.35	7.96
Oleic	282.47	4	
Camphoric (+)	200.23	4.72	5.83
Isobutyric	88.11	4.84	

Chiral salt formers are in italic
Chiral salt formers with an asterisk (*) available as enantiomer
Class 2: Low toxicity and good tolerability

solvates and is described elsewhere (Good and Rodríguez 2008; Variankaval et al. 2006).

Historically, solvent-mediated crystallization frequently has been used to optimize solid-state phase, since this method translates to a preferable scalable process. If limited quantities of compound are available and experimental time constraints are high, thermal crystallization methods may be used as a first pass (Chen et al. 2005). These two techniques now will be described in detail.

Table 7.3c List of class 3 acid salt formers

Name	Mw	p*K*a1	p*K*a2	p*K*a3
Hydrobromic	80.91	−6		
Nitric	63.01	−1.32		
Trifluoroacetic	114.02	0.23		
Saccharin	183.19	1.31		
Acetic, 2,2-dichloro-	128.94	1.35		
Acesulfame	163.15	*2*		
Camphor-10-sulfonic (+)	232.29	*2.17*		
Salicylic	138.12	2.97	13.82	
Salicylc, 4-amino-	153.14	3.25	*10*	*3.5*
Mandelic (DL)	152.15	3.37		
Formic	46.02	3.75		
Benzoic, 4-acetamido-	179.17	*4.3*		
Cinnamic	148.16	4.40		
Caproic	116.16	4.8		
Undec-10-enoic	184.28	*4.9*		

Chiral salt formers are in italic
Chiral salt formers with an asterisk (*) available as enantiomer
Class 3: Less advisable for general use, but used in the past

7.3.3 *Solvent-Mediated Crystallization Screening*

Solvent mediation (crystallization from solvents) is a commonly used method for generating crystalline forms. Solvent-mediated crystallization methods include exploration of differing crystallization conditions such as solvent composition, temperature, pressure, supersaturation, and humidity.

7.3.3.1 Slurry Ripening Crystallization Methods

Slurry/ripening techniques which mimic larger scale processes are an effective first step in the phase discovery/optimization process. This method simply involves suspending solids in a suitable recrystallization solvent above the solubility limit. Achieving saturation in a given solvent system may provide an environment in which a crystalline phase will have a lower solubility limit than the corresponding amorphous form. Although a given crystalline phase results in a material with lower free energy than an amorphous material, various packing arrangements may exist which have even lower free energy and solubility. Hence, slurry conversions between polymorphic phases utilizing non-solvating solvent systems are a useful way to generate more stable crystalline phases of a molecule. As indicated in Fig. 7.5, the higher solubility phase (equilibrium defined by Ksp1) dissolves due to the driving force created by the propagation of the lower solubility phase (defined by Ksp2) creating a continuous drive to replace the molecules in solution. As the

Table 7.4 List of basic salt formers

Name	Mw	p*K*a1	p*K*a2	p*K*a3	Class
Potassium hydroxide	56.10	14			1
Sodium hydroxide	40.00	14			1
Magnesium hydroxide	58.32	14	11.4		1
Calcium hydroxide	74.10	12.6	11.57		1
Choline hydroxide	121.18	11			1
Ammonia *volatile*	17.03	9.27			1
*Arginine (L)**	174.20	13.2	9.09	2.18 A	1
*Lysine (L)**	146.19	10.79	9.18	2.16 A	1
t-Butylamine	73.14	10.68			1
Glucamine, N-methyl-	195.21	8.03			1
Hydrabamine *low solubility, sec. amine*	596.97	*11.92*	*11.32*		2
Diethylamine *secondary amine*	73.14	10.93			2
Benzathine *low solubility, sec. amine*	240.35	*9.99*	*9.39*		2
Ethanol, 2-(diethylamino)—*nitrosamines*	117.19	*9.58*			2
Pyrrolidine, 1-(2-hydroxyethyl)-	115.18	*9.44*			2
Benethamine low solubility, *sec. amine*	211.30	*9.41*			2
Deanol	89.14	*8.83*			2
Tromethamine	121.14	8.02			2
Morpholine, 4-(2-hydroxyethyl)—*prodrugs*	131.17	*7.39*			2
Zinc hydroxide	99.40	14	9.64		3
Ethylenediamine	60.10	10.09	7.00		3
Piperazine	86.14	9.82	5.68		3
Ethanolamine (2-aminoethanol)	61.08	9.50			3
Diethanolamine (2,2′-iminobis(ethanol))	105.14	*8.96*			3
Triethanolamine (2,2′,2″-nitrilotris(ethanol))	149.19	7.82			3
Imidazole (1H-)	68.08	6.95			3

Chiral salt formers are in italic
Chiral salt formers with an asterisk (*) available as enantiomer

higher free energy phase is consumed, the lower energy and more stable phase will continue to form until equilibrium is achieved. Essentially, the solubility difference between the phases under the experimental conditions drives the dissolution of the less stable and more soluble phase (Cardew and Davey 1985).

The risk of a new more stable phase appearing can increase with scale as API is produced with fewer impurities available to poison initiating nucleation events. Likewise, the appearance of a more stable phase makes it more difficult to reproduce a previously observed metastable phase.

Initiating nucleation by varying solvent systems is common in slurry-ripening experiments. Examples of solvent systems with differing dielectric constants, hydrogen bonding potential, and crystallization temperature ranges have been utilized to aid in the prediction of API solubility (Chen and Song 2004; Tung et al. 2008). The objective is to use a system where nucleation probability of a new

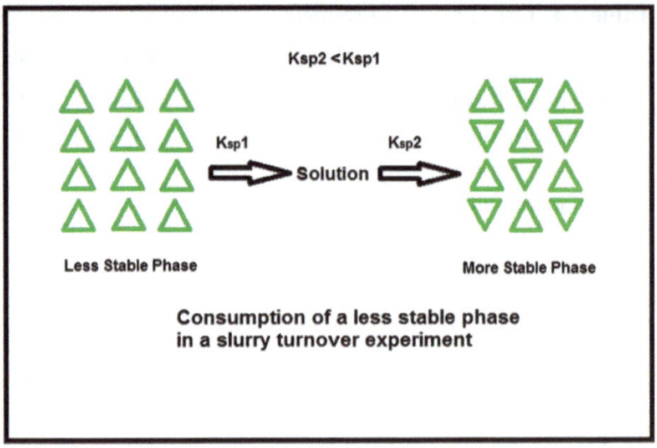

Fig. 7.5 Consumption of a higher solubility phase in a slurry turnover experiment

crystalline phase has the highest chance of success. Prediction of the equilibrium solubility concentration of an API in a variety of solvent systems allows for selection of the optimum systems where the balance of solubility and supersaturation can occur on very small quantities of material. Furthermore, systems with poor solubility may be less likely to produce large enough quantities of initiating nucleation events to generate measurable quantities of a new phase on the time scale of an experiment.

Many other variables have the ability to impact primary nucleation of a crystalline material. Studies have established that varying temperature, agitation, levels of impurities and their makeup, and solid surface characteristics can have a remarkable impact on initiating nucleation events and subsequent propagation. Wilhelm Ostwald first theorized that initiating nucleation events in solution would result in the discovery of the most stable crystalline phase last (Nyvlt 1995). Conceptually, this is due to the irreversible nature of metastable phases to transform into more thermodynamically stable phases. More stable states also have the greatest difference in free energy from the initial state. The large difference may preferentially favor initiating nucleation of a less stable phase not as far removed in free energy from the initial material. One can imagine the transformation of a less stable phase to a more stable phase, but unless an enantiotropic relationship exists (discussed in Sect. 7.4) where there is a transition temperature above which the stable-phase relationship exchanges, converting to the metastable phase from the more stable phase would be thermodynamically opposed.

7.3.3.2 Evaporation and Temperature Control Crystallization Methods

Using evaporation to change the quantity of solvent relative to the compound is another way in which supersaturation can be achieved, enabling nucleation and propagation of new crystalline materials. Similarly, supersaturation can be achieved for compounds which have variability in their saturation concentration as a function of temperature (Fig. 7.6).

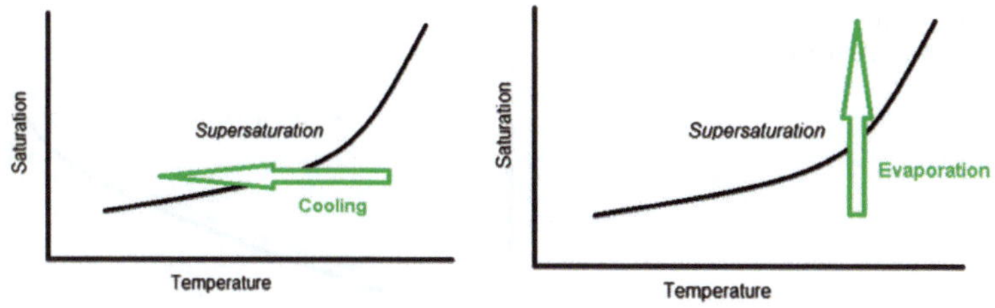

Fig. 7.6 Graphs showing the effect of cooling (Δ temperature, *left*) and evaporation (Δ solvent volume, *right*) to achieve supersaturation

Evaporation as a first-pass technique is typically pursued with a compound being fully dissolved in the solvent system. Solubilization can be achieved by equilibration with time or is commonly done by gently heating and agitating the material in solution. Hot filtration of remaining solids enables the removal of potential nucleation-hindering materials or low-solubility impurities which may poison a desired nucleation event (Myerson 2002). Nucleation sites remain an important crystallization initiation point. Techniques such as etching the glass vessel, addition of seeds of a crystalline material of similar molecular structure, or even the introduction of inert external nucleation materials like activated carbon may be used to cause primary crystallization.

Thermal instability of a compound is important with methods involving heat, as heating and dissolving reactive organic compounds in organic solvents can generate degradants. We will discuss transitions between crystalline phases which are a result of temperature variation later in this chapter. When the objective is to nucleate a new crystalline phase, it is commonplace to utilize a similar technique to that described in the evaporation experiment where filtration at elevated temperatures can remove potential nucleation poisons. Temperature in these systems can be manipulated in many ways, but achieving the goal of a slightly supersaturated system in which nucleation can be initiated remains the same. When scaling material quantities (discussed in Sect. 7.4), seeding a heated mixture with a desired stable crystalline phase at the supersaturation point can avoid generation of metastable phases upon cooling as illustrated in Fig. 7.7.

7.3.3.3 Antisolvent Precipitation and Antisolvent Vapor Diffusion Crystallization Methods

Adding a miscible non-solubilizing solvent (antisolvent) to a solution of compound dissolved in a solubilizing solvent reduces the saturation solubility in the new mixed system and may create a supersaturated environment (Fig. 7.8). Since small changes in the volume of the antisolvent can have a striking effect on compound solubility, antisolvent addition is typically done dropwise with agitation to ensure mixing. Using this method, it is typical that at a certain point, solids precipitate in an uncontrolled manner. This favors formation of crystalline phases

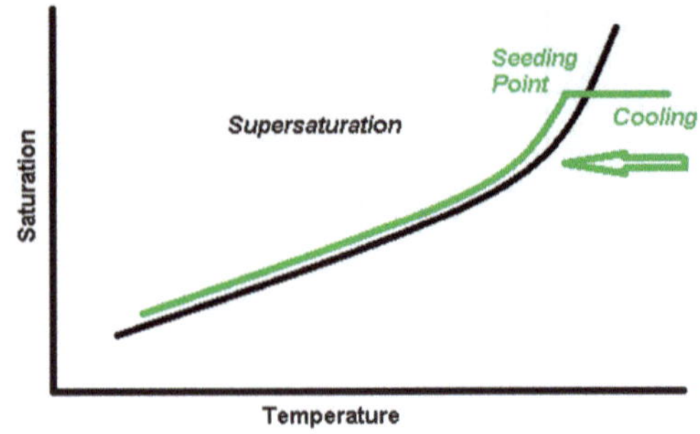

Fig. 7.7 Utilizing seeding- and temperature-dependent solubility

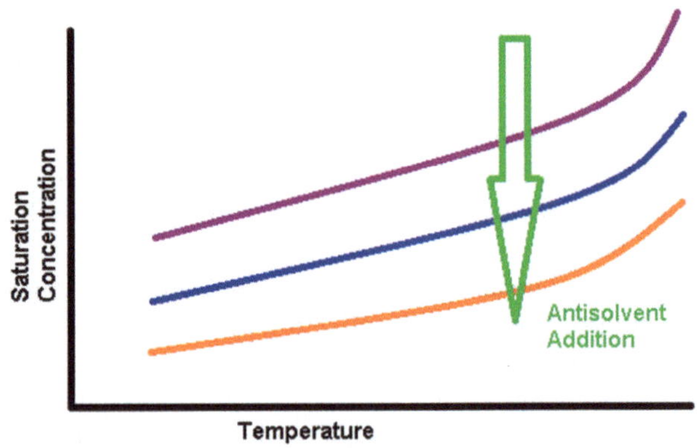

Fig. 7.8 Utilizing antisolvent addition to impact the solubilizing ability of the solvent

with a small free energy difference from the higher energy amorphous material. Hence this process typically results in higher energy phases or solvates rather than more stable anhydrous phases.

Imagine slowing down the antisolvent addition process to a deliberately and controlled addition of the antisolvent. On scale, control over this process would be quite difficult, but at small scale, this experiment can be mimicked by addition of the antisolvent as a vapor to the solvent and dissolved compound. Hence, simply exposing the molecule in solution to the vapor of antisolvent in an enclosed environment can cause crystallization. Vapor-induced crystallization methods typically include water vapor, individual or mixed organic solvent vapors, or a combination of the two. Antisolvent diffusion and similar methods also intrinsically have a higher likelihood of generating solvated crystalline phases when compared to other methods like thermal annealing. However, the ability to probe very small quantities of material which may result in a marked purity upgrade and generation of seed makes this an attractive nondestructive early-development-phase discovery technique. Experimentally, the molecule is dissolved in a solvent system in which it has high solubility. Then this combination is added to a container possessing vapor of a solvent which the molecule has lower solubility in. If the antisolvent is miscible in the initial solvent system, over time the solvating properties of that system will change as it approaches equilibrium, preferentially driving past the point of supersaturation and creating a precipitant which, hopefully, is crystalline.

7.3.4 Humidity-Mediated Crystallization Methods

In some cases, solvated phases have the ability to rearrange once the solvent is removed or replaced in the crystalline lattice. A common example of this effect is seen in the presence or absence of water vapor. Discovery of a more stable crystalline anhydrous or hydrate phase resulting from rearrangement in the crystal packing of a transient solvate can be quite valuable. Methods as simple as placing a solvated crystalline phase in a vacuum oven with air or nitrogen purge have produced stable and viable anhydrous phases and desolvates suitable for development.

There are a number of methods used to control the relative humidity in the environment around a sample. Inorganic salt chambers which regulate the humidity in the enclosure are a simple way to expose a material to a controlled humidity environment. Readily available salts such as sodium chloride or bromide, magnesium chloride or nitrate, lithium chloride, and potassium chloride or acetate when submerged in water in a chamber at a given temperature will result in a defined relative humidity in the enclosed environment (Table 7.5).

For fine control over humidity, dynamic vapor sorption (DVS) instrumentation is commonly used. This technique not only allows for subtle humidity manipulation of gasses flowing over a sample, but it also measures quantity and speed of the solvent's uptake. When crystalline hydrates form, the water molecules may occupy space in the crystalline lattice which is accessible to external humidity. This result will be manifested as a step in the water vapor sorption isotherm which typically can be used to calculate the stoichiometry of the resulting hydrate (Fig. 7.9). If a phase absorbs water via channels, the level of hydration has a tendency to be linear with regard to experimental humidity. Although the DVS technique does not define that a new material is in fact crystalline, it can be used as a guide to generate the materials which can be analyzed by visual microscopy, variable relative humidity XRPD, and Raman techniques discussed later in this chapter. DVS as a characterization method is also useful for determining if deliquescence of a crystalline salt form occurs should the counterion have a very high affinity for water. If the deliquescence point is at a low enough humidity, it is advisable to identify an alternative phase.

Table 7.5 Percent relative humidity generated in headspace by water-saturated inorganic salts in the terrestrial and biorelevant temperature range (adapted from Wexler and Hasegawa 1954; Greenspan 1977)

Temperature	5 °C	20 °C	25 °C	40 °C	50 °C
Lithium chloride	11	11	11	11	11
Potassium acetate		23	23		
Magnesium chloride	34	33	33	32	31
Magnesium nitrate	59	54	53	48	45
Sodium bromide	64	59	58		
Sodium chloride	76	75	75	75	74
Potassium chloride	88	85	84	82	81
Potassium nitrate	96	95	94	89	85
Potassium sulfate	98	98	97	96	96

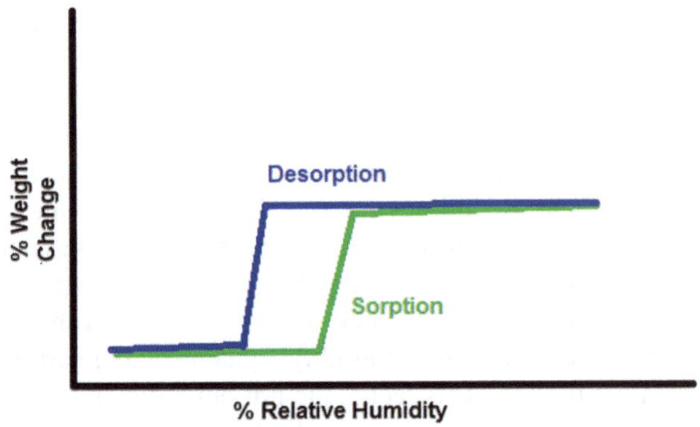

Fig. 7.9 Typical dynamic vapor sorption/desorption isotherm for an anhydrate to stoichiometric hydrate transition

7.3.5 *Temperature-Mediated Crystallization Methods*

The glass transition temperature (Tg) determined via thermal analysis of an amorphous material defines the temperature at which the material changes properties from a brittle solid. There are several methods which can be used to define this temperature, but a simple and commonly utilized technique is via differential scanning calorimetry (DSC). This instrument measures the difference in the quantity of heat required to increase the temperature of a sample vs. a reference. It can also be utilized to determine a host of temperature-induced transitions like phase transitions, including endothermic or exothermic transitions and fusion/melting point of a crystalline material along with the corresponding heat capacity. An exothermic transition noted by this technique may be indicative of a crystallization event.

Using a technique such as DSC to detect transitions due to crystallization is a simple, material-sparing method to discover crystalline phases and generate valuable information on them. Typically only a few mgs of material are required. Via instrument control, one can impart control over heating rate, cooling rate, sweeping gasses, and on some systems even modulate the temperature rapidly with precise control. It is typically not recommended to perform open solvent-mediated experiments inside the instrument, but the introduction of pans which can be hermetically sealed along with the ubiquitous use of the complementary technique of thermogravimetric analysis (TGA) makes the instruments a mainstay for defining material properties of pharmaceuticals.

Once information has been gathered on the various thermal events, additional control techniques aimed at phase discovery may be undertaken. Methods in which to vary temperature aimed at generating new crystalline phases include temperature cycling near the Tg, controlled or quench cooling from the melt, sublimation, and annealing. In some cases, an exothermic event may occur above the Tg which indicates crystallization. If an event is observed, one may manipulate the material above that temperature and slowly cool it attempting to isolate crystalline seed for further study. In a minority of cases, multiple events may occur in the study, corresponding to several crystalline forms with which to follow up. As discussed with solvent-mediated crystallization, thermal methods may only need to be used for the initial generation of a new crystalline phase. Once a particular crystalline phase has come into existence, it may no longer require the originating system to generate more of that phase.

7.3.6 High-Throughput Solvent-Mediated Phase Screening

Discussed previously, different types of experiments will have a tendency to create crystalline phases which are inherently more or less stable based on the dynamics of the crystallization. Solid-state experiments like thermal cycling, cooling from the melt, and sublimation are simplistic experimentally, but unfortunately the process is quite difficult to miniaturize and reproduce in a high-throughput fashion. Solvent-mediated experiments on the other hand are highly amenable to automation due to the rise in automation and apparatus capable of manipulating caustic chemicals. As noted in Fig. 7.10, these experiments are also more likely to form stable crystalline phases as crystallization times can be on the scale of hours and days. The resulting materials can also be further manipulated by solid-state experiments like exposure to solvent vapor, humidity, and altered temperature for extended periods of time to both discover new phases and understand stability.

The simplicity of both solvent- and temperature-mediated crystallizations coupled with advances in robotics used for preparation, manipulation, and characterization has led to the development of rapid "high -throughput" protocols for crystallization screening (Morissette et al. 2004). Here, solvent-mediated phase discovery methods are organized into a matrix of crystallization experiments commonly referred to as crystallization screens via workflows like those seen in Fig. 7.11. Advances in automation enabled faster preparation using minimal compound and more sensitive analysis, providing the ability to generate and analyze small quantities of many different crystalline "leads." Thus, the experimental

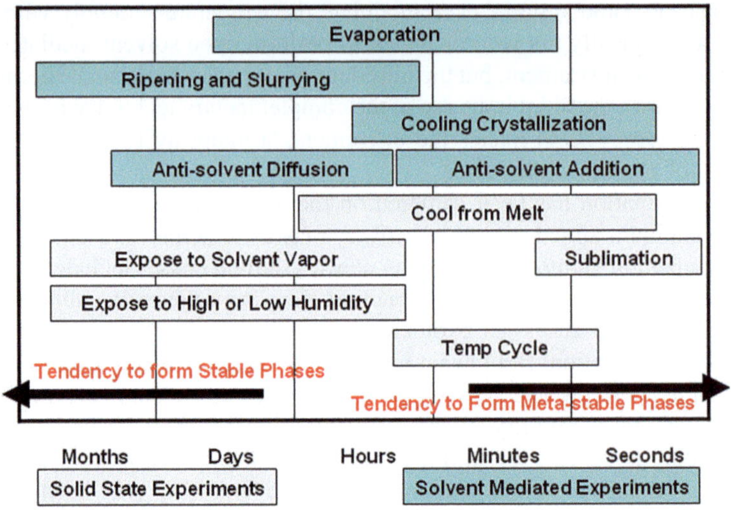

Fig. 7.10 Typical experimental timelines for discovery of new crystalline phases (some data from Anderton 2007)

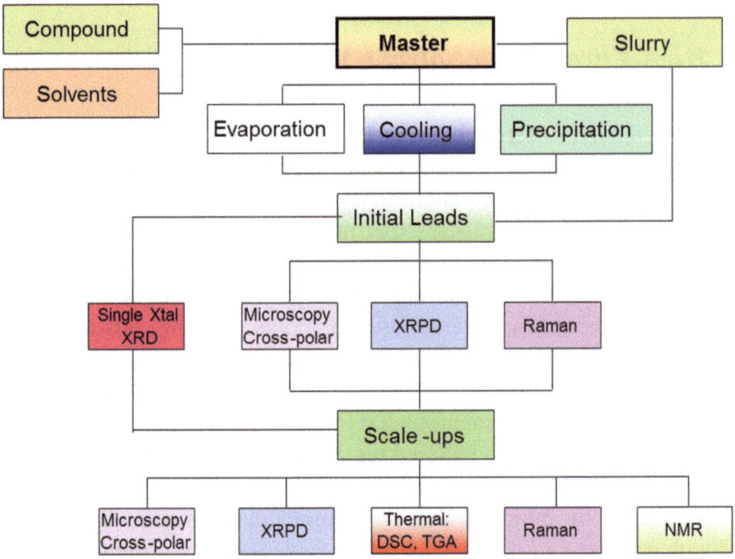

Fig. 7.11 Generic workflow for matrix crystallization experimentation (some data from http://www.freeslate.com/life-science-applications/small-molecule/preformulation)

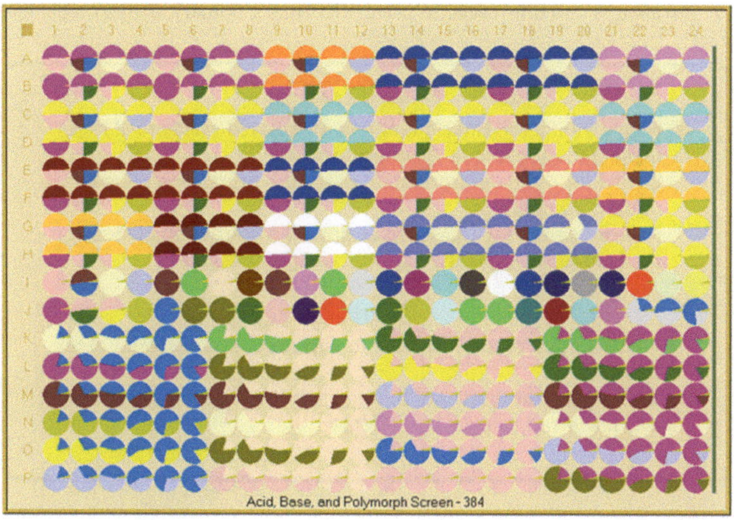

Fig. 7.12 Example of a 384-well screening design matrix aimed at discovery of acid and base salts, co-crystals, and polymorphs. Different colors represent different components within the well

crystallization matrix has evolved to the standardized well plate format popularized in biologics screening and is capable of handling 96, 384 (Fig. 7.12), and higher concurrent solvent crystallization experiments. Specialized solid- and liquid-handling robots using tailored matrix protocols designed to target crystallization of preferred counterions and solvent systems were developed to handle the sample throughput. Temperature, evaporation, humidity, and antisolvent controls also were incorporated along with variable agitation methods. Advances in computing have led to ease in deconvolution of the resulting analysis datasets further enabled by computerized categorization.

Solution-mediated high-throughput experiments in the 1–5 mg range can be a time-, compound-, and resource-sparing way to discover viable crystalline phases as compared to manual methods. Besides the resource savings, this method also attempts to mimic, although on very small scale, chemical processing conditions which may be utilized on scale to generate batches. A targeted set of experiments with only a few hundred mgs of material has the potential to yield information on solvent incompatibilities, crystal morphology, and processing temperature dependence. If the use of a nondestructive analysis technique is utilized (such as XRPD), the material can be incubated for short periods of time at various temperature and humidity ranges and reanalyzed for an early understanding of a phase's physical stability along with the potential to form new phases from solid-to-solid transitions and humidity-mediated processes.

Table 7.6 Advantages and drawbacks to using XRPD as a primary analysis technique for HT screening

Advantages

- Highly useful for small quantities of isolated solids (1-2 mgs) making the technique amenable for HT
- Useful for rapid differentiation of unique crystalline phases via pattern comparison
- Powder sample will present all possible diffraction planes for analysis and responsive to wet sample analysis

Drawbacks

- Isomorphous phases difficult to differentiate by powder x-ray analysis
- Sample preparation may be required to mitigate sample height and preferred orientiation effects

7.3.7 X-Ray and Raman Analysis Screening Characterization Methods

The small quantities of material generated by automated preparation methods require nondestructive and rapid analytical methods for initial evaluation. Evolving technology has pushed the speed of gathering XRPD and Raman measurements to minutes or seconds. Furthermore, sample stages designed for incorporation of template-based multi-well plates along with position-selectable software enable rapid hands-off analysis of high-throughput sample quantities (Kojima et al. 2006). The nondestructive nature of these techniques also allows for complementary analysis while sparing the crystalline seed for future use in scale-up experiments. Automated Raman and XRPD instrumentation for high-throughput analysis is now commercially available with XRPD arguably being the preferred first-pass HT technique due to advantages over other techniques (Table 7.6).

As an analytical technique, XRPD diffraction yields information which is derived from the crystal packing arrangement of the molecule. Planes of electron density interact with a single-wavelength X-ray beam and produce a unique pattern which can be used to distinguish one phase from another. Hence, the "fingerprint" generated by XRPD (as noted in Fig. 7.13) of a particular crystalline form is easily differentiated from differing packing arrangements even of the same molecule as is the case with polymorphs.

Raman analysis is a vibrational spectroscopic technique which uses scattering of a tunable incident beam of radiation (typically in the visible or NIR band) to produce photons with different wavelengths emitted post-sample interaction. This method allows for highly differentiated signals from very small quantities of material and the ability to detect small spectrum differences. These differences can be attributed to particular atoms in a molecule and allow for an understanding

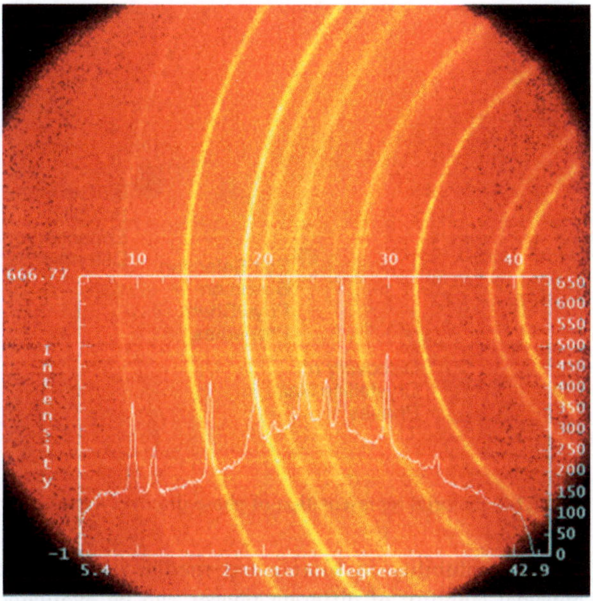

Fig. 7.13 Characteristic diffraction rings as seen on a plate detector in 2-D space and corresponding peak-averaged XY pattern

of likely intra- and intermolecular interactions. Raman analysis can have advantages over X-ray techniques when studying solvates, salt, and co-crystal forms that crystallize in identical crystallographic space groups with similar packing parameters but have differing cosolvents, counterions, etc. (Saesmaa et al. 1990). Unfortunately, this technique has a higher likelihood of sample alteration or destruction through localized heating by the tunable incident beam. Comprehensive descriptions of these techniques can be found in various publications (Jenkins and Snyder 1996).

There is some debate over the best first-pass characterization technique. However, it is clear that utilized together, Raman and XRPD datasets can be quite definitive for differentiating between phases. The interpretation of the large datasets generated becomes an issue and a keen eye is required to pick up on the subtle differences between spectra or patterns. Mixtures of metastable and stable crystalline phases and the presence of transient solvates or co-crystals are all examples of areas where the analyst's judgment is often required over computer-based data classification.

7.4 Selection of a Lead Development Phase

7.4.1 Definitive Characterization of Select Lead Phases

Once phase screening studies have been completed and there are multiple phases to choose from, it is appropriate to collect further appropriate experimental information in order to select between possible development phases. Here key attributes

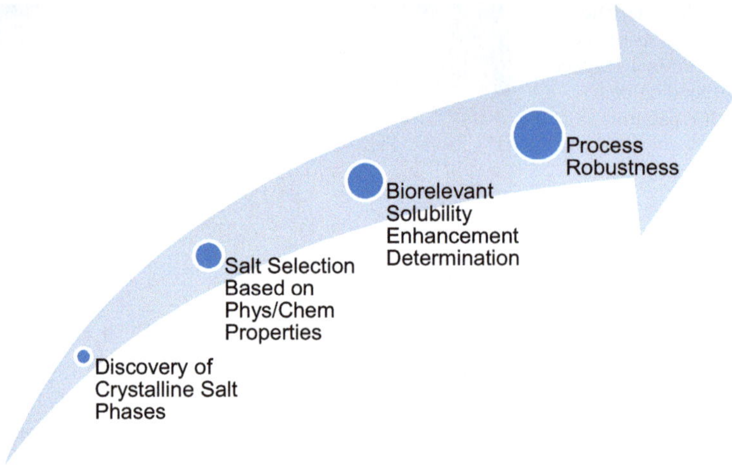

Fig. 7.14 Pictorial representation of phase discovery, evaluation, and selection process

such as solubility in biorelevant media, degree of crystallinity, and chemical/physical stability of different phases can be evaluated and compared in order to select the preferred phase for development. One way to visualize the key attributes that differentiate phases is shown in Fig. 7.14.

7.4.2 Scaling Crystalline Material for Detailed Characterization

Before we discuss more detailed phase characterization, it is worth briefly commenting on scaling select leads to have enough material to evaluate. Should a suitable phase be discovered during screening studies, it may be used to seed a larger scale supersaturated solution via solvent-mediated methods. If a new phase truly is viable, even if it was found initially by a solvent-free thermal method, it is likely to propagate in a batch-scale processing environment if appropriately added to a saturated solution. Even crystals isolated from a solvent system that is not suitable for human consumption can be used as seeds, with the understanding that a more innocuous solvent system must be used for batches intended for the clinic.

Controlled-temperature solvent-mediated techniques with seeding are favored when generating larger quantities of a crystalline phase. Even with limited knowledge of saturation concentration at differing temperatures and how that changes with temperature, quantities of a desired crystalline phase may be generated reproducibly. Heating and solution agitation imparted to a slurry solution above the saturation concentration can be followed by seeding and slow cooling. This method will first generate a bed of nucleation points of the desired crystalline phase, providing sites for propagation. The subsequent cooling will facilitate continued precipitation from solution as the desired phase and prolonged agitation is likely to ensure conversion to a single thermodynamically stable phase. One caveat with this technique is that different crystalline phases may be stable at differing temperatures. As discussed later in this section, enantiotropic polymorphs have a conversion temperature which should be utilized as a temperature limit if the relationship is understood in advance.

Once a stable phase is identified, it is unlikely to change unless a new, more stable phase under the storage or experimental processing conditions emerges. It is atypical that thorough phase screening protocols will fail to turn up stable phases at given storage and processing conditions. Furthermore, the risk of negative impacts on solubility due to form change by switching between closely related thermodynamically related crystalline phases is far less than what would be observed going from amorphous material to a stable crystalline phase.

Any crystal phase used for seeding in a crystallization process will usually achieve a significant purification upgrade. Precipitates from antisolvent techniques may not be crystalline, but may have a different impurity profile. In this case, the resultant mixture may be further manipulated via complementary material control techniques such as temperature cycling to generate a crystalline phase via nucleus generation at a temperature-related supersaturation point.

7.4.3 Biorelevant Solubility Evaluation

Solubility can be measured by various high-throughput and manual methods, and this topic has been well described in the literature over the years (Serajuddin et al. 1986). High-throughput methods are generally employed during the lead optimization stages of discovery, where the solubility is determined via precipitation from a solvent stock solution using robotic methods. However, the precipitated phase is generally not characterized in this method. For phase discrimination, solubility determination starting from a solid of known phase obviously is required. It is prudent to study each phase's solution behavior in several buffers at different pH values and also in simulated gastrointestinal media (FeSSIF/FaSSIF; Dressman and Reppas 2000). Salts with varied counterions can have significantly different solubility behavior and can be rank ordered for preference. Absolute risks for poor oral absorption as related to solubility and dose can be estimated using the concept of maximum absorbable dose (MAD, Hilgers 2003). Solubility and stability in potential preclinical formulation vehicles also should be evaluated. This topic is discussed briefly later in the upcoming formulation section.

7.4.4 Thermal and Hygroscopicity-Related Phase Stability

A thorough assessment of physicochemical properties of an assortment of salts and phases typically begins with studies of the temperature- and humidity-related stability, since these are two highly variable conditions to which a drug substance or product may be subjected. Towards this end, melting point, solvation/desolvation behavior, and transition temperatures between differing polymorphic phases all should be studied.

7.4.4.1 Thermal Stability

DSC and thermogravimetric analysis (TGA) are two common thermal analytical techniques utilized to study and differentiate between different crystalline phases (Craig and Reading 2006). A common type of DSC has dual heaters for the pan holding the sample and a standard. Typically, the sample is heated at a linear rate, held isothermally, or cycled (modulated) throughout the heating range. The heat flow to or from a sample is measured and observations of exothermic and endothermic events (i.e., melting, crystallization, or T_g) along with changes in heat capacity can be reproducibly made.

A TGA instrument measures weight loss of a sample as a function of increasing temperature. This method is most useful for determining if solvents are associated with a compound and the temperature where a compound decomposes. Coupling a TGA instrument with in-line techniques capable of analyzing the resulting emissions throughout the temperature range (e.g., Mass Spec) can provide qualitative determinations of components that are liberated upon heating.

In the case of polymorphism, the relationship between true polymorphs with differing unit cells may be classified as monotropic or enantiotropic. This polymorphic relationship is further defined in Figs. 7.15, 7.16, and 7.17. For enantiotropic systems, the significance of this solid-state phase transition between discovered polymorphic phases is paramount in early development as the selected processing temperature will directly impact the isolated crystal form in the case of enantiotropic polymorphic phases. In these cases, thermodynamic stability of the polymorphs depends on the temperature at a given pressure.

It is important to remember that the kinetic stability of convertible phases also may play a role in conversion determinations and impact the ability to predict storage behavior. In the case of metastable phases, either phase may be suitable depending on the processing or stability conditions chosen. This topic has been described in a variety of publications and a recommendation for an in-depth guide to differentiating between these phases utilizing thermal methods of analysis can be found in Craig's chapter of pharmaceutical applications of DSC (Craig and Reading 2006).

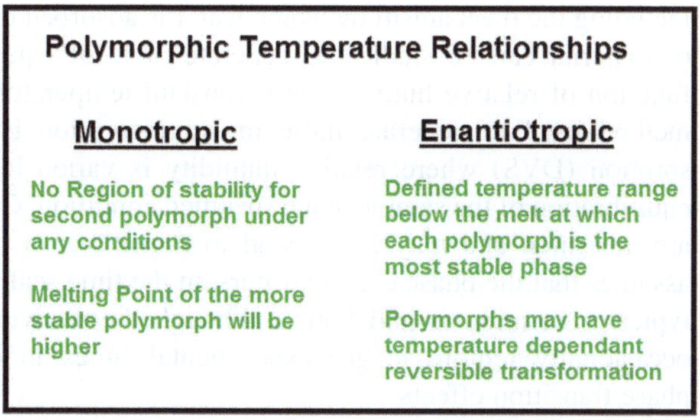

Fig. 7.15 Monotropic and enantiotropic polymorphic relationships

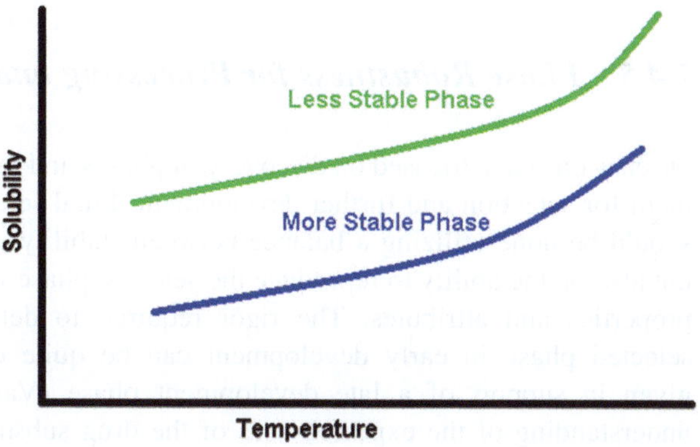

Fig. 7.16 Monotropic relationship between polymorphs

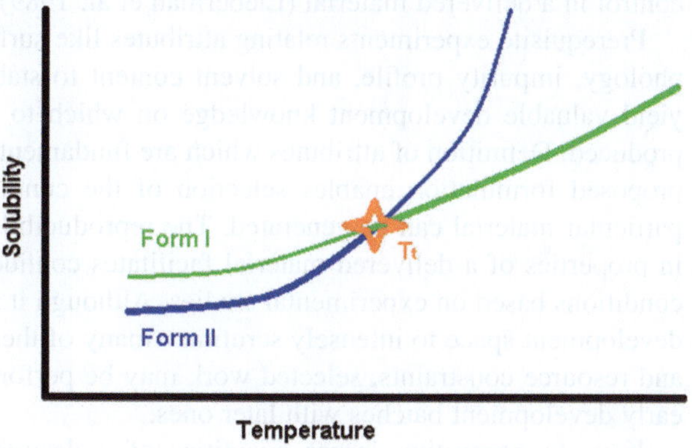

Fig. 7.17 Enantiotropic relationship between polymorphs with a transition temperature (Tt) below which form II is the more stable phase

7.4.4.2 Hygroscopicity Liabilities

Dramatic effects of water-solid interactions may occur under relatively high-humidity conditions, such as deliquescence as previously mentioned. Hence, it is important to gain an understanding of the impact of water on a given phase or salt, including the mechanism by which water is adsorbed or desorbed. Small quantities of material can be studied to generate plots of equilibrium water content as a function of relative humidity at a constant temperature. A common instrumental method used to generate these moisture-sorption isotherms is dynamic vapor sorption (DVS) where relative humidity is varied in step fashion, followed by equilibration of the sample at the specified condition. Changes in the sample weight are measured and may correspond to transitions in physical state. This method assumes that the phase change occurs on the time scale of the experiment which is typically in minutes and hours, although thermodynamically favored processes occasionally require longer experimental times in order to overcome kinetic phase transition effects.

7.4.5 Phase Robustness for Processing and Storage

Discussions have focused on discovery of phases and initial analysis to differentiate them for selection and further development. Final selection of a preclinical phase should be done utilizing a balance between stability and processability concerns, but also on the ability to reproduce the selected phase with control over the desired properties and attributes. The rigor required to determine processability of a selected phase in early development can be quite different than the attention given in support of a late development phase (Variankaval et al. 2008). An understanding of the expected uses of the drug substance along with the storage and formulation processing requirements can guide the choice of attributes which to control in a delivered material (Lieberman et al. 1989).

Prerequisite experiments relating attributes like surface area, particle size, morphology, impurity profile, and solvent content to stability and performance will yield valuable development knowledge on which to build as larger batches are produced. Definition of attributes which are fundamental to the performance of the proposed formulation enables selection of the condition ranges under which a particular material can be generated. The reproducibility imparted by consistency in properties of a delivered material facilitates confidence in selection of storage conditions based on experimental studies. Although it may be difficult in the early development space to intensely scrutinize many of these attributes due to material and resource constraints, selected work may be performed focused on correlating early development batches with later ones.

Keys to generating larger quantities of a desired crystalline phase include understanding phase relationships between known crystalline phases providing nucleation sites, and enabling growth to desired shape and size. There are many ways to vary temperature, with two main goals comprising the technique. The first goal is to promote nucleation at as many sites as possible, providing a large quantity of crystalline seeds on which bulk crystalline solids can propagate. The second goal is to promote growth on this seed bed which sustains the packing of more molecules on top of each other in an ordered fashion, thus removing impurities and creating crystals of similar size, shape, and properties.

Knowledge of the saturation concentration of a compound at a given starting temperature and how the corresponding change with a drop in temperature can enable quantities of the desired crystalline phase to be generated reproducibly. Chromatographic experiments which consume minimal amounts of material may be performed to determine these concentrations. Aspects of large-scale manufacturing can be paralleled in an effort to create material which has uniform and reproducible particle properties which are likely to be sought after for their tendency to flow through process train equipment without sticking to surfaces or segregate from other formulation components.

7.5 Preclinical Formulations as Related to API Phase and Physicochemical Properties

7.5.1 Preclinical Formulation Options as Related to Drug Phase

Now that the rationale and tactics for identifying a suitable API phase have been described, a short section on how to approach preclinical formulations in the context of phase is in order. Even the most desirable crystalline phases cannot progress without adequate oral absorption in preclinical toxicology studies, The key area of focus often is enhancing the oral absorption of poorly soluble compounds, regardless of the existing solid-state phases, and various enabling formulation options can be engaged for enhancing oral absorption (Kwong et al. 2011; Mass et al. 2007). One way to envision formulation options in relation to complexity is shown in Fig. 7.18.

When a suspension is used, micronization is a relatively simple and well-established means of increasing dissolution rate and reducing absorption variability. As particle size is further reduced, nanoparticles ("nanos," particle size below 1 μm) can be prepared for maximizing dissolution rate (Merisko-Liversidge et al. 2003). Small-scale methods for making nanoparticles such as media milling and high-pressure homogenization are becoming increasingly common in a discovery setting. Generally, the addition of a surfactant stabilizer is required to keep the nanoparticles from agglomerating. Regardless of particle size, suspension formulations should be characterized at least by light microscopy, so as to have some understanding of what was dosed.

Although a simple solution or suspension formulation is always preferred, poor solubility or the need for increased exposure often requires us to move towards more complex enabled formulation options as shown in Fig. 7.18. There is a wealth of information available on the use of various enabling excipients such as surfactants, cosolvents, or lipid/emulsion formulations (Porter et al. 2007; Kommuru

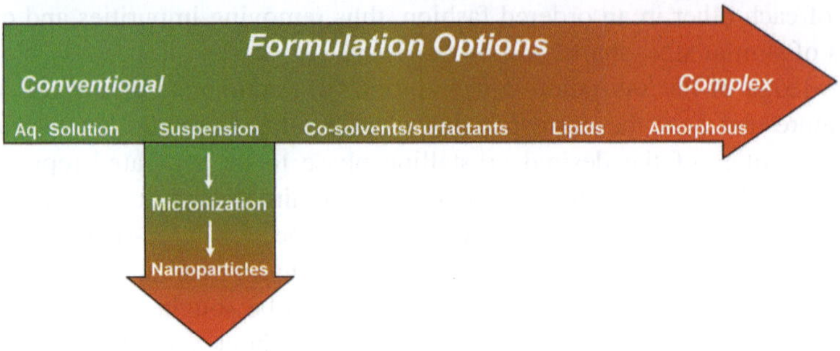

Fig. 7.18 Some preclinical formulation options in relation to complexity

et al. 2001) which all can be very effective for enhancing the absorption of insoluble compounds. It should be noted, however, that care must be taken in selecting the formulations used for in vivo behavioral models, where a poorly tolerated enabled formulation can confound observations from the study. Furthermore, although tremendous improvement in oral exposure can be gained through the use of enabled formulations, such measures cannot fix poor intrinsic API properties such as low cell membrane permeability (often seen initially via *low* permeability in the in vitro Caco 2 cell model) or high in vivo clearance rates due to extensive metabolism.

For concentrated solution formulations, the likelihood that the API will "crash" out once the formulation enters the GI tract is a well-known caveat. This is not always insurmountable, since a precipitate may be reasonably soluble such as an amorphous phase or nanoparticles. Regardless, the potential impact of this precipitation process can be evaluated by in vitro means, where the formulation is sequentially diluted into simulated gastric fluid followed by a second dilution into FaSSIF. Assay of the final concentration of the supernatant solution after removal of any precipitates allows for a series of formulations to be rank ordered for solubility enhancement in the GI tract. This gives a reasonable prediction of the formulations' in vivo performance and a rationale for formulation selection.

Although we have described many desirable attributes of a robust crystalline API phase, the more soluble amorphous phase is often attractive for enhancing drug solubility. However, it is generally preferred to disperse the amorphous drug into an amorphous polymer (i.e., a "solid solution" in a cellulosic polymer) in order to prevent crystallization Kaushal et al. 2004). The practical methods of choice for the preparation of amorphous drug-polymer dispersions are spray drying or hot melt extrusion. Although it is fairly common in academic and development settings, the routine use of amorphous drug dispersions in a discovery setting is less mature, partly due to the relatively large quantities of drug required to make the dispersion. As the preparative methods increase in efficiency (and decrease to mg scale), amorphous dispersions now are finding increased utility well before entry into development.

Once a suitable API phase and a functional preclinical formulation are identified, the chemical and physical stability of the API in the formulation must be well

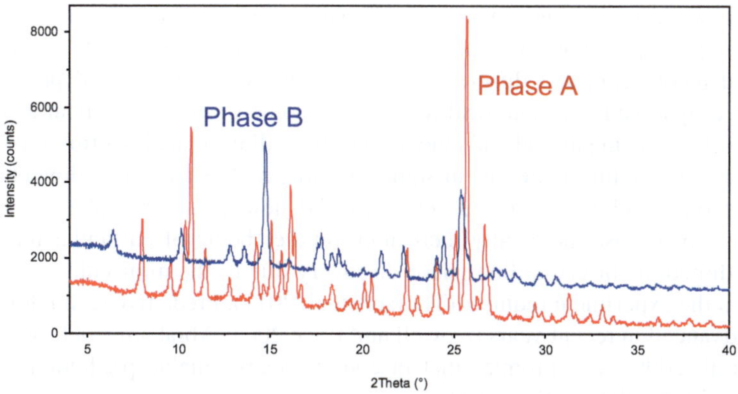

Fig. 7.19 XRPD patterns showing the formation of a new polymorph (phase conversion from A → B)

understood and carefully monitored. Although the API chemical stability requirement is obvious, phase changes often are more subtle but can be just as deleterious to the formulation's performance. This is exemplified in Fig. 7.19, where powder X-ray diffraction of API solids isolated from a suspension formulation clearly showed that after several hours, the initial phase of the API converted to a new polymorph (phase A → B). As is often the case, the new previously unobserved phase was less soluble (more stable) than the starting phase and adequate exposure could no longer be achieved.

As mentioned previously, this scenario supports the strategy of performing at least some polymorph screening before any significant preclinical safety studies start, in order to build confidence that a stable API phase is in hand and adequate exposure can be maintained.

7.6 Conclusion

This chapter has provided a rationale and some methods for the early identification of a solid-state phase that is suitable for clinical development, where lessons learned in the pharmaceutical industry as a consequence of catastrophic phase change have put phase selection at the forefront of pharmaceutical materials science research. Although generation of viable phases and a full understanding of their associated physicochemical properties can be a daunting task, these efforts constitute a worthwhile path towards de-risking potential phase-related development variables and can ultimately save time and resource. As new drugs' physicochemical properties trend towards poor solubility and subsequent oral absorption, early phase optimization will only continue to grow in importance and practice.

If developments of the recent past are any indication of the future state of phase screening in early development, then the drive to utilize smaller quantities of material while increasing efficiency and analysis information are sure to advance. Technology improvements in spectral pattern recognition are likely to provide more detailed recognition and differentiation between newly discovered phases when analyzed by advancing analytical techniques. Some of these analytical techniques can now be done in parallel and innovations have allowed analysis times to become faster while enabling increases in signal resolution. The advent of flow chemistry has percolated ideas of in situ screening for new crystalline phases, with the evolution of robust chemically inert microfluidics being of particular interest for future directions of crystalline phase discovery due to the miniscule quantities of material the experiment requires (Thorson et al. 2011). Predictive technologies also have advanced in recent years (Nowell and Price 2005; Abraham and Probert 2006; Dey et al. 2006) with promise that in coming years reliable prediction of stable phases will translate to the bench.

Finally, changes in development paradigms and availability of relatively inexpensive and resource-sparing future screening techniques may enable additional comprehensive screening analyses to be completed on ever-decreasing quantities of material. This type of advancement would enable hundreds, even thousands, of experiments to progress in parallel at very early stages, potentially on long time scales enabling generation of stable phases earlier in the development life cycle of a drug.

References

Abraham NL, Probert MIJ (2006) A periodic genetic algorithm with real-space representation for crystal structure and polymorph prediction. Phys Rev B 23(22):224104–224110

Anderton C (2007) A valuable technique for polymorph screening. Am Pharm Rev 10:34–40

Bak A et al (2008) The co-crystal approach to improve the exposure of a water-insoluble compound: AMG 517 sorbic acid co-crystal characterization and pharmacokinetics. J Pharm Sci 97(9):3942–3956

Bastin RJ, Bowker MJ, Slater BJ (2000) Salt selection and optimisation procedures for pharmaceutical New chemical entities. Org Proc Res Dev 4(5):427–435

Bauer J, Spanton S, Henry R, Quick J, Dziki W, Porter W, Morris J (2001) Ritonavir: an extraordinary example of conformational polymorphism. Pharm Res 18(6):859–866

Berge SM, Bighley LD, Monkhouse DC (1977) Pharmaceutical salts. J Pharm Sci 66(1):1–19

Bernstein J (2002) Polymorphism in molecular crystals. IUCr monographs on crystallography, vol 14. http://books.google.com/books/about/Polymorphism_in_Molecular_Crystals.html?id=Rg97h45jtIkC

Byrn S, Pfeiffer RR, Stowell JG (1999) Solid-state chemistry of drugs, 2nd edn. SSCI, Inc., West Lafayette

Cardew PT, Davey RJ (1985) The kinetics of solvent-mediated phase transformations. Proc R Soc A398:415–428

Chemburkar SR et al (2000) Dealing with the impact of ritonavir polymorphs on the late stages of bulk drug process development. Org Process Res Dev 4(5):413–417

Chen C-C, Song Y (2004) Solubility modeling with a non-random Two-liquid segment activity coefficient model. Ind Eng Chem Res 43:8354

Chen S, Guzei IA, Yu L (2005) New polymorphs of ROY and New record for coexisting polymorphs of solved structures. J Am Chem Soc 127:9881–9885

Craig D, Reading M (2006) Thermal analysis of pharmaceuticals. CRC, Boca Raton, FL
Dey A, Pati NN, Desiraju GR (2006) Crystal structure prediction with the supramolecular synthon approach: experimental structures of 2-amino-4-ethylphenol and 3-amino-2-naphthol and comparison with prediction. Cryst Eng Comm 8:751–755
Dressman JB, Reppas C (2000) In vitro-in vivo correlations for lipophilic, poorly water-soluble drugs. Eur J Pharm Sci 11:S73–S80
DiMasi JA, Feldman L, Seckler A, Wilson A (2010) Trends in risks associated with new drug development: success rates for investigational drugs. Clin Pharmacol Ther 87:272–277
Florence AJ, Johnston A, Price SL, Nowell H, Kennedy AR, Shankland N (2006) An automated parallel crystallisation search for predicted crystal structures and packing motifs of carbamazepine. J Pharm Sci 95:1918–1930
Gibson M (2009) Pharmaceutical preformulation and formulation: a practical guide from candidate drug selection to commercial dosage form, 2nd edn. Interpharm CRC, Boca Raton, FL
Good D, Rodriguez N (2008) Solubility advantage of pharmaceutical cocrystals. Crystal Growth Design 9:2252–2264
Gould PL (1986) Salt selection of basic drugs. Int J Pharm 33:201–217
Greenspan L (1977) Humidity fixed points of binary saturated aqueous solutions. J Res National Bureau of Standards—A 81A(1977):89–96
Hancock BC, Parks M (2000) What is the true solubility advantage for amorphous pharmaceuticals? Pharm Res 17(4):397–404
Higgins J (2009) Perspectives on preformulation programs at the discovery-development interface. Am Pharm Rev 12:42–46
Hilgers A (2003) Predicting oral absorption of drugs: a case study with a novel class of antimicrobial agents. Pharm Res 20(8):1149–1155
Hilfiker R (2006) Polymorphism in the pharmaceutical industry. Wiley, Weinheim
Hulme AT, Price SL (2007) Toward the prediction of organic hydrate crystal structures. J Chem Theory Comput 3:1597–1608
Jenkins R, Snyder RL (1996) Introduction to powder X-ray diffractometry. Chemical analysis: a series of monographs on analytical chemistry and its applications, vol 138, Wiley Hoboken, NJ
Kaushal A, Gupta P, Bansal A (2004) Amorphous drug delivery systems: molecular aspects, design and performance. Crit Rev Ther Drug Carrier Syst 21:133–193
Kerns EH, Di L (2008) Drug-like properties: concepts, structure design and methods. Academic, London
Kojima T, Onoue S, Nurase N, Katoh F, Mano T, Matsuda Y (2006) Crystalline form information from multiwell plate salt screening by use of raman spectroscopy. Pharm Res 24:806–812
Kommuru TR, Gurley B, Khan MA, Reddy IK (2001) Self-emulsifying drug delivery systems (SEDDS) of coenzyme Q10: formulation development and bioavailability assessment. Int J Pharm 212:233–246
Kwong E, Higgins J, Templeton AC (2011) Strategies for bringing drug delivery tools into discovery. Int J Pharm 412:1–7
Lancaster RW et al (2006) Racemic progesterone: predicted in silico and produced in the solid state. Chem Commun 4921–4923
Lieberman HA, Lachman L, Schwartz JB (1989) Pharmaceutical dosage forms: tablets, vol 1. Marcel Dekker, Inc., New York, pp 1–73
Lipinski CA, Lombardo R, Dominy BW, Feeney PJ (1997) Experimental and computational approaches to estimate solubility and permeability in drug discovery and development settings. Adv Drug Deliv Rev 23:3–25
Mass J, Kamm W, Hauck G (2007) An integrated early formulation strategy-from Hit evaluation to pre-clinical candidate profiling. Eur J Pharm Biopham 66:1–10
Meenan PA, Anderson SR, Klug D (2002) The influence of impurities and solvents on crystallization. In: Myerson AS (ed) Handbook of industrial crystallization, 2nd edn. Butterworth-Heinemann, Boston, pp 67–100

Merisko-Liversidge E, Liversidge G, Cooper E (2003) Nanosizing: a formulation approach for poorly-water-soluble compounds. Eur J Pharm Sci 18:113–120

Morissette SL, Almarsson O, Peterson M, Remenar J, Read M, Lemmo A, Ellis S, Cima M, Gardener C (2004) High throughput crystallisation: polymorphs, co-crystals and solvates of pharmaceutical solids. Adv Drug Deliv Rev 56:275–300

Morissette SL, Soukasene S, Levinson D, Cima MJ, Almarsson O (2003) Elucidation of crystal form diversity of the HIV protease inhibitor ritonavir by high-throughput crystallization. Proc Natl Acad Sci 100(5):2180–2184

Myerson AS (2002) Handbook of industrial crystallization, 2nd edn. Butterworth-Heinemann, Woburn

Nakanishi T et al (1998) Synthesis of NK109, an anticancer benzo[c]phenanthridine alkaloid. J Org Chem 63:4235

Neumann MA, Perrin MA (2005) J Phys Chem B 109:15531–15541

Newman AW et al (2008) Chaper 14. Salt and cocrystal form selection in preclinical development handbook. Wiley-Interscience, Hoboken, pp 455–481

Nowell H, Price SL (2005) Validation of a search technique for crystal structure prediction of flexible molecules by application to piracetam. Acta Crystallogr Sect A: Struct Sci B61:558–568

Nyvlt J (1995) The Ostwald rule of stages. Cryst Res Technol 30:443–449

Ouvard C, Price SL (2004) Cryst Growth Des 4:1119–1127

Palucki M, Higgins J, Templeton A, Kwong E (2010) Early optimization of the solid state phase and Pre-clinical toxicology formulation for potential drug candidates. J Med Chem 53 (16):5897–5905

Porter CJ, Trevakis NL, Charman WN (2007) Lipids and lipid-based formulations: optimizing the oral delivery of lipophilic drugs. Nat Rev Drug Discov 6:231–248

Saesmaa T, Makela T, Tannienen VP (1990) Physical studies on the benzathine and embonate salts of some blactam antibiotics. Part 1. X-Ray powder diffractometric study. Acta Pharm Fenn 99:157

Serajuddin A, Sheen P, Mufson D, Bernstein D, Augustine MA (1986) Preformulation study of a poorly water-soluble drug, alpha-pentyl-3-(2-quinolinylmethoxy)benzenemethanol: selection of the base for dosage form design. J Pharm Sci 75(5):492–496

Stahl PH, Wermuth CG (2011) Handbook of pharmaceutical salts—properties, selection and use (2nd Revised Edition). Int Union of Pure and Applied Chemistry, Wiley

Stephenson GA, Aburub A, Woods TA (2011) Physical stability of salts of weak bases in the solid-state. J Pharm Sci 100:1607–1617

Thorson MR et al (2011) A microfluidic platform for pharmaceutical salt screening. Lab Chip 11:3829–3837

Tung HH, Tabora J, Variankaval N, Bakken D, Chen C-C (2008) Prediction of pharmaceutical solubility Via NRTL-SAC and COSMO-SAC. J Pharm Sci 97(5):1813–1820

Variankaval N, Wenslow R, Murray J, Hartman R, Helmy R, Kwong E, Clas S, Dalton C, Santos I (2006) Preparation and solid-state characterization of nonstoichiometric cocrystals of a phosphodiesterase-IV inhibitor and L-tartaric acid. Cryst Growth Des 6(3):690–700

Variankaval N, Cote A, Doherty MF (2008) From form to function: crystallization of active pharmaceutical ingredients. AIChE J 54(7):1682–1688

Wexler A, Hasegawa S (1954) Relative humidity-temperature relationships of some saturated salt solutions in the temperature range 0 °C to 50 °C. J Res Nat Bur Stand (US) 53(1):19–26

Chapter 8
Efficient Laboratory Methods to Assess Risk and Design Formulations

Stephen R. Byrn and Roy J. Haskell

The first chapter in this section outlined a strategy to reduce risk in early development of solid oral dosage forms. The second chapter summarized strategies for drug candidate phase optimization in discovery space. This chapter outlines later stage screening and formulation design during the development process. In particular, this chapter outlines a strategy to reduce risk in formulation design by determining solid state properties and then incorporating solid forms with known properties into a formulation that does not change that form.

The first step in assessing and reducing risk is understanding whether a solid material for formulation and development exists in a crystalline or amorphous state or mixtures. Figure 8.1 illustrates the crystalline state as a perfectly ordered solid with molecules (circles) packed in an orderly array. Figure 8.1 illustrates an amorphous material as a disordered material with only short-range order. Crystalline materials give an X-ray diffraction pattern because Bragg planes exist in the material (see Fig. 8.2). Amorphous materials do not give a diffraction pattern (see Fig. 8.2). Of course, there are many interesting cases where a pharmaceutical material shows an intermediate degree of order falling somewhere between the highly ordered crystalline state and the disordered amorphous state. From a thermodynamic point of view, crystalline materials are more stable but the rate of transformation of amorphous materials to crystalline materials can be highly variable (Taylor et al. 2010).

Additionally, as outlined in the second chapter in this section, crystals of pharmaceutical materials can exist in different forms. These solid-state modifications of a compound are referred to as crystalline forms. When differences between

S.R. Byrn (✉)
Department of Industrial and Physical Pharmacy, Purdue University, 575 Stadium Mall Drive, West Lafayette, IN 47906, USA
e-mail: sbyrn@purdue.edu

R.J. Haskell
Discovery Pharmaceutics, Bristol-Myers Squibb, Wallingford, CT, USA

A.C. Templeton et al. (eds.), *Discovering and Developing Molecules with Optimal Drug-Like Properties*, AAPS Advances in the Pharmaceutical Sciences Series 15,
DOI 10.1007/978-1-4939-1399-2_8

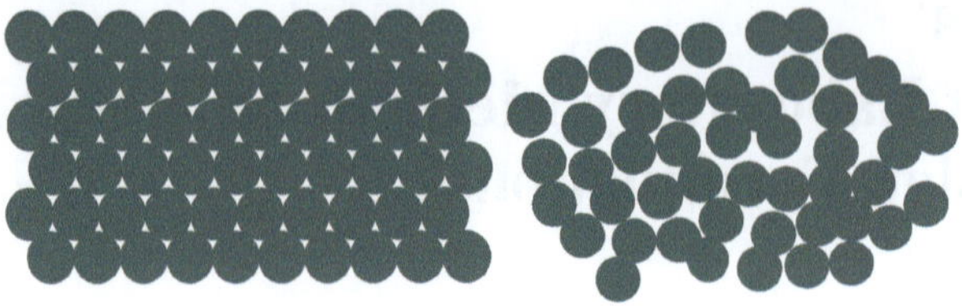

Fig. 8.1 Idealized view of crystalline (*left panel*) and amorphous (*right panel*) amorphous material. In this two-dimensional figure the molecule are viewed as *circle*

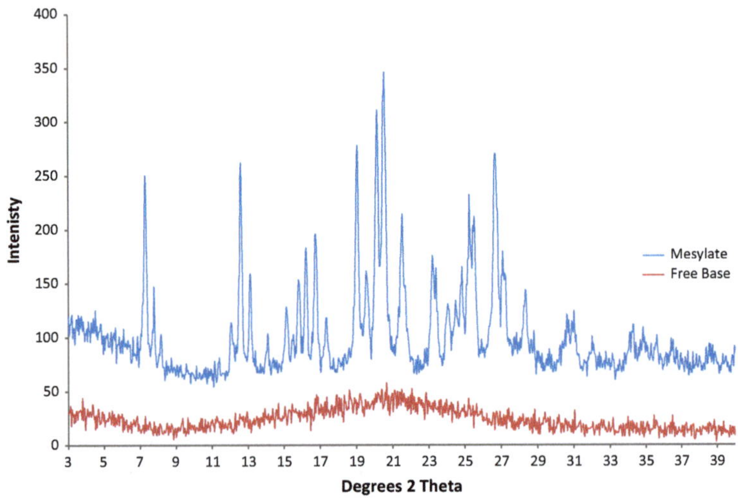

Fig. 8.2 X-ray diffraction pattern of two samples, crystalline, and amorphous

early batches of a substance are found, the use of the term "form" is particularly helpful since it allows subsequent, more accurate, description of a given variant batch (i.e., polymorph, solvate, habit, or amorphous material). The term pseudopolymorphism is applied frequently to designate solvates. These solid state modifications have different physical properties.

To put these definitions into a practical context, consider two cases (aspirin and paroxetine hydrochloride) in which a compound was crystallized for subsequent formulation and different-shaped crystals resulted in each experiment. Although sometimes dramatically different shapes were obtained upon changing solvents for the various crystallizations, the final interpretations in the two cases are different. For aspirin, X-ray powder diffraction showed that all crystals regardless of shape had the same diffraction pattern. Thus, the different shaped crystals are termed crystal habits. For paroxetine hydrochloride, the different shaped crystals had different X-ray powder diffraction patterns. Subsequent analysis showed that the crystals did not contain solvent. Thus, these different crystals are polymorphs. (Note that hydrates of paroxetine hydrochloride also exist.)

Another important step in reducing risk during formulation is to understand the size and shape of the crystalline form to be formulated. Crystals of a pharmaceutical material from different sources can vary greatly in their size and shape. Typical particles in different samples may resemble, for example, needles, rods, plates, prisms, etc. Such differences in shape are collectively referred to as differences in morphology and particle size. Lack of control of particle size represents a risk to control of critical drug product properties for formulation.

8.1 Risk in Selecting Amorphous Materials for Incorporation into Drug Product

Selecting amorphous materials for a final formulation has substantial risk but also substantial gain. Amorphous materials have no long range order and are thermodynamically metastable as pointed out in Chap. 2 of this series. An amorphous solid is characterized by a unique glass transition temperature T_g, the temperature at which it changes from a glass to a rubber. When T rises above T_g, the rigid solid can flow and the corresponding increase in molecular mobility can result in crystallization or increased chemical reactivity of the solid. Several historic papers describe some additional details of amorphous materials. Pikal and coworkers at Eli Lilly showed that amorphous materials can also have reduced chemical stability (Pikal et al. 1977) and Fukuoka showed amorphous materials had a tendency to crystallize (Fukuoka et al. 1991). Nevertheless, in some cases, amorphous forms were historically used as products. An excellent example is novobiocin (Mullins and Macek 1960) which exists in a crystalline and an amorphous form. The crystalline form is poorly absorbed and does not provide therapeutic blood levels; in contrast, the amorphous form is readily absorbed and is therapeutically active. Further studies show that the solubility rate of the amorphous form is 70 times greater than the crystalline form in 0.1 N HCl at 25 °C when particles <10 mm are used. Table 1.7 (Haleblian 1975) presents data for the plasma levels of novobiocin's amorphous and crystalline forms and for sodium novobiocin, which also gives detectable plasma levels, but is chemically unstable in solution. Amorphous materials are of tremendous interest in current development regimes. More information will be presented on amorphous formulations below.

It is possible to summarize the risks involved in selecting amorphous materials as the final form:

1. Lower purity
2. Less physically and chemically stable

   Crystalline hydrate > anhydrous crystal > amorphous

3. More hygroscopic

From this list it is clear that crystalline materials are generally more desirable and impart lower risk unless they are so insoluble that they cannot be used as medicines.

8.2 Risk in Selecting Solvates and Cocrystals for Incorporation into Drug Product

In some cases selecting a solvate as a final form involves risk. Solvates contain solvents regularly incorporated into the crystal lattice. When the solvent is water the solid form is called a hydrate. Solvates and hydrates can, in some cases, lose solvent at temperatures near room temperature. In such cases there is a substantial risk of solid form changes during storage.

Solvates and hydrates do not have the same composition as unsolvated materials. Solvates and hydrates are sometimes referred to as pseudomorphs or solvatomorphs. Interestingly it is possible for solvates and hydrates to be polymorphic. In such a case one has polymorphic solvates. Kuhnert Brandstatter in her 1971 book showed photomicrographs of 16 solvated forms of estradiol.

Figure 8.3 shows the crystal structure of caffeine monohydrate. The crystal of caffeine is built up by stacking the layers shown in Fig. 8.3 on top of each other. Thus, the hydrate molecules are in tunnels in this solid form.

It is important to note that the FDA has defined polymorphs as "different crystalline forms of the same drug substance. This may include solvation or hydration products (also known as pseudopolymorphs) and amorphous forms. Per the current regulatory scheme, different polymorphic forms are considered the same active ingredients." Thus, for purposes of registration, scientists are directed to define polymorphs more broadly to include amorphous forms, solvates, and hydrates.

Cocrystals, that is, two component crystals, are another solid material of interest. Like solvates, the new crystalline structure imparts different properties including solubility, stability, and mechanical properties to the material. Of special interest are cocrystals with altered solubility of stability. Figure 8.4 shows the crystal structure of a cocrystal of 2-methoxy-4-nitrophenol-4-(dimethylamino)pyridine

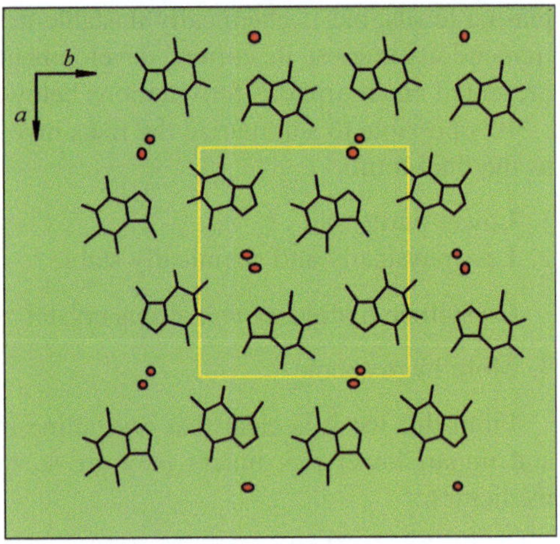

Fig. 8.3 Projection of the crystal structure of caffeine hydrate on the *ab* plane

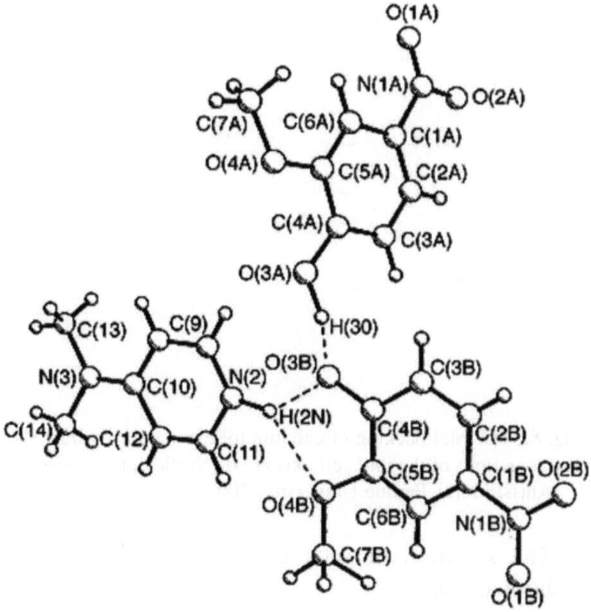

Fig. 8.4 Crystal structure of a cocrystal (2-methoxy-4-nitrophenol-4-(dimethylamino)pyridine (2:1).). The unit cell parameters are $a = 6.880$, $b = 38.40$, $c = 8.454$ and the space group is $Pna2_1$

(2:1) (Huang et al. 1997). The FDA has recently released a draft guidance defining cocrystals as "Solids that are crystalline materials composed of two or more molecules in the same crystal lattice." Currently, there are no or few cocrystalline materials on the market. Thus, the development of a cocrystalline form may impart some risk.

8.3 Risk in Selecting Salts for Incorporation into Drug Product

Pharmaceutical salts are substances formed by a reaction of an acid and a base. When a carboxylic acid reacts with an amine a salt is typically formed. However, the degree of proton transfer can vary

$$RCOOH + H_2N - R' \rightarrow RCOO^- \cdots\cdots\cdots H_3N^+ - R$$

depending on the acidity and basicity of the reacting groups.

Figure 8.5 shows the crystal structure of calcium tolfenamate trihydrate. It is clear that the unit cell is composed of regions containing mostly hydrocarbon functional groups and regions containing polar functionalities. This type of crystal packing is typical for salts.

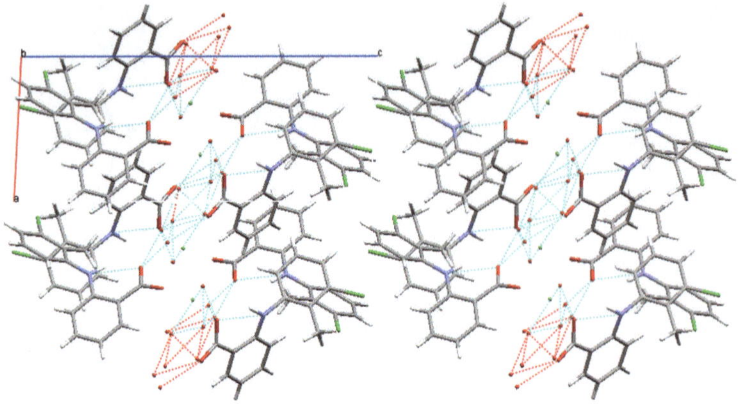

Fig. 8.5 Crystal packing of calcium tolfenamate trihydrate showing hydrogen bonding network. The directions of the unit cell axis are (*a*) vertical, (*c*) across, and (*b*) out of the plane of the paper (F. Atassi Ph.D., Purdue University, 2007)

The selection of salts as the final crystal form does not typically impart additional risk.

8.4 Using Decision Trees as a Guide for Reducing Risk

In 1995 Byrn, Pfeiffer, Ganey, Poochikian, and Hoiberg from Purdue University and the FDA published a paper using decision trees to describe a strategy to reduce risk by identifying the best solid form early in development (Byrn et al. 1995). In this way it is possible to ensure uniformity of solid form in clinical trials and resolve solid state issues before critical stages of development. The decision trees also suggested appropriate analytical methods for control. This is important since appropriate analytical methods reduce risk by providing reliable information. Four decision trees were presented: polymorphs, hydrates/solvates, desolvated solvates, and amorphous forms. This 1995 paper described a typical postdiscovery screen as follows:

> Solvents should include those used in the final crystallization steps and those used during formulation and processing and may also include water, methanol, ethanol, propanol, isopropanol, acetone, acetonitrile, ethyl acetate, hexane and mixtures if appropriate. New crystal forms can often be obtained by cooling hot saturated solutions or partly evaporating clear saturated solutions. The solids produced are analyzed using X-ray diffraction and at least one of the other methods. In these analyses, care must be taken to show that the method of sample preparation (i.e. drying, grinding) has not affected the solid form.

In the late 1990s the International Committee on Harmonization used a similar decision tree approach to describe how final specifications for the solid form in drug substances (API) and drug product should be determined. In this case, risk is reduced by developing appropriate specifications. Several decision trees were

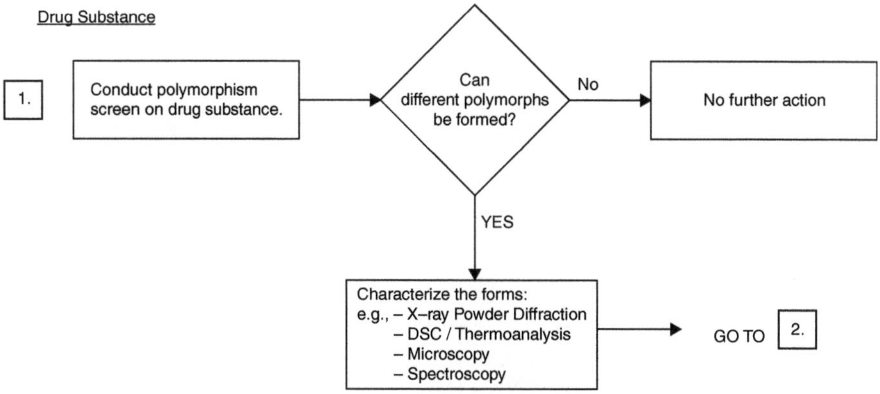

Fig. 8.6 ICH Q6A question 1 on polymorphs: can different polymorphs be formed?

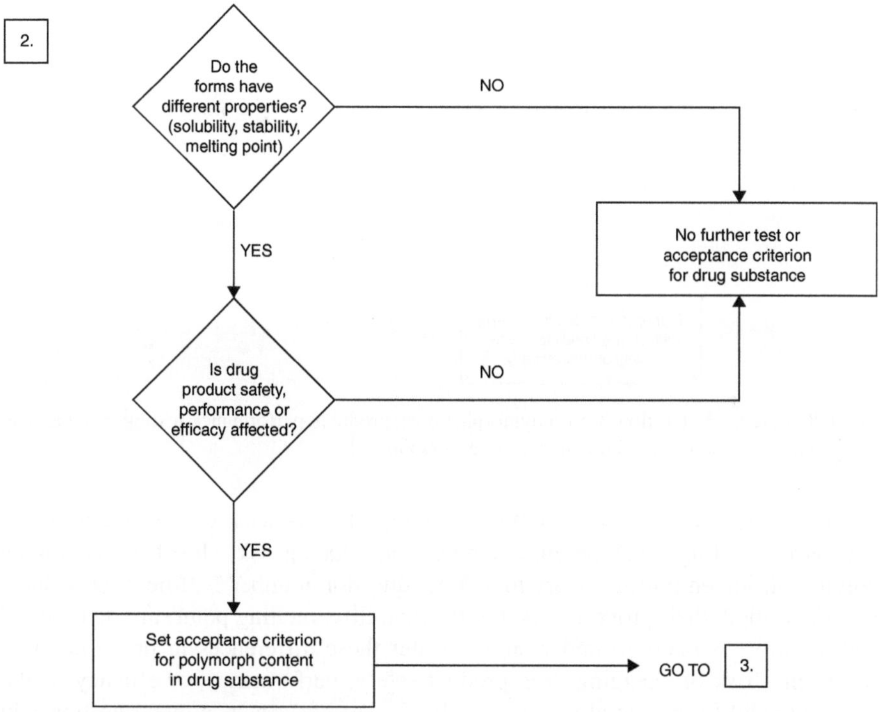

Fig. 8.7 ICH Q6A question 2 on polymorphs: do the forms have different properties (solubility, stability, melting point)

presented in the ICH Q6A document including decision trees on particle size and polymorphs. The ICH utilized the broadened definition of polymorphs that includes hydrates, solvates, and amorphous forms. The ICH decision trees on polymorphs are divided into three questions as shown in Figs. 8.6, 8.7, and 8.8.

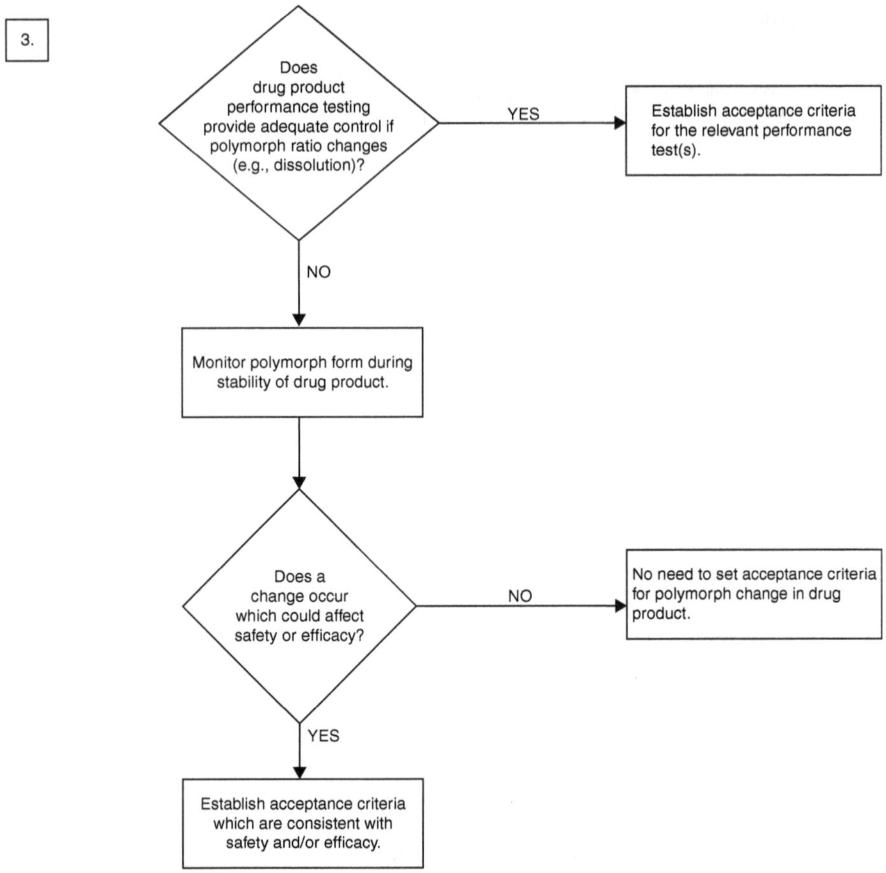

Fig. 8.8 ICH Q6A question 3 on polymorphs: does products performance testing provide adequate control if polymorph ratio changes (e.g. dissolution)?

These three decision trees outline a strategy that is widely used today during postdiscovery drug development as a means of reducing risk. Most firms conduct a polymorph screen postdiscovery to address question number 1. If new forms have been identified, their properties (solubility, stability, melting point) are determined and an effort is made to understand whether these differences in properties will represent a risk of changing drug product safety, performance, or efficacy. If the different solid forms can affect safety, efficacy, or performance then question 3 in the decision tree (Fig. 8.1) is addressed by determining whether drug product testing can detect changes in ratios of these forms. Additionally, the ratios of forms are monitored during stability studies to make sure changes that affect performance, safety, or efficacy do not occur. Using this strategy it is possible to reduce risk by finding a developable solid form rapidly.

8.5 Additional Steps to Reduce Risk in Designing a Formulation

Simultaneously with screening for the solid form, a number of questions need to be addressed: (a) What is the "intrinsic stability" of the compound? (b) What is the likely dose? (c) How soluble is the compound/formulation? (d) How well is the compound absorbed? (e) What is its BCS class? (f) What are its solid state properties and stability? (g) How well will the powder flow? (h) Is moisture an issue? and (i) What is the likely design, composition, and manufacturing procedure of the formulation/product?

As has already been discussed the solubility of the compound is a critical quality important for specifications and development. The solubility of a solid substance is the concentration at which the solution phase is *in equilibrium with a given solid phase at a stated temperature and pressure.* Under these conditions, the solid is neither dissolving nor continuing to crystallize. Note that the definition implies the presence of a specific solid phase. Once determined under the stated conditions, however, we can talk about the "solubility" of a given phase (e.g., a specific polymorph or pseudopolymorph) as a quantity, even in the absence of that solid phase. The Gibbs phase rule specifies that at constant temperature and pressure a two-phase system, such as solid drug and solvent, should achieve a well-defined, invariant concentration given sufficient time (i.e., at equilibrium). Thus, the term, equilibrium solubility, which is also referred to as the thermodynamic solubility. For the two-phase system to be at equilibrium, the chemical potential of both components must be equal to each other. If this is not the case, then depending upon the relative values, the system will experience a thermodynamic driving force toward dissolution or precipitation. Note also that while the rule confirms the ability of a compound to have a particular solubility under defined conditions, it says nothing about what actual value of solubility is eventually attained.

Use of the term "equilibrium" in connection with crystallizing systems requires clarification. When a substance exists in more than one crystal form, that is, when other polymorphs are possible, only the *least soluble* of these at a *given* temperature is considered the most physically stable form at that temperature, all others are considered to be metastable forms. In given cases, a solution of a substance may be in apparent equilibrium with one of these metastable phases for a long time, in which case, the system is in metastable equilibrium and is expressing the thermodynamic solubility of *that* solid form.

The most easily conceptualized driver of solubility is the manner which the compound interacts with the solvent. The more extensively a compound interacts with the solvent molecules at the expense of the solvent interacting with itself, the greater the enthalpic contribution toward having more drug solution at equilibrium. The entropic considerations are more complex. The increase in disorder brought about by spreading a drug molecule from an immobile solid into a fluid is opposed by the loss of disorder experienced by the solvent molecules as they are displaced by the presence of a drug compound. It is the relative contribution of these two

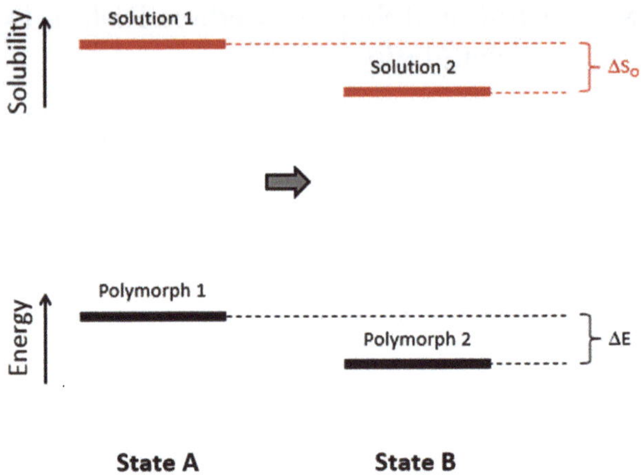

Fig. 8.9 Solubility and energy diagram of two forms

mechanisms that determines if dissolution is entropically favored or not. Equally important to the solvent–drug interactions in solution are the drug–drug interactions that must be disrupted in order for a compound to be removed from the solid form in the first place. Because the strength of these interactions in the solid state depends upon the orientation of the compositional atoms with respect to each other, the chemical potential represented by a given solid is determined by the particular polymorph in which it exists. Since, as noted earlier, the relative chemical potential of the solid and solution defines the solubility, the choice of solid form will determine the measured value of this critical parameter.

Figure 8.9 is useful in describing supersaturation phenomena and definitions and their relationship to solubility. Supersaturation is the amount of compound in solution in excess of that present under equilibrium conditions. However, in the same sense that equilibrium solubility has little meaning in the absence of a defined solid form, some supersaturation scenarios require knowing the solid phase present. For example, when the chemical potential represented by Solution 2 is equal to that of Polymorph 2, the amount of dissolved drug can be considered to be the equilibrium solubility. It can also be considered the most stable situation involving the least risk since Polymorph 2 is the lowest energy solid in which the compound can exist at a specified temperature and pressure. Similarly, the amount of drug in Solution 1 can be considered to be the equilibrium solubility of Polymorph 1. It is important to note that this represents a higher order of risk since while Solution 1 is saturated with respect to Polymorph 1, it is supersaturated with respect to Polymorph 2. Were a third solid form of even higher energy to exist, call it Polymorph 3, then the solution at the equilibrium solubility of this form would be supersaturated with respect to both Polymorph 2 and Polymorph 1. In all cases, the possibility of a high energy polymorph converting to one of lower energy represents the source of risk.

There is another scenario deserving of consideration and that is when there is more compound in solution than consistent with the equilibrium solubilities of

Polymorphs 1 and 2, yet there is no solid simultaneously present. Setting aside the possibility that the hypothesized Polymorph 3 is present, but only in a state difficult to observe, (i.e., subvisible particles), this level of drug can be considered the kinetic solubility of the compound. Such a solution, which can be prepared by a shift in pH, dilution from water-miscible solvents, or dissolution of a salt, is likely to be highly unstable toward precipitation and represent the highest risk of all. It is important to note that unlike equilibrium solubility, the measured kinetic solubility of a compound may be highly dependent on the manner in which the experiment was conducted (Alsenz and Kansey 2007).

8.5.1 Experimental Considerations for Measuring Solubility

As noted earlier, the solubility is affected by the solids and solvents employed, and while its theoretical value is completely defined, it is ultimately determined through experiment. Thus, the manner in which the solubility of a compound is evaluated can also have an effect if not on its absolute value then on the practitioner's ability to understand it. There are several important aspects to consider.

Time: The equilibrium solubility with respect to any solid form can only be determined when sufficient time has been allowed to elapse to ensure saturation. The surrogate for knowing if this has occurred is to make a measurement at multiple time points after combining drug and solvent and then assume that a lack of change in value indicates the attainment of equilibrium.

Separation: Assessing the quantity of drug in solution necessitates separating the solid material, be it undissolved or precipitated, from the supernatant before analysis. Thus, the degree to which the separation is "complete" will affect the experimental value. In many cases filtration, frequently using membranes possessing 0.22 μm diameter pores, is employed. However, the arbitrariness of result so obtained can be addressed by considering the fact that the prevalence of such filters is due primarily to their ability to remove microbes as a means of sterilization. If microbes happened to be of a larger or smaller dimension, then estimates of solubility using this approach would average correspondingly higher or lower since greater or smaller amounts of small drug particles could be mistaken for molecularly dispersed compound. A similar argument can be made with respect to centrifugation since the size cutoff produced will depend on the choice of the force and duration applied.

Sorption: Any surface that contacts the experimental materials, particularly after the aforementioned separation step, is a possible source of compound loss that will inevitably lead to an underestimate of solubility. This is of particular concern for low solubility compounds where the sorption of even nanograms of compound can represent loss of most of the dissolved material. The practical solution to this problem is paying extreme attention to detail—bordering on paranoia—and successfully duplicating results across labs.

Starting Material: Compound can be introduced into the solvent in the form of a solid or as a solution. In the former case, the chemical potential of the solution rises to match that of the solid whereas in the latter case the chemical potential of the solution drops over time. Starting with solution is a perfectly acceptable way of introducing compound to the solvent if the solid form present during separation is unambiguously identified. Indeed, methods that start with dissolved compound, such as potentiometric titrations or light scattering, are those the least subject to artifact (Box et al 2006; Lindfors et al 2006; Ilevbare and Taylor 2013). In all cases sufficient time must be allowed for measured values to reach a steady state.

The most significant practical aspect of the above is that there is an extreme degree of risk associated with any experimental value reported for solubility in the absence of a thorough understanding of the solid phase with which the solution is in equilibrium. All other things being equal, a higher energy solid state will lead to a higher measured value for solubility, whereas lower solubilities are obtained from lower energy solids. However, by its very nature, a higher energy solid is thermodynamically unstable with respect to those at lower energy. The situation is described in Fig. 8.9. Two polymorphs, 1 and 2, are specified to differ from each other by a given amount of energy, ΔE, with the latter being more stable than the former. Under constant experimental conditions, the water solubility of the nonionized drug, ΔS_O, will differ by a value corresponding to that same amount of energy, with the higher solubility being associated with the higher energy solid. Should Polymorph 1 be used to generate a saturated solution and the equilibrium solid present convert to Polymorph 2 during the experiment, then the amount of drug in solution will drop. It is important to note, however, that the time scale over which this decrease occurs is not easily predicted (Ozaki et al 2012), but the effect on risk is clear. A formulation prepared from Polymorph 2 has the potential to generate significantly less bioavailability than the same formulation prepared from Polymorph 1. The risk is even greater when transitioning from discovery to development since the amorphous materials (e.g., films, pastes, etc,) frequently used in discovery will likely have ΔE values in substantial excess of the solids employed later in the preclinical space. Any biopharmaceutical assessment of a compound must therefore also try to quantify the extent of these risks. Note also, however, that the utility of amorphous dispersions to enhance bioavailability is based on leveraging this same effect to good purpose, with the difference being that particular formulation emphasis is placed on kinetically stabilizing the enabling solid with respect to conversion to low energy states.

Risk in pharmaceutical development increases as the solubility decreases. In certain cases it is impossible to develop a product if the solubility is too low. Risk in development also increases with an increase in the number of solid forms. It can require substantial effort to develop a formulation if multiple forms exist. In some cases the cost of developing a solid with multiple equilibrating forms can be prohibitive.

The solubility and permeability are combined to determine the BCS class (Table 8.1). BCS class I drugs dissolve easily and are easily transported into the blood stream because they are highly permeable with respect to the membranes in

Table 8.1 BCS classification system

BCS classification	Solubility	Permeability
BCS class I	High	High
BCS class II	Low	High
BCS class III	High	Low
BCS class IV	Low	Low

the GI tract. BCS class III and IV drugs have poor permeability and are generally difficult to develop. BCS class II drugs are of the greatest importance for formulation/medicine design because the structure of the solid, the formulation, and many other factors are likely to have a significant effect on bioavailability and ultimately safety, performance, and efficacy. Several important drugs that are widely prescribed are BCS Class 2 including: atorvastatin calcium, celecoxib, efavirenz, irbesartan, lopinavir, medroxyprogesterone acetate, raloxifene hydrochloride, simvastatin, and warfarin sodium. Of the marketed drugs nearly 70 % are in BCS Class I or II with 31 % being in BCS Class II. It has been estimated that as high as 80 % of the drugs under development are BCS Class II. Risk increases as you move from BCS Class I to BCS Class II. Drugs with poor permeability (BCS Classes III and IV) are even harder to develop and have high risk of failure.

Particle size, like polymorphism, is one of the most critical aspects of solid state chemistry and the design of formulations. Failure to control particle size represents one of the greatest risks in development. The incorrect particle size can cause a change in the rate of dissolution and affect safety, efficacy, and performance. The ICH Q6A document on specifications makes it clear that particle size of APIs (drug substances) must be controlled especially if they are poorly soluble.

Likewise, failure to control polymorphism represents one of the greatest risks to development. Screening, as outlined earlier, is an approach often used to reduce this risk by finding the best solid form, and a manufacturing process. Screening can reduce the risk of process failure and lead to a reliable process for manufacturing. During the later stages of development complete screening studies are recommended. In these studies, a wide range of solvents are used including solvents used or of possible use in the manufacturing process, solvents spanning a wide range of polarities, and solvents having various functionalities. Typically, a list of 60 or more solvents is used to select the best solvents for these experiments. Solvents containing water (such as ethyl acetate water) are also used in this screen. In addition, melt-based crystallization on a microscope is recommended. In this regard, the scientists at Innsbruck, Austria have used melt recrystallization as a method for polymorph screening for more than 70 years. In a more recent example, Yu used non-solvent-based melt crystallization methods to discover new forms of ROY (2005). Additionally, grinding is used in an attempt to prepare new forms. Likewise crash cooling out of solvents and crystallization of amorphous materials are also used in an attempt to obtain new forms.

If a supersaturated solution is created it is important to screen for a crystallization inhibitor (Vandecruys et al. 2007). As indicated, screening is typically done for polymorphs including amorphous forms, salts, cocrystals, and nanoparticulate formulations. Unfortunately, all important forms are not found upon screening, and instances of late appearing important forms still occur. Thus, screening can reduce but not eliminate risk.

In addition to screening and selection of the best solid form and optimization/control of API and drug product, additional experiments are also carried out again to reduce risk. These experiments include determination of the partition coefficient (logP). This reflects the hydrophobicity of the drug and can be useful in determining the BCS class. The solubility of all available forms is determined as well as the degree of precipitation of any solid forms. Since solubility can depend on solid form, the solubility is typically determined in aqueous buffers, organic solvents, surfactants, and perhaps cyclodextrins and lipids. The solution and solid state stability of the API is determined under stress conditions including extreme pH, temperature light, and humidity. This provides information on the intrinsic chemical stability of the system and this knowledge is critical in formulation development. The pKa is also determined or calculated. This provides important information on the acidity/basicity of the material.

Once the solid form has been selected based on the above screening experiments, the stability of that form is determined under stress and accelerated conditions. This provides important information on how to handle that particular form and the risk of transformation. The dissolution properties of this form are also monitored. This provides important information on what might happen in the GI tract.

8.6 Reduction of the Risk of Failure Due to Poor Solubility

For poorly soluble compounds, salt formation and the formation of amorphous forms can reduce the risk of failure. Salt formation was discussed earlier and in the previous chapter. Rapid dissolution rates can be achieved by some salts. In fact, for some pharmaceutical salts, such as sodium phenytoin, a solubility enhancement of about 1,000,000 is achieved. This clearly shows the desirability of finding salt forms and explains why a very large number of drugs are developed as salts. Berge et al. (1977) summarized the approaches to forming salts.

> Choosing the appropriate salt . . . can be a very difficult task, since each salt imparts unique properties to the parent compound.
>
> Salt-forming agents are often chosen empirically. Of the many salts synthesized, the preferred form is selected by pharmaceutical chemists primarily on a practical basis: costs of raw materials, ease of crystallization, and percent yield. Other basic considerations include stability, hydroscopicity and flowability of the resulting bulk drug. Unfortunately, there is no reliable way of predicting the influence of a particular salt series on the behaviour of the parent compound. Furthermore, even after many salts of the same basic agent have been prepared, no efficient screening techniques exist to facilitate selection of

the salt most likely to exhibit the desired pharmacokinetic, solubility, and formulation profiles. (Berge at 1.)

The number of salt forms available to a chemist is large ... Various salts of the same compound often behave quite differently because of physical, chemical, and thermodynamic properties they impart to the parent compound. For example, a salt's hydrophobicity and high crystal lattice energy can affect dissolution rate and, hence, bioavailability. (Berge at 2.)

Salt formation is a means of altering the physical, chemical, and biological characteristics of a drug without modifying its chemical structure. Clearly, the salt form can have a dramatic influence on the overall properties of the parent compound. At present, selecting a salt form that exhibits desired combination of properties is a difficult semi-empirical choice. (Berge at 16.)

Berge and the previous chapter lists 80 acids for forming salts with drug compounds that are basic and 21 bases for forming salts with drug compounds that are acidic. In part because hydrochloric acid naturally occurs in the stomach, that acid is by far the most frequently used to make drug salts, accounting for almost half of the salts approved by the FDA. This is despite the fact that due to the "common ion effect," hydrochloride salts tend to be less soluble in the stomach. Nevertheless, it is a nearly universal practice among chemists tasked with making a salt of a basic drug compound to try to make the hydrochloride salt first. When the hydrochloride salt is unacceptable for one reason or another, there is no universal second choice; the list is very long.

A viable alternative to salt formation, especially in cases where a salt cannot be formed, is to develop an amorphous form/formulation. There are several products containing amorphous forms on the market including Kaletra and sporanox. A recent review reports that amorphous formulations can result in as much as an 82× increase in bioavailability (Newman et al. 2012). Law and coworkers (2003) reported greatly enhanced plasma concentrations of amorphous ritonavir over crystalline material. Ritonavir is one of the components in Kaletra. Figure 8.10 shows the results of studies of a 2:1 HPMC-P:itraconazole dispersion in dogs. Clearly, the amorphous dispersion results in a large increase in bioavailability.

In order to find the best amorphous form and formulation an amorphous screen is conducted. First a brief study of the solubility of the solid form and polymers is carried out to find a solvent that dissolves both the polymer and drug. Mixtures of methylene chloride and ethanol are particularly attractive but other solvents of interest include acetone, acetonitrile, methanol, isopropanol, THF, ethyl acetate, and pentane. An amorphous dispersion screen is carried out using a 1:1 ratio and 1:2 ratio of the drug and polymer. Pure drug is used as a control. The drug and polymer are mixed in a minimum amount of solvent and the solvent is evaporated as fast as possible using a rotary evaporator with a heated bath on the solvent flask and a good vacuum. Solids are scraped out, dried further in a vacuum oven, ground to a powder, and analyzed by XRPD. The following polymers are typically used: PVP (2 molecular weights), HPMC, HPMC-AS, HPMC-P, crospovidone,

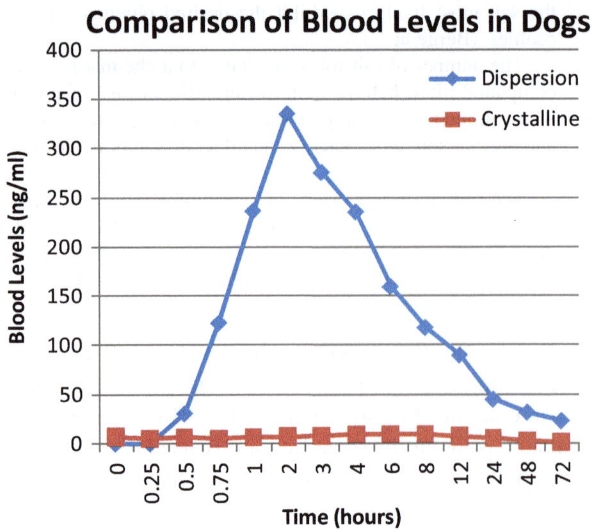

Fig. 8.10 Comparison of the bioavailability of a 2:1 HPMC-P:Itraconazole dispersion in dogs

polymethacrylates (two), PEG, and pluronic. The dispersions are analyzed by XRPD. Amorphous dispersions are stored at 40 °C/75 % RH for 3 days and reanalyzed by XRPD to verify stability.

Our lab and collaborators have also developed an amorphous screen that can be carried out in the levitated drop apparatus available at Argonne National Laboratory (see Benmore chapter). In this study, the drug-polymer solution is levitated in drops. The evaporating levitated drop/particle is analyzed every 30 s with high energy X-rays at the synchrotron X-ray source. This analysis provides instant feedback on the crystallinity of the evaporated drop/particles. It is also possible to carry out pair distribution function analysis of the diffraction pattern produced during the evaporation. More details are provided in the chapter by Chris Benmore appearing later in this book section.

Shah and coworkers (Hu et al. 2013) have suggested another approach to forming amorphous compounds involving precipitation of amorphous solids from organic solvents. This method is particularly useful for compounds with high melting points, making melt-extrusion difficult, and low solubility in volatile solvents, making spray drying or rotary evaporation difficult. In this method drug and polymer are codissolved in water miscible solvents such as DMA, DMSO, or DMF and added to cold acidic water such as 0.01 N HCl. A variation of this method uses acidic polymers which can form salts or associations with weak bases or hydrogen bonding accepting sites. They suggest using, for example, 3 antisolvent/water ratios, 3 concentrations of drug, and three polymers (see above) for an initial 27 experiments. They suggest working on the 100 mg scale for large scale and they also have suggested a microplate screening method using 96 well plates at the 0.1 mg scale and an X-ray powder diffractometer capable of transmission analysis.

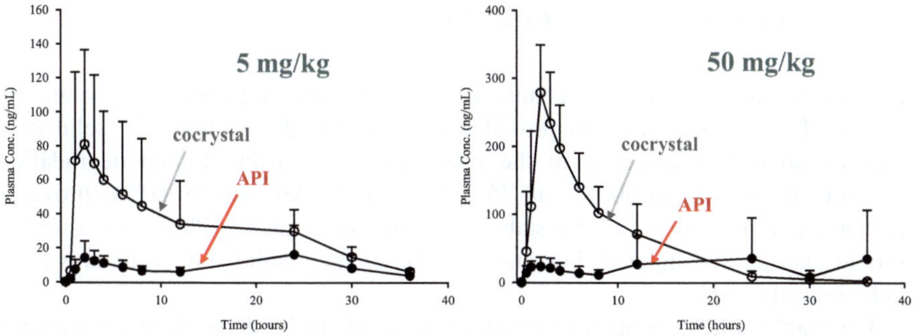

Fig. 8.11 Enhanced bioavailability of a cocrystal over the parent drug (modified from McNamera et al. 2006)

Another alternative to salt formation or amorphous formation is cocrystal formation. Figure 8.11 shows that a cocrystal enhanced the bioavailability of an amide containing API by about 4×.

Amorphous and cocrystalline formulations are classified as supersaturated drug delivery systems as discussed earlier. Crystallization inhibitors are sometimes used to prevent premature crystallization for these formulations. The screening method Brewster and coworkers (Vandecruys et al. 2007) suggested is as follows: The drug is dissolved in DMF typically at 100 mg/mL. In a second vessel was placed 0.01 N HCl and 2.5 % of the crystallization inhibitor candidate (polymer or compound). The DMF solution is added dropwise to a stirred solution of the crystallization inhibitor until a precipitate is just noticed visually. The solution is then filtered and the concentration determined by HPLC. The pH is also measured since it could have an effect on the solubility of some drugs. A control experiment is done without the excipient present in the water. The ability to inhibit precipitation is then determined by the ratio of the concentration of the drug attained in the precipitation experiment to the control. In some cases solubility ratios of >10,000 were attained. Additionally, the stability of the supersaturated solution was assessed. Good crystallization inhibitors are incorporated in the formulation. It is interesting to note that the first steps in this method are quite similar to those used by Shah and coworkers (Hu et al. 2013).

In addition to solubility/dissolution rate, the solid form can influence a number of other properties important for formulation including: milling, blending, tableting, dry filling, suspension formulation, and lyophilization. Transformations to other forms can also occur during these processes.

8.7 Reducing Risk Due to Instability

Initially, the intrinsic stability is analyzed using functional group analysis and the outline of solid state chemistry provided in Chap. 1 in this section. The solid state stability should be assessed in the phase of choice using forced degradation experiments (usually at 80 °C and 75 °C/40 % RH). Analysis by HPLC provides immediate information on the stability of the solid phase. Photolytic stability should also be investigated. For unstable phases, new solid state structures need to be investigated.

For stable solids, excipient compatibility needs to be done. A few excipients from each class are mixed with the drug in a 1:1 ratio and 5 % moisture is added. The mixture is then stressed at 80 °C for 7 days. HPLC analysis reveals instability and in subsequent formulation the incompatible excipients are excluded.

In addition, it is useful to determine solution stability of the drug in buffers of relevance to the body (pH 1, 6.8 and 7.4). Typically the solution is heated at 50 °C for 3–6 days and analyzed by HPLC. Solution degradation suggests that wet granulation should not be used to prepare the final formulation.

8.8 Formulation Design to Reduce Risk

The best approach to formulation design is to follow the old adage "Just do something."

The first choice for reduced risk early formulations is a powder in a capsule formulation. This can be prepared precisely by a modern dosator machine. This formulation requires verifying that the capsule does not cause degradation or physical transformation of the solid form. It has been estimated that over 90 % of the solid materials can be accurately filled into capsules using a dosator. Additionally, a simple dissolution study can verify that there is good release of the drug from this simple formulation.

If an amorphous dispersion is selected for development, it is quite routine to fill this dispersion directly into capsules. We have found that, in some cases, the polymer dispersion and the capsule form a plug during dissolution. In such cases, it is necessary to utilize a disintegrant to break up the dispersion; thus forming a powder + disintegrant in capsule formulation.

In the event that a powder in a capsule does not work, or a larger scale formulation is needed, a wet granulation or direct compression formulation and manufacturing method provides the approach with the lowest risk. Wet granulations are particularly preferred if a thorough analysis of the solid state chemistry indicates there is little possibility of transformation during the formulation process. For these formulations a diluent, binder, disintegrant, and lubricant is typically added. In some cases glidants and flow promoters are also added. For wet granulations, a super disintegrant is sometimes used but it should be remembered that the sodium

Table 8.2 Common tablet excipients

Diluents	Binders
Lactose USP	Acacia
Lactose USP, anhydrous	Cellulose derivatives
Lactose USP, spray-dried	Gelatin
Directly compressible starches	Glucose
Hydrolyzed starches	Polyvinylpyrrolidone (PVP)
Microcrystalline cellulose NF	Starch, paste
Other cellulose derivatives	Starch, pregelatinized
Dibasic calcium phosphate dihydrate NF	Sodium alginate and alginate derivatives
Mannitol USP	
Sorbitol	Sorbitol
Sucrose USP powder	Tragacanth
Sucrose-based materials	
Calcium sulfate dihydrate NF	
Dextrose	
Disintegrants	Lubricants
Starch	Stearic acid
Starch derivatives	Stearic acid salts
Clays	Stearic acid derivatives
Cellulose	Talc
Cellulose derivatives	Polyethylene glycols
Alginates	Surfactants
PVP, cross-linked	Waxes
Croscarmellose sodium	
Sodium starch glycolate	
Pregelatinized starch	
Glidants and Flow Promoters	
Silica derivatives	
Talc	
Cornstarch	

superdisintegrants can react with ionic bases. Table 8.2 lists common tablet excipients.

Next, the lowest risk approach is to make a trial formulation. A typical small-scale method of manufacture by wet granulation is shown below:

For amorphous dispersions of drug in polymer, wet granulation is not recommended and probably not needed. Instead the polymer–drug dispersion is blended with a superdisintegrant and a lubricant and perhaps a filler such as MCC or lactose and compressed into tablets or filled into capsules. If the drug and/or polymer is ionic it is safest to use a nonionic disintegrant. For early studies the drug–polymer dispersion can be filled directly into capsules using a dosator.

The tablets produced are analyzed for dissolution compared to an equal dose of drug (API) powder. Typically both simulated gastric fluid and simulated intestinal fluid are used. It is easiest to use an in situ probe to determine relative dissolution rate. But HPLC is also recommended especially in cases where aggregation of the drug will cause artifacts in UV analysis.

The tablets are also analyzed by X-ray diffraction and HPLC to make sure the manufacturing process does not cause physical or chemical changes. The API is used as a control. In addition, binary and tertiary mixture of API and excipients are made and analyzed for changes using X-ray diffraction and HPLC. Finally, it is recommended that the tablets be subjected to some degree of forced degradation conditions to rule out physical changes or chemical degradation under stress conditions.

At this point, the risk for development of this formulation is assessed. The tablets produced are stressed at 40 °C/75 % RH for a week and reanalyzed by HPLC and X-ray diffraction. If no changes are detected a full scale early stage forced degradation study is carried out using exposure for a month at stress conditions. The stressed tablets are again tested by X-ray diffraction, HPLC, and dissolution. After these studies another risk assessment is done and the process is repeated until an acceptable formulation and manufacturing method is achieved.

8.9 Conclusion

In conclusion, this chapter has reviewed strategies to deal with risk in formulation design. Initially, risk of choosing the incorrect form is addressed, subsequently approaches to overcome the barrier to poor solubility including salt formation and the use of amorphous dispersions are outlined. The risk of these various approaches is also briefly addressed. Finally the incorporation of the appropriate solid form or dispersion into a formulation is discussed.

References

Alsenz J, Kansey A (2007) High throughput solubility measurement in drug discovery and development. Adv Drug Deliv Rev 59:546–567

Berge SM et al (1977) Pharmaceutical Salts. J Pharm Sci 66:1–19

Box KJ, Volgyi G et al (2006) Equilibrium versus kinetic measurements of aqueous solubility, and the ability of compounds to supersaturate in solution—a validation study. J Pharm Sci 95 (6):1298–1307

Byrn S, Pfeiffer R, Ganey M, Hoiberg C, Poochikian G (1995) Pharmaceutical solids: a strategic approach to regulatory considerations. Pharm Res 12:945–954

Fukuoka E, Makita M, Nakamura Y (1991) Glassy state of pharmaceuticals. V. Relaxation during cooling and heating of glass by differential scanning calorimetry. Chem Pharm Bull 39:2087–2090

Haleblian JK (1975) Characterization of habits and crystalline modifications of solids and their pharmaceutical applications. J Pharm Sci 64:1269–1288

Hu Q, Choi DS, Chokshi H, Shah N et al (2013) Highly efficient miniaturized coprecipitation screening (MiCoS) for amorphous solid dispersion formulation development. Int J Pharm 450:53–62

Huang K-S, Britton D, Etter MC, Byrn SR (1997) A novel class of phenol-pyridine co-crystals for second harmonic generation. J Mater Chem 7(5):713–720

Ilevbare GA, Taylor LS (2013) Liquid–liquid phase separation in highly supersaturated aqueous solutions of poorly water-soluble drugs: implications for solubility enhancing formulations. Cryst Growth Des 13:1497–1509

Kuhnert-Brandstätter M (1971) Thermomicroscopy in the analysis of pharmaceuticals. Pergamon, New York, NY

Law D, Schmitt ED, Kennan I et al (2004) Ritonavir–Peg 8000 amorphous solid dispersions. J Pharm Sci 93(3):563–570

Lindfors L, Forssen S, Skantze P et al (2006) Amorphous drug nanosuspensions. 2. Experimental determination of bulk monomer concentrations. Langmuir 22:911–916

McNamara DP, Childs SL, Giordano J et al (2006) Use of a glutaric acid cocrystal to improve oral bioavailability of a low solubility API. Pharm Res 23:1888–1897

Mullins JD, Macek TJ (1960) Some pharmaceutical properties of novobiocin. J Am Pharm Assoc Sci Ed 49:245–248

Newman A, Knipp G, Zografi G (2012) Assessing performance of amorphous solid dispersions. J Pharm Sci 101:1355–1377

Ozaki S, Minamisono T et al (2012) Supersaturation–nucleation behavior of poorly soluble drugs and its impact on the oral absorption of drugs in thermodynamically high-energy forms. J Pharm Sci 101(1):214–222

Pikal MJ, Lukes AL, Lang JE (1977) Thermal decomposition of amorphous β-lactam antibacterials. J Pharm Sci 66:1312–1316

Taylor LA, Van Eerdenbrugh B, Bard JA (2010) Crystallization tendency of active pharmaceutical ingredients following rapid solvent evaporation—classification and comparison with crystallization tendency from undercooled melts. J Pharm Sci 99:3826–3838

Vandecruys R, Peeters J, Verreck G, Brewster MC (2007) Use of a screening method to determine excipients which optimize the extent and stability of supersaturated drug solutions and application of this system to solid formulation design. Int J Pharm 342:168–175

Chapter 9
Advanced X-Ray Analytical Methods to Understand Structure, Properties, and Risk

C.J. Benmore

9.1 Solubility and Bioavailability

The structural interactions and inter-solubilities of drug and polymer molecules strongly influence both the stability and overall solubility of the resulting medicines. Understanding phase relationships between a drug and its polymer host is essential to optimize clinical drug formulations and understand the risk for crystallization. In this chapter we use both structural and levitation studies to help investigate the drug–polymer, drug–drug and polymer–polymer interactions in solutions and amorphous mixtures. Understanding these factors will help to generate and interpret phase relationships for the systems of interest. The results are interpreted using x-ray pair distribution function analysis, in terms of atom–atom drug–polymer interactions between the host and solute components, using the example of the itraconazole–PVP system. The motivation is to develop a more fundamental glass science basis for designing amorphous drug systems (Gao 2008; Willart and Descamps 2008) and for estimating the risk for crystallization.

Amorphous materials are produced in ways that inhibit molecules from arranging in a crystalline lattice. Nucleation is the initial process in crystallization, whereby molecules arranged in an ordered structure influence neighboring molecules to convert to a similar or lower energy state. The absence of any seed crystallites in the amorphous matrix is therefore extremely important for phase stability. It follows that material produced which is not completely amorphous will behave differently over time to a material that contains even a fraction of a percent of crystalline matter. Completely amorphous materials are much more likely to remain in that state over a longer time period. Heterogeneous nucleation normally occurs at sites on surfaces in contact with the material. These external surfaces are

C.J. Benmore (✉)
X-ray Science Division, Advanced Photon Source, Argonne National Laboratory, 9700S. Cass Avenue, Lemont, IL 60439, USA
e-mail: benmore@anl.gov

A.C. Templeton et al. (eds.), *Discovering and Developing Molecules with Optimal Drug-Like Properties*, AAPS Advances in the Pharmaceutical Sciences Series 15,
DOI 10.1007/978-1-4939-1399-2_9

eliminated in containerless experiments. Homogeneous nucleation on the other hand usually takes place upon supercooling or superheating the solution, and occurs spontaneously and randomly without preferential sites.

The ultimate goal of this x-ray analysis is to characterize and design systems that are highly soluble in body fluid. While stable amorphous drug–polymer products are needed to have a long shelf life, they must also be easily dissolved in body fluids to produce high absorption of the active pharmaceutical ingredient. Less stable systems would be expected to provide relatively higher solubility but they would have a greater risk of crystallization during storage. Thermodynamic measurements on phase-pure drug compositions have indicated that many amorphous compositions are far from equilibrium (Baird et al. 2010, 2012). The departure from equilibrium increases with the difficulty of glass formation. Spray dried materials can often result in particles that contain some degree of crystalline material, due to particles coming in contact with the walls of the vessel during the drying process, causing heterogeneous nucleation. Acoustic levitation provides the operator with more control over the droplet and completely prevents any surface interaction over the time of the drying process. This synthesis route can produce materials that are both very far from equilibrium, still have a long shelf life, and have a reduced risk of crystallization.

9.2 Pair Distribution Function Analysis Using High-Energy X-Rays

X-ray diffraction from polycrystalline powders of organic molecules exhibits Bragg peaks associated with long-range ordering. These sharp peaks may be accurately indexed with a lattice and the positional and thermal displacement parameters corresponding to a specific space group. Amorphous materials produced by solvent evaporation or glasses quenched from the liquid state have an inherently disordered structure on intermediate length scales and the absence of long-range order, resulting in a broad oscillating diffraction pattern (Elliott 1991). Absolute structural determination of these disordered materials is considerably more difficult than crystals, although the short-range interactions, which govern the packing between adjacent molecules, may still be well defined in some cases. Organic glass formers often have complex structures due to the changes in molecular conformation and intermolecular bonding, which can lead to variable or different orientational correlations than those observed in the crystalline state. The degree of disorder and structural variability lends itself to ambiguity in structure determination, and generally several techniques are required to obtain a clear picture of the structure of a particular glass. Nonetheless, of all the techniques used to characterize the structure of amorphous and glassy materials (x-ray and neutron diffraction, EXAFS's, NMR, and Raman spectroscopy), x-ray diffraction with pair distribution function analysis (PDF) has remained one of the most popular and commonly used methods (Benmore 2012). Warren (1934) first pioneered the Pair Distribution

Function (PDF) technique in the 1930s, using x-rays to study noncrystalline forms of matter. With the advent of third generation synchrotrons and the availability of high fluxes of high-energy x-rays (>60 keV) the PDF technique has become widespread in the past decade. High-energy x-ray beamlines at major facilities around the world, e.g., the Advanced Photon Source (USA), Spring-8 (Japan), and the European Synchrotron Radiation Facility (France), are now able to routinely perform measurements on low-Z organic amorphous and glassy materials (Benmore 2012). PDF has been used to study the structural conformation and interactions of molecules in the liquid and glassy state for decades (see overview in Benmore (2012)), but until recently there has been very little application to amorphous pharmaceuticals.

A schematic of a synchrotron PDF measurement is shown in Fig. 9.1. There are several advantages for using high-energy synchrotron x-rays over conventional (laboratory based) x-ray diffraction techniques: (1) the experiments are much faster, lasting from seconds to minutes, (2) structure factors can be measured out to much higher momentum transfers (Q) at smaller scattering angles (2θ), which in turn leads to higher real space resolution, (3) attenuation and multiple scattering effects are negligible for small samples, i.e., typically ~1 mm^3 (Neuefeind et al. 1996). This is because the photoelectric absorption decreases as ~E^{-3}.

The measured x-ray pair distribution function is obtained via a Fourier transform of the measured and corrected diffraction pattern. The PDF represents an average of all the interactions within the bulk material, and is defined as the probability of finding an atom at a distance r away from a central atom as illustrated in Fig. 9.2. In this sense it is a general technique that can be equally applied to materials that are in

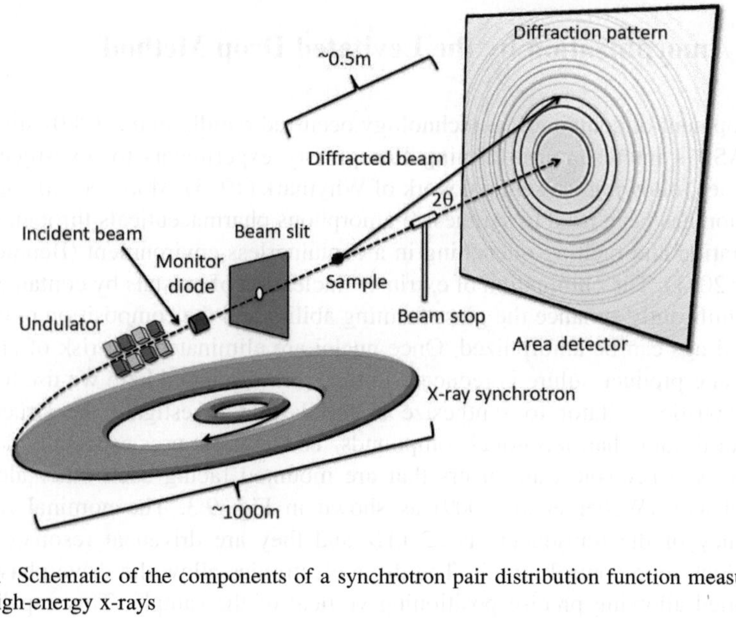

Fig. 9.1 Schematic of the components of a synchrotron pair distribution function measurement using high-energy x-rays

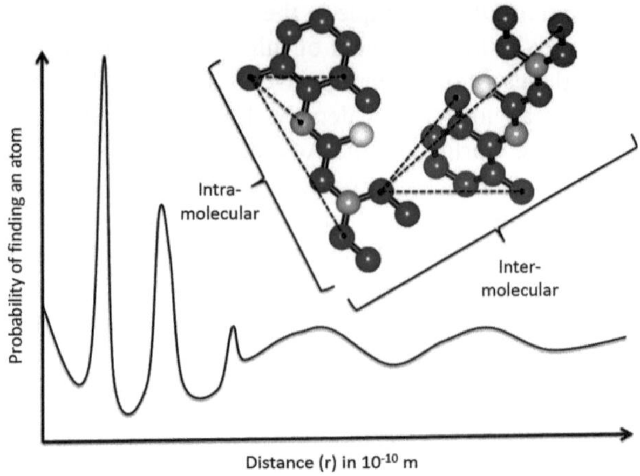

Fig. 9.2 A schematic of a Fourier transform of the x-ray diffraction pattern, yielding the Pair Distribution Function (PDF). The PDF describes the probability of finding an atom at a particular distance *r*, which contains information on both intra- and intermolecular interactions

the polycrystalline, nanocrystalline, amorphous, glassy, or liquid states. For small organic molecules it is useful to try and separate the scattering pattern into the contributions arising from intramolecular and intermolecular scattering; although this is not trivial for many pharmaceuticals where overlap between intra- and intermolecular correlations complicates the deconvolution process. In common with most methods for investigating the structure of disordered materials the x-ray PDF method is most powerful when interpreted along with information from other methods, e.g., NMR and computational modeling.

9.3 Amorphization by the Levitated Drop Method

Development of containerless technology occurred rapidly in the 1980s stimulated by NASA's interest in performing low gravity experiments to investigate fluid properties following the seminal work of Whymark (1975). More recently, acoustic levitation has been used to synthesize amorphous pharmaceuticals through solvent evaporation and melting quenching in a containerless environment (Benmore and Weber 2011). The elimination of extrinsic nucleation of crystals by container walls can significantly enhance the glass forming ability and the composition range over which drugs can be amorphized. Once nuclei are eliminated, the risk of crystallization and product failure is reduced. In the work discussed here we use a single-axis acoustic levitator to synthesize materials and investigate the structure of nonequilibrium pharmaceutical compounds. The levitator uses especially designed piezo-driven acoustic transducers that are mounted facing each other along the vertical axis (Weber et al. 2009) as shown in Fig. 9.3. The nominal resonant frequency of the transducers is 22 kHz and they are driven at resonance by a phase lock loop control circuit. The drive electronics allow the drive phase to be controlled allowing precise positioning vertical of the sample. This capability is

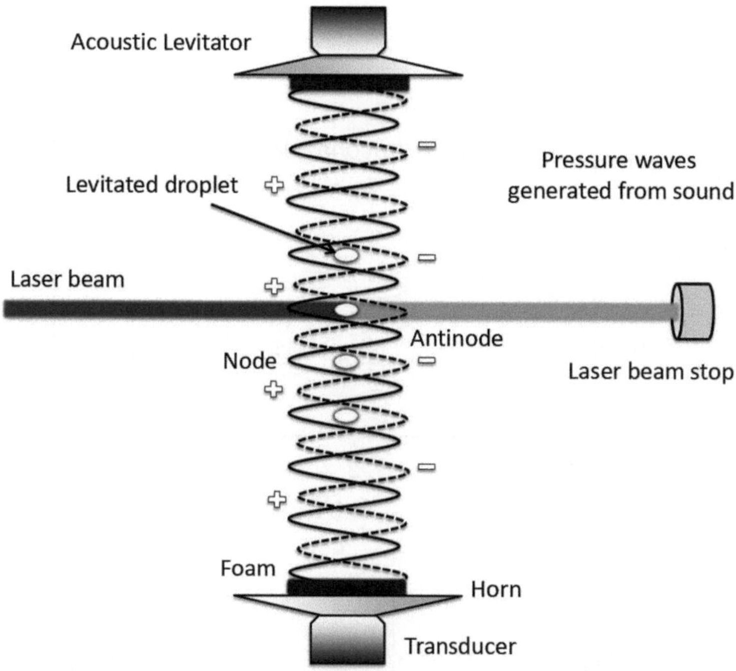

Fig. 9.3 Schematic of the main components of a single-axis acoustic levitator used to levitate droplets using sound. Two opposing sound waves generated by the transducers create regions of varying pressure and nodes where materials can be suspended. Heating or melting can be achieved by local heating with a low power (10 W) CO_2 laser

particularly important to align samples with the x-ray beam for in situ measurements of the structure (Weber et al. 2012).

The solvent used to prepare the materials is an important aspect of the amorphization process. Also, the solubility, evaporation rate, and drop size that can be levitated (determined by the density and surface tension of the liquid) influence the way in which the liquid drug–polymer mixtures will behave. An example of the way in which details of the crystallization process occurs from a drug–polymer mixture in solution is captured using x-ray diffraction is given in Fig. 9.4a, b. Figure 9.4c compares the evaporation rate of droplets of ethanol, acetone, and isopropanol, which have similar densities and surface tensions.

9.4 X-Ray Structure Factor Determination: Theory

X-ray scattering theory has been covered comprehensively in many excellent texts and for brevity the reader is referred to the literature (Klug and Alexander 1974; Warren 1990). This section will concentrate on the extraction of a pseudo-nuclear

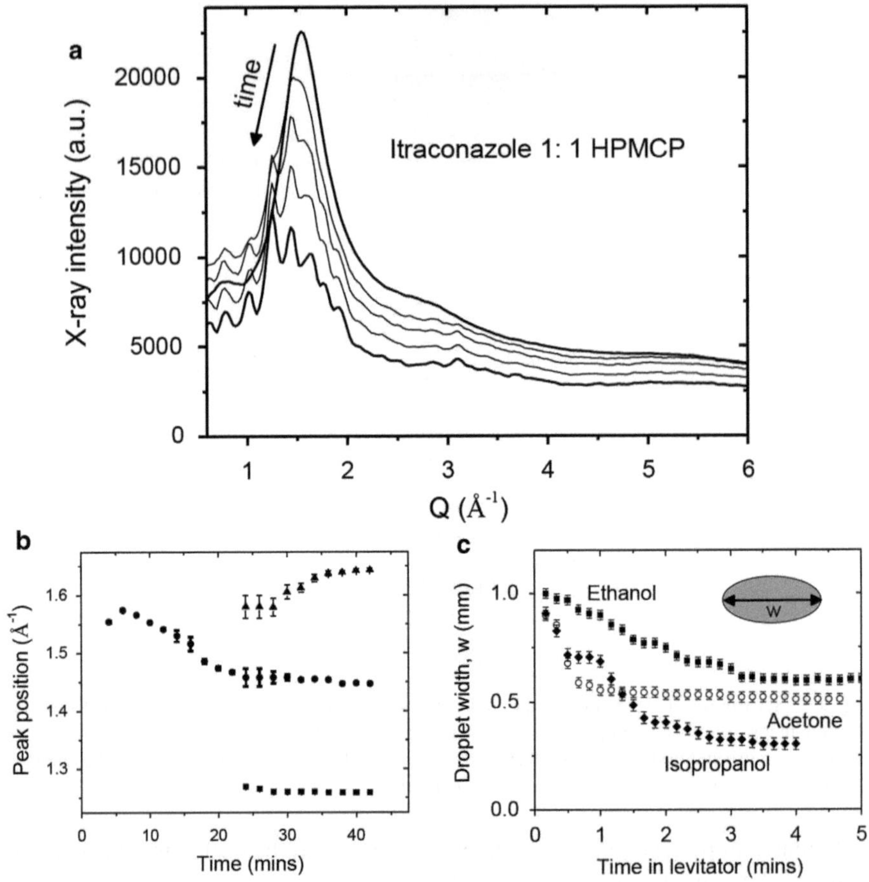

Fig. 9.4 (**a**) Crystallization of a mixture of itraconazole and hydroxypropylmethyl cellulose phthalate from a solution of ethanol-methylene chloride observed over time using x-ray diffraction. (**b**) Shows the evolution of peak positions as the solvent evaporates and crystallites form. (**c**) Evaporation of levitated droplets of ethanol (*squares*), acetone (*circles*), and isopropanol (*diamonds*) under ambient conditions, monitored by the change in flattened droplet width (w) as a function of time. Initial droplet widths ranged from 1 to 2 mm and were normalized to an initial droplet size of ~1 mm for comparison purposes

pair distribution function from the measured scattered x-ray intensity of the electron clouds. There are many different formalisms for the total structure factor $S(Q)$ as a function of momentum transfer Q (Keen 2001), but experimentally, they all derive from the measured elastic contribution of the scattered x-ray intensity $I(Q)$ that represents the electron–electron interactions. X-ray form factors $f(Q)$ represent the scattering from the electron cloud and their amplitude increases with atomic number, Z. Practically, the decrease in $f(Q)$ with increasing Q limits the maximum value of Q that it is obtainable with x-rays for low-Z (organic) materials. The nonspherical shape of the electron cloud in liquids or glasses with few electrons, such as hydrogenous or light element materials, suggests a redistribution of charge needs to be taken into account (Soper 2007). However, this effects the shape of the $S(Q)$ only at low Q-values, typically $Q \leq 1$ Å^{-1}, and generally the spherical atom approximation is adequate for obtaining an accurate PDF. In order to obtain a pseudo-atomic function it is necessary to divide by a so-called sharpening function. The most common formalism is to divide by the average scattering $\langle f^2(Q)\rangle$ to yield the pseudo-nuclear structure factor $S(Q)$,

$$S(Q) - 1 = \frac{I(Q) - \left(\sum_{i=1}^{n} f_i^2(Q)\right) - C(Q)}{f^2(Q)} \tag{9.1}$$

where i and j represent the different atomic species in the material. The average scattering is given by $\langle f^2(Q)\rangle = [\sum_{i,j=1}^{n} c_i f_i(Q)]^2$ and $C(Q)$ represents the Compton scattering (see Fig. 9.5a). The measured $S(Q)$ is most commonly expressed using the Faber–Ziman formalism (Faber and Ziman 1965) as the sum of the x-ray weighted element specific partial structure factors $S_{ij}(Q)$,

$$S(Q) - 1 = \left(\frac{\sum_{i,j} c_i c_j f_i(Q) f_j(Q)\left[S_{ij}(Q) - 1\right]}{\langle f^2(Q)\rangle}\right) \tag{9.2}$$

$S(Q)$ is measured over a finite range in reciprocal space from $Q_{\min}$ and $Q_{\max}$. Of these two limits the $Q_{\max}$ often has the most noticeable effect on the Fourier transform to extract the PDF, and measurements on high-energy beamlines at advanced synchrotron sources enable measurements out to the highest $Q_{\max}$ values. Ultimately there are several consistency checks, which can be made to fully analyzed x-ray diffraction data. Namely, $S(Q)$ tends to unity at high-Q values, and for homogeneous liquids the normalized $I(Q)$ tends to the isothermal compressibility as Q tends to zero. This behavior is described by the equation,

$$I(Q = 0) = \rho k_B T \chi_T Z \tag{9.3}$$

where Z is the total number of electrons in a single molecule, k_B is the Boltzmann constant, T is the absolute temperature, and χ_T is the isothermal compressibility. For homogeneous glasses this equation is generally valid at conditions corresponding to the glass transition temperature where the liquid structure is frozen in.

The pseudo-nuclear pair distribution function $G(r)$ is obtained via a Sine Fourier transformation of $Q \cdot S(Q)$ using,

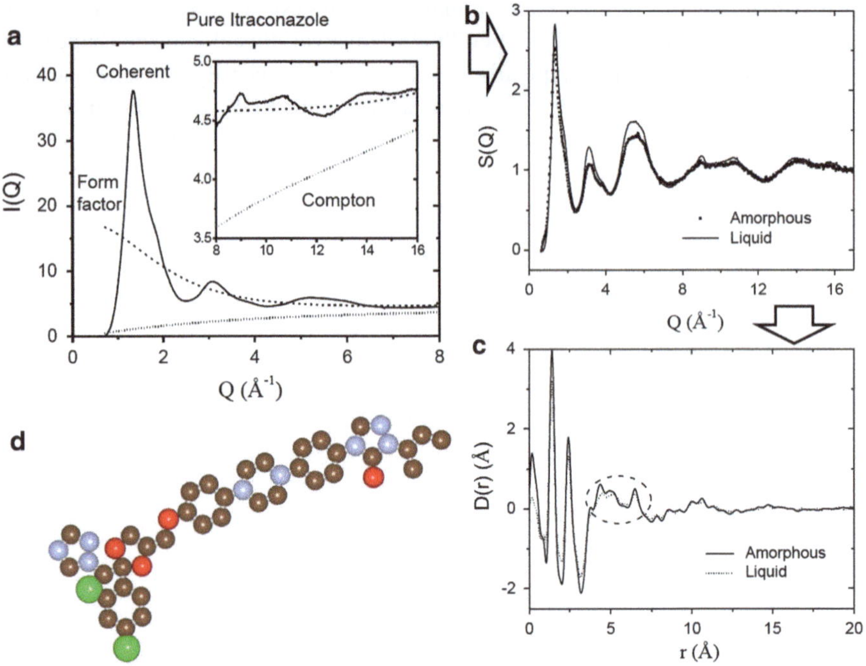

Fig. 9.5 An example of the pair distribution function analysis of pure itraconazole measured in the acoustic levitator using high-energy x-ray diffraction. (**a**) The background corrected x-ray intensity, *I*(*Q*), normalized to the "self scattering" which comprises the Compton scattering and the molecular form factor based on the independent atom approximation (**b**). The x-ray structure factors, *S*(*Q*), of amorphous and liquid itraconazole extracted from *I*(*Q*). (**c**) A Fourier transform of *S*(*Q*) yields the differential pair distribution function *D*(*r*) with the bulk density removed. The small but measureable changes between the amorphous and liquid state most noticeably appear as a sharpening of the peaks and in the 4–7 Å region corresponding to several atom–atom nearest neighbor distances. (**d**) Shows the molecular structure of itraconazole

$$G(r) - 1 = \frac{1}{2\pi^2 r\rho} \int_{Q\min}^{Q\max} Q[S(Q) - 1]M(Q, \Delta) \sin(Qr) \mathrm{d}Q \tag{9.4}$$

where ρ is the atomic (or molecular) number density in Å^3. A Lorch (Lorch 1969) or other modification function $M(Q, \Delta)$ is often used to minimize the oscillations generated during the Fourier transform over a finite Q-range (defined by the $Q_{\max}$ limit), where the parameter Δ describes the average width in real space. Another commonly used real space representation includes the *differential* pair distribution function, $D(r)$, which removes the bulk density, emphasizes the longer-r correlations,

$$D(r) = 4\pi\rho r\,[G(r) - 1] \tag{9.5}$$

9.5 Experiment: The Example of Pure Itraconazole

Itraconazole is a potent antifungal drug with a relatively poor dissolution rate and low bioavailability after oral administration. Modulated differential scanning calorimetry measurements have revealed two endothermic transitions at 74 and 90 °C, and Six et al. (2001) have suggested a chiral nematic mesophase forms upon cooling from the melt, followed by limited rotational mobility within the molecule. It has also been observed that upon cooling from the melt a color change occurs from transparent to white at 90 °C corresponding to the first endothermic peak (Tarnacka et al. 2013). Based on FTIR measurements these authors (Tarnacka et al. 2013) find homeotropic alignment of the rigid rod-like molecular cores and suggest a nematic phase exists below 90 °C and a smetic A phase below 74 °C. Previous x-ray diffraction studies have demonstrated a strong temperature dependence of the diffuse scattering signal of the principal peak (Atassi et al. 2013), and that variable degrees of ordering occur at low-Q values on samples quenched from different temperatures (Tarnacka et al. 2013).

An illustration of Eqs. (9.1), (9.3), (9.4), and (9.5) is shown below for the case of x-ray diffraction and PDF analysis for pure itraconazole. The high-energy x-ray experiments were performed on beamlines 6-ID-D and 11-ID-C at the Advanced Photon source using incident energies of 100 keV and 115 keV respectively. Essentially, after the standard x-ray source (polarization, Q-calibration), sample (attenuation, multiple and Compton scattering, form factor), and detector (dark current, oblique incidence, etc.) dependent corrections have been applied, the $S(Q)$ can be extracted using Eq. (9.1), see Fig. 9.5a. Figure 9.5b shows the x-ray structure factor for both liquid and amorphous itraconazole, produced by laser heating in the acoustic levitator (as illustrated in Fig. 9.3). The x-ray PDF obtained by a Sine Fourier transform shown in Fig. 9.5c primarily shows sharp C–C distances at low-r, although small changes can be observed mainly with the growth of peaks at 4.4, 5.0, 6.5, and 10.6 Å between the liquid and amorphous states. These correspond primarily to second and higher atom–atom nearest neighbor arrangements and indicate increased ordering either within the molecule (see Fig. 9.5d) or due the alignment of adjacent rod-like molecules.

For molecular liquids it is useful to separate the total structure factor into the contributions arising from intramolecular and intermolecular scattering; $S(Q) = S_{\text{intra}}(Q) + S_{\text{inter}}(Q)$, where the intramolecular contribution is given by a zeroth order spherical Bessel function and the root mean square vibrational amplitude for each atom–atom pair. The intermolecular function usually decays rapidly at high-Q values except for any short-range interactions. For many of the molecular drugs studied the molecules contain >50 atoms, so the initial parameters are obtained by fits to the crystal structure and the Debye–Waller thermal factors σ_{ij} are estimated within the empirical algorithm,

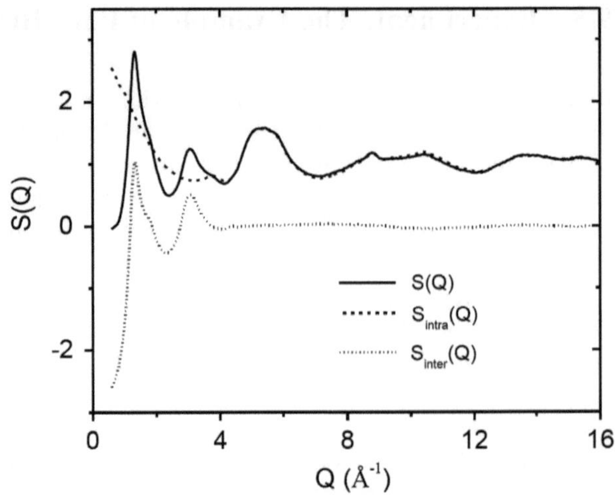

Fig. 9.6 The measured structure factor for itraconazole (*solid line*) is broken down into its intramolecular $S_{\text{intra}}(Q)$ (*dashed line*) and intermolecular $S_{\text{inter}}(Q)$ (*dotted line*) components

$$S_{\text{intra}}(Q) = \sum_{i,j=1}^{m} \frac{Nc_i c_j f_i(Q) f_j(Q)}{2} \sin c\left(Qr_{ij}\right) \exp\left(-Q^2 \sigma_{ij}\right) \tag{9.6}$$

Details are given in Benmore et al. (2013) and Mou et al. (in preparation). An example of the intramolecular fitting process for itraconazole is shown in Fig. 9.6.

9.6 Itraconazole–PVP Mixtures

In order to reduce the risk of crystallization, scientists have been interested in preparing molecular mixtures of drugs and polymers since the early work of Chiou and Riegelman (1971). Poor solubility of the amorphous drug in polymer within a solid dispersion may result in concentrated drug domains, which have a higher risk of crystallization than a molecularly dispersed system during long-term storage (Mura et al. 1998). Even a mixture containing nanocrystalline and amorphous domains would be at a risk to crystallize and a heterogenous system would still show a single T_g (Pham et al. 2010). For this reason, the ability of the PDF method to determine the presence of a molecular mixture and the absence of nanocrystalline and amorphous domains would be an important step forward in assessing the risk of crystallization. Put another way, the ability to determine that a molecular mixture existed in dispersions would provide a powerful method to assess the risk of crystallization. In this section we explore the application of the PDF method to understanding the structure of the polymer–itraconazole systems.

Amorphous screens allow a broad range of materials to be prepared in different ways and with different drug-binder ratios. Here, saturated solutions of a matrix of itraconazole–polymer mixtures were made by dissolving the materials in a stirred ethylene-methylene chloride solution held in a temperature-controlled bath. The

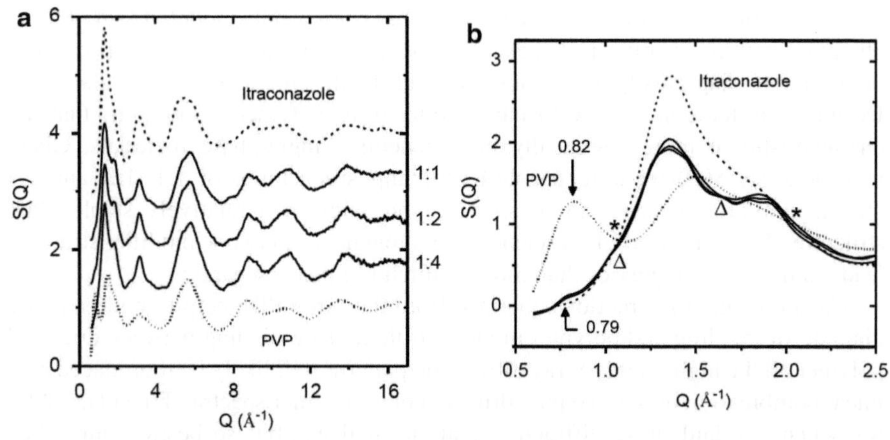

Fig. 9.7 (**a**) The measured x-ray structure factors of itraconazole-polyvinylpyrrolidine (PVP) for composition ratios of 1:1, 1:2, 1:4 together with the pure components shown at the *top* and *bottom*. (**b**) Zoom-in of the low-*Q* variation as a function of composition. *Asterisks* denote crossover points between the pure species; *triangles* denote isosbestic points within the 1:1 to 1:4 compositional range (see text)

resulting solution was filtered to remove undissolved material and inserted into the levitator using a syringe or a calibrated automatic pipette. As the solvent evaporates, the sample shrinks in size and reaches a stable end point after several minutes. One of the main differences between acoustic levitation and the commonly used spray drying process in pharmaceutical production is droplet size. The droplet size of samples prepared by levitation is typically ~0.5–1.0 mm in diameter, leaving particles sizes of ~300–500 μm after the solvent evaporates. In spray drying the particle sizes are typically 100–200 μm. This means the drug–polymer particles produced by spray drying have a larger surface area but preliminary tests suggest that the bulk structure remains the same. Advantages of levitated drop method are that they provide a unique way to access metastable and nonequilibrium forms of materials that cannot be produced by other methods. The first step is to survey a matrix of different drug–polymer compositions after the solvent has evaporated from the acoustically levitated droplets. The measured x-ray structure factors are shown in Fig. 9.7.

The x-ray scattering patterns reveal that pure itraconazole can be produced in the amorphous form using this method, although repeat experiments showed some traces of crystallinity in the spectra. The itraconazole:PVP ratios of 1:1, 1:2, and 1:4 yield an almost identical diffraction pattern with very little variation with composition, including a very small pre-peak at $Q_1 \sim 0.79$ Å^{-1}, which is close to the intense first sharp diffraction peak in PVP at 0.82 Å^{-1}. The first sharp diffraction peak at position Q_1 in the x-ray diffraction $S_X(Q)$ is associated with the existence of intermediate or medium range order in glass with a periodicity of $2\pi/Q_1$, although its origin is still controversial. Medium range order in glasses has been defined as covering the region ~5–20 Å (Elliott 1991).

The role of drug–polymer phase behavior has been widely modeled using Flory–Huggins solution theory (Hancock and Zografi 1993), to help develop pseudo-binary phase diagrams by taking into account the dissimilarity in molecular sizes. Desirable systems are likely to have strong drug–polymer interactions that are characteristic of more chemically homogeneous, single-phase materials. Given that the x-ray $S(Q)$ for itraconazole:PVP composition ratios of 1:1, 1:2, and 1:4 are all similar, this tends to indicate the presence of a relatively stable glass structure. As in classical glass science, a homogeneous network-modifier structure leads to more stable glasses than those that tend to phase separate.

The atom–atom interactions obtained from the x-ray PDF help to build a phase diagram of the drug and polymer mixture. If there is no binding between drug and polymer all the high-energy x-ray diffraction patterns will likely be reproduced by a linear combination of the two pure drug and pure polymer spectra (Tse et al. 2005). So, when overlaid these diffraction patterns will exhibit isosbestic points, i.e., points that remain constant as a function of composition. If there is a strong x-ray sensitive interaction between drug and polymer the linear combination test will not work. This is the case in the Itraconazole:PVP system as can be seen in Fig. 9.7b at $Q \sim 1.04$ and 1.92 Å^{-1} denoted by asterisks. However, within the compositional range 1:1 to 1:4 isosbestic points do exist at $Q \sim 1.0$ and ~1.65 Å^{-1}, suggesting a region of phase stability in the system.

9.7 The Pair Distribution Function

Pair distribution function analysis has recently been used to investigate the structure of pure phase pharmaceuticals (Benmore et al. 2013; Mura et al. 1998; Atassi et al. 2010). The next step is to apply this method to a drug-binder system. Here, representative differential pair distribution functions for the Itraconazole–PVP system are shown in Fig. 9.8. The low-r peaks at 1.4 and 2.4 Å are shifted to lower distances in the 1:1 mixture compared to the pure components (most noticeably PVP), and maybe due to additional chemical pressure compressing the molecules in the mixture. In addition there is a strong nonlinear behavior in the region of the leading edge of peak at 3.6 Å. A more detailed interpretation of this region is described in terms of separating out the drug–polymer interactions from the drug–drug and polymer–polymer correlations later in this chapter.

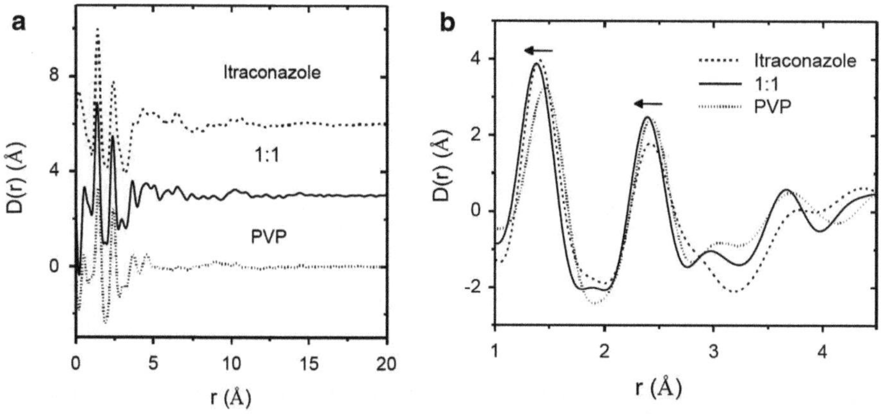

Fig. 9.8 (**a**) The x-ray pair distribution functions of Itraconazole-polyvinylpyrrolidine (PVP) for a composition ratio of 1:1 together with the pure components, obtained by the Fourier transformation of the corresponding curves in Fig. 9.7a. (**b**) Zoom-in of the low-*r* peak position changes as a function of composition. We note that the drug–polymer mixture exhibits shorter atom–atom correlations than those in either of the pure components

9.8 Extracting the Drug–Polymer Interactions at the Atomic Level

The scattering pattern is undoubtedly complex but can be broken down into three components: drug–drug, drug–polymer, and polymer–polymer interactions. For high content >50 % polymer containing samples, either a molecular model or the intramolecular drug–drug contribution may allow us to extract the drug–polymer interactions, provided the drug is dispersed randomly in the polymer and not clustered in regions with significant intermolecular drug–drug interactions. A proposed structural hypothesis to guide the PDF data analysis of medicines is that the drug–polymer mixtures that show the shortest and most intense intermolecular atom–atom contacts are most likely to exhibit the highest dissolution rate and bioavailability, because this will signify the strongest interaction (e.g., evidence of a bond) between the drug and polymer molecules, which keeps the system amorphous. In this case, strong interactions between the drug and polymer will be reflected by peaks in the PDF, i.e., the short intermolecular atom–atom peaks will be sharper compared to, say, broad non-bonded (van der Waals) interactions (Benmore et al. 2013). The main parameters to be identified in the drug–polymer PDF are therefore the nearest neighbor intermolecular peak heights and peak widths. This approach is in part supported by the studies of Wegiel (2013) and Taylor who showed that the shifts of the IR OH and C=O groups in some polyphenol systems could be correlated with stability of the glasses.

Various methods or molecular modeling programs can be used to determine the atom–atom distances and separate the intramolecular and intermolecular drug and polymer interactions. Here we concentrate on an experimental approach for

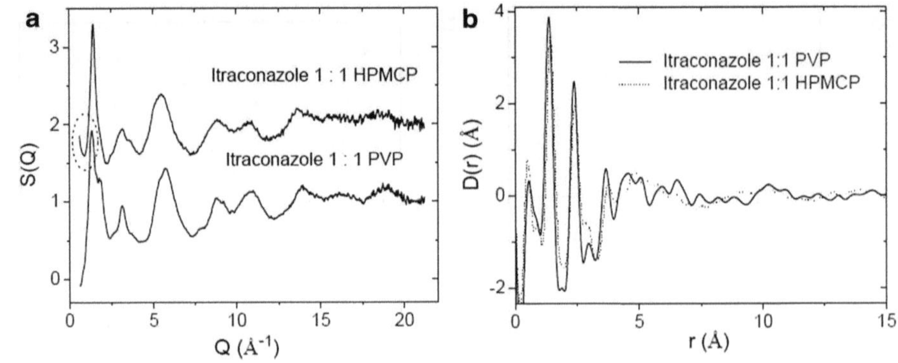

Fig. 9.9 (**a**) The measured x-ray structure factors of equal mixtures of itraconazole-polyvinylpyrrolidine (PVP) and Itraconazole-hydroxypropylmethyl cellulose phthalate (HPMCP). The rise at low-Q in the HPMCP spectra indicates density fluctuations within the material due to heterogeneity on the nanometer scale. (**b**) The corresponding pair distribution functions indicate the PVP mixture is more structured than the HPMCP mixture

extracting this information. As pointed out earlier, a useful test to see if there is a significant difference in the linear scaling of the interactions of the structural components of an amorphous drug–polymer system is by measuring the system as a function of composition using x-ray PDF. If a nonlinear progression of the structure factor is observed, this suggests that the nature of the drug–polymer and/or drug–drug, polymer–polymer interactions has been altered in some way, e.g., crystallization of one or more of the components, a molecular conformation change, or an indication of some region of phase stability between the drug and polymer has been reached. This is an important finding as it suggests the structures maybe unstable due to competing forces and the compositional range has to be narrowed to find a stable or metastable structural region. However, narrowing the compositional range also makes the nature of the drug–polymer interaction much harder to extract. In addition, if significant scattering intensity is observed at the lowest-Q values >2 Å, this is an indication that there are significant density fluctuations and possible phase separation over the range of a few nanometers and possibly longer distances (see Fig. 9.9).

Since a linear progression of the structure factor was found in the Itraconazole: polymer system over the compositional range 1:1 to 1:4 (see Fig. 9.7b), this suggests that the atom–atom structural interactions of the drug–drug, drug–polymer and polymer–polymer most likely scale linearly with composition within this range (Tse et al. 2005). If this assumption is true, and the observation of isosbestic points lead to an essentially linear variation of the partial structure factors with composition over this region, the drug–polymer interactions can be extracted from a series of diffraction patterns using the following method.

First we consider a drug comprising elements A and B, and a polymer comprising elements X and Y. For a 1:1 drug-polymer mixture the composition will be

$A_1B_1X_1Y_1(Q)$. Using the Faber–Ziman formalism (Faber and Ziman 1965) this leads to a total structure factor $S_{1:1}(Q)$ given by,

$$\begin{aligned}\langle F_{1:1}\rangle^2 S_{1:1}(Q) =& 4f_A^2(Q)[S_{AA}(Q)-1]+4f_B^2(Q)[S_{BB}(Q)-1]+4f_X^2(Q)[S_{XX}(Q)-1]\\ &+4f_Y^2(Q)[S_{YY}(Q)-1]+8f_A(Q)f_B(Q)[S_{AB}(Q)-1]\\ &+8f_A(Q)f_X(Q)[S_{AX}(Q)-1]+8f_A(Q)f_Y(Q)[S_{AY}(Q)-1]\\ &+8f_B(Q)f_X(Q)[S_{BX}(Q)-1]+8f_B(Q)f_Y(Q)[S_{BY}(Q)-1]\\ &+8f_X(Q)\end{aligned} \tag{9.7}$$

where the fractions represent the concentration (c) part of the weighting factors, which are given by c_α^2 for like atoms and $2c_\alpha c_\beta$ for different atom types α and β. The factors $f_\alpha(Q)$ represent the Q dependent x-ray atomic weighting factors and $\langle F_{1:1}\rangle^2=\langle\sum c_\alpha f_\alpha(Q)^2\rangle$ the average scattering of a particular drug–polymer composition. $S_{\alpha\beta}(Q)$ are the atom–atom partial structure factors that we intend to separate. In this case $S_{AA}(Q)$, $S_{BB}(Q)$, $S_{AB}(Q)$ represent both intra- and intermolecular drug–drug interactions. $S_{XX}(Q)$, $S_{YY}(Q)$, $S_{XY}(Q)$ represent both intra- and intermolecular polymer–polymer interactions. $S_{AX}(Q)$, $S_{AY}(Q)$, $S_{BX}(Q)$, $S_{BY}(Q)$ represent ONLY the intermolecular drug–polymer interactions. Similar expressions can be written for other compositions, i.e., $\langle F_{1:2}\rangle^2 S_{1:2}(Q)$ and $\langle F_{1:4}\rangle^2 S_{1:4}(Q)$. The predominantly intermolecular drug–polymer interactions can then be extracted using the formula,

$$\begin{aligned}\Delta S^{\text{inter}}_{\text{drug-polymer}}(Q) =& \langle F_{1:1}\rangle^2 S_{1:1}(Q)+\langle F_{1:4}\rangle^2 S_{1:4}(Q)-2\langle F_{1:2}\rangle^2 S_{1:2}(Q)-WS^{\text{intra}}_{\text{drug}}(Q)\\ =& f_A(Q)f_X(Q)[S_{AX}(Q)-1]+f_A(Q)f_Y(Q)[S_{AY}(Q)-1]\\ &+f_B(Q)f_X(Q)[S_{BX}(Q)-1]+f_B(Q)f_Y(Q)[S_{BY}(Q)-1]\end{aligned} \tag{9.8}$$

where $WS^{\text{intra}}_{\text{drug}}(Q)$ represents the weighted signal from the intramolecular drug–drug interaction. For high polymer contents the intermolecular drug–drug interactions are assumed to be small relative to the total, such that the $\Delta S^{\text{inter}}_{\text{drug-polymer}}(Q)$ difference function contains information relating mainly to intermolecular drug–polymer interactions.

9.9 Example: The Itraconazole $C_{35}H_{38}N_8O_4Cl_2$: $[C_6H_9N_1O_1]_n$ PVP System

As in most pharmaceutical-binder materials the most dominant elements are carbon and hydrogen. For itraconazole (denoted I) carbon $c_C^I=0.402$ and hydrogen $c_H^I=0.437$ (representing 84 % of the total atoms) and for PVP (denoted P, represents 88 % in total) where carbon $c_C^P=0.353$ and hydrogen $c_H^P=0.529$. Despite their abundance, hydrogen correlations can often be neglected because they scatter

Table 9.1 The atom–atom weighting factors for the drug–polymer interaction for each of the three compositions together with the difference function $\Delta S^{\text{inter}}_{\text{drug-polymer}}(Q)$

Atom–atom interaction		Faber–Ziman weighting factors for Itraconazole:PVP (%)			
Drug	Polymer	1:1	1:2	1:4	$\Delta S_{\text{drug-polymer}}{}^{\text{inter}}(Q)$
C	C	**33.2**	**21.1**	**21.1**	**56.9**
C	H	9.0	10.5	10.5	−5.2
C	N	7.1	8.2	8.2	−4.0
C	O	8.1	9.4	9.4	−4.6
H	C	6.6	7.6	7.6	−3.8
H	H	0.8	1.0	1.0	−0.5
H	N	1.3	1.5	1.5	−0.7
H	O	1.5	1.7	1.7	−0.8
N	C	9.7	11.2	11.2	−5.6
N	H	2.4	2.8	2.8	−1.4
N	N	0.9	1.1	1.1	−0.5
N	O	2.2	2.5	2.5	−1.2
O	C	5.5	6.4	6.4	−3.2
O	H	1.4	1.6	1.6	−0.8
O	N	1.1	1.3	1.3	−0.6
O	O	0.6	0.7	0.7	−0.4
Cl	C	5.9	6.8	6.8	−3.4
Cl	H	0.5	1.7	1.7	−5.1
Cl	N	1.1	1.3	1.3	−0.7
Cl	O	1.3	1.5	1.5	−0.7

It can be seen that carbon–carbon interactions dominate the x-ray scattering function

x-rays so weakly, i.e., $Z_C = 6$, $Z_H = 1$ and $f_H(Q) \rightarrow 0$ rapidly with increasing Q (we note that $f_\alpha(Q=0) = Z_\alpha$). Water is a perfect example of this, as the x-ray PDF is essentially a measurement of the oxygen–oxygen correlations (Soper 2007). Conversely, although Cl has $Z_{Cl} = 17$ the concentration in itraconazole is small $c^I_C = 0.023$. This essentially results in the x-ray scattering from the itraconazole–PVP system to be dominated by carbon-related interactions, i.e., the backbones of the molecules (see Table 9.1).

So for brevity, here we just write the correlations from the dominant C–C relations. The total structure factor with composition 1:1 is dominated by itraconazole–PVP interactions $S_{IP}(Q)$,

$$S_{1:1}(Q) = 0.296 \times [S_{II}(Q) - 1] + 0.476 \times [S_{IP}(Q) - 1] + 0.228 \times [S_{PP}(Q) - 1] \tag{9.9}$$

The total structure factor with compositions 1:2 and 1:4 is dominated by an increasing weighting factor of PVP–PVP interactions $S_{PP}(Q)$,

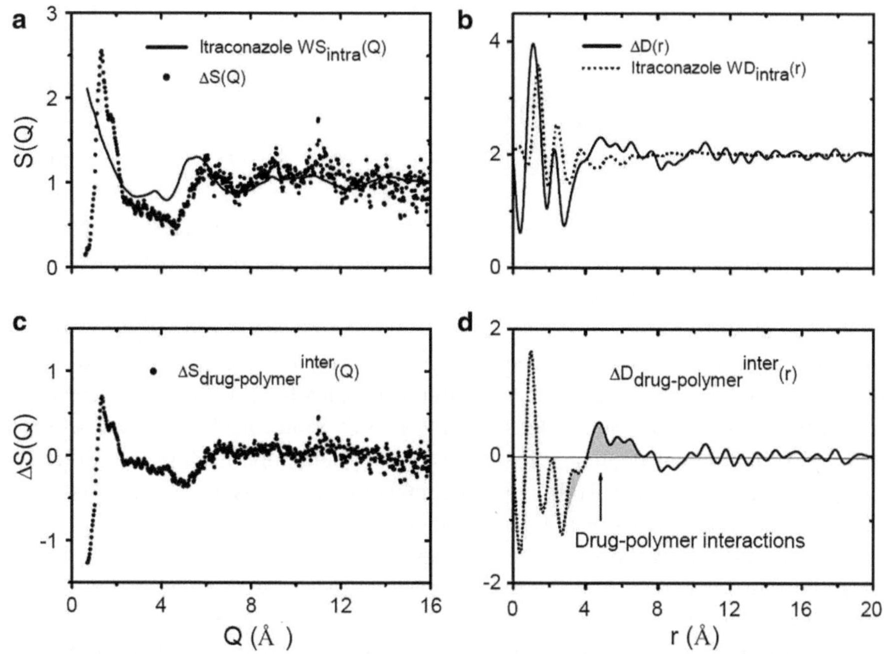

Fig. 9.10 (**a**) The difference structure factor, $\Delta S(Q)$, which is dominated by the carbon–carbon, itraconazole–PVP interactions and itraconazole–itraconazole interactions (*solid circles*). The weighted intramolecular diffraction pattern of pure itraconazole is represented by $WS_{\mathrm{drug}}^{\mathrm{intra}}(Q)$ (*line*). (**b**) The corresponding differential pair distribution functions $\Delta D(r)$ (*solid line*) compared to $WD_{\mathrm{drug}}^{\mathrm{intra}}(r)$ (*dashed line*). Both curves were Fourier transformed at a $Q_{\max} = 10\,\text{Å}^{-1}$ (where the point scatter masks any meaningful oscillation in $\Delta S(Q)$), and a Lorch modification function applied. (**c**) The difference between the curves in **a** is shown as *circles* and relates to the estimated intermolecular drug–polymer interaction $\Delta S_{\mathrm{drug-polymer}}^{\mathrm{inter}}(Q)$. (**d**) The difference between the curves in **b** is shown below and relates to the intermolecular drug–polymer interaction $\Delta D_{\mathrm{drug-polymer}}^{\mathrm{inter}}(r)$ (see text). The *shaded region* represents the region of nearest C–C intermolecular drug–polymer interactions

$$\begin{aligned}S_{1:2}(Q) &= 0.168 \times [S_{\mathrm{II}}(Q) - 1] + 0.299 \times [S_{\mathrm{IP}}(Q) - 1] + 0.533 \times [S_{\mathrm{PP}}(Q) - 1]\\ S_{1:4}(Q) &= 0.059 \times [S_{\mathrm{II}}(Q) - 1] + 0.208 \times [S_{\mathrm{IP}}(Q) - 1] + 0.732 \times [S_{\mathrm{PP}}(Q) - 1]\end{aligned} \tag{9.10}$$

By adding spectra 1:1 and 1:4 and subtracting twice the 1:2 pattern, we extract a function dominated by itraconazole–PVP interactions with some additional itraconazole–itraconazole interactions, which we approximate to $S_{\mathrm{drug}}^{\mathrm{intra}}(Q)$. Unfortunately, it is problematic to precisely subtract the intramolecular itraconazole–itraconazole interactions in this case (extracted from the diffraction pattern of the pure drug in Fig. 9.6) because the intramolecular bond distances shift to lower-r values in the mixture, as shown previously in Fig. 9.8. Therefore in Fig. 9.10a we compare $\Delta S(Q)$ to the appropriately scaled function $WS^{inter}_{drug}(Q)$, whereby the drug–polymer interaction is the difference between the two curves.

$$\Delta S(Q) = \Delta S^{inter}_{drug\text{-}polymer}(Q) + WS^{intra}_{drug}(Q)$$
$$= 2[S_{1:1}(Q) + S_{1:4}(Q) - 2S_{1:2}(Q)]/3 = [S_{IP}(Q) - 1] \quad (9.11)$$

The $\Delta S^{inter}_{drug\text{-}polymer}(Q)$ function, shown in Fig. 9.10c, is dominated by carbon–carbon interactions (57 %) which are 11 times larger than the next largest atom–atom correlations (see Table 9.1).

The peak shifts around 1.0 and 2.2 Å in Fig. 9.10b are due to slightly shorter intramolecular bond lengths that occur when the drug and polymer are mixed, and do not subtract correctly when the molecular itraconazole–itraconazole interactions are removed. The main intermolecular (C–C) correlation therefore occurs between 4 and 7 Å (and maybe even at 3.5 Å) between drug and polymer (this is denoted by the shaded region of Fig. 9.10d). No appreciable interaction is observed beyond this distance within the errors of the measurement.

Putting aside the validity of the assumption that the isosbestic points infer an essentially linear variation of the partial structure factors with composition, at this point it is appropriate to discuss the experimental error analysis associated with PDF measurements. While the scatter of measured data points in Q-space gives a clear indication of the random error associated with $S(Q)$ and $\Delta S(Q)$ (which increases with increasing Q-values) the extent of the errors is not so clear in real space. In general, distances at lowest-r in $D(r)$, say <2 Å, are the region most prone to long-wavelength systematic errors in the measurement or Fourier transform procedure and also inadequacies in the spherical electron cloud approximation. At higher distances, an estimation of the extent of uncertainty can be gleaned from the magnitude of the transform ripples that penetrate through the data. It should also be noted that since $\Delta S(Q)$ is a double difference function most of the systematic errors associated with the measurement cancel. Finally, some intermolecular itraconazole–itraconazole interactions may remain in the $\Delta S^{inter}_{drug\text{-}polymer}(Q)$ function despite the samples being polymer rich. We note that in pure itraconazole the short-range intermolecular interactions are centered on a primary peak at 4.4 Å, which would only partially account for the measured intensity labeled "drug–polymer interactions" in Fig. 9.10b.

Figure 9.11 shows possible intermolecular C–C interactions between itraconazole and PVP to illustrate the findings of the differential pair distribution analysis. For a full interpretation of the PDF data, computer simulation or modeling techniques are required. Classical molecular dynamics simulations provide a powerful method for understanding the structure and interaction of drug–polymer systems and can straightforwardly predict the measured x-ray PDF by moving atoms within a simulation box according to predefined interatomic potentials (Allen and Tildesley 1993). Similarly inverse methods such as Empirical Potential Structure Refinement have been specifically designed such that the experimental

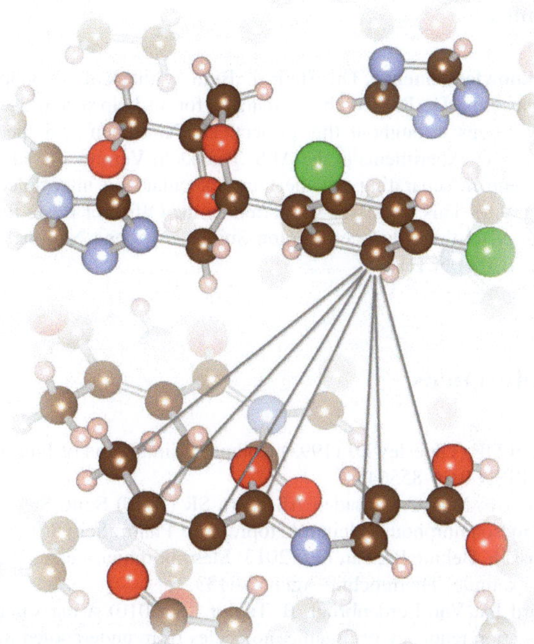

Fig. 9.11 A schematic of the molecular interaction depicts representative carbon–carbon itraconazole–PVP interactions between the molecular backbones

PDF data constrains the three-dimensional model in order to obtain a chemically realistic structure that agrees with the x-ray measurement (Soper 2005).

9.10 Future

From an experimental standpoint the efficacy of the x-ray PDF method hinges on obtaining a database of information on several drug–polymer systems to characterize and evaluate the variability of the interaction upon mixing. Furthermore, the structural data needs to be correlated with solubility and the shelf life of the medicine to find and predict structure-property trends that may lead to highly stable products. In terms of the methodology presented, evaluating many more compositions within the phase diagram, to determine regions of crystallinity and phase separation, would enhance the usefulness of the x-ray PDF technique. Consequently, this would enable us to employ a more robust matrix analysis of the type described by Neuefeind and Benmore (2009). Moreover, a greater number of measurements would enable the extraction of a more reliable drug–polymer pair distribution function, provided the linear behavior of the partial structure factors is indeed proven to be a sound approximation inferred by the observation of isosbestic points.

Acknowledgements Dr. Stephen Byrn is thanked for helpful discussions and review of the manuscript. Dr. Rick Weber is thanked for his help with the levitator experiments and stimulating discussions throughout this project. Dr. Xin Chen is helped with the sample preparation and levitation experiments at the APS. Thanks to Veronika Kondev for Fig. 9.4c and Qiushi Mou for his help in separating out the intramolecular and intermolecular contributions for itraconazole (Fig. 9.6). This work was supported by the Office of Basic Energy Sciences, U.S. Department of Energy, at the Advanced Photon Source, Argonne National Laboratory under contract number DE-AC02-06CH1135.

References

Allen MP, Tildesley DJ (1993) Computer simulation of liquids. Oxford University Press, Oxford. ISBN 0-19-855645-4

Atassi F, Mao C, Masadeh AS, Byrn SR (2010) Solid State characterization of amorphous and mesoamorphous calcium ketoprofen. J Pharm Sci 99(9):3684

Atassi F, Behme RJ, Patel PJ (2013) Mesomorphous versus traces of crystallinity: the itraconazole example. Thermochim Acta 574:133

Baird JA, Van Eerdenbrugh B, Taylor LS (2010) A classification system to assess the crystallization tendency of organic molecules from undercooled melts. J Pharm Sci 99:3787

Baird JA, Santiago-Quinonez D, Rinaldi C, Taylor LS (2012) Role of viscosity in influencing the glass forming ability of organic molecules from the undercooled melt state. Pharm Res 29:271

Benmore CJ (2012) A review of high-energy X-ray diffraction from glasses and liquids. ISRN Materials Science, Vol. 852905. http://dx.doi.org/10.5402/2012/852905

Benmore CJ, Weber JKR (2011) Amorphization of molecular liquids of pharmaceutical drugs by acoustic levitation. Phys Rev X 1:011004

Benmore CJ, Weber JKR, Tailor AN, Cherry B, Yarger JL, Mou Q, Weber W, Neuefeind J, Byrn SR (2013) Structural characterization and aging of glassy pharmaceuticals made using acoustic levitation. J Pharm Sci 102:1290

Chiou WL, Riegelman S (1971) Pharmaceutical applications of solid dispersion systems. J Pharm Sci 60:1281–1301

Elliott SR (1991) Medium range structural order in covalent amorphous solids. Nature 354:445

Faber TE, Ziman JM (1965) A theory of the electrical properties of liquid metals. III. The resistivity of binary alloys. Philos Mag 11:153. doi:10.1080/14786436508211931

Gao P (2008) Amorphous pharmaceutical solids: characterization, stabilization, and development of marketable formulations of poorly soluble drugs with improved oral absorption. Mol Pharm 5:903

Hancock BC, Zografi G (1993) The use of solution theories for predicting water vapor absorption by amorphous pharmaceutical solids: a test of the Flory–Huggins and Vrentas models. Pharm Res 10(9):1262

Keen DA (2001) A comparison of various commonly used correlation functions for describing total scattering. J Appl Cryst 34:172

Klug HP, Alexander LE (1974) X-ray diffraction procedures: for polycrystalline and amorphous materials, 2nd edn. Wiley Interscience Publication, New York. ISBN 0471493694, 9780471493693

Lorch E (1969) Neutron diffraction by germania, silica and radiation-damaged silica glasses. J Phys C 2(2):229

Mou Q, Benmore CJ, Yarger JL (Submitted to J. Appl. Cryst.)
Mura P, Faucci MT, Manderioli A, Furlanetto S, Pinzauti S (1998) Thermal analysis as a screening technique in preformulation studies of picotamide solid dosage forms. Drug Dev Ind Pharm 24 (8):747–756
Neuefeind J, Benmore CJ (2009) Formalism for the determination of structural isotope effects with neutrons. Nucl Instrum Meth A 600(1):257
Neuefeind J, Zeidler MD, Poulsen HF (1996) The atomic and electronic structure of liquid N-methylformamide as determined from diffraction experiments. Mol Phys 87(1):189
Pham TN, Watson SA, Edwards AJ, Chavda M, Clawson JS, Strohmeier M, Vogt FG (2010) Analysis of amorphous solid dispersions using 2D solid-state NMR and 1H T1 relaxation measurements. Mol Pharm 7(5):1667–1691
Six K, Verreck G, Peeters J, Binnemans K, Berghmans H, Augustijns P, Kinget R, Van den Mooter G (2001) Investigation of thermal properties of glassy itraconazole: identification of a monotropic mesophase. Thermochim Acta 376:175
Soper AK (2005) Partial structure factors from disordered materials diffraction data: an approach using empirical potential structure refinement. Phys Rev B 72(10):1
Soper AK (2007) Joint structure refinement of X-ray and neutron diffraction data on disordered materials: application to liquid water. J Phys Condens Matter 19(33):335206
Tarnacka M, Adrjanowicz K, Kaminski E, Grzybowska K, Kolodziejczyk K, Wlodarczyk P, Hawelek L, Garbacz G, Kocot A, Paluch M (2013) Molecular dynamics of itraconazole at ambient and high pressure. Phys Chem Chem Phys 15:20742
Tse JS, Klug DD, Guthrie M, Tulk CA, Benmore CJ, Urquidi J (2005) Investigation of the intermediate- and high-density forms of amorphous ice by molecular dynamics calculations and diffraction experiments. Phys Rev B 71:214107
Warren BE (1934) The diffraction of X-rays in glass. Phys Rev 45(10):657
Warren BE (1990) X-ray diffraction. Courier Dover Publications, New York. ISBN 0486663175, 9780486663173
Weber JKR, Rey CA, Neuefeind J, Benmore CJ (2009) Acoustic levitator for structure measurements on low temperature liquid droplets. Rev Sci Instrum 80:083904
Weber JKR, Benmore CJ, Tumber SK, Tailor AN, Rey CA, Taylor LS, Byrn SR (2012) Acoustic levitation—recent developments and emerging opportunities in biomaterials research. Eur Biophys J 41:397
Wegiel LA (2013) Stability of amorphous drug polymer dispersions, impact of molecular drug polymer interactions. Ph. D. Thesis, Purdue University, 89(125): 128
Whymark RR (1975) Acoustic field positioning for containerless processing. Ultrasonics 13(6): 251
Willart JF, Descamps M (2008) Solid state amorphization of pharmaceuticals. Mol Pharm 5:905

Part III
Use of Physicochemical Properties for Preclinical Formulation Selection and Early Clinical Formulations

Chapter 10
Performance and Characterization of Amorphous Solid Dispersions: An Overview

Grace Ilevbare, Patrick Marsac, and Amitava Mitra

10.1 Introduction

Advances in drug discovery techniques such as use of multiple high-throughput activity screening and combinatorial chemistry favor complex chemical structures with high pharmacological potency. However, these advances have led to selection of drug candidates with poor drug-like properties (e.g., highly lipophilic and poorly water soluble) that face significant problems later in drug development (Venkatesh and Lipper 2000). Recent analyses show a consistent increase in the number of water insoluble drug candidates that are in clinical development across the pharmaceutical industry, with some reports suggesting that up to 90 % of the new chemical entities in development can be classified as water insoluble compounds (i.e., BCS II and BCS IV) (Benet and Wu 2006). These types of molecules often show low oral bioavailability and saturation of exposure with increasing dose and present significant challenges to pharmaceutical scientists to develop formulations which can provide adequate bioavailability.

Solubility plays a central role, along with permeability and dissolution rate, in drug absorption after oral dosing (Martinez and Amidon 2002; Sugano 2012b). Drug absorption from the gastrointestinal (GI) tract is a highly complex process and is dependent on numerous factors such as drug physicochemical properties, formulation characteristics and GI physiological factors. For the purpose of this chapter, a simplified schematic of the oral absorption process is shown in Fig. 10.1. As seen in

G. Ilevbare
Discovery Pharmaceutical Sciences, Merck & Co., Kenilworth, NJ, USA

P. Marsac (✉)
Formulation Sciences, Merck & Co., 770 Sumneytown Pike, West Point, PA 19486-0004, USA
e-mail: Patrick_Marsac@merck.com

A. Mitra
Biopharmaceutics, Merck & Co., West Point, PA, USA

A.C. Templeton et al. (eds.), *Discovering and Developing Molecules with Optimal Drug-Like Properties*, AAPS Advances in the Pharmaceutical Sciences Series 15,
DOI 10.1007/978-1-4939-1399-2_10

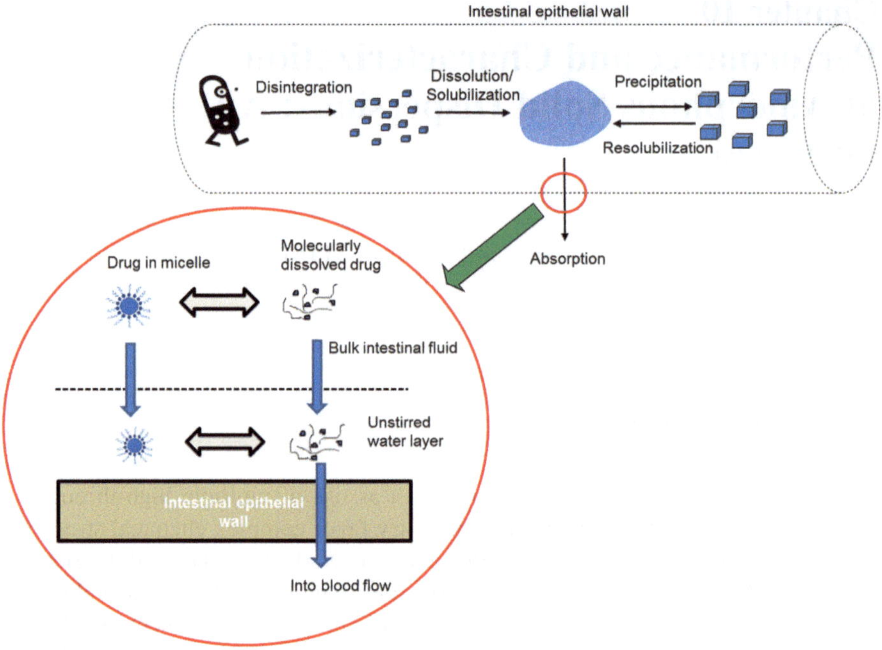

Fig. 10.1 A simplified schematic presentation of the drug absorption process after oral administration

the figure, drug absorption after oral dosing is a sequential process, which can be described as follows:

1. The drug product disintegrates to form small solid particles.
2. These particles then undergo dissolution to release solubilized drug. The solubilized drug is usually a mixture of molecularly dissolved drug and drug molecules bound to the bile salt micelles present in the small intestinal fluid.
3. The solubilized drug then migrates closer to the intestinal epithelial wall due to mixing and convection of the bulk intestinal fluid, and further diffuses into the unstirred water layer (UWL) adjacent to the epithelial cells.
4. Finally, the molecularly dissolved drug permeates across the intestinal epithelial wall and gets absorbed.

Once the solubility limit is reached in the small intestinal fluid, no more dissolution will take place until the dissolved drug gets absorbed. Hence, for low solubility molecules achieving complete solubilization of the dose can become a significant challenge. It is also possible that the dissolved drug can precipitate in the intestinal fluid. The precipitate may subsequently redissolve depending on the solubility of the form and its dissolution kinetics. Hence it becomes imperative that formulations are developed which can transiently increase solubility and also maintain supersaturation for a physiologically relevant time period so as to enable adequate absorption and bioavailability.

Scientists are able to influence absorption of poorly soluble molecules by crystal form selection, manipulating drug properties (e.g., salts, cocrystals, and amorphous forms) and formulation engineering. Most of these approaches boost absorption by solubilizing the drug and also maintaining the drug in solution state by decreasing the rate of crystallization, thus prolonging the supersaturated state. Crystalline solids are mostly used in pharmaceutical drug formulations because of their physical and chemical stability. However, the superior stability of crystalline solids can have a significant adverse impact on dissolution and solubility, especially for BCS Class II and IV compounds. Low solubility is a significant barrier to the effective delivery of therapeutic agents, consequently high energy forms such as amorphous solids can be utilized to improve solubility and delivery. Amorphous solids hold the promise of superior performance (Serajuddin 1999) because of the inherently higher energy state of active pharmaceutical ingredients (API) over the crystalline counterpart.

However, a higher energy alone does not guarantee higher bioavailability. Amorphous materials have unique challenges as compared to other more common approaches for increasing bioavailability due to an increased risk of crystallization to a more stable crystalline form. In a recent review of 40 research papers, the dissolution and bioavailability of amorphous formulations relative to reference formulations was examined (Newman et al. 2012). Improved bioavailability was observed in ~82 % of the systems studied leaving ~18 % of the systems demonstrating similar or lower values (Newman et al. 2012). Therefore, rendering a material amorphous does not guarantee success and as will be described in this and the two subsequent chapters, the product must be engineered to achieve superior performance. Engineering superior products requires an understanding of how the method of manufacture and the functional excipients impact the fate of the product during storage and after administration. The latter demands an understanding of how the material behaves as it travels through the gastrointestinal tract and is absorbed through the gut wall.

Performance is defined by demonstration of superior bioavailability. However, demonstrating superior bioavailability over traditional formulation approaches early in development may or may not inherently imply successful demonstration of other, perhaps longer term, performance requirements. First, there is a need to demonstrate adequate physical stability. This requires selection of formulation and process conditions that yield a product that is completely amorphous. Much more difficult to demonstrate is the avoidance of amorphous–amorphous phase separation over the lifetime of the product. Second, amorphous formulations are most often more chemically labile and avoiding chemical degradation can be more challenging than the corresponding crystalline material. Third, the manufacture of the solid dispersion intermediate and downstream processing to the final product should allow for robust manufacturability and scale-up. Finally, the development and characterization of amorphous products carry unique challenges which may or may not have regulatory precedence and hence proactive discussion between the Sponsor and Agency is required to enable successful approval of these products.

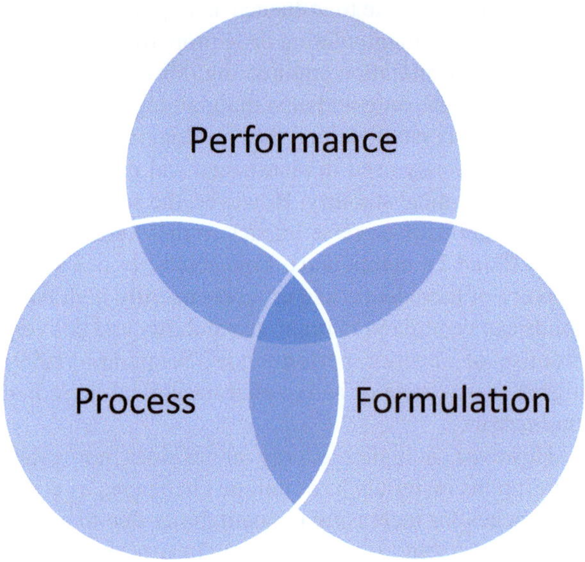

Fig. 10.2 The process and the formulation together give the desired performance. The challenge is to engineer the product with the desired performance characteristics so as to achieve superior stability and bioavailability

In this chapter, performance will be defined and concepts and measures of performance peculiar to amorphous systems will be highlighted. Failure modes and other phenomena unique to amorphous materials, such as crystallization and phase separation will be discussed. In the following chapters, the two most common methods of production, spray drying and hot-melt extrusion, will be outlined in detail with an emphasis on how the formulation and processing impact the observed performance. It will be shown that performance is dictated by the combination of formulation and processing and these concepts are interrelated as illustrated in Fig. 10.2. Not all formulations will provide the desired performance; some formulations may provide the required performance but cannot be manufactured in a robust fashion. Alternatively, although some processes may yield an amorphous material, the performance required may not be realized. Furthermore, some processes may give the performance required but may not be robust. Engineering to achieve the ideal performance requires a simultaneous optimization of process and formulation.

10.2 Drug Solubility, Dissolution Rate, and Permeability and Methods of Quantification

Solubility and dissolution are interrelated properties of a drug molecule. Solubility is the thermodynamic parameter measured by the concentration in solution of a compound in equilibrium with excess solid compound. Solubility is dependent on drug (e.g., p*K*a, lipophilicity, as well as intermolecular forces in the solid state) and solvent properties (e.g., pH, polarity, as well as solvent interactions) (Avdeef 2001; Martin 1993; Yalkowsky 1999). For crystalline compounds, solubility is defined as follows (Yalkowsky 1999):

$$\ln(x)_{\text{solute}} = -\frac{\Delta H_{\text{f}}}{R}\left(\frac{1}{T} - \frac{1}{T_{\text{m}}}\right) - \frac{\Delta C_p}{R}\left(1 - \frac{T_{\text{m}}}{T} + \ln\frac{T_{\text{m}}}{T}\right) - \ln\gamma \tag{10.1}$$

where $\ln(x)_{\text{solute}}$ is the natural log of mole fraction solubility, ΔH_{f} is the molar heat of fusion of the crystal, R is the universal gas constant, T is the temperature of interest, T_{m} is the melting temperature, and ΔC_p is the heat capacity difference between the crystal and the liquid. The activity coefficient, γ accounts for nonideal solute–solvent interactions. For details on derivation of Eq. (10.1), the reader is encouraged to consult previously published comprehensive texts on the topic (Yalkowsky 1999; James 1986; Martin 1993). Based on Eq. (10.1), factors such as higher melting point, higher heat of fusion, and lower solution temperature will translate into lower solubility. It is difficult to accurately model solubility without experimentation; careful attention to experimental detail is necessary to obtain high quality solubility data. Solubility becomes the rate-limiting step to absorption when the concentration of dissolved drug in the small intestine reaches the equilibrium solubility limit of the drug in the intestinal fluid, so that solid drug can no longer dissolve. In other words, solubility-limited absorption occurs when the dose is so high that the gut fluids are saturated with the drug(s) and therefore demonstrate a less than proportional increase in AUC with dose. Formulation approaches to enhance solubility such as using amorphous drug, or salt forms of the drug could be used to increase absorption in such cases.

Along with solubility, intestinal permeability is another important parameter that governs oral drug absorption. Permeation across the intestinal membrane is a sequential process of permeation through the unstirred water layer (UWL) adjacent to the intestinal wall, followed by permeation across the intestinal epithelial cell membrane into the enterocytes (Lennernas 1998; Sugano 2009a, b). Mathematically, the effective intestinal permeability is expressed as (Sugano 2009a):

$$\frac{1}{P_{\text{eff}}} = \left(\frac{1}{P_{\text{ep}}} + \frac{1}{P_{\text{UWL}}}\right)\frac{1}{\text{FE}} \tag{10.2}$$

Here, FE is the surface area expansion coefficient of the fold (plicate) structures in the small intestine, P_{eff} is the effective intestinal permeability, P_{ep} is the epithelial cell membrane permeability given by:

$$\log P_{\text{ep}} = 1.1(\log P) - 5.63 \tag{10.3}$$

P_{UWL} is the unstirred water layer permeability:

$$P_{\mathrm{UWL}} = \frac{D_{\mathrm{eff,UWL}}}{h_{\mathrm{eff}}} + P_{\mathrm{wc}} \tag{10.4}$$

$D_{\mathrm{eff,UWL}}$ is the effective diffusion coefficient in the UWL, h_{eff} is the effective thickness of UWL, and P_{wc} is the UWL permeability coefficient by water convection. As is seen from Eqs. (10.2)–(10.4), the effective intestinal permeability is a function of partitioning of the drug substance into UWL adjacent to the intestinal wall and permeation into the enterocyte of the intestinal wall. Equation (10.3) shows that the epithelial cell permeability depends on the log P of the compound. Although this might indicate that increase in log P would result in proportional increase in permeability, this is not the case. High log P compounds would have a tendency to sequester in the intestinal epithelial cells and also would have limited diffusion across the UWL due to their high lipophilicity. This has been demonstrated experimentally in permeability studies using Caco-2 cell monolayers, which has clearly shown an upper limit in molecule lipophilicity above which there is minimal to no gain in permeability (Camenisch et al. 1998).

Oral absorption becomes permeability limited when drug permeation across the intestinal epithelia is slow compared to the dissolution rate. Manipulating the formulation typically cannot mitigate this type of absorption limitation. Published reports have suggested the use of permeability enhancers in the formulation to transiently increase the permeability of drugs across the intestinal wall by disrupting the lipid bilayer structure of the epithelial membrane, opening tight junctions (Aungst 2000, 2012; Goldberg and Gomez-Orellana 2003). However, the use of permeation enhancers is limited and, to date, there are no approved oral drug products that use permeation enhancers, primarily because of toxicity concerns, i.e., once the intestinal membrane has been permeabilized to facilitate absorption of the drug, any xenobiotic and/or microorganism might also get absorbed (Goldberg and Gomez-Orellana 2003; Swenson and Curatolo 1992).

Dissolution rate is commonly expressed by Noyes–Whitney equation:

$$\mathrm{DR} = \frac{\mathrm{d}Q}{\mathrm{d}t} = DA\left(\frac{C_{\mathrm{S}} - C_{\mathrm{b}}}{h}\right) \tag{10.5}$$

where DR is dissolution rate, D is the diffusion coefficient, A is the solid surface area, C_{s} is solubility, C_{b} is bulk solution concentration and h is the diffusion layer thickness. Based on Eq. (10.5), dissolution rate will increase with increasing surface area, diffusivity, and drug solubility. It is also evident that dissolution rate will decrease with increasing thickness of the aqueous boundary and the concentration of drug already in solution. Absorption can become dissolution rate limited when the dissolution rate of the formulation is much slower than the permeation rate, so the dissolved drug is immediately absorbed and the saturation solubility in the intestinal fluid is never reached. In this scenario, the rate of drug absorption is determined by the rate of dissolution and drug particle size reduction will be effective in enhancing absorption.

As indicated above and in Fig. 10.1 oral absorption is controlled by dissolution, diffusion, permeability, solubility and precipitation (crystallization). The importance of solubility, dissolution, and permeability is further illustrated by Eq. (10.6):

$$F_{\text{absorption}} = C_{\text{b}} \times P \times \text{SA} \tag{10.6}$$

where $F_{\text{absorption}}$, the mass transport of free dissolved API molecules across the intestinal membrane, P is the permeability, C_{b}, the free concentration of drug in the unstirred water layer, and SA, the surface area available for absorption (Sjögren et al. 2013). Surface area in the human GI is highest in the small intestine and permeability generally decreases as a function of distance down the GI tract (He et al. 1998; Sjögren et al. 2013).

Due to the complexities of drug product development process, the ability to accurately predict oral drug absorption during drug product development is becoming more important given the time and resource challenges facing the pharmaceutical industry. To this end, physiologically based pharmacokinetic (PBPK) modeling enables the prediction of the plasma concentration versus time profiles and PK parameters in human based on preclinical in vitro and in vivo data and can thus provide a valuable resource to support decision making at various stages of the drug development process. Hence PBPK modeling is now routinely conducted to guide formulation development and is also used by regulatory agencies during review of new drug applications (NDA) (Huang et al. 2013; Kostewicz et al. 2013; Zhang et al. 2011). These models strive to mimic the drug absorption process from the gastrointestinal (GI) tract and predict the plasma concentration time profiles based on the physicochemical properties of the drug, formulation characteristics, GI physiology and dose. Briefly, the models mathematically describe the drug dissolution process in the GI tract and subsequent absorption as the dissolved drug passes through the small intestine and the colon (Agoram et al. 2001). Changes in dissolution rate as a result of changes in the formulation, pH dependent solubility, absorption in the colon, gut/liver metabolism, effect of transporters, and other variables may also be accounted for in these models. A detailed discussion of the absorption modeling approaches is out of scope for this chapter and interested readers are referred to the references cited above. Nevertheless, the use of simpler calculations can provide a rapid assessment of the biopharmaceutical profile of the drug and guide formulation approaches in the early stages of drug development (Oh et al. 1993; Rohr 2006). These include dimensionless numbers such as dose number (DO), which is a ratio of dose concentration (i.e., highest dose in 250 mL dosing vehicle) over the solubility in GI fluid as shown in Eq. (10.7) below (Oh et al. 1993; Rohr 2006).

$$\text{DO} = \frac{\left(\frac{\text{Dose}}{\text{Volume}}\right)}{\text{Solubility}} \tag{10.7}$$

Based on this equation, as DO becomes greater than 1 the dose will exceed the solubilization capacity of the medium and absorption will become solubility limited. In this scenario, formulation approaches to enhance solubility (i.e., attain supersaturation) such as by using the amorphous form of the drug through solid dispersion technology, use of a salt form, cyclodextrin complexation, and lipid formulation may aid in enhancing absorption. The DO concept primarily gives an indication of potential for solubility limited absorption. For example, solubility of the crystalline form of itraconazole in simulated gastric fluid (SGF) has been reported to be 2.6 μg/mL and undetectable in fasted state simulated intestinal fluid (FaSSIF) (Taupitz et al. 2013). The calculated DO based on SGF solubility would be 154 at a dose of 100 mg, indicating that most of the dose will not dissolve in the stomach if the crystalline drug was dosed and thus there is a risk of poor bioavailability. However, when formulated as a complex in cyclodextrin, itraconazole solubility as high as 1.9 mg/mL was achieved in SGF (Taupitz et al. 2013). This would result in a DO of 0.2 at 100 mg indicating that the dose will be completely soluble in stomach (assuming that there is no dissolution rate limitation). This concept can be effectively used in assessing risk of solubility limited absorption of molecules and whether solubilizing formulation technologies would be needed to achieve adequate bioavailability.

Another dimensionless number commonly used is the dissolution number (DN) which is the ratio of the small intestinal transit time to the time required for complete dissolution (Eq. 10.8) (Oh et al. 1993; Rohr 2006). A $DN > 1$ suggests that the time for complete dissolution is faster than the small intestinal transit time, hence the full dose may be available in solution for absorption. On the other hand, a $DN < 1$ (i.e., dissolution is slower than transit time), indicates the possibility of a dissolution rate limited absorption and in such cases approaches such as drug particle size reduction could be used to increase the dissolution rate of the formulation.

$$\mathrm{DN} = \frac{\left(\frac{\pi R^2 L}{Q}\right)}{\left(\frac{r^2 \rho}{3DC_S}\right)} \tag{10.8}$$

Here, R is the radius of the small intestine (cm), L is the small intestinal length (cm), Q is the volume flow rate in the GI tract (cm^3/min), C_S is the saturation solubility of drug in the small intestinal medium (mg/mL), D is the diffusion coefficient of the drug (cm^2/s), and ρ is the density of the drug (g/cm^3). A dimensionless number that incorporates both permeability and solubility can also be used to assess the biopharmaceutical profile of the drug candidate, one such concept is the Maximum Absorbable Dose (MAD):

$$\mathrm{MAD} = P_{\mathrm{eff}} \times \mathrm{solubility} \times \mathrm{SISA} \times \mathrm{SITT} \tag{10.9}$$

where, P_{eff} is the effective intestinal permeability, SISA the small intestinal surface area, and SITT the small intestinal transit time. However, in practice due to the uncertainty in the in vitro or in situ measurements of P_{eff} using systems such as

Caco-2 cell monolayer or rat intestinal perfusion, the utility of MAD in routine risk assessment may be limited. This dimensionless number is similar to the $F_{\text{absorption}}$ concept as described in Eq. (10.6).

10.3 The Supersaturation Concept and Amorphous Solubility

The use of supersaturating dosage forms, such as amorphous solids, has long been recognized as a good formulation approach to increase absorption and bioavailability of poorly soluble drugs, since the increased intraluminal drug concentration can result in enhanced drug flux across the intestinal membrane (Brouwers et al. 2009). The advantages of formulating drugs as amorphous forms have been demonstrated by a number of studies (Bollag et al. 2010; Goldberg et al. 1966; Hancock and Parks 2000; Joshi et al. 2004; Kennedy et al. 2008; Konno et al. 2008; Kwong et al. 2011; Law et al. 2001, 2004; Leuner and Dressman 2000; Six et al. 2004). As noted above, the performance of amorphous solids as a supersaturating dosage form depends on the choice of formulation and processing conditions to yield a pure amorphous solid with no or minimal crystallization of the amorphous drug during storage and upon dosing. In dissolution studies of supersaturating dosage forms, API solution concentrations much higher than those achieved using the corresponding thermodynamically stable crystalline form are often observed, indicating that supersaturated solutions are being generated, as illustrated in Fig. 10.3. Typically, these increases in solubility will translate into an increase in oral absorption if the enhanced concentrations can be maintained long enough for absorption to occur. The longevity of the supersaturation achieved during dissolution of the amorphous form is of critical importance and is dictated by the physiological conditions.

Amorphous solids lack long range order characteristic of crystalline materials, so for amorphous solids, Eq. (10.1) reduces to:

$$\ln(x)_{\text{solute}} = -\ln\gamma \tag{10.10}$$

(Hancock and Parks 2000; Hoffman 1958), the ratio of amorphous to crystalline solubility $\left(\frac{\sigma_{\text{amorphous}}}{\sigma_{\text{crystalline}}}\right)$ can be written as:

$$\frac{\sigma_{\text{amorphous}}}{\sigma_{\text{crystalline}}} = \exp\left(\frac{\Delta G_{\text{a-c}}}{RT}\right) \tag{10.11}$$

where $\Delta G_{\text{a}-\text{c}}$, the difference in the free energy of the crystalline and amorphous material can be estimated using the Hoffman equation (Hoffman 1958):

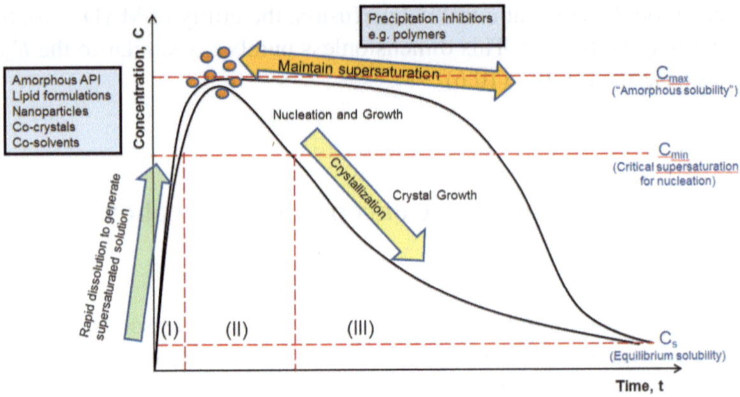

Fig. 10.3 A schematic of drug concentration versus time profile illustrating the concept of achieving and maintaining supersaturation in the gastrointestinal tract (adapted and revised based on a figure from Brouwers et al. 2009). The figure also shows a representation of LaMer's diagram depicting stages of nucleation and crystal growth with change in concentration as a function of time: region I before nucleation, region II of nucleation and growth, and region III of growth analogous to a typical dissolution profile that is usually obtained for ASDs of poorly water soluble drugs. (LaMer and Dinegar 1950; Sarode et al. 2014)

$$\Delta G_{\mathrm{a-c}} = \frac{\Delta H_{\mathrm{f}}}{RT}\left(\frac{\Delta T \cdot T}{{T_{\mathrm{m}}}^{2}}\right) \tag{10.12}$$

The Hoffman equation was derived for temperatures between T_g (glass transition temperature) and T_m (melting temperature) under the assumption that the enthalpy of the supercooled liquid and crystal vary linearly with temperature. The enthalpy difference between the supercooled liquid and the crystalline materials was represented by a linear function. The Hoffman equation takes into consideration changes in enthalpy and entropy of fusion as a function of temperature, but does not require explicit knowledge of the heat capacity (Hoffman 1958). More accurately, the heat capacity of the amorphous and crystalline materials can be measured using differential scanning calorimetry (DSC) (Murdande et al. 2010a, b). The aforementioned assumption (temperature range used will be between T_g and T_m) made by Hoffman may affect the accuracy of Eq. (10.12) for predicting the theoretical solubility of an amorphous solid. Most dissolution experiments are performed at room temperature (~25 °C) or physiological body temperature (37 °C), which are both lower than typical glass transition temperatures for many amorphous pharmaceutical compounds. If the experimental temperature is close enough to the glass transition temperature, then the value generated by Eq. (10.12) is an acceptable estimate of the energy difference between the amorphous and crystalline state (Hoffman 1958). Although the Hoffman equation may provide a reasonable estimation of ΔG_{a-c}, it does not take into consideration the degree of ionization of the amorphous and crystalline materials and the reduced activity of the amorphous solid resulting from water sorption. Thus, more accurately, the theoretical solubility advantage of an amorphous material (Eq. 10.13) (Ilevbare and Taylor 2013) can be estimated using the experimentally determined crystalline solubility (C_{eq}), the free energy difference between the crystalline and amorphous forms ($\Delta G_{\text{a}\rightarrow\text{c}} = \mu_2^{\text{a}} - \mu_2^{\text{c}}$), the activity of the amorphous solute saturated with water $\left(\text{e}^{[-I(\text{a}_2)]}\right)$, the fraction ionized for an amorphous solute (α^{a}) and the fraction ionized for the crystalline solute (α^{c}) as shown in the following equation (Murdande et al. 2010a, b):

$$C_{\text{amorphous}} = C_{\text{eq}} \times \text{e}^{[-I(\text{a}_2)]} \times \text{e}^{\left[\frac{\mu_2^{\text{a}} - \mu_2^{\text{c}}}{RT}\right]} \times \left(\frac{1 - \alpha^{\text{c}}}{1 - \alpha^{\text{a}}}\right) \qquad (10.13)$$

$\text{e}^{[-I(\text{a}_2)]}$ can be calculated from the moisture sorption profile of the amorphous solid and represents the reduction in the thermodynamic activity due to the presence of moisture. Experimentally, the number of moles of water absorbed per mole of amorphous solid as a function of relative humidity is determined and this data is used to estimate the water content at a relative humidity of 100 (water activity of 1). The Gibbs–Duhem equation is then used to calculate the activity of the amorphous drug. The third term is the change in standard state chemical potential on transforming the crystal to the dry amorphous state. The last term on the right hand side in the above equation represents differences in the ionization states in solutions equilibrated with amorphous versus crystalline forms. This ionization term is equal to 1 if the API is unionizable or is present in a buffered solution where ionization is minimal.

In summary, the amorphous solubility advantage can be described well using thermodynamic principles. The theoretical enhancement in solubility can be several orders of magnitude (Hancock and Parks 2000). However, when comparing predicted solubility values for several amorphous drugs with experimentally reported values in the literature, there is typically a poor correlation, especially for moderately and fast crystallizing compounds. This inconsistency has been attributed, in part, to the rapid crystallization to a lower solubility drug form from the amorphous material upon contact with dissolution fluids or crystallization from the supersaturated solution generated upon dissolution of the amorphous solid (Hancock and Parks 2000; Konno et al. 2008).

The thermodynamic driving force for crystallization is created by the presence of a supersaturated solution. The driving force for crystallization is the difference in chemical potential ($\Delta\mu$) of a molecule in the supersaturated and saturated solution (Mullin 2001b).

$$\Delta\mu = kT\ln\left(\frac{c}{c_{\text{s}}}\right) \qquad (10.14)$$

where k is the Boltzmann constant, T is the absolute temperature, c is the solute concentration and, c_s is the equilibrium solute concentration at equilibrium temperature, T.

For dilute solutions, the supersaturation ratio, S, is given by the ratio of concentrations (Ilevbare et al. 2012c):

$$S = \frac{c}{c_s} \tag{10.15}$$

The magnitude of S is an important factor in determining how long supersaturation can be maintained. A system is always stable (i.e., no crystallization will occur) when $c - c_s < 0$. A system is considered to be metastable when $c - c_s > 0$; the concentration is above the equilibrium solubility and an increase in bioavailability can be observed since spontaneous crystallization is improbable. Under these conditions, nucleation, precipitation, deposition and growth of a new phase are possible. Spontaneous nucleation occurs when $c - c_s \gg 0$.

Shown in Fig. 10.3 is an illustration of a typical concentration–time profile (LaMer's diagram) for an amorphous solid depicting stages of nucleation and crystal growth (LaMer and Dinegar 1950; Sarode et al. 2014). During Stage 1, solution concentration increases steadily with time until the equilibrium solubility, C_s, of the drug is attained and exceeded. If the rate of dissolution is slower than the rate of crystallization from the bulk amorphous solid, then solution concentration may not exceed C_s or may exceed C_s, but C_{min} will not be attained. Spontaneous nucleation occurs when a critical supersaturation is attained, C_{min}. Solution concentrations between C_{min} and C_s can be maintained for extended periods of time because there is an insufficient driving force to overcome the energy barrier to nucleation. During stage II, nucleation and growth occur simultaneously and if the dissolution rate of the amorphous solid dispersion (ASD) is slower, the concentration drops below C_{min}, thereby stopping the nucleation process. However, the growth of already formed nuclei still continues, further reducing the concentration to C_s during stage III. C_{max} is the maximum solution concentration that can be achieved during the dissolution of an amorphous solid; often, it can also be described as the amorphous "solubility." The kinetics and mechanism of nucleation and crystal growth are highly dependent on concentration and may be difficult to control. It may be possible to prolong the level of supersaturation obtained during stage II by incorporating effective nucleation and/or crystal growth inhibitor(s) in the formulation. This observation is of great interest since it means that the solution concentration can be increased and maintained with a low risk of crystallization. Guzmán et al. (2007) described the generation and stabilization of a metastable supersaturated state using the "spring and parachute" model (Fig. 10.3). Higher energy forms such as amorphous solids that transiently increase solubility relative to that of the stable crystalline form were likened to the "spring." The higher energy state of amorphous solids can translate into an increase in solubility and drug absorption, but rapid crystallization back to the more stable and less soluble crystalline form can limit this benefit (Guzmán et al. 2007). Additives that inhibit crystallization were described as "parachutes"; they maintain supersaturation for an extended time period. The enhanced and prolonged solubilization could lead to further increases in absorption and bioavailability.

10.4 Crystallization Inhibition by Polymeric Additives

The use of amorphous solid dispersions has become a well-known strategy for inhibiting crystallization, as the amorphous solid dispersions can have increased physical stability over neat amorphous material. The supersaturated solutions generated from the dissolution of amorphous solids may lead to an increase in absorption compared to that of a saturated solution if supersaturation can be maintained for a physiologically relevant type period. However, due to the inherent thermodynamic instability of the supersaturated state, which will eventually lead to drug crystallization, prolonged maintenance of supersaturation in the gastrointestinal tract may be difficult (Ilevbare et al. 2012c). It has been demonstrated that the addition of polymeric additives, even at very low concentrations, can be used to inhibit crystallization from supersaturated solutions (Alonzo et al. 2010; Dai et al. 2007; Femi-Oyewo and Spring 1994; Gao et al. 2009; Lindfors et al. 2008; Raghavan et al. 2001; Simonelli et al. 1970; Vandecruys et al. 2007). If one or both of the crystallization processes can be retarded or inhibited, supersaturation may be maintained for a physiologically relevant time period, leading to enhanced absorption. Polymeric additives that have been investigated in this context include polyvinyl pyrrolidone (PVP) (Alonzo et al. 2010; Femi-Oyewo and Spring 1994; Gao et al. 2009; Lindfors et al. 2008; Simonelli et al. 1970; Vandecruys et al. 2007), polyethylene glycol (PEG) (Dai et al. 2007; Gao et al. 2009), hydroxypropyl methyl cellulose (HPMC) (Alonzo et al. 2010; Raghavan et al. 2001), and hydroxypropyl methyl cellulose acetate succinate (HPMCAS) (Alonzo et al. 2010; Dai et al. 2007; Femi-Oyewo and Spring 1994; Gao et al. 2009; Raghavan et al. 2001; Vandecruys et al. 2007). Although the inhibition of crystallization from aqueous supersaturated solutions of drugs by polymeric additives has been extensively reported, little is known about which polymer properties influence their inhibitory potential. It has been proposed that maintenance of supersaturation through the addition of polymeric additives is the result of intermolecular interactions in solution (hydrogen bonding) and/or steric hindrance of recrystallization. Raghavan et al. attributed the decreased crystallization rate of hydrocortisone acetate in polymer solutions to attractive forces between the polymer and the drug, e.g., by hydrogen bonding (Raghavan et al. 2001). Hasegawa et al. (1988) investigated ability of CMEC (carboxymethylethyl cellulose) to inhibit crystallization of nifedipine, griseofulvin and spironolactone from supersaturated solutions. The authors noted that the adsorption of CMEC on the surface of nifedipine crystals was the most important factor in inhibiting crystallization from supersaturated solution. Ziller and Rupprecht also reported that inhibition is caused by polymer molecules occupying adsorption sites on the crystal face (Ziller and Rupprecht 1988). However, in all of the examples above, it is difficult to discriminate between effects on nucleation and crystal growth. The following section focuses on the effect of polymers on nucleation and crystal growth.

10.4.1 Nucleation

The formation of crystals in liquid solutions begins with nucleation. In solution crystallization, nucleation is of practical importance in pharmaceutical systems since it plays an important role in determining the final crystal structure, form and size distribution properties (Erdemir et al. 2009). Drug delivery challenges concerning the concentration threshold above which crystallization of an active pharmaceutical ingredient (API) occurs upon administration are related to the kinetic stability of supersaturated solutions, which is in turn regulated by the nucleation mechanisms and kinetics (Rodríguez-hornedo and Murphy 1999). Thus, optimum control over crystallization from supersaturated solutions is facilitated by understanding the fundamentals of nucleation kinetics and the mechanism of inhibition by additives.

Classical nucleation theory (CNT) is the simplest and most widely used theory that describes the nucleation process. Although CNT was originally derived for condensation of a vapor into a liquid, it has been applied to explain precipitation of crystals from supersaturated solutions. CNT assumes that clusters evolve in size by a sequence of molecular additions until they reach a critical size. While CNT allows for the estimation of the size of the critical nucleus and nucleation rate (Eq. 10.4), it does not provide any information about the structure of aggregates or pathways leading from the solution to the solid crystal (Erdemir et al. 2009; Schuth 2001; Vekilov 2004).

$$J = K_0 \exp\left(-\frac{\Delta G_{\mathrm{crit}}}{k_{\mathrm{B}}T}\right) \tag{10.16}$$

Where, K_0 is the pre-exponential kinetic factor:

$$K_0 = N_0 v_0 \tag{10.17}$$

N_0 is the solubility of the material expressed as the number of molecules of the crystallizing phase in a unit volume, and v_0 is the frequency of atomic or molecular transport at the nucleus–liquid interface. Equation (10.4) may be rewritten as:

$$J = N_0 v_0 \exp\left(-\frac{16\pi {\gamma_{12}}^3 \Omega^2}{3(k_{\mathrm{B}}T)^3(\ln S)^2}\right) \tag{10.18}$$

Here γ is the interfacial energy per unit area between the crystallization medium and the nucleating cluster, Ω is the volume of a molecule inside the crystal, k_B is Boltzmann's constant, and T is temperature.

In the past decade, a line of simulations, theories and experimental studies suggests that nucleation of solids from solution does not proceed through the classical nucleation model but follows more complex routes. Alternatively, the two-step nucleation model (Erdemir et al. 2009; Vekilov 2004) (which was initially proposed for protein crystallization) proposes that a sufficient-sized disordered cluster of solute molecules forms first, followed by reorganization of that cluster into an ordered structure. The applicability of the two-step mechanism to both macromolecules and small organic molecules has been demonstrated, suggesting that this mechanism may be important for crystallization processes from solutions (Vekilov 2004). Molecular dynamics simulations also generate a two-step nucleation process (Anwar et al. 2009). The organization time of the cluster has been proposed to be the rate-limiting step of the nucleation process, due to the observed increase in organization time of appropriate lattice structures with greater molecular complexity of the solute. An illustration of the two-step nucleation model is shown in Fig. 10.4 (Erdemir et al. 2009).

Determination of the induction time, also known as the nucleation time, is a well-established method of characterizing crystal formation kinetics (Cerdeira et al. 2004; Kuldipkumar et al. 2006; Raghavan et al. 2001; van der Leeden et al. 1993; Verdoes et al. 1992). After the creation of a supersaturated solution, a certain induction time may elapse prior to the formation of a detectable amount of the new crystalline phase (Kuldipkumar et al. 2006; Sohnel and Mullin 1988). However, it is not possible to directly measure nucleation time, because the nuclei

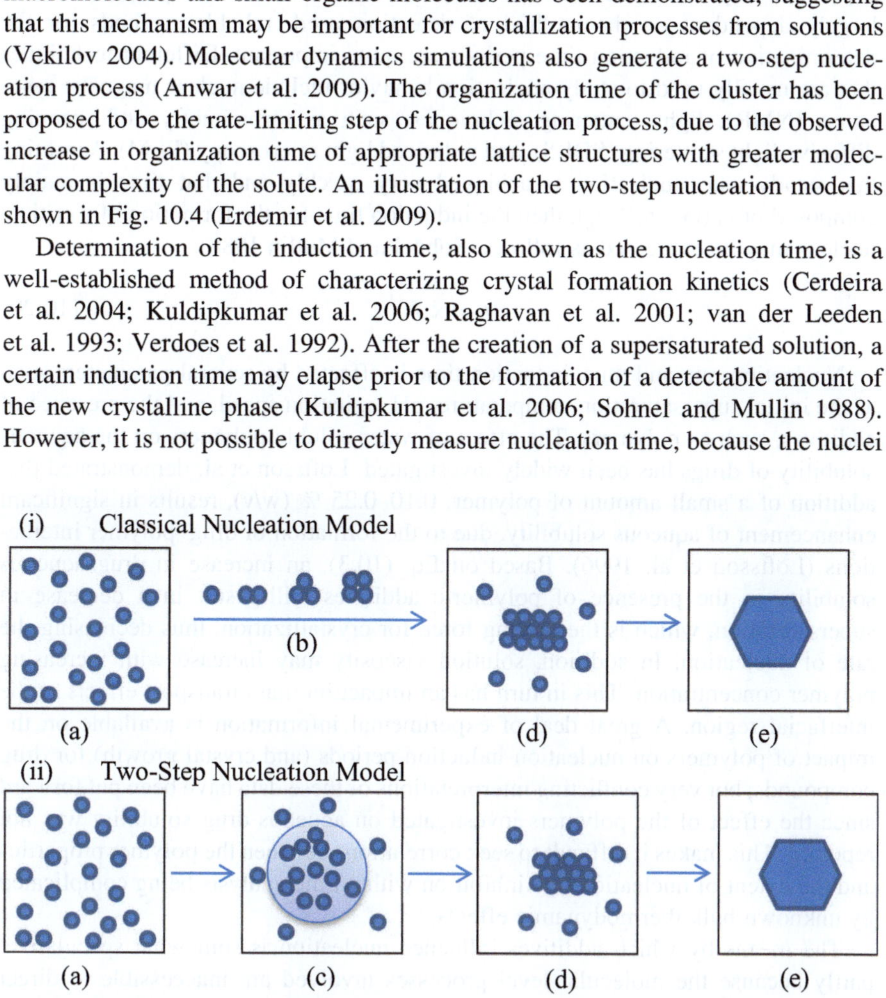

Fig. 10.4 Alternative pathways leading from solution to solid crystal: (a) supersaturated solution; (b) ordered subcritical cluster of solute molecules, proposed by classical nucleation theory; (c) liquid-like cluster of solute molecules, dense precursor proposed by two-step nucleation theory; (d) ordered crystalline nuclei; (e) solid crystal

formed can only be detected after they have grown to an experimentally detectable size. Therefore, the experimental nucleation time, t_{ind}, can be defined as the sum of the time for critical nucleus formation (true nucleation time, t_n) and growth to detectable size, t_g (Sohnel and Mullin 1988).

$$t_{ind} = t_n + t_g \tag{10.19}$$

This time is an experimental observable and is a measure of the ability of the solution to remain supersaturated, that is, in metastable equilibrium. Since the experimentally measured induction time is not a fundamental property and is largely dependent on the method of detection used, reliable methods for the determination of induction time periods are quite important. Different techniques that are usually used to study nucleation behavior include nuclear magnetic resonance (NMR), light scattering techniques, such as turbidimetry and scanning diffusive light scattering (SDLS), and polarized light microscopy (PLM). Assuming that steady-state nucleation is achieved very quickly and that time is mainly composed of t_n (i.e., $t_n \gg t_g$), then the induction time for the formation of a critical nucleus may be expressed as follows (Sohnel and Mullin 1988):

$$t_{ind} \propto J^{-1} \tag{10.20}$$

Nucleation rate and thus, induction time is affected by several parameters such as the initial supersaturation, temperature, pH, agitation speed, and the presence of additives, such as polymers. The effect of water-soluble polymers on the aqueous solubility of drugs has been widely investigated. Loftsson et al. demonstrated that addition of a small amount of polymer, 0.10–0.25 % (w/v), results in significant enhancement of aqueous solubility, due to the formation of drug–polymer interactions (Loftsson et al. 1996). Based on Eq. (10.3), an increase in drug aqueous solubility in the presence of polymeric additives will result in a decrease in supersaturation, which is the driving force for crystallization, thus decreasing the rate of nucleation. In addition, solution viscosity may increase with increasing polymer concentration. This in turn has an impact on mass transport effects in the interfacial region. A great deal of experimental information is available on the impact of polymers on nucleation-induction periods (and crystal growth) for drug compounds, but very conflicting interpretations of these data have been put forward since the effect of the polymers investigated on aqueous drug solubility was not reported. This makes it difficult to seek correlations between the polymer properties and the extent of nucleation rate inhibition without the analysis being complicated by unknown bulk thermodynamic effects.

The means by which additives influence nucleation is somewhat speculative, partly because the molecular-level processes involved are inaccessible to direct experimental observations. Based on the two step nucleation model, it can be assumed that polymers most likely affect nucleation kinetically by hindering the reorganization of a cluster of solute molecules into an ordered structure, which is proposed to be the rate-limiting step of the two-step nucleation model. An effective nucleation inhibitor should thus interact favorably with the solute (Anwar et al. 2009). A strong affinity of the polymer for the solute, relative to that for the solvent and other polymer molecules in solution, not only ensures that the polymer molecules interact with the solute aggregates, but also leads to disruption of the reorganization of the solute clusters (Anwar et al. 2009). This will impact the ability of the polymer to mix with drug pre-nucleation aggregates which in turn would be expected to influence the effectiveness of the polymer as a nucleation inhibitor. Important intermolecular interactions between a polymer and drug molecule include specific interactions such as hydrogen bonding and nonspecific interactions such as hydrophobic interactions. Molecular simulations on a model system also suggest that the relative strength of the pairwise intermolecular interactions between solute, solvent, and additive, as well as size factors, influences the impact of the additive on nucleation kinetics (Anwar et al. 2009). In a structure–property study, it was found that the effectiveness of the various polymers in inhibiting nucleation depended on the hydrophobicity of the polymer relative to that of the model drug compounds—celecoxib, efavirenz, and ritonavir (Ilevbare et al. 2012c). The hydrophobicity of the polymer most likely influences the ability of the polymer to form polymer–solute interactions relative to polymer–solvent and polymer–polymer interactions. Polymer–solute interactions would be expected to hinder the reorganization of a cluster of solute molecules into an ordered crystal structure (Ilevbare et al. 2012c).

10.4.2 Crystal Growth

Crystal growth commences after the formation of stable nuclei. Crystal growth depends on external factors (temperature, supersaturation, solvent, and presence of impurities), and internal factors (three-dimensional crystal structure, molecular bonds, and crystal defects) (Mullin 2001b; Rodríguez-hornedo and Murphy 1999; Sangwal 2007). Crystal growth occurs by the diffusion of molecules from the bulk/ liquid phase to the crystal surface, followed by the incorporation or integration of the molecules into the crystal lattice. At the surface, molecules become organized into the space lattice through an adsorbed layer. Therefore, crystal growth mechanisms fall into two main categories: volume diffusion and surface integration controlled processes. If the diffusion to the crystal surface is the slowest step, growth is diffusion controlled. Diffusion controlled growth is not applicable if there is no dependence of growth rate on hydrodynamic conditions such as flow rate or stirring rate (Sangwal 2007). If the integration of the molecules into the crystal lattice is the rate limiting step, then growth is considered to be integration rate limited. However, neither the diffusion step nor the integration step will happen unless the solution is supersaturated. The expressions for diffusion and integration controlled processes are given in Eqs. (10.9) and (10.10), respectively (Garside 2002).

$$R_G = k_d\left(\frac{C_\infty - C_I}{1 - w_\infty}\right) \tag{10.21}$$

$$R_G = k_r(C_I - C^*)^r \tag{10.22}$$

where R_G is the overall mass growth rate of a crystal, k_d and k_r are the mass transfer rate coefficient and integration rate coefficient, respectively, C_I is the intermediate concentration between the bulk and the crystal surface, C_∞ is the concentration in the bulk solution, C^* is the equilibrium concentration, w_∞ is the bulk solution concentration as a mass fraction and r is the reaction order. When the rate of diffusion and integration are equal, the overall growth rate is (Garside 2002):

$$R_G = k_G(C_\infty - C^*)^g \tag{10.23}$$

where k_G is the growth rate constant and g is the reaction order.

Depending on the extent of supersaturation, layer growth or continuous growth may occur (Rodríguez-hornedo and Murphy 1999; Sangwal 2007). Layer growth occurs at low supersaturations and can be further divided into two subcategories by which steps are forms: screw dislocation and two-dimensional nucleation (Mullin 2001b; Sangwal 2007). The continuous growth model is applicable at higher supersaturations when the crystal–solution interface at the molecular level is rough, and there are many defects or potential growth sites. In this case, growth proceeds isotropically resulting in nonfaceted crystals and responds to very small changes in the growth driving chemical potential (supersaturation) (Rodríguez-hornedo and Murphy 1999).

It has been demonstrated that a polymer that is an effective nucleation inhibitor may not have a similar impact on the inhibition of crystal growth, especially at high supersaturations (Abbou Oucherif et al. 2013; Alonzo et al. 2012; Ilevbare et al. 2012b, c, d). Alonzo et al. observed that preventing the formation of crystals (nucleation) in supersaturated solutions of felodipine through the addition of a polymeric crystallization inhibitor (hydroxypropylmethyl cellulose (HPMC)) was a more effective strategy for maintaining supersaturation than inhibiting crystal growth, since crystal growth could not be inhibited at high supersaturation conditions representative of that generated by the amorphous formulation (Alonzo et al. 2012). However, preventing nucleation may not always be possible and inhibition of growth may be necessary to maintain supersaturation and thus bioavailability. For example, the amorphous formulation may contain seed crystals resulting from the manufacturing process, which will lead to rapid de-supersaturation unless effective crystal growth inhibitors have been included in the formulation. Small traces of crystalline material can thus potentially have a significant impact on the extent and duration of supersaturation. In aqueous solution, crystal growth inhibition is typically characterized by measuring the rate of de-supersaturation in the presence of seed crystals, with and without pre-dissolved polymers (Ilevbare et al. 2012d). When seeds are added to a supersaturated solution, the excess solute can be consumed by two mechanisms (Ilevbare et al. 2012b): (1) the seed crystals can grow to a larger size; (2) new additional crystals can be generated by primary and/or secondary nucleation, resulting in a distribution of particles of a smaller size than the added seed crystals. Therefore, it is important to determine the predominant mechanism of desupersaturation during seeded-desupersaturation experiments.

It is generally accepted that for an additive to inhibit growth when added at a low concentration at which there are no effects of solution viscosity or polymer concentration on equilibrium solubility, adsorption of the inhibitor to the growing surface is an important factor. The coverage of the surface sites available for adsorption is characterized by the parameter, θ (Sangwal 2007). The number of sites, n_{max}, available for adsorption per unit area of a surface for a given set of growth conditions is constant. At a given concentration of impurity, c_i, the coverage of adsorption sites, θ_{eq}, may be expressed by:

$$\theta_{eq} = \frac{n_{ad}}{n_{max}} \tag{10.24}$$

where n_{ad} is the number of adsorption sites occupied per unit area on the surface at a particular temperature. At $c_i = 0$, the available adsorption sites on the surface are vacant, that is, $n_{ad} = 0$ and $\theta_{eq} = 0$. θ_{eq} increases as c_i increases, reaching 1 when all adsorption sites are occupied ($n_{ad} = n_{max}$) (Sangwal 2007). However, complete coverage, $\theta_{eq} = 1$, is not considered to be a necessary condition for the complete prevention of crystal growth (Kubota and Mullin 1995).

Polymer molecules inhibit the introduction of drug molecules from solution into the crystal lattice by occupying growth sites and thereby acting as a mechanical barrier (Hasegawa et al. (1988). Several factors, such as hydrophobicity (Tian et al. 2007; Zimmermann et al. 2009), electrostatic interaction (Pan et al. 2001; Tjipangandjara and Somasundaran 1991), and hydrogen bonding between the adsorbate and the interfacial species (Raghavan et al. 2001; Somasundaran and Huang 2000; Somasundaran and Krishnakumar 1997) have been found to contribute to the adsorption process. In a structure–property study, Ilevbare et al. found that the hydrophobicity of the polymer, as well as the presence of ionized functional groups, were important in determining the effectiveness of a polymer as a crystal growth inhibitor for three model compounds—celecoxib, efavirenz, and ritonavir (Ilevbare et al. 2012d, 2013a). Using FTIR spectroscopy, the authors also demonstrated that the formation of specific interactions was in part responsible for the crystal growth inhibition of ritonavir by poly(*N*-isopropylacrylamide) (Ilevbare et al. 2012d, 2013a) and efavirenz by PVP (Ilevbare et al. 2013a). Specific intermolecular interactions (such as hydrogen bonding) may be important in cases in which polymer and the crystallizing drug have accessible donor and acceptor groups.

As highlighted above, crystal growth rate/mechanism is highly dependent on supersaturation, so it is expected that the effectiveness of the polymers as crystal growth rate inhibitors will also be dependent on supersaturation. It has been demonstrated that growth is inhibited to a large extent at low supersaturation, but only marginally at supersaturations corresponding to the amorphous "solubility" (Alonzo et al. 2012; Ilevbare et al. 2012a, c). A change of growth mechanism from (slow) layer growth to (fast) continuous growth (Fig. 10.4) may be responsible for the much lower effectiveness of polymers in inhibiting growth at very high supersaturations. At very high supersaturations, growth proceeds by a random addition of solute molecules to the crystal surface and it is likely that there are always favorable growth sites available even in the presence of adsorbed polymer. In other words, at high supersaturations, the density of interaction of the adsorbed polymer with the crystal surface is insufficient to block every potential growth site and so crystals will grow rapidly (Ilevbare et al. 2013a).

Lastly, it is important to consider the effect of multiple additives on crystal growth from solution since combinations of additives (polymers and surfactants) are often used in pharmaceutical products to improve the delivery of poorly water-soluble active pharmaceutical ingredients (API). Additive interactions have not been widely studied and may promote or inhibit crystallization (nucleation and crystal growth) in an unpredictable manner, which in turn has an impact on the extent and duration of supersaturation. In the absence of additive effects on either supersaturation or solution viscosity, and assuming adsorption of additive molecules on the surface of crystal is a required step for modification of growth to occur, the following scenarios are possible for binary additives (Ilevbare et al. 2012b): (1) Both additives adsorb independently at the interface and may either interact with different surface adsorption sites or compete for adsorption sites (Ghodbane and Denoyel 1997; Wesley et al. 1999); (2) If both additives adsorb at the surface and they also interact to form solution complexes, they may adsorb as a complex directly from solution or they may adsorb individually. In this case, there may be competition between the additives and the complex for adsorption sites (Ghodbane and Denoyel 1997; Otsuka et al. 1996). By performing seeded-desupersaturation experiments, the effect of a series of polymer–polymer and polymer–surfactant combinations on crystal growth inhibition of ritonavir was investigated (Ilevbare et al. 2012b). The majority of the polymer–polymer combinations investigated had a synergistic effect on crystal growth inhibition, that is, the combination of two polymers was more effective in inhibiting crystal growth compared to either of the individual polymers. In contrast, no shortening of nucleation-induction times was observed when the polymer combinations were used, with the polymers actually inhibiting nucleation (Ilevbare et al. 2012b). The effectiveness of the polymer–polymer combinations was ascribed to the formation of inter-polymer complexes through hydrophobic interactions that adsorb and interact favorably with the crystallizing solute and/or interaction of individual polymers at different adsorption sites. The acceleration of crystal growth in the presence of polymer–surfactant combinations was attributed to weakened interactions between the polymer and the surface of the crystallizing solute brought about by the presence of surfactant molecules. Based on these observations, careful evaluation of the impact of combinations of additives on crystallization behavior is recommended in order to optimize the performance of supersaturating dosage forms (Ilevbare et al. 2012b).

10.4.3 Impact of Solution Crystallization Inhibition on Performance

So far the discussion has focused on understanding the impact of polymers on crystallization inhibition and the mechanism by which inhibition occurs. For supersaturating dosage forms, such as amorphous solid dispersions, the ability of a polymeric additive to inhibit solution crystallization determines the success of the formulation for drug delivery. Although the in vivo conditions of the gastrointestinal tract are extremely difficult to simulate in vitro, perhaps an understanding of the changes in the thermodynamic properties of a drug during dissolution, supersaturation, and crystallization in comparative dissolution studies may be helpful for appreciating the potential enhancement of absorption in vivo by polymeric amorphous solid dispersions.

It has been demonstrated that amorphous solid dispersions provide improved physical stability over neat amorphous material and enhanced oral bioavailability for poorly soluble compounds over their crystalline counterparts. For example, Law et al. studied the absorption of poly(ethylene glycol) (PEG)-ritonavir amorphous solid dispersions with different drug loadings using beagle dogs as an animal model (Law et al. 2004). Amorphous solid dispersions containing 10–30 % drug exhibited significant increases in exposure over crystalline drug. The 10 % amorphous solid dispersion exhibited increases of 22- and 13.7-fold in AUC and C_{max}, respectively; this increase in solution concentration was higher than that reported for pure amorphous ritonavir (Law et al. 2001). Similarly, an animal bioavailability study in dogs was performed to compare a capsule containing 1:2 (w/w) itraconazole: HPMC-P (hydroxypropylmethyl cellulose phthalate) amorphous solid dispersion, prepared by spray-drying, to those filled with crystalline itraconazole (Engers et al. 2008). The itraconazole:HPMC-P amorphous solid dispersion exhibited significantly higher C_{max} and bioavailability than the crystalline material alone. The C_{max} was approximately 17- to 35-fold higher compared to crystalline material, while the systemic exposure, was 511–1,911 % of that observed from crystalline itraconazole.

In addition to improving bioavailability for poorly soluble compounds in normal fasted state, amorphous solid dispersions can also mitigate the effect of food on pharmacokinetics of these drugs. It is well known that co-dosing with food can sometimes dramatically affect pharmacokinetics of drugs, commonly known as food effect (Fleisher et al. 1999). For example, typically for a BCS class II drug (low solubility, high permeability) an increase in AUC and/or C_{max} is expected with food, whereas for a BCS class IV drug (low solubility, low permeability) food can either increase or decrease AUC and/or C_{max}. Since a food effect can have significant implications on drug product development (Lentz 2008), solubility enhancing formulations are now being increasingly utilized to minimize or mitigate the changes in PK when the product is dosed with food. Since these formulations can increase solubility and dissolution rate of the drug in the GI tract and thus increase their bioavailability, this may minimize the effect of food on PK. Lopinavir/ritonavir solid dispersion tablets made using melt extrusion technology showed minimal effect on AUC and C_{max} when dosed with high fat or moderate fat meal, as compared to dosing in fasted state (Klein et al. 2007). In contrast, the soft gelatin capsule formulation showed 62 % and 32 % increase for lopinavir, and 44 % and 28 % for ritonavir, in AUC and C_{max} respectively, when dosed with a moderate fat meal as compared to fasted state.

10.5 Dissolution and Amorphous Nanoparticle Formation: Effect on Performance

Dissolution of a solid dispersion depends on the proportional quantities and properties of both the drug and carrier (Corrigan 1985). Craig et al. presented a simplified model of the process by which drug particles are released during dissolution (Craig 2002). This model assumes that there is a highly concentrated polymer layer at the dissolving surface (at least at high polymer content) through which the drug must pass before it is released into the bulk solution phase. The two main mechanisms for drug release from solid dispersions that have been described in the literature are carrier drug-release and carrier-controlled release (Albers et al. 2009; Craig 2002; Tho et al. 2010; Vasconcelos et al. 2007). For carrier drug-release, the drug is molecularly dispersed within the concentrated polymer layer and the rate-limiting step for dissolution of drug is the release of polymer itself. For carrier-controlled release, the dissolution of the drug particles into the polymer-rich layer is a slow process. These two release mechanisms may occur simultaneously if the drug particles are partially dissolved in the polymer-rich diffusion layer before being released. It is important to note that at low drug concentrations, the release mechanism could be either carrier drug-release or carrier-controlled (Craig 2002). Although a moderately hydrophobic polymer may be a very effective crystallization inhibitor for a given drug compound, it may have a negative impact on the rate of drug release from the solid dispersion at high polymer concentrations. More hydrophilic polymers, such as PVP, can be potentially combined with moderately hydrophobic polymers in amorphous matrices to improve drug release rates. Therefore, by judicious selection of polymer combinations, it may be possible to substantially delay crystallization during supersaturation generated by dissolution of the amorphous solid while concurrently improving formulation properties (Ilevbare et al. 2012b).

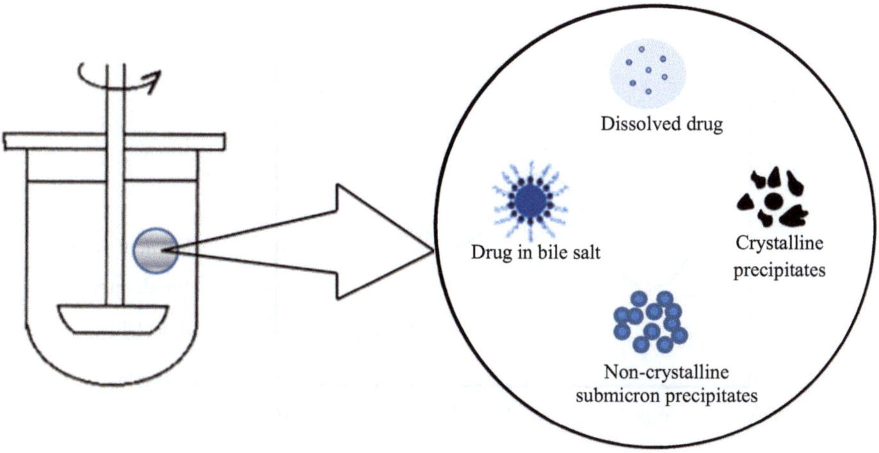

Fig. 10.5 Species formed during dissolution of amorphous solid dispersions in simulated intestinal fluid

Shown in Fig. 10.5 is an illustration of the various "structures" a drug compound can exist as during dissolution of an amorphous solid dispersion in simulated fluid. A drug compound may completely dissolve in solution (free-drug in solution), be trapped in the hydrophobic core of bile salt micelles, form crystalline precipitates or noncrystalline submicron colloidal dispersions from supersaturated solution. Often times when amorphous solid dispersion is dissolved in aqueous media, the resultant solution is observed to be cloudy, characteristic of the presence of a second light scattering phase. This observation is consistent with studies in which supersaturating dosage forms, consisting of amorphous solid dispersions, are found to have a rapid dissolution rate (Alonzo et al. 2011; Hsieh et al. 2012; Ilevbare and Taylor 2013; Karavas et al. 2006), and submicron particles are formed in situ upon dispersion in the aqueous media (Ilevbare et al. 2013b). In recent studies, it has been demonstrated that when sufficiently high supersaturated solutions are generated in aqueous media, poorly water-soluble compounds can form noncrystalline (amorphous) precipitates (Hsieh et al. 2012; Ilevbare et al. 2013b; Ilevbare and Taylor 2013). This observation is consistent with the study by Friesen et al., where it was found that colloidal particles began to form for an amorphous solid dispersion containing a poorly water soluble drug only at high supersaturation (Friesen et al. 2008). Using the solvent-shifting (co-solvent) method, Ilevbare et al. determined the concentration of colloidal nanoparticle formation for eight hydrophobic poorly water-soluble drug molecules. This concentration was found to be consistent for a given medium and temperature, is predictable, and can be related to the theoretical amorphous "solubility" (Ilevbare et al. 2013b). Although numerous researchers have demonstrated that amorphous solid dispersions are capable of generating nanosuspensions of drug—submicron colloidal dispersions of nearly pure drug particles—when dissolved in aqueous solution (Aisha et al. 2012; Alonzo et al. 2011; Frank et al. 2012; Friesen et al. 2008; Tachibana and Nakamura 1965),

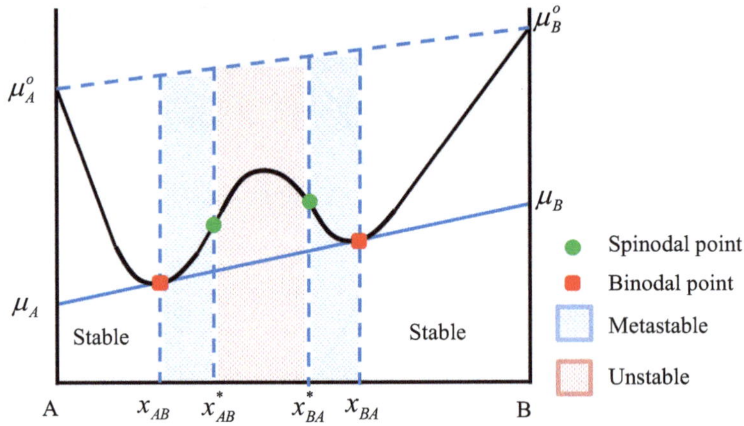

Fig. 10.6 Free energy diagram illustrating phase separation. Stable, unstable and metastable refer to the stability of the liquid phases

the formation mechanism and the factors influencing the size and stability of these species are not well understood. Based on experimental observations, as well as results presented in the literature, the behavior of compounds showing amorphous precipitation and displaying extended supersaturation, relative to the crystalline equilibrium solubility, can be explained by considering the liquid-liquid phase separation (LLPS) phenomenon (Hsieh et al. 2012; Ilevbare and Taylor 2013; Raina et al. 2012). The "liquid" undergoing phase separation from the bulk solution phase is an amorphous supercooled liquid (perhaps alternatively an amorphous glass), since the process is occurring at a temperature below the melting point of the crystalline material. The term "oiling out" is often used in place of LLPS if the compound is above its melting point. However, many compounds have a melting point lower than room temperature, and hence exist as a liquid or "oil" at room temperature.

The thermodynamic basis for LLPS has been presented in several papers (Hsieh et al. 2012; Ilevbare and Taylor 2013; Lafferrere et al. 2004). The criterion for stability of a homogeneous solution is based on the free energy between the mixed and unmixed states ΔG_{mix}. Figure 10.6 shows a free energy of mixing versus composition (x) diagram for two components A (amorphous drug) and B (water) at constant temperature. For a partially miscible system the free energy of mixing is expressed as:

$$\Delta G_{\text{mix}} = x_{\text{A}}\left(\mu_{\text{A}} - \mu_{\text{A}}^{\circ}\right) + x_{\text{B}}\left(\mu_{\text{B}} - \mu_{\text{B}}^{\circ}\right) \tag{10.25}$$

where x_{A} and x_{B} are the mole fraction of the individual components and μ_{A} and μ_{B} are the chemical potentials of the components and the sign $^{\circ}$ indicates the standard state. The free energy of the system can be reduced by separating into phases with compositions x_{AB} and x_{BA}; these points are called the binodal points. Both phases, x_{AB} and x_{BA}, have equal chemical potentials and are in equilibrium. Starting from a

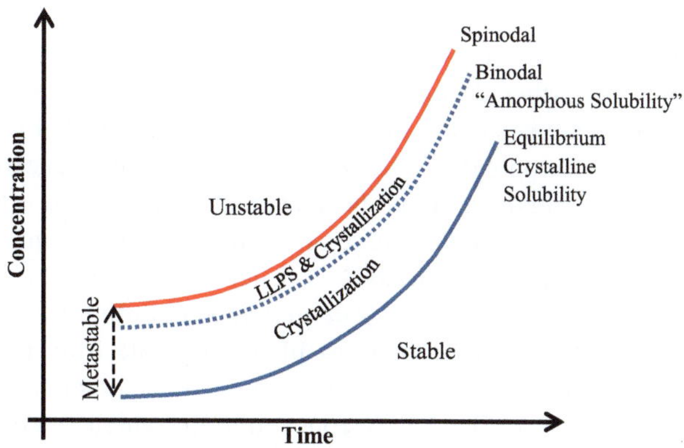

Fig. 10.7 Concentration-time diagram showing the metastable zone widths of LLPS and crystallization

dilute solution (B) and adding drug into the solvent, drug will dissolve leading to an increase in concentration. During this process, the solution is homogeneous (stable), the process is thermodynamically favored (negative ΔG_{mix}), and the chemical potentials of all components in the system are lower for the solution compared to the corresponding physical mixture. At a composition of x_{BA}, further addition of drug will not lead to an increase in concentration because chemical potential increases beyond this point. Rather, a new drug-rich phase of composition x_{AB} will form, which is in equilibrium with x_{BA}. For this hypothetical amorphous drug–water system, the binodal points represent the theoretical onset of LLPS, and corresponds to the upper-limit solution concentration (amorphous solubility), x_{BA}, that can be generated by the solid in the absence of any crystallization and the maximum water content of the drug rich phase, x_{AB} (Raina et al. 2012). In Fig. 10.6, the inflection points, x^*_{AB} and x^*_{BA} of the free energy curve are known as the spinodal points.

To evaluate the crystallization tendency of a system, a concentration versus temperature phase diagram similar to Fig. 10.7 is typically used. As mentioned above, Fig. 10.6 describes phase behavior at constant temperature. Figure 10.7 can be derived from free energy versus composition curves at different temperatures. Shown in Fig. 10.7 are the crystalline equilibrium solubility, the coexistence (binodal/amorphous solubility) and the spinodal curves. For a compound with a melting point higher than the experimental temperature, the binodal point x_{BA}, can be used to construct the coexistence (amorphous solubility) curve. The coexistence curve represents the concentration above which the drug-rich noncrystalline phase will begin to form; it is the LLPS limit relative to temperature. Below the equilibrium solubility curve a system is always stable (i.e., no crystallization will occur), while above the spinodal curve, the solution is unstable and there is no free energy barrier for the nucleation of a new phase, that is, spontaneous crystallization will occur. A system is considered to be metastable between the equilibrium solubility curve and spinodal curves; an increase in bioavailability may be observed since spontaneous crystallization is unlikely. It is apparent that the coexistence curve falls within the metastable region for crystallization and thus, LLPS will not be observed if the kinetics of crystal nucleation is rapid relative to LLPS. The metastable region can be further divided into two regions as depicted in Fig. 10.7 and these two regions are separated by the coexistence curve. Below the coexistence curve only crystallization can occur, while above the coexistence curve, either crystallization or LLPS can occur. It has been demonstrated that the distance been the spinodal and coexistence (binodal) curve is small for organic compounds (Ilevbare and Taylor 2013; Lafferrere et al. 2004; Veesler et al. 2003).

The colloidal species observed upon dissolution of the rapidly dissolving amorphous solid dispersions most likely originate from LLPS, since amorphous solid dispersions dissolve to generate supersaturated drug solutions (Ilevbare et al. 2013b). The LLPS concentration, which is comparable to the amorphous "solubility," represents the upper limit of solution concentration of "free" drug that can be achieved following the dissolution of an amorphous solid dispersion; exceeding the LLPS concentration leads to the formation of a colloidal phase of submicron size (Ilevbare et al. 2013b). The supersaturation behavior of drug compounds is highly correlated to their inherent crystallization tendency (Hsieh et al. 2012; Van Eerdenbrugh et al. 2010). Thus, it is expected that the nature of the species formed upon dissolution of amorphous solid dispersions will depend on the crystallization tendency of the API present in the amorphous solid dispersion. As noted above, LLPS will only occur if the kinetics of crystal nucleation is slow relative to the LLPS process (Ilevbare and Taylor 2013). Poorly water soluble drug compounds that undergo the LLPS process and thereby yield noncrystalline colloidal particles have been shown to belong to the group of slow crystallizing compounds (Hsieh et al. 2012; Ilevbare et al. 2013b; Ilevbare and Taylor 2013). Using a suitable polymer, the rate of crystallization of intermediate or fast crystallizing compounds can be significantly delayed and therefore, amorphous blends of these types of compounds with water soluble crystallization inhibitors may generate noncrystalline particulates upon dissolution. It has also been surmised that the duration of time the generated particulates remain amorphous in aqueous solution may be related to the ability to form stable amorphous formulations that possess long shelf-life stability (Brick et al. 2003; Friesen et al. 2008; Hsieh et al. 2012) and, when dissolved, yield supersaturated solutions that can be maintained for a physiologically relevant time period.

Colloidal nanoparticles generated during dissolution of a given amorphous solid dispersion are most likely important contributors to the bioavailability enhancement often seen with this supersaturating dosage form (Frank et al. 2012). Serajuddin et al. hypothesized that colloidal nanoparticles could dissolve during GI transit, thus increasing bioavailability (Serajuddin 1999). Nanoparticles are well known to improve dissolution rate and bioavailability due to their high surface area to volume ratio. As described in Eq. (10.5), dissolution rate is directly proportional to the surface area (A) of the dissolving drug particle and inversely proportional to the diffusion layer thickness (h). Since particle size reduction increases A and decreases h, nanoparticles can provide significant increase in the dissolution rate. The effect of particle size on drug solubility, dissolution rates and corresponding drug absorption has been well established both theoretically and experimentally (Fincher 1968; Hintz and Johnson 1989; Kaneniwa and Watari 1978; Kaneniwa et al. 1978; Lu et al. 1993; Waterman and Sutton 2003). The increase in solubility by particle size reduction can be estimated using the Ostwald–Freundlich equation (Kelvin equation) (Sugano 2012a):

$$\frac{S_{0,r_\mathrm{p}}}{S_{0,\infty}} = \exp\left(\frac{2\gamma v_\mathrm{m}}{r_\mathrm{p}RT}\right) \tag{10.26}$$

where S_{0,r_p} and $S_{0,\infty}$ are the solubility of particles with radius r_p and that of an infinitely large particle (larger than several micrometers), respectively, γ is the interfacial tension between the solid surface, the fluid and v_m is the molar volume of a drug, R is the gas constant and T is the absolute temperature.

Another hypothesis suggests that the nanoparticles can increase bioavailability by increasing effective permeability (Sugano 2010). This theory hypothesizes that the nanoparticles by the virtue of their small size can diffuse into the UWL, thus decreasing the effective thickness of the UWL thereby increasing the P_UWL (Eq. 10.4) and in effect increasing the effective permeability (Eq. 10.2) (Sugano 2010). Permeability studies with nanosuspensions of budesonide and spironolactone in Caco-2 cell monolayers, showed that the nanosuspensions had significantly higher flux than the saturated solution formulation of these compounds (Lenhardt et al. 2008). Since in these studies the nanosuspension was compared to a saturated solution, dissolution rate enhancement can be ruled out as possible explanation for the higher flux observed with the nanosuspensions. Although the exact mechanism for increase in flux with the nanosuspension is not known, this data does show that the nanoparticles should be increasing bioavailability by some mechanism other than increase in dissolution rate and solubility. Clearly, these results highlight the complexities of the mechanism by which these novel solubilizing formulations such as amorphous solid dispersion and nanosuspensions enhance bioperformance and this has presumably driven recent research focused on the fate of these oral dosage forms post administration.

In order to take advantage of the properties of colloidal dispersions, the submicron particles must be stabilized. Colloidal particles may undergo coalescence since colloidal particles are generated through the formation of highly supersaturated solutions. Polymers have been used to stabilize colloidal dispersions (Aisha et al. 2012; Douroumis and Fahr 2007; Friesen et al. 2008; Moribe et al. 2006; Tachibana and Nakamura 1965) and can also serve as crystallization inhibitors (Ilevbare et al. 2012c, d). It has been proposed that the size and stability of colloidal particles depends on solution conditions (such as pH, buffer and ionic strength) and the structure and properties of the molecules including the number of polar atoms, number of hydrogen bond donors and acceptors, conformational flexibility and presence of ionized groups (Ilevbare et al. 2013b; Frenkel et al. 2005). Analysis of aggregate size distributions for NNRTI APIs under various physiological conditions demonstrated dependence of aggregation size on solution pH (Frenkel et al. 2005) at constant ionic strength, aggregate size was found to increase with increasing pH (decreasing extent of ionization). This dependence of particle size on drug and/or additive ionization, and thus on pH, may have important implications for drug absorption since pH varies along the GI tract.

10.6 Solid-State Failure Modes: Crystallization and Amorphous–Amorphous Phase Separation

In addition to manipulating the performance of the amorphous API upon administration, additives are introduced with the intention of manipulating the performance of the solid state. Specifically, polymers, surfactants, and/or other excipients are selected with the intention of avoiding solid-state failure modes—most notably crystallization and amorphous–amorphous phase separation. Engineering the amorphous dispersion often requires that the drug, the polymer, and any other additives, are miscible. That is, in order to manipulate the thermodynamic driving force for crystallization and the molecular mobility of an amorphous API, the manufacturing process must incorporate additives on a molecular level. Therefore, an understanding of miscibility in the solid state and the impact which mixing has on those properties which mitigate physical failure risks are very important in the design of amorphous solid dispersions. In this section, a brief introduction of solid-state physical failure modes is provided and many important tools and techniques used to directly measure failure or to measure those factors which facilitate failure will be outlined.

10.6.1 Thermodynamics Drives and Mobility Facilitates Physical Failure

Crystallization from the solid state can be related to the thermodynamic driving force for the formation of a critical size nucleus and subsequent crystal growth, as well as the associated activation energies for each process (Bhugra and Pikal 2008; Fisher et al. 1949; Gutzow and Schmelzer 1995; Mullin 2001a). In the absence of a polymer, the thermodynamic driving force for crystallization of an API is easily measured using calorimetric techniques. The presence of a miscible polymer will act to reduce the thermodynamic driving force for crystallization and various tools and techniques have been developed to quantify the degree to which the chemical potential of an API is reduced (Marsac et al. 2009, 2006; Marsac and Taylor 2008). Of course thermodynamics alone cannot predict the kinetics of crystallization and the molecular mobility of the API also plays an important role (Bhugra and Pikal 2008; Ediger et al. 1996; Mullin 2001a). Most often, physical stability risk is considered to be low when the material is stored well below the temperature where cooperative motions are minimal (i.e., below the glass transition temperature). However, examples exist whereby the T_g alone does not predict physical stability (Vyazovkin and Dranca 2007), nucleation may be unavoidable during the production of amorphous materials (Baird et al. 2010; Van Eerdenbrugh et al. 2010), and crystal growth rates may exceed those expected based on diffusion controlled growth by orders of magnitude at temperatures near the glass transition temperature (Sun et al. 2008; Wu and Yu 2006; Hikima et al. 1995; Ishida et al. 2007). These results highlight the importance of measuring sub-T_g motions and, as with cooperative motions, polymers may only impact local motions which occur well below the glass transition temperature if miscibility is achieved. Various tools and techniques provide access to motions which occur over different timescales (Bhattacharya and Suryanarayanan 2009; Ediger et al. 1996). Linking molecular motions to physicochemical phenomena remains an area of active research and will be discussed within the framework of the tools outlined below. One approach is to measure the temperature dependence of a specific motion and demonstrate that the activation energy observed is coincident with the activation energy of a physical phenomenon such as crystallization. If this link can be made, the motion may be related to the observed phenomena. This approach, and derivatives of this approach, have been the source of active research in the pharmaceutical literature aimed at predicting shelf life of amorphous systems (Alie et al. 2004; Aso et al. 2000, 2004; Bhugra et al. 2006, 2007, 2008a, b; Caron et al. 2010; Greco et al. 2012; Korhonen et al. 2008).

The theoretical constructs used to describe phase separation of an amorphous drug from aqueous media can also be used to describe phase separation of a binary drug polymer system. Further, the criteria associated with the compositional dependence of the free energy of mixing two amorphous materials dictates the mechanism of phase separation, namely, separation via spinodal decomposition or via nucleation and growth. Similar to the arguments made for crystallization from the amorphous phase, kinetic restrictions in the solid state may not allow for realization of amorphous–amorphous phase separation.

In principle, measuring amorphous–amorphous phase separation is simple. A property of the mixture is measured and compared to the same property of the materials from which the mixture is made. This could include any number of properties such as the temperature dependence of the enthalpy (the heat capacity), the difference in density, optical properties, interaction of the material with various perturbations such as mechanical stress, application of an electrical potential across the material, or introduction of radiation which discriminates between mixtures and individual components as the result of interactions which might occur between the various materials over length-scales of molecular order. However, in practice, measurements can be exceedingly difficult since amorphous materials often have properties which are not as well defined as crystalline materials and amorphous–amorphous phase separation is not often expressed as discrete phases of well-defined composition. It is left to the scientist to decide which tools best discriminates between an amorphous single phase system and an amorphous system which exhibits compositional heterogeneity.

10.7 Tools and Techniques Used to Measure Solid-State Stability

In additional to the polymer and other excipients which may reside in the amorphous solid dispersion, functional excipients such as disintegrants, lubricants, compaction aids, flow aids, among others may be added to the oral dosage form—each dilutes the analytical response associated with the API. Further, each of these components has unique morphology, particle size distribution, and often convoluted analytical signatures. Therefore, detecting a change in the solid state of the drug in multicomponent systems is complex. Most often, several techniques are used simultaneously and each may provide a unique perspective on the mechanism of failure.

The ability of a particular technique to discriminate between the desired solid-state form and a failed system is a function of how well the analytical signature of each can be discriminated. In general, crystallization in an amorphous matrix can often be detected quite well given the highly ordered nature of crystalline solids. In contrast, amorphous–amorphous phase separation is much more difficult to detect. This is largely the result of the greater distribution of local environments which are present in an amorphous material relative to the crystalline counterpart. As a consequence, the response of the material to most external perturbations will be more diffuse in nature. For instance, vibrational frequencies of chemical bonds will show broader response to infrared radiation and nuclear magnetic moments of an atom will display broader peaks as a result of the faster free inductive decay. Further complicating our ability to discriminate between two amorphous phases is the fact that amorphous–amorphous phase separation is often not well defined. Specifically, amorphous–amorphous phase separation most often will not present as two discrete phases of specific composition dictated by the thermodynamics of mixing. Rather, it more often may occur as a distribution of compositions. The latter may be the result of kinetically arrested phase separation, incomplete mixing during processing, or spatial-temporal environmental differences within particulates during storage and handling. This can result in phase separation expressed as a continuum of compositions.

Characterization of amorphous solid dispersions provides an understanding of the link between the process, the formulation, and measures of performance. First, consider the solid dispersion intermediate. The amorphous solid dispersion must be characterized to demonstrate the lack of crystalline material. Moreover, realization of the utility of each functional excipient placed in the solid dispersion most often requires demonstration of molecular level mixing. In the early stages of development, selection of the solid dispersion formulation components, measures of crystallization tendency and homogeneity are conducted as a function of process and formulation. Next, the solid dispersion is incorporated into a solid oral dosage form and solid-state characterization of the powder properties, amenable to downstream processing, is conducted. Formulation selection necessarily includes excipients which allow for adequate production of tablets or capsules and consistent release of the API from the dosage form. The crystallization tendency of the API must also be measured in the final dosage form since solid-state interactions may alter the crystallization kinetics as compared to those measured in the amorphous solid dispersion. All the while, specific measurements may be conducted which link fundamental properties of the solid dispersion to the crystallization rates, for instance, molecular mobility and thermodynamic properties. Finally, changes in the material which are undetectable may be indirectly measured via stressing the system so as to force failure. That is, if a material is prepared in a consistent fashion, it may often fail in a consistent fashion. In this section of the chapter, techniques which may be applied to understand amorphous materials will be outlined with a focus on the ability of each tool to detect failure modes and phenomena peculiar to amorphous solid dispersions.

10.7.1 Differential Scanning Calorimetry

Differential scanning calorimetry (DSC) is perhaps the most widely used technique applied to amorphous solid dispersions. In this technique, the energy input associated with heating the material can be measured along with any thermal transition such as a melting point or recrystallization. This technique has been used to measures many properties of amorphous materials including the presence of crystals, glass forming tendency, molecular mobility, crystallization tendency, and crystallization rate. The technique also allows for access to the thermodynamic properties of mixtures.

Most notably, DSC is used to demonstrate miscibility in multicomponent amorphous solid dispersions. The glass transition temperature of each material is compared to the glass transition temperature of the mixture. If the T_gs of each component are unique and if the T_g shows a strong compositional dependence near the target composition of the formulation, DSC may be a useful technique to demonstrate bulk homogeneity or lack thereof. For instance, mixtures of indomethacin, citric acid, and PVP were shown to contain a miscibility gap using DSC—ternary systems may be of particular relevance when considering addition of surfactants, pH modifying agents, or other additives into solid dispersion systems. Although indomethacin was shown to be completely miscible with PVP and citric acid, a miscibility gap was found to exist between citric acid and indomethacin. Interestingly, this miscibility gap could be "closed" with the appropriate addition of PVP (Lu and Zografi 1998). Also of great importance is the glass transition temperature of the solid dispersion system as a function of water content. Water acts to reduce the glass transition temperature of solid dispersion systems, increase the thermodynamic driving force for crystallization, and may facilitate crystallization and/or amorphous phase separation. Therefore, as a matter of standard practice, solid dispersions are placed at variable relative humidity, the amount of water at each relative humidity is measured by dynamic vapor sorption, and the depressed glass transition temperature is measured using hermetically sealed DSC pans. The glass transition temperature at any water activity can then be compared with the expected water activity at any storage condition.

Although DSC remains the workhorse for demonstrating the phase behavior of amorphous materials, several challenges may occur when utilizing DSC in practice. In some instances, the change in heat capacity across the glass transition temperature is not often easily accessed. This could be the result of an inherently small change in heat capacity when passing through the glass transition temperature. Alternatively, if a multicomponent system possesses a large distribution in local environments, the glass transition temperature may be inherently broad. Further, if the material contains a large distribution in molecular weights or structural variants, as is the case of some polymers and surfactants, the glass transition temperature may occur over a broad range. Moreover, it has been reported that the length-scales of phase separation required for detectability may be on the order of 30 nm (Newman et al. 2008). Complicating the situation further the sample may be impacted by simply the act of conducting the measurement, as with any non-isothermal technique. For instance, Marsac et al. showed remixing of felodipine with poly(vinyl pyrrolidone) (PVP) during a DSC experiment when cycling from a temperature below T_g to progressively higher temperatures above T_g (Marsac et al. 2010). Perhaps most importantly, solid-state phase amorphous separation may not occur as discrete phases but rather as compositional gradients across the material making this mode of failure very difficult to detect with great certainty. In short, all of these possibilities exist and it may be difficult to completely deconvolute observed phenomena which result from compositional heterogeneity, inherent properties of the individual components, and results which manifest strictly as a consequence of the measurement itself. Therefore, the scientist is often required to complement DSC measurements with orthogonal measurements.

DSC analysis is also a useful tool to identify thermodynamic properties of the API. As was noted above, the heat capacity of the crystalline and amorphous materials can be coupled with the enthalpy of fusion to determine the difference in free energy between the crystalline and amorphous materials and therefore the solubility ratio of the two (with additional corrections as noted prior) (Hancock and Parks 2000; Murdande et al. 2010a, b, 2011a, b). In binary systems, DSC analysis can also be used to measure the strength of interactions between a drug and a polymer near the melting point of the drug. Specifically, if melting point depression is observed, by definition the chemical potential of the drug must be reduced in the molten phase and therefore the drug and the polymer must be miscible in the molten phase (Marsac et al. 2006, 2009). Careful interpretation of results is required since the ability to access the thermodynamic endpoint is a function of many properties which dictate the kinetics of the measurement. Therefore, it is useful to approach the liquidous line from several directions in order to improve the accuracy and confidence (Mahieu et al. 2013; Tian et al. 2013; Sun et al. 2010; Tao et al. 2009). Nevertheless, with respect to miscibility, these approaches provide experimental measures for understanding not only whether mixing occurred (as indicated by a single T_g) but also provide an insight into the strength of interactions between the components in the amorphous solid dispersion. The importance of this approach in understanding hot-melt extrusion processing space will be discussed further in the chapter focused on hot-melt extrusion.

Molecular mobility measurements have also been conducted using DSC and provide access to motions occurring on the order of >100 s. Specifically, the amorphous material is held at a temperature below the T_g for a period of time during which the material will undergo structural relaxation. Subsequent heating through the T_g leads to an increase in enthalpy back to the metastable state (i.e., the supercooled liquid) and manifests as an endotherm in the DSC thermogram coincident with the glass transition temperature (Hancock and Shamblin 2001; Shamblin et al. 2000; Shamblin and Zografi 1998). The resulting data can be fit to the Kohlrausch–Williams–Watts Equation, for instance, to determine an average relaxation time and the distribution of timescales of relaxation (Mao et al. 2006). Correlating relaxation times as measured by DSC with crystallization tendency has proven to be very difficult as reflected by numerous studies in the literature (Gupta et al. 2004; Kakumanu and Bansal 2002; Matsumoto and Zografi 1999; Shamblin and Zografi 1998) and several drawbacks to this approach are noted (Johari et al. 2005; Kawakami and Pikal 2005; Luthra et al. 2008). Additional studies using DSC will be discussed in the HME chapter as applied to crystallization tendency from the melt.

10.7.2 X-Ray Powder Diffraction

Alongside DSC, X-ray diffraction remains a workhorse for detecting crystalline API in an amorphous matrix. This is the result of X-ray diffraction offering excellent discrimination between amorphous and crystalline forms. As defined by Bragg's Law, when the difference between the incident and diffracted waves is an integral multiple of the wavelength of X-rays, constructive interference occurs and gives rise to a Bragg diffraction pattern. As described by Eq. (10.27), n is an integral, λ is the wavelength of the incident radiation, d is the interplanar distance between crystal planes and Θ is the angle between the incident radiation and the crystal plane.

$$n\lambda = 2d \sin \Theta \tag{10.27}$$

Alternatively, amorphous materials will not give constructive interference and instead will yield what is often termed an "amorphous halo." Examples showing the utility of X-ray diffraction for characterization of amorphous materials will not be reviewed here but the reader is referred to several references for additional information (Nunes et al. 2005; Surana and Suryanarayanan 2000; Yamada and Suryanarayanan 2006, 2007).

More recently, X-ray diffraction methods have been applied to amorphous materials with the intention of understanding the degree of mixing on a length-scale which is perhaps difficult to measure using traditional DSC methods. Specifically, the "amorphous halo" (i.e., the sum total of the Bragg + diffuse scattering intensities) may be used to produce a pair distribution function (PDF) via Fourier transformation. The resulting PDF then describes the probability of finding two atoms separated by a certain interatomic distance. Miscible systems (i.e., single phase systems containing multiple components) will contain nearest neighbor distances unique from the corresponding single component systems. Consider, for instance, a binary API-polymer system and the "pure curve resolution method" as outlined by Newman et al. (2008). When the PDF of the binary is unique from the weighted average of the individual PDFs, the material may be interpreted as a single phase miscible system. Alternatively, when the PDF of the binary can be represented by a weighted average of the PDFs of the individual components, the material can be interpreted as having two amorphous phases. Using this approach, Newman et al. studied three systems: dextran-poly(vinylpyrrolidone), trehalose-dextran and indomethacin-poly(vinylpyrrolidone). The dextran-poly (vinylpyrrolidone) system showed two glass transition temperatures as measured by differential scanning calorimetry and the PDF of the binary was very similar to a weighted average of the two components. The indomethacin-PVP binary displayed a single T_g and the PDF could not be described by a weighted average of the individual components. Most strikingly, the trehalose-dextran system displayed a single T_g as measured by DSC but the PDF of the binary could not be described by a weighted average of the two component PDF functions suggesting phase separation on a length-scale not captured by DSC. In summary, XRPD coupled with PDF analysis potentially provides useful information when taken in consideration with orthogonal techniques.

10.7.3 Vibrational Spectroscopy

Vibrational spectroscopy has proven very useful in understanding the molecular level nature of amorphous solid dispersions and may serve as a practical tool for characterizing phase behavior. Included in this category are Raman spectroscopy, Mid-IR, and Near-IR approaches. These approaches can be used to show both crystallization (Rumondor and Taylor 2010a; Shah et al. 2006) and amorphous–amorphous phase separation within an amorphous matrix (Marsac et al. 2010; Padilla et al. 2011; Rumondor et al. 2011). Vibrational spectroscopy is also often used to describe the influence of intermolecular interactions on stability of amorphous materials (Khougaz and Clas 2000; Tang et al. 2002a, b; Tantishaiyakul et al. 1999; Taylor and Zografi 1997). Further, spectroscopic tools can often be used on-line to measure properties of dispersions during manufacturing (Saerens et al. 2011, 2012).

Infrared spectroscopy, in some instances, can be a useful tool to discriminate between a single phase and a two phase amorphous binary system. That is, if interactions between the species in the solid dispersion exist, differences between the vibrational spectra of physical mixtures and mixed systems may be observed using vibrational spectroscopy (Rumondor et al. 2009, 2011; Rumondor and Taylor 2010b). For instance, felodipine and PVP were prepared as amorphous solid dispersions demonstrating a single glass transition temperature (Marsac et al. 2010). The IR spectrum clearly distinguished between drug–drug hydrogen bond interactions and drug–polymer hydrogen bond interactions. Upon exposure to high relative humidity, drug–polymer interactions were interrupted in favor of drug–drug interactions. However, crystallization was not observed and therefore, these results provide molecular level evidence of the bulk observation of two glass transition temperatures. Of course, the technique becomes more difficult to apply to systems with complex or weak hydrogen bonding patterns and so may not be applied generally.

Raman spectroscopy is perhaps the most useful in characterizing amorphous products in multicomponent systems. This is because Raman provides an excellent ability to discriminate active ingredients from common excipients. Further, although the impact of sorbed water may influence Raman spectra, water is not directly observed (as is the case in IR spectroscopy) and therefore qualitative measurements can be made at ambient conditions without the need for rigorous removal of water. In one example, it was shown that two solid dispersions prepared by HME at identical compositions gave a single glass transition temperature (Qian et al. 2010). However, the physical stability was found to be a function of the processing condition. This was explained by differences in compositional homogeneity (or lack thereof) as measured by confocal Raman spectroscopy. In another example, Raman spectroscopy was used for the detection of crystalline ibipinabant (Sinclair et al. 2011). Solid dispersions were prepared with PVP and sodium lauryl sulfate (SLS) via spray drying and formulated into a tablet matrix at a level of 1 wt % ibipinabant. Coupled with multivariate analysis, a limit of detection of 5 wt% crystalline was achieved corresponding to 0.05 wt% crystalline ibipinabant in the finished tablet (Sinclair et al. 2011). In another study, Raman mapping was used to understand amorphous–amorphous phase separation in freeze dried samples of dextran and PVP (Padilla et al. 2011). Amorphous phase separation was measured by Raman and the results were found to be comparable to DSC and XPRD (see PDF discussion above) results with some differences observed at compositions near the boundary of the phase separation. In this publication, the authors note the exceedingly long run times associated with Raman mapping approaches. Recently, nonlinear spectroscopic approaches have shown promise to increase efficiency of spectral collection and may demonstrate great utility in the future (Strachan et al. 2011).

10.7.4 Broadband Coherent Anti-Stokes Raman Scattering

Broadband Coherent Anti-Stokes Raman Scattering (BCARS) can be categorized as a nonlinear approach and proves to exhibit the discrimination achieved with traditional Raman scattering while offering superior spatial resolution and speed. For instance, Hartshorn et al. demonstrated that BCARS could be used to discriminate between two polymorphs of indomethacin, the so-called α and γ, within a tablet matrix (Hartshorn et al. 2013). Further, the ability to quickly and specifically map (i.e., collecting Raman spectrum over all relevant frequencies) facilitated detection of an unexpected phase undetected with bulk measurements. The changes in the spectra of the unexpected phase were shown to be consistent with a small amount of amorphous indomethacin. In general, "deliberate subsampling" (i.e., mapping) allows for detection of minor components within a multicomponent system. However, obtaining representative sampling requires the speed which may only be offered by multi-photon spectroscopy.

10.7.5 Second Harmonic Generation and Two Photon Fluorescence

Other types of nonlinear spectroscopy which are showing great promise in the pharmaceutical arena include second-harmonic generation and two photon fluorescence. These approaches have proven particularly useful in detecting crystalline materials in an amorphous matrix. Second harmonic generation (SHG) was applied to evaluate the nucleation and crystal growth within thin films of griseofulvin and chloropropamide and a limit of detection of 1 part per ten billion by volume and an estimated crystalline domain size on the order of ~90 nm (Wanapun et al. 2010) was demonstrated. Note that this length scale offers access to detection of crystalline material via SHG prior to detection via optical microscopy. Further, SHG offers the ability to detect crystallinity in a turbid matrix as was demonstrated in powdered amorphous griseofulvin—of course this is very difficult if not impossible with optical microscopy (Wanapun et al. 2011). Moving to a binary system, naproxen was prepared as an amorphous dispersion with HPMCAS and SHG was shown to provide superior limits of detection as compared with Raman and XRPD while results agreed well at compositions where all three methods were able to detect crystalline Naproxen (Kestur et al. 2012). Having shown superior detection limits in thin films and solid dispersion drug polymer mixtures, the approach was applied to common excipients—lactose, mannitol, and Avicel among others—so as to understand the utility of SHG in multicomponent systems (Toth et al. 2012). Although some excipients showed SHG signal using an infrared incident radiation source, Toth et al. showed that ultraviolet radiation exhibited better SHG discrimination between excipients and crystalline griseofulvin. Further, the authors coupled UV-SHG with UV-two photon excited fluorescence and demonstrated four orders of magnitude stronger signal for crystalline griseofulvin as compared to the most active excipient, lactose.

10.7.6 Solid-State Nuclear Magnetic Resonance Spectroscopy

Solid-state nuclear magnetic resonance spectroscopy (ssNMR) involves the excitation of specific nuclei using a high powered radiofrequency pulse sequence in the presence of a static magnetic field (Berendt et al. 2006; Tishmack et al. 2003). Excited nuclei relax to the static magnetic field producing a free induction decay (FID). A Fourier transformation of the FID yields the observed spectra and the resulting signal is directly related to the number of like nuclei present in the sample. That is, NMR is an inherently quantitative approach making it particularly appealing when multiple forms of an API co-exist. Moreover, the approach is particularly useful for molecules containing NMR sensitive nuclei which reside only in the API. For instance, ^{19}F is often found in APIs but not excipients offering the opportunity for excellent discriminating power in multicomponent systems. Wenslow used ^{19}F ssNMR to successfully discriminate between the amorphous and crystalline forms of a muscarinic M3 receptor antagonist after formulation into a tablet at 2 wt% (Wenslow 2002), highlighting both the sensitivity and the ability to easily discriminate in dilute multicomponent mixtures. With respect to mobility measurements, ssNMR offers access to molecular motions ranging from 10^{-11} to 10^{3} s via relaxation measurements and 10^{-5} to 10^{-10} cm^2/s via diffusion measurements (Ediger et al. 1996). The crystallization tendency of nifedipine and phenobarbital has been related to the molecular mobility of each as measured by ^{1}H NMR relaxation times (Aso 2000; Aso 2001). Specifically, the temperature dependence of molecular mobility and the temperature dependence of measures of physical instability were correlated suggesting a linkage between the two.

In addition to the conventional use of 1D ssNMR experiments to detect a specific API form in the presence of other API forms, 2D ssNMR correlation techniques and 1H T1 relaxation methods have emerged as powerful tools for the structural analysis of amorphous solid dispersions (Pham et al. 2010). Heteronuclear-detected 1H T1 measurements were shown to detect phase separation in amorphous solid dispersions. In the same manner, CP-HETCOR methods were successful utilized to demonstrate glass formation in amorphous dispersions. 1H-13C CPMAS HETCOR experiments rely on spin diffusion between the drug and the polymer. In most cases the 1H chemical shift resolution is sufficient since the polymers used do not contain aromatic protons. If the drug molecule contains fluorine, the 1H-19F CP-HETCOR experiment can be employed which is more sensitive in detecting spin diffusion than the 1H-13C version. Another useful approach is the evaluation of signal intensity variations caused by spin diffusion to proton-rich polymers which can prove the presence of a homogeneous dispersion in the absence of adequate resolution in the proton dimension. The relaxation and HETCOR experiments can be supported with a range of other ssNMR methods, including the 1H DQ-BABA, 19F RFDR, and 2H experiments. The 1H DQ-BABA and 19F RFDR experiments detect H-H and F-F intermolecular proximities, respectively. All these ssNMR methods to confirm the formation of amorphous dispersions have the advantage that they do not require the preparation of reference blends and physical mixtures.

10.7.7 *Dynamic Mechanical Analysis*

Dynamic Mechanical Analysis (DMA) has found great utility in the polymer industry but fewer publications have shown the use of this approach for understanding amorphous solid dispersion systems. However, DMA should be easily translated into the pharmaceutical arena and offer another option to evaluate the phase behavior of multicomponent amorphous systems. Most often the material of interest is exposed to a sinusoidal mechanical stress, σ, and the resulting strain, ε, is measured with t representing time, ω the frequency, and δ is the phase lag between the stress and the strain.

$$\sigma = \sigma_0 \sin (t\omega + \delta) \tag{10.28}$$

$$\varepsilon = \varepsilon_0 \sin (t\omega) \tag{10.29}$$

The storage modulus represents the elastic component of the response and is simply $(\sigma_0/\epsilon_0)\cos\delta$ and the viscous component of the response, or the loss modulus, is simply $(\sigma_0/\epsilon_0)\sin\delta$. The ratio of the storage modulus to the loss modulus, or $\tan\delta$, is a measure of the viscoelasticity of the material. As an amorphous material approaches the glass transition, the storage modulus decreases while the loss modulus increases giving rise to a peak in the $\tan\delta$ curve. This peak provides a natural point to assign as the glass transition temperature. According to the time–temperature superposition principle, relaxation times having shorter timescales can be accessed either at higher temperatures or high frequencies. Alternatively, relaxation times having longer timescales may be accessed at lower temperatures or lower frequencies. Having the ability to probe the sample over both the temperature scale and the frequency scale is part of method development and the signal should be optimized accordingly. However, as with any non-isothermal approach, the impact of the measurement on the sample should be considered during method development.

Carpenter et al. demonstrated that DMA was able to show well defined glass transition temperatures in lyophilized protein and protein excipient mixtures and results compared well with high ramp rate DSC results and dilatometry measurements (Carpenter et al. 2009). Further, the authors demonstrated that DMA could readily detect phase separation in a lysozyme and trehalose mixture (Carpenter et al. 2009). This was concluded from the observation of two discrete drops in the elastic modulus and two clear peaks in $\tan\delta$ over the temperature range studied. DMA was also shown to detect two phases within a mixture of select nonreactive polymer and photo-initiated free radical methacrylate systems. Specifically, two peaks in the $\tan\delta$ were clearly detected in select binary systems as compared to a single peak in the photo-initiated free radical methacrylate system in the absence of the secondary polymer (Szczepanski et al. 2012). Park et al. showed that DMA was useful to understand the phase behavior in hydrogels made of poly(vinyl) alcohol (PVA) and methylcellulose (MC) systems prepared using glutaraldehyde in the presence of HCl (Park et al. 2001). In another study, Karabanova compared the phase separation of polyurethane and poly(2-hydroxyethyl methacrylate) semi-interpenetrating polymer networks prepared by both photopolymerization and thermopolymerizaton (Karabanova et al. 2008). It has been previously shown that the two systems are immiscible (Karabanova et al. 2004). However, DMA results clearly demonstrated subtle differences in the tanδ between the samples prepared by the two distinct methods and this, in part, helped to explain exceptional difference in performance of the two materials (Karabanova et al. 2008). Additional examples showing the utility of DMA in understanding the phase behavior of complex (i.e., multicomponent) systems can be consulted for further explanation (Jyotishkumar et al. 2009; Tsuji and Ikada 1996). Moreover, although not considered a primary tool for measuring noncooperative motions in the pharmaceutical arena, secondary relaxation processes are often measured in the polymer sciences using DMA. Such studies are generally focused on correlation with measures of mechanical integrity and can show the complex relationship between secondary relaxations and composition in multicomponent systems. For instance, Poirier-Brulez et al., showed the impact of water and sucrose on secondary relaxations in starch based glasses (Poirier-Brulez et al. 2006). It is only a matter of time before this technique is leveraged more effectively in developing amorphous solid dispersion systems.

10.7.8 Dielectric Analysis and Thermally Stimulated Current

Dielectric Analysis (DEA) and Thermally Stimulated Current (TSC) operate under the principle that dipoles with sufficient mobility will respond to an external electric field. Most generally, the governing equations which describe dielectric analysis (DEA) are similar to those described above for DMA only the elastic and viscous component of the response can be replaced by the dielectric permittivity and the dielectric loss associated with the respective oscillating perturbations (i.e., mechanical or electrical). Alternatively, DEA can also be conducted in the time domain in which case the response to a potential is measured as a function of time as opposed to frequency (Feldman et al. 1992; Bhugra et al. 2008a). Although DEA can be used to detect crystallization from an amorphous matrix (Alie et al. 2004; Bhardwaj et al. 2013; Bhardwaj and Suryanarayanan 2011, 2012a, b; Bhattacharya and Suryanarayanan 2009; Bhugra et al. 2007, 2008b; Dantuluri et al. 2011) and amorphous–amorphous phase separation (Power et al. 2007), it is a particularly appealing tool for measuring molecular motions in amorphous materials since it provides access to both cooperative and noncooperative (local or secondary relaxations) motions. That is, motions of a vast number of timescales may be accessed with DEA (10^{-11} to 10^4 s) (Ediger et al. 1996). Various models can be applied to dielectric data which capture the functional dependence of the dielectric response on frequency, time, or temperature (Jonscher 1983). Fitting the data to these models and/or application of appropriate curve resolution to de-convolute various overlapping motions provides an insight into the temperature and frequency dependence of each mode of motion. The timescales of physical instability can then be measured with the aim of finding a link between specific modes of molecular motion and crystallization tendency (Alie et al. 2004; Bhardwaj et al. 2013; Bhardwaj and Suryanarayanan 2011, 2012a, b; Bhattacharya and Suryanarayanan 2009; Bhugra et al. 2007, 2008b; Dantuluri et al. 2011).

TSC provides access to timescales between 25 and 3,000 s and can detect exceptionally low currents (Correia et al. 2001). In one manifestation of the TSC experiment, the material of interest may be held at a high temperature, exposed to an electric field, quenched (the temperature is reduced while maintaining the static electric field), and then reheated in the absence of the electric field. Upon reheating, different modes of mobility are activated based on the temperature and the resulting sample current is detected by the instrument to give information regarding these different mobility modes. This type of experimental design is termed thermally stimulated depolarization current (TSDC). Alternatively, the material may simply be heated in the presence of an electrical potential, termed thermally stimulated polarization current (TSPC) as the current being measured is the result of the sample polarization. In another experiment, called thermal windowing, a similar approach to the TSDC experiment is taken. However, the current is only applied over a small temperature range and incremental changes in the temperature range are made in subsequent experiments. This approach enables access to distributions of relaxations which occur only over a small temperature range and the resulting data can be used to understand the activation energy associated with specific relaxation processes (Shmeis et al. 2004a, b). TSC has proven to be useful in detecting small amounts of amorphous material within a largely crystalline matrix (Galop and Collins 2001) and has also been used as a tool to detect amorphous–amorphous phase separation (Shmeis et al. 2004a, b). For instance, Schmeis et al. prepared solid dispersions of a model compound and PVP. At some concentrations, although DSC showed a single glass transition temperature, TSC displayed multiple events which the authors attributed to a limit in the miscibility between the drug and the polymer. Further, the authors note that TSC was able to distinguish between modes of motion which demonstrate only small changes in the heat capacity and therefore are difficult to detect by DSC. Interestingly, the authors were able to demonstrate two alpha type transitions in the pure amorphous API as well as the alpha transition and a sub-T_g beta-relaxation in the polymer. Continued research in dielectrics is motivated by the high sensitivity of these techniques and potential to predict shelf life of amorphous pharmaceutical products by linking measures of mobility to crystallization kinetics.

10.7.9 Atomic Force Microscopy

Atomic force microscopy (AFM) can be used in various configurations and coupled with multiple techniques for the desired application. Most routine approaches include operation in either contact mode or tapping mode. As the name would imply, contact mode consists of holding the tip of the probe in constant contact with the sample. The force is maintained constant and the associated cantilever deflection is recorded as the probe is scanned across the sample surface. This is accomplished using a photodiode array that captures the position of a laser reflected off of the top surface of the cantilever. A piezoelectric scanner is used to both raster the tip across the surface as well as move the cantilever in the *z*-direction in response to changes in surface topography. Tapping mode significantly reduces issues with damage of the probe and the sample and is therefore used more often than contact mode for soft organic pharmaceutical materials. In tapping mode, the amplitude and frequency of cantilever oscillation are set and as the tip travels about the surface, changes in the surface topography lead to differences between the set-point and measured amplitude. In addition, changes in the phase angle of the oscillation can be observed and mapped in 2D as a result of changes in the mechanical properties of the material, the interaction between the material and the probe, and/or changes in friction between the material and the probe among others differences. Further, coupling with thermal probes and spectroscopic probes remains an area of great interest for characterizing pharmaceutical systems. The ability to overlay high resolution AFM images with calorimetric or chemical information holds the promise of providing an additional insight into failure modes in amorphous solid dispersions.

AFM can be used to understand both crystallization and phase separation of amorphous materials (Marsac et al. 2010; Lauer et al. 2013; Meeus et al. 2013; Price and Young 2004; Qi et al. 2011; Yang et al. 2013). In one study, solid dispersions prepared by spin casting felodipine with polyvinylpyrrolidone showed featureless topographical and phase images. After exposure to high relative humidity, films became visually cloudy and the corresponding AFM images showed a clear increase in surface roughness and phase shift. Further exposure to high relative humidity led to the development of well-defined morphology consistent with crystallization and was verified by birefringence under cross-polars using optical microscopy (Marsac et al. 2010). Application of AFM has also been demonstrated on materials manufactured via spray drying and hot melt extrusion. For instance, crystallization on the surface of single particles of spray dried amorphous lactose was detected with AFM. Interestingly, the particles became "smooth" when first exposed to high relative humidity, presumably the result of the plasticizing effect of moisture, followed by a roughening of the surface as a result of crystallization of lactose (Mahlin et al. 2004). Scanning electron microscopy images confirmed the coarsening or roughening of the particles surface post crystallization. In a follow up study, amorphous dispersions of lactose were prepared with PVP and analyzed with AFM (Mahlin et al. 2006). The authors noted that the spray dried solid dispersions were enriched in PVP on the surface as measured by X-ray photoelectron spectroscopy (see below for more discussion on this technique). A higher relative humidity was required before crystallization could be detected. Clearly, coupling multiple techniques proved useful in this study. It is also interesting to note that amorphous lactose particles prepared with PVP showed very different morphology as compared to those prepared in the absence of PVP. This work highlights the impact of both formulation and process on performance. In another study, AFM was used not only as a screening tool, but to understand the impact of the HME process on the resulting solid dispersion (Lauer et al. 2013). Specifically, it was shown that solid dispersions prepared by solvent casting and subsequently heated to drive off residual solvent and allow for high temperature thermal homogenization showed greater homogeneity as compared to a sample extruded at 140 °C. However, when the same sample was extruded at 160 °C, AFM showed similar results as the solvent cast material. This result highlights again the process-formulation-performance interplay. In the same study, the authors note the potential utility of AFM to distinguish between mechanisms of phase separation: spinodal decomposition and phase separation via nucleation and growth. Finally, AFM has been shown to demonstrate utility in multicomponent systems. Yang et al. note that immiscible polymer blend formulations may offer superior advantages over the corresponding miscible binaries (Yang et al. 2013). Characterization of ternaries made of Eudragit E PO and PVP-VA amorphous solid dispersions with felodpine was made possible using a combination of DSC, attenuated total reflectance FT-IR and AFM (Yang et al. 2013). Indeed, complementary characterization tools will continue to become more important as the community develops more sophisticated formulations.

10.7.10 X-Ray Photoelectron Spectroscopy

Surface analysis may be important for characterization of particulates that are produced in a way which allows for compositional heterogeneity as a function of position within the particle. For instance, the spray drying process and precipitation processes may be susceptible to spatial heterogeneity as will be discussed below and in Chap. 12. Perhaps also of concern but not addressed in detail in this text is production via amorphous precipitation (Shah et al. 2012; Sertsou et al. 2002; Kislalioglu et al. 1991). In the amorphous precipitation approach, mechanisms which may facilitate compositional heterogeneity throughout the particle include differences in the solubility profile as a function of solvent–antisolvent composition between the components of the system, degree and extent of mixing during the very dynamic particle formation process, among others. It is reasonable to consider the impact of surface properties on the performance of the resulting particles. Typically solid dispersions are prepared with hydrophobic API. High concentrations of API on the surface may compromise wetting during the dissolution process and increase physical stability risk. Alternatively, if a hydrophilic polymer or a surfactant preferentially resides on the surface, the impact may be superior wettability. Given the importance in linking surface properties to performance, methods which provide an insight into the surface properties of particulates are important to explore. These approaches have found use in other arenas (Faldt et al. 1993; Jones et al. 2013; Kim et al. 2003, 2009) and more recently, the pharmaceutical industry is demonstrating the importance of surface analysis (Dahlberg et al. 2008; Meeus et al. 2013).

X-ray photoelectron spectroscopy (XPS) can be used to quantitatively detect almost all atomic species with the exception of hydrogen and helium and generally this approach probes several to tens of atomic layers on the surface of the sample—depending upon the specifics of the equipment, radiation source, and the nature of the solid. XPS can be used as a microscopic approach with *x–y* resolution on the order of about 20 μm (for typical pharmaceutical systems). Alternatively, XPS can be used to analyze large surface areas (i.e., multiple particles simultaneously). If the latter approach is chosen, it is important to couple XPS results with an understanding of the morphology of the particles. For instance, as will be discussed in more detail in the next chapter, spray dry particles may take on the morphology of hollow spheres, shriveled raisins, or shattered egg shells, among others. If the composition on the interior and exterior surface of a particle is different, an average of the two may be observed when multiple particles are analyzed simultaneously. Therefore, the morphology and the area of illumination must be considered together.

In XPS, an incident photon ionizes an atom and produces an ejected electron. The principle of XPS can be described by Eq. (10.30) where K.E. is the kinetic energy of the ejected electron, h is Plank's constant, v is the frequency of the incident electron, and B.E. is the binding energy of the ejected electron.

$$\text{K.E.} = hv - \text{B.E.} \tag{10.30}$$

Since each atom has unique energy levels associated with each electronic orbital, the binding energy is unique to each type of atom. The intensity of various peaks will depend on the number of atoms present, the type of orbital (different orbitals may have different degrees of interaction), and the impact of various other processes which are beyond the scope of discussion here. Examples whereby this approach has been used for pharmaceutical systems include the recent work by Meeus et al. whereby XPS showed that the surface composition of PVP and PLGA in microspheres prepared by spray-drying was a function of the humidity and thermal history of the particles (Dahlberg et al. 2008; Meeus et al. 2013). In another example, Dahlberg et al. compared solid dispersions made with two model API systems and three polymers. It was found that solid dispersions prepared of the more hydrophobic API generally displayed higher surface composition (Dahlberg et al. 2008). Further, the authors also note that the rate of solvent removal and the solvent system used to prepare the solid dispersion may impact the degree to which the heterogeneity is observed. As described below, the particle formation process is a function of multiple physical phenomena such as solubility, diffusivity, surface activity, and phase behavior of the multicomponent system, among others. The rate limiting process is difficult to determine given the fast dynamics of particle formation. However, the results presented above demonstrate the complex interplay between the formulation, the process, and the resulting particles produced. This, in turn, may impact the performance.

As demonstrated above, XPS measurements can be used to show the surface composition in multicomponent pharmaceutical systems. However, it remains to be seen whether surface heterogeneities translate into heterogeneities throughout the particle which may impact bioavailability. For this, additional characterization tools should be considered which allow for complete structural and chemical characterization of single particles. In the next sections, additional techniques will be discussed which may meet this need.

10.7.11 Solution Calorimetry

Solution calorimetry has been used to measure the heat of mixing polymer systems (Brunacci et al. 1994; Righetti et al. 2002; Weeks et al. 1977). Recently, this approach has been extended to amorphous solid dispersion systems and offers the ability to measure the phase behavior of drug–polymer systems (Marsac and Taylor 2008). In this approach, the heat of solution of a physical mixture is compared with the heat of solution of a solid dispersion. A solvent is chosen which readily dissolves all of the species which make up the solid dispersion. This solution serves as a thermodynamic reference state from which the heat of mixing the species in the solid dispersion can be calculated through the application of Hesse's Law as outlined in the equation below. The larger the difference between the heat of solution of the solid dispersion and the heat of solution of a physical mixture at the same composition, the greater the thermodynamic driving force for mixing. Therefore, there is a greater the reduction in the chemical potential of the drug in the presence of the polymer.

$$\Delta H_{\text{drug-polymer}} = \left(x_{\text{drug}}.\Delta H_{\text{drug}} + x_{\text{polymer}}.\Delta H_{\text{polymer}}\right) - \Delta H_{\text{dispersion}} \tag{10.31}$$

Since the heat of solution is conducted on glassy materials, an excess enthalpy over the supercooled liquid is inherent to the system. This can be accounted for by including the appropriate heat capacity terms in the equation above which capture the enthalpy difference between the glassy material and the supercooled liquid at the temperature of interest. The solution calorimetry approach is particularly attractive because it allows for a room temperature measurement of the enthalpy of mixing. Further, this approach is very appealing since it allows for measurement of the sum total of all interactions which may occur between the species in the dispersion. In contrast, spectroscopic methods, such as IR or ssNMR, only provide an insight into interactions which manifest as a change in distribution and strength of the respective analytical signatures. The approach is particularly exciting because it allows for access to the thermodynamic properties of the amorphous dispersion and, perhaps more importantly, a comparison of the enthalpy of mixing and the enthalpy of crystallization. The corresponding free energy of mixing can then be calculated from a model describing the entropy of mixing. In practice, this translates to a room temperature measurement of the chemical potential of the drug in the solid dispersion and therefore the solubility of the drug in the polymer.

10.8 Manufacturing Single Phase Multicomponent Amorphous Materials

As described above, the influence of additives on the performance of the API most often requires consistent generation of a single phase multicomponent amorphous material and application of multiple characterization tools enables development of a robust amorphous solid dispersion. Compositional fluctuations may compromise the fate of the dispersion when exposed to the gut and/or the propensity for physical failure on stability (i.e., crystallization). Consider the two most common methods of preparation of amorphous solid dispersions, namely, hot-melt extrusion and spray drying. In the case of hot-melt extrusion, the starting materials are most often crystalline active, an amorphous polymer, and perhaps other additives such as a surfactant. The combination of distributive (breaking of particles), dispersive (mixing on a particulate level), and thermal homogenization (thermodynamically driven mixing) must be such that the individual components reach a homogeneous end state. If the two materials are miscible and the temperature which the material experiences is above the liquidous line (see the Chap. 11), then given sufficient time and energy input, a homogeneous single phase multicomponent endpoint should be reached. However, if the energy input is insufficient—as defined by inadequate shear, insufficient time for inter-diffusion of the various species, etc.—then a distribution in composition may persist by the time the material exits the die. In a similar vein, the spray drying process includes the preparation of a feed solution which is subsequently atomized into a spray dry chamber. The process of evaporating the carrier solvent may lead to compositional heterogeneities (see the Chap. 12). That is, as the solvent front recedes, a compositional gradient will promote diffusion or "remixing" back into the bulk droplet. Depending on the relative rates of diffusion of the dissolved species and the rate of evaporation, a radial concentration distribution in the resulting particle may occur. Moreover, some of the dissolved species may be surface active and certainly each material will have a different solubility in the carrier solvent/co-solvent system. Further, a phase envelope may exist in either process as dictated by the thermodynamics of mixing the species which make up the solid dispersion system. Other processes whereby amorphous materials are isolated may also lead to gradual compositional heterogeneities—only the route by which these are realized may be unique. Consider the precipitation of amorphous materials via high shear anti-solvent methods

(Sertsou et al. 2002; Shah et al. 2012). The rate of precipitation of each of the species and local variations in the solvent composition may promote the development of inhomogeneity within the particulates. Regardless, the end result is the same—compositional heterogeneity may result from the kinetics of the processes by which the materials are produced relative to the kinetics of the mixing (i.e., in the case of hot melt extrusion) or the kinetics of solvent removal (i.e., in the case of spray drying or high shear precipitation). In either case, it is noted that heterogeneity may occur regardless of the thermodynamic end-state and although it is important to understand the thermodynamics of mixing materials, the manufacture of amorphous solid dispersions must be controlled so as to allow for access to the end-state. The tools and techniques outlined in this chapter provide an insight into the link between processing, formulation, and performance. In the subsequent two chapters, a detailed discussion of the two most widely used manufacturing processes for generating amorphous materials is outlined in more detail.

References

Abbou Oucherif K, Raina S, Taylor LS, Litster JD (2013) Quantitative analysis of the inhibitory effect of HPMC on felodipine crystallization kinetics using population balance modeling. CrystEngComm 15:2197–2205

Agoram B, Woltosz WS, Bolger MB (2001) Predicting the impact of physiological and biochemical processes on oral drug bioavailability. Adv Drug Deliv Rev 50:S41–S67

Aisha AFA, Ismail Z, Abu-Salah KM, Majid AMSA (2012) Solid dispersions of α-mangostin improve its aqueous solubility through self-assembly of nanomicelles. J Pharm Sci 101:815–825

Albers J, Alles R, Matthée K, Knop K, Nahrup JS, Kleinebudde P (2009) Mechanism of drug release from polymethacrylate-based extrudates and milled strands prepared by hot-melt extrusion. Eur J Pharm Biopharm 71:387–394

Alie J, Menegotto J, Cardon P, Duplaa H, Caron A, Lacabanne C, Bauer M (2004) Dielectric study of the molecular mobility and the isothermal crystallization kinetics of an amorphous pharmaceutical drug substance. J Pharm Sci 93:218–233

Alonzo D, Zhang GZ, Zhou D, Gao Y, Taylor L (2010) Understanding the behavior of amorphous pharmaceutical systems during dissolution. Pharm Res 27:608–618

Alonzo DE, Gao Y, Zhou D, Mo H, Zhang GG, Taylor LS (2011) Dissolution and precipitation behavior of amorphous solid dispersions. J Pharm Sci 100:3316–3331

Alonzo DE, Raina S, Zhou D, Gao Y, Zhang GGZ, Taylor LS (2012) Characterizing the impact of hydroxypropyl methyl cellulose on the growth and nucleation kinetics of felodipine from supersaturated solutions. Cryst Growth Des 12:1538–1547

Anwar J, Boateng PK, Tamaki R, Odedra S (2009) Mode of action and design rules for additives that modulate crystal nucleation. Angew Chem Int Ed 48:1596–1600

Aso Y, Yoshioka S et al (2000) Relationship between the crystallization rates of amorphous nifedipine, phenobarbital, and flopropione, and their molecular mobility as measured by their enthalpy relaxation and H-1 NMR relaxation times. J Pharm Sci 89(3):408–416

Aso Y, Yoshioka S et al (2001) Explanation of the crystallization rate of amorphous nifedipine and phenobarbital from their molecular mobility as measured by C-13 nuclear magnetic resonance relaxation time and the relaxation time obtained from the heating rate dependence of the glass transition temperature. J Pharm Sci 90(6):798–806

Aso Y, Yoshioka S, Kojima S (2004) Molecular mobility-based estimation of the crystallization rates of amorphous nifedipine and phenobarbital in poly(vinylpyrrolidone) solid dispersions. J Pharm Sci 93:384–391

Aungst BJ (2000) Intestinal permeation enhancers. J Pharm Sci 89:429–442

Aungst BJ (2012) Absorption enhancers: applications and advances. AAPS J 14:10–18

Avdeef A (2001) Physicochemical profiling (solubility, permeability, charge state). Curr Top Med Chem 1:277–354

Baird JA, Van Eerdenbrugh B, Taylor LS (2010) A classification system to assess the crystallization tendency of organic molecules from undercooled melts. J Pharm Sci 99:3787–3806

Benet LZ, Wu CY (2006) Using a biopharmaceutics drug disposition classification system to predict bioavailability and elimination characteristics of new molecular entities. North Jersey Drug Metabolism Discussion Group, Somerset, NJ

Berendt RT, Sperger DM, Isbester PK, Munson EJ (2006) Solid-state NMR spectroscopy in pharmaceutical research and analysis. Trends Anal Chem 25:977–984

Bhardwaj SP, Suryanarayanan R (2011) Subtraction of DC conductivity and annealing: approaches to identify Johari-Goldstein relaxation in amorphous trehalose. Mol Pharm 8:1416–1422

Bhardwaj SP, Suryanarayanan R (2012a) Molecular mobility as an effective predictor of the physical stability of amorphous trehalose. Mol Pharm 9:3209–3217

Bhardwaj SP, Suryanarayanan R (2012b) Use of dielectric spectroscopy to monitor molecular mobility in glassy and supercooled trehalose. J Phys Chem B 116:11728–11736

Bhardwaj SP, Arora KK, Kwong E, Templeton A, Clas S-D, Suryanarayanan R (2013) Correlation between molecular mobility and physical stability of amorphous itraconazole. Mol Pharm 10:694–700

Bhattacharya S, Suryanarayanan R (2009) Local mobility in amorphous pharmaceuticals—characterization and implications on stability. J Pharm Sci 98:2935–2953

Bhugra C, Pikal MJ (2008) Role of thermodynamic, molecular, and kinetic factors in crystallization from the amorphous state. J Pharm Sci 97:1329–1349

Bhugra C, Shmeis R, Krill SL, Pikal MJ (2006) Predictions of onset of crystallization from experimental relaxation times I-correlation of molecular mobility from temperatures above the glass transition to temperatures below the glass transition. Pharm Res 23:2277–2290

Bhugra C, Rambhatla S, Bakri A, Duddu SP, Miller DP, Pikal MJ, Lechuga-Ballesteros D (2007) Prediction of the onset of crystallization of amorphous sucrose below the calorimetric glass transition temperature from correlations with mobility. J Pharm Sci 96:1258–1269

Bhugra C, Shmeis R, Krill SL, Pikal MJ (2008a) Different measures of molecular mobility: comparison between calorimetric and thermally stimulated current relaxation times below T-G and correlation with dielectric relaxation times above T-G. J Pharm Sci 97:4498–4515

Bhugra C, Shmeis R, Krill SL, Pikal MJ (2008b) Prediction of onset of crystallization from experimental relaxation times. II. Comparison between predicted and experimental onset times. J Pharm Sci 97:455–472

Bollag G, Hirth P, Tsai J, Zhang J, Ibrahim PN, Cho H, Spevak W, Zhang C, Zhang Y, Habets G, Burton EA, Wong B, Tsang G, West BL, Powell B, Shellooe R, Marimuthu A, Nguyen H, Zhang KYJ, Artis DR, Schlessinger J, Su F, Higgins B, Iyer R, D'Andrea K, Koehler A, Stumm M, Lin PS, Lee RJ, Grippo J, Puzanov I, Kim KB, Ribas A, Mcarthur GA, Sosman JA, Chapman PB, Flaherty KT, Xu X, Nathanson KL, Nolop K (2010) Clinical efficacy of a RAF inhibitor needs broad target blockade in BRAF-mutant melanoma. Nature 467:596–599

Brick MC, Palmer HJ, Whitesides TH (2003) Formation of colloidal dispersions of organic materials in aqueous media by solvent shifting. Langmuir 19:6367–6380

Brouwers J, Brewster ME, Augustijns P (2009) Supersaturating drug delivery systems: the answer to solubility-limited oral bioavailability? J Pharm Sci 98:2549–2572

Brunacci A, Pedemonte E, Cowie JMG, Mcewen IJ (1994) The thermodynamics of mixing of polystyrene and poly(alpha-methylstyrene) from a calorimetric viewpoint. Polymer 35:2893–2896

Camenisch G, Alsenz J, Van De Waterbeemd H, Folkers G (1998) Estimation of permeability by passive diffusion through Caco-2 cell monolayers using the drugs' lipophilicity and molecular weight. Eur J Pharm Sci 6:313–319

Caron V, Bhugra C, Pikal MJ (2010) Prediction of onset of crystallization in amorphous pharmaceutical systems: phenobarbital, nifedipine/PVP, and phenobarbital/PVP. J Pharm Sci 99:3887–3900

Carpenter J, Katayama D, Liu L, Chonkaew W, Menard K (2009) Measurement of T-G in lyophilized protein and protein excipient mixtures by dynamic mechanical analysis. J Ther Anal Calorim 95:881–884

Cerdeira M, Candal RJ, Herrera ML (2004) Analytical techniques for nucleation studies in lipids: advantages and disadvantages. J Food Sci 69:R185–R191

Correia NT, Alvarez C, Ramos JJM, Descamps M (2001) The beta-alpha branching in D-sorbitol as studied by thermally stimulated depolarization currents (TSDC). J Phys Chem B 105:5663–5669

Corrigan OI (1985) Mechanisms of dissolution of fast release solid dispersions. Drug Dev Ind Pharm 11:697–724

Craig DQM (2002) The mechanisms of drug release from solid dispersions in water-soluble polymers. Int J Pharm 231:131–144

Dahlberg C, Millqvist-Fureby A, Schuleit M (2008) Surface composition and contact angle relationships for differently prepared solid dispersions. Eur J Pharm Biopharm 70:478–485

Dai W-G, Dong LC, Shi X, Nguyen J, Evans J, Xu Y, Creasey AA (2007) Evaluation of drug precipitation of solubility-enhancing liquid formulations using milligram quantities of a new molecular entity (NME). J Pharm Sci 96:2957–2969

Dantuluri AKR, Amin A, Puri V, Bansal AK (2011) Role of alpha-relaxation on crystallization of amorphous celecoxib above T-G probed by dielectric spectroscopy. Mol Pharm 8:814–822

Douroumis D, Fahr A (2007) Stable carbamazepine colloidal systems using the cosolvent technique. Eur J Pharm Sci 30:367–374

Ediger MD, Angell CA, Nagel SR (1996) Supercooled liquids and glasses. J Phys Chem 100:13200–13212

Engers D, Teng J, Jimenez-Novoa J, Gent P, Hossack S, Campbell C, Ivanisevic I, Templeton A, Byrn S, Newman AW (2008) Screening and animal bioavailability studies on itraconazole amorphous dispersions. AAPS Annual Meeting, Atlanta, GA

Erdemir D, Lee AY, Myerson AS (2009) Nucleation of crystals from solution: classical and two-step models. Acc Chem Res 42:621–629

Faldt P, Bergenstahl B, Carlsson G (1993) The surface coverage of fat on food powders analyzed by ESCA (electron-spectroscopy for chemical-analysis). Food Struct 12:225–234

Feldman YD, Zuev YF, Polygalov EA, Fedotov VD (1992) Time domain dielectric-spectroscopy—a new effective tool for physical-chemistry investigation. Colloid Polym Sci 270:768–780

Femi-Oyewo MN, Spring MS (1994) Studies on paracetamol crystals produced by growth in aqueous solutions. Int J Pharm 112:17–28

Fincher JH (1968) Particle size of drugs and its relationship to absorption and activity. J Pharm Sci 57:1825–1835

Fisher JC, Hollomon JH, Turnbull D (1949) Rate of nucleation of solid particles in a subcooled liquid. Science 109:168–169

Fleisher D, Li C, Zhou Y, Pao LH, Karim A (1999) Drug, meal and formulation interactions influencing drug absorption after oral administration—clinical implications. Clin Pharmacokinet 36:233–254

Frank KJ, Westedt U, Rosenblatt KM, Holig P, Rosenberg J, Magerlein M, Fricker G, Brandl M (2012) The amorphous solid dispersion of the poorly soluble ABT-102 forms nano/microparticulate structures in aqueous medium: impact on solubility. Int J Nanomedicine 7:5757–5768

Frenkel YV, Clark AD, Das K, Wang Y-H, Lewi PJ, Janssen PAJ, Arnold E (2005) Concentration and pH dependent aggregation of hydrophobic drug molecules and relevance to oral bioavailability. J Med Chem 48:1974–1983

Friesen DT, Shanker R, Crew M, Smithey DT, Curatolo WJ, Nightingale JAS (2008) Hydroxypropyl methylcellulose acetate succinate-based spray-dried dispersions: an overview. Mol Pharm 5:1003–1019

Galop M, Collins GL (2001) Thermally stimulated currents observed in pharmaceutical products. Thermochim Acta 367:37–41

Gao P, Akrami A, Alvarez F, Hu J, Li L, Ma C, Surapaneni S (2009) Characterization and optimization of AMG 517 supersaturatable self-emulsifying drug delivery system (S-SEDDS) for improved oral absorption. J Pharm Sci 98:516–528

Garside JEA et al (2002) Measurement of crystal growth and nucleation rates. Icheme, Rugby, UK

Ghodbane J, Denoyel R (1997) Competitive adsorption between non-ionic polymers and surfactants on silica. Colloids Surf A Physicochem Eng Asp 127:97–104

Goldberg M, Gomez-Orellana I (2003) Challenges for the oral delivery of macromolecules. Nat Rev Drug Discov 2:289–295

Goldberg AH, Gibaldi M, Kanig JL (1966) Increasing dissolution rates and gastrointestinal absorption of drugs via solid solutions and eutectic mixtures III: experimental evaluation of griseofulvin—succinic acid solid solution. J Pharm Sci 55:487–492

Greco S, Authelin J-R, Leveder C, Segalini A (2012) A practical method to predict physical stability of amorphous solid dispersions. Pharm Res 29:2792–2805

Gupta P, Kakumanu VK, Bansal AK (2004) Stability and solubility of celecoxib-PVP amorphous dispersions: a molecular perspective. Pharm Res 21:1762–1769

Gutzow I, Schmelzer J (eds) (1995) The vitreous state: thermodynamics, structure, rheology, and Crystallization and crystallization. Springer, Berlin, Gmbh & Co. Kg. Book

Guzmán HR, Tawa M, Zhang Z, Ratanabanangkoon P, Shaw P, Gardner CR, Chen H, Moreau J-P, Almarsson Ö, Remenar JF (2007) Combined use of crystalline salt forms and precipitation inhibitors to improve oral absorption of celecoxib from solid oral formulations. J Pharm Sci 96:2686–2702

Hancock BC, Parks M (2000) What is the true solubility advantage for amorphous pharmaceuticals? Pharm Res 17:397–404

Hancock BC, Shamblin SL (2001) Molecular mobility of amorphous pharmaceuticals determined using differential scanning calorimetry. Thermochim Acta 380:95–107

Hartshorn CM, Lee YJ, Camp CH Jr, Liu Z, Heddleston J, Canfield N, Rhodes TA, Hight Walker AR, Marsac PJ, Cicerone MT (2013) Multicomponent chemical imaging of pharmaceutical solid dosage forms with broadband cars microscopy. Anal Chem 85:8102–8111

Hasegawa A, Taguchi M, Suzuki R, Miyata T, Nakagawa H, Sugimoto I (1988) Supersaturation mechanism of drugs from solid dispersions with enteric coating agents. Chem Pharm Bull 36:4941–4950

He Y-L, Murby S, Warhurst G, Gifford L, Walker D, Ayrton J, Eastmond R, Rowland M (1998) Species differences in size discrimination in the paracellular pathway reflected by oral bioavailability of poly(ethylene glycol) and D-peptides. J Pharm Sci 87:626–633

Hikima T, Adachi Y, Hanaya M, Oguni M (1995) Determination of potentially homogeneous-nucleation-based crystallization in O-terphernyl and an interpretation of the nucleation-enhancement mechanism. Phys Rev B 52:3900–3908

Hintz RJ, Johnson KC (1989) The effect of particle size distribution on dissolution rate and oral absorption. Int J Pharm 51:9–17

Hoffman JD (1958) Thermodynamic driving force in nucleation and growth processes. J Chem Phys 29:1192–1193

Hsieh YL, Ilevbare GA, Van Eerdenbrugh B, Box KJ, Sanchez-Felix MV, Taylor LS (2012) pH-induced precipitation behavior of weakly basic compounds: determination of extent and duration of supersaturation using potentiometric titration and correlation to solid state properties. Pharm Res 29:2738–2753

Huang S-M, Abernethy DR, Wang Y, Zhao P, Zineh I (2013) The utility of modeling and simulation in drug development and regulatory review. J Pharm Sci 102:2912–2923

Ilevbare GA, Taylor LS (2013) Liquid-liquid phase separation in highly supersaturated aqueous solutions of poorly water-soluble drugs: implications for solubility enhancing formulations. Cryst Growth Des 13:1497–1509

Ilevbare GA, Liu H, Edgar KJ, Taylor LS (2012a) Inhibition of solution crystal growth of ritonavir by cellulose polymers—factors influencing polymer effectiveness. CrystEngComm 14:6503–6514

Ilevbare GA, Liu H, Edgar KJ, Taylor LS (2012b) Effect of binary additive combinations on solution crystal growth of the poorly water-soluble drug, ritonavir. Cryst Growth Des 12:6050–6060

Ilevbare GA, Liu H, Edgar KJ, Taylor LS (2012c) Maintaining supersaturation in aqueous drug solutions: impact of different polymers on induction times. Cryst Growth Des 13:740–751

Ilevbare GA, Liu H, Edgar KJ, Taylor LS (2012d) Understanding polymer properties important for crystal growth inhibition—impact of chemically diverse polymers on solution crystal growth of ritonavir. Cryst Growth Des 12:3133–3143

Ilevbare GA, Liu H, Edgar KJ, Taylor LS (2013a) Impact of polymers on crystal growth rate of structurally diverse compounds from aqueous solution. Mol Pharm 10:2381–2393

Ilevbare GA, Liu H, Pereira J, Edgar KJ, Taylor LS (2013b) Influence of additives on the properties of nanodroplets formed in highly supersaturated aqueous solutions of ritonavir. Mol Pharm 10:3392–3403

Ishida H, Wu T, Yu L (2007) Sudden rise of crystal growth rate of nifedipine near T-G without and with polyvinylpyrrolidone. J Pharm Sci 96:1131–1138

James KC (1986) Solubility and related properties. Marcell Decker, New York

Johari GP, Kim S, Shanker RM (2005) Dielectric studies of molecular motions in amorphous solid and ultraviscous acetaminophen. J Pharm Sci 94:2207–2223

Jones JR, Prime D, Leaper MC, Richardson DJ, Rielly CD, Stapley AGF (2013) Effect of processing variables and bulk composition on the surface composition of spray dried powders of a model food system. J Food Eng 118:19–30

Jonscher AK (1983) Dielectric relaxation in solids. Chelsea Dielectrics Press Limited, London

Joshi HN, Tejwani RW, Davidovich M, Sahasrabudhe VP, Jemal M, Bathala MS, Varia SA, Serajuddin ATM (2004) Bioavailability enhancement of a poorly water-soluble drug by solid dispersion in polyethylene glycol-polysorbate 80 mixture. Int J Pharm 269:251–258

Jyotishkumar P, Koetz J, Tiersch B, Strehmel V, Ozdilek C, Moldenaers P, Haessler R, Thomas S (2009) Complex phase separation in poly(acrylonitrile-butadiene-styrene)-modified epoxy/4,4′-diaminodiphenyl sulfone blends: generation of new micro- and nanosubstructures. J Phys Chem B 113:5418–5430

Kakumanu VK, Bansal AK (2002) Enthalpy relaxation studies of celecoxib amorphous mixtures. Pharm Res 19:1873–1878

Kaneniwa N, Watari N (1978) Dissolution of slightly soluble drugs. IV. Effects of particle size of sulfonamides on in vitro dissolution rate and in vivo absorption rate, and their relation to solubility. Chem Pharm Bull 26:813–826

Kaneniwa N, Watari N, Iijima H (1978) Dissolution of slightly soluble drugs. V. Effect of particle size on gastrointestinal drug absorption and its relation to solubility. Chem Pharm Bull 26:2603–2614

Karabanova LV, Boiteux G, Gain O, Seytre G, Sergeeva LM, Lutsyk ED (2004) Miscibility and thermal and dynamic mechanical behaviour of semi-interpenetrating polymer networks based on polyurethane and poly(hydroxyethyl methacrylate). Polym Int 53:2051–2058

Karabanova LV, Boiteux G, Seytre G, Stevenson I, Lloyd AW, Mikhalovsky SV, Helias M, Sergeeva LM, Lutsyk ED, Svyatyna A (2008) Phase separation in the polyurethane/poly(2-hydroxyethyl methacrylate) semi-interpenetrating polymer networks synthesized by different ways. Polym Eng Sci 48:588–597

Karavas E, Georgarakis E, Bikiaris D (2006) Felodipine nanodispersions as active core for predictable pulsatile chronotherapeutics using PVP/HPMC blends as coating layer. Int J Pharm 313:189–197

Kawakami K, Pikal MJ (2005) Calorimetric investigation of the structural relaxation of amorphous materials: evaluating validity of the methodologies. J Pharm Sci 94:948–965

Kennedy M, Hu J, Gao P, Li L, Ali-Reynolds A, Chal B, Gupta V, Ma C, Mahajan N, Akrami A, Surapaneni S (2008) Enhanced bioavailability of a poorly soluble VR1 antagonist using an amorphous solid dispersion approach: a case study. Mol Pharm 5:981–993

Kestur US, Wanapun D, Toth SJ, Wegiel LA, Simpson GJ, Taylor LS (2012) Nonlinear optical imaging for sensitive detection of crystals in bulk amorphous powders. J Pharm Sci 101:4201–4213

Khougaz K, Clas SD (2000) Crystallization inhibition in solid dispersions of MK-0591 and poly (vinylpyrrolidone) polymers. J Pharm Sci 89:1325–1334

Kim EHJ, Chen XD, Pearce D (2003) On the mechanisms of surface formation and the surface compositions of industrial milk powders. Drying Technol 21:265–278

Kim EHJ, Chen XD, Pearce D (2009) Surface composition of industrial spray-dried milk powders. 2. Effects of spray drying conditions on the surface composition. J Food Eng 94:169–181

Kislalioglu MS, Khan MA, Blount C, Goettsch RW, Bolton S (1991) Physical characterization and dissolution properties of ibuprofen—eudragit coprecipitates. J Pharm Sci 80:799–804

Klein CE, Chiu Y-L, Awni W, Zhu T, Heuser RS, Doan T, Breitenbach J, Morris JB, Brun SC, Hanna GJ (2007) The tablet formulation of lopinavir/ritonavir provides similar bioavailability to the soft-gelatin capsule formulation with less pharmacokinetic variability and diminished food effect. J Acquir Immune Defic Syndr 44:401–410

Konno H, Handa T, Alonzo DE, Taylor LS (2008) Effect of polymer type on the dissolution profile of amorphous solid dispersions containing felodipine. Eur J Pharm Biopharm 70:493–499

Korhonen O, Bhura C, Pikal MJ (2008) Correlation between molecular mobility and crystal growth of amorphous phenobarbital and phenobarbital with polyvinylpyrrolidone and L-proline. J Pharm Sci 97:3830–3841

Kostewicz E, Aarons L, Bergstrand M, Bolger M, Galetin A, Hatley O, Jamei M, Lloyd R, Pepin X, Rostami-Hodjegan A, Sjögren E, Tannergren C, Turner D, Wagner C, Weitschies W, Dressman J (2013) PBPK models for the prediction of in vivo performance of oral dosage forms. Eur J Pharm Sci 57:300–321

Kubota N, Mullin JW (1995) A kinetic model for crystal growth from aqueous solution in the presence of impurity. J Cryst Growth 152:203–208

Kuldipkumar A, Kwon GS, Zhang GGZ (2006) Determining the growth mechanism of tolazamide by induction time measurement. Cryst Growth Des 7:234–242

Kwong AD, Kauffman RS, Hurter P, Mueller P (2011) Discovery and development of telaprevir: an NS3-4A protease inhibitor for treating genotype 1 chronic hepatitis C virus. Nat Biotechnol 29:993–1003

Lafferrere L, Hoff C, Veesler S (2004) Study of liquid-liquid demixing from drug solution. J Cryst Growth 269:550–557

Lamer VK, Dinegar RH (1950) Theory, production and mechanism of formation of monodispersed hydrosols. J Am Chem Soc 72:4847–4854

Lauer ME, Siam M, Tardio J, Page S, Kindt JH, Grassmann O (2013) Rapid assessment of homogeneity and stability of amorphous solid dispersions by atomic force microscopy-from bench to batch. Pharm Res 30:2010–2022

Law D, Krill SL, Schmitt EA, Fort JJ, Qiu Y, Wang W, Porter WR (2001) Physicochemical considerations in the preparation of amorphous ritonavir–poly(ethylene glycol) 8000 solid dispersions. J Pharm Sci 90:1015–1025

Law D, Schmitt EA, Marsh KC, Everitt EA, Wang W, Fort JJ, Krill SL, Qiu Y (2004) Ritonavir–PEG 8000 amorphous solid dispersions: in vitro and in vivo evaluations. J Pharm Sci 93:563–570

Lenhardt T, Vergnault G, Grenier P, Scherer D, Langguth P (2008) Evaluation of nanosuspensions for absorption enhancement of poorly soluble drugs: in vitro transport studies across intestinal epithelial monolayers. AAPS J 10:435–438

Lennernas H (1998) Human intestinal permeability. J Pharm Sci 87:403–410

Lentz KA (2008) Current methods for predicting human food effect. AAPS J 10:282–288

Leuner C, Dressman J (2000) Improving drug solubility for oral delivery using solid dispersions. Eur J Pharm Biopharm 50:47–60

Lindfors L, Forssén S, Westergren J, Olsson U (2008) Nucleation and crystal growth in supersaturated solutions of a model drug. J Colloid Interface Sci 325:404–413

Loftsson T, Fri∂riksdóttir H, Gu∂mundsdóttir TK (1996) The effect of water-soluble polymers on aqueous solubility of drugs. Int J Pharm 127:293–296

Lu Q, Zografi G (1998) Phase behavior of binary and ternary amorphous mixtures containing indomethacin, citric acid, and PVP. Pharm Res 15:1202–1206

Lu ATK, Frisella ME, Johnson KC (1993) Dissolution modeling: factors affecting the dissolution rates of polydisperse powders. Pharm Res 10:1308–1314

Luthra SA, Hodge IM, Pikal MJ (2008) Investigation of the impact of annealing on global molecular mobility in glasses: optimization for stabilization of amorphous pharmaceuticals. J Pharm Sci 97:3865–3882

Mahieu A, Willart J-F, Dudognon E, Danede F, Descamps M (2013) A new protocol to determine the solubility of drugs into polymer matrixes. Mol Pharm 10:560–566

Mahlin D, Berggren J, Alderborn G, Engstrom S (2004) Moisture-induced surface crystallization of spray-dried amorphous lactose particles studied by atomic force microscopy. J Pharm Sci 93:29–37

Mahlin D, Berggren J, Gelius U, Engstrom S, Alderborn G (2006) The influence of PVP incorporation on moisture-induced surface crystallization of amorphous spray-dried lactose particles. Int J Pharm 321:78–85

Mao C, Chamarthy SP, Byrn SR, Pinal R (2006) A calorimetric method to estimate molecular mobility of amorphous solids at relatively low temperatures. Pharm Res 23:2269–2276

Marsac PJ, Taylor LS (2008) Thermodynamics of mixing drugs and polymers. AAPS National Meeting 10(S2):T2447

Marsac PJ, Shamblin SL, Taylor LS (2006) Theoretical and practical approaches for prediction of drug-polymer miscibility and solubility. Pharm Res 23:2417–2426

Marsac PJ, Li T, Taylor LS (2009) Estimation of drug-polymer miscibility and solubility in amorphous solid dispersions using experimentally determined interaction parameters. Pharm Res 26:139–151

Marsac PJ, Rumondor ACF, Nivens DE, Kestur US, Stanciu L, Taylor LS (2010) Effect of temperature and moisture on the miscibility of amorphous dispersions of felodipine and poly (vinyl pyrrolidone). J Pharm Sci 99:169–185

Martin A (1993) Physical pharmacy. Lea & Febiger, Philadelphia, PA

Martinez MN, Amidon GL (2002) A mechanistic approach to understanding the factors affecting drug absorption: a review of fundamentals. J Clin Pharmacol 42:620–643

Matsumoto T, Zografi G (1999) Physical properties of solid molecular dispersions of indomethacin with poly(vinylpyrrolidone) and poly(vinylpyrrolidone-co-vinylacetate) in relation to indomethacin crystallization. Pharm Res 16:1722–1728

Meeus J, Scurr DJ, Amssoms K, Davies MC, Roberts CJ, Van Den Mooter G (2013) Surface characteristics of spray-dried microspheres consisting of PLGA and PVP: relating the influence of heat and humidity to the thermal characteristics of these polymers. Mol Pharm 10:3213–3224

Moribe K, Pongpeerapat A, Tozuka Y, Yamamoto K (2006) Drug nanoparticle formation from drug/HPMC/SDS ternary ground mixtures. Pharmazie 61:97–101

Mullin JW (2001a) Crystallization. Reed Educational and Professional Publishing Ltd., Oxford

Mullin JW (2001b) Solutions and solubility, 4th edn. Butterworth-Heinemann, Oxford

Murdande SB, Pikal MJ, Shanker RM, Bogner RH (2010a) Solubility advantage of amorphous pharmaceuticals: I. A thermodynamic analysis. J Pharm Sci 99:1254–1264
Murdande SB, Pikal MJ, Shanker RM, Bogner RH (2010b) Solubility advantage of amorphous pharmaceuticals: II. Application of quantitative thermodynamic relationships for prediction of solubility enhancement in structurally diverse insoluble pharmaceuticals. Pharm Res 27:2704–2714
Murdande SB, Pikal MJ, Shanker RM, Bogner RH (2011a) Aqueous solubility of crystalline and amorphous drugs: challenges in measurement. Pharm Dev Technol 16:187–200
Murdande SB, Pikal MJ, Shanker RM, Bogner RH (2011b) Solubility advantage of amorphous pharmaceuticals, Part 3: is maximum solubility advantage experimentally attainable and sustainable? J Pharm Sci 100:4349–4356
Newman A, Engers D, Bates S, Ivanisevic I, Kelly RC, Zografi G (2008) Characterization of amorphous API:polymer mixtures using X-ray powder diffraction. J Pharm Sci 97:4840–4856
Newman A, Knipp G, Zografi G (2012) Assessing the performance of amorphous solid dispersions. J Pharm Sci 101:1355–1377
Nunes C, Mahendrasingam A, Suryanarayanan R (2005) Quantification of crystallinity in substantially amorphous materials by synchrotron X-ray powder diffractometry. Pharm Res 22:1942–1953
Oh DM, Curl RL, Amidon GL (1993) Estimating the fraction dose absorbed from suspensions of poorly soluble compounds in humans—a mathematical-model. Pharm Res 10:264–270
Otsuka H, Esumi K, Ring TA, Li JT, Caldwell KD (1996) Simultaneous adsorption of poly (N-vinyl-2pyrrolidone) and hydrocarbon/fluorocarbon surfactant from their binary mixtures on hydrophilic/hydrophobic silica. Colloids Surf A Physicochem Eng Asp 116:161–171
Padilla AM, Ivanisevic I, Yang Y, Engers D, Bogner RH, Pikal MJ (2011) The study of phase separation in amorphous freeze-dried systems. Part I: Raman mapping and computational analysis of XRPD data in model polymer systems. J Pharm Sci 100:206–222
Pan Z, Campbell A, Somasundaran P (2001) Polyacrylic acid adsorption and conformation in concentrated alumina suspensions. Colloids Surf A Physicochem Eng Asp 191:71–78
Park JS, Park JW, Ruckenstein E (2001) Thermal and dynamic mechanical analysis of PVA/MC blend hydrogels. Polymer 42:4271–4280
Pham TN, Watson SA, Edwards AJ, Chavda M, Clawson JS, Strohmeier M, Vogt FG (2010) Analysis of amorphous solid dispersions using 2D solid-state NMR and H-1 T-1 relaxation measurements. Mol Pharm 7:1667–1691
Poirier-Brulez F, Roudaut G, Champion D, Tanguy M, Simatos D (2006) Influence of sucrose and water content on molecular mobility in starch-based glasses as assessed through structure and secondary relaxation. Biopolymers 81:63–73
Power G, Vij JK, Johari GP (2007) Dielectric relaxation and crystallization of nanophase separated 1-propanol-isoamylbromide mixture. J Chem Phys 127
Price R, Young PM (2004) Visualization of the crystallization of lactose from the amorphous state. J Pharm Sci 93:155–164
Qi S, Belton P, Nollenberger K, Gryczke A, Craig DQM (2011) Compositional analysis of low quantities of phase separation in hot-melt-extruded solid dispersions: a combined atomic force microscopy, photothermal Fourier-transform infrared microspectroscopy, and localised thermal analysis approach. Pharm Res 28:2311–2326
Qian F, Huang J, Zhu Q, Haddadin R, Gawel J, Garmise R, Hussain M (2010) Is a distinctive single T-G a reliable indicator for the homogeneity of amorphous solid dispersion? Int J Pharm 395:232–235
Raghavan SL, Trividic A, Davis AF, Hadgraft J (2001) Crystallization of hydrocortisone acetate: influence of polymers. Int J Pharm 212:213–221
Raina SA, Zhang GGZ, Alonzo DE, Wu J, Zhu D, Catron ND, Gao Y, Taylor LS (2012) Enhancements and limits in drug membrane transport using supersaturated solutions of poorly water-soluble drugs. Cryst Growth Des 12(3):1538–1547

Righetti MC, Cardelli C, Scalari M, Tombari E, Conti G (2002) Thermodynamics of mixing of poly(vinyl chloride) and poly(ethylene-co-vinyl acetate). Polymer 43:5035–5042
Rodríguez-Hornedo N, Murphy D (1999) Significance of controlling crystallization mechanisms and kinetics in pharmaceutical systems. J Pharm Sci 88:651–660
Rohr B (2006) Biopharmaceutics modeling and the role of dose and formulation on oral exposure. In: Borchardt RT, Kerns EH, Hageman MJ, Thakker DR, Stevens JL (eds) Optimizing the drug like properties of leads in drug discovery. Springer, New York
Rumondor ACF, Taylor LS (2010a) Application of partial least-squares (PLS) modeling in quantifying drug crystallinity in amorphous solid dispersions. Int J Pharm 398:155–160
Rumondor ACF, Taylor LS (2010b) Effect of polymer hygroscopicity on the phase behavior of amorphous solid dispersions in the presence of moisture. Mol Pharm 7:477–490
Rumondor ACF, Ivanisevic I, Bates S, Alonzo D, Taylor LS (2009) Evaluation of drug-polymer miscibility in amorphous solid dispersion systems. Pharm Res 26:2523–2534
Rumondor ACF, Wikstroem H, Van Eerdenbrugh B, Taylor LS (2011) Understanding the tendency of amorphous solid dispersions to undergo amorphous-amorphous phase separation in the presence of absorbed moisture. AAPS PharmSciTech 12:1209–1219
Saerens L, Dierickx L, Lenain B, Vervaet C, Remon JP, De Beer T (2011) Raman spectroscopy for the in-line polymer-drug quantification and solid state characterization during a pharmaceutical hot-melt extrusion process. Eur J Pharm Biopharm 77:158–163
Saerens L, Dierickx L, Quinten T, Adriaensens P, Carleer R, Vervaet C, Remon JP, De Beer T (2012) In-line NIR spectroscopy for the understanding of polymer-drug interaction during pharmaceutical hot-melt extrusion. Eur J Pharm Biopharm 81:230–237
Sangwal K (2007) Additives and crystallization process: from fundamentals to applications. Wiley, Hoboken, NJ
Sarode AL, Wang P, Obara S, Worthen DR (2014) Supersaturation, nucleation, and crystal growth during single- and biphasic dissolution of amorphous solid dispersions: polymer effects and implications for oral bioavailability enhancement of poorly water soluble drugs. Eur J Pharm Biopharm 86:351–360
Schuth F (2001) Nucleation and crystallization of solids from solution. Curr Opinion Solid State Mater Sci 5:389–395
Serajuddin ATM (1999) Solid dispersion of poorly water-soluble drugs: early promises, subsequent problems, and recent breakthroughs. J Pharm Sci 88:1058–1066
Sertsou G, Butler J, Hempenstall J, Rades T (2002) Solvent change co-precipitation with hydroxypropyl methylcellulose phthalate to improve dissolution characteristics of a poorly water-soluble drug. J Pharm Pharmacol 54:1041–1047
Shah B, Kakumanu VK, Bansal AK (2006) Analytical techniques for quantification of amorphous/crystalline phases in pharmaceutical solids. J Pharm Sci 95:1641–1665
Shah N, Sandhu H, Phuapradit W, Pinal R, Iyer R, Albano A, Chatterji A, Anand S, Choi DS, Tang K, Tian H, Chokshi H, Singhal D, Malick W (2012) Development of novel microprecipitated bulk powder (MBP) technology for manufacturing stable amorphous formulations of poorly soluble drugs. Int J Pharm 438:53–60
Shamblin SL, Zografi G (1998) Enthalpy relaxation in binary amorphous mixtures containing sucrose. Pharm Res 15:1828–1834
Shamblin SL, Hancock BC, Dupuis Y, Pikal MJ (2000) Interpretation of relaxation time constants for amorphous pharmaceutical systems. J Pharm Sci 89:417–427
Shmeis RA, Wang ZR, Krill SL (2004a) A mechanistic investigation of an amorphous pharmaceutical and its solid dispersions, Part I: a comparative analysis by thermally stimulated depolarization current and differential scanning calorimetry. Pharm Res 21:2025–2030
Shmeis RA, Wang ZR, Krill SL (2004b) A mechanistic investigation of an amorphous pharmaceutical and its solid dispersions, Part II: molecular mobility and activation thermodynamic parameters. Pharm Res 21:2031–2039
Simonelli AP, Mehta SC, Higuchi WI (1970) Inhibition of sulfathiazole crystal growth by polyvinylpyrrolidone. J Pharm Sci 59:633–638

Sinclair W, Leane M, Clarke G, Dennis A, Tobyn M, Timmins P (2011) Physical stability and recrystallization kinetics of amorphous ibipinabant drug product by Fourier transform Raman spectroscopy. J Pharm Sci 100:4687–4699

Six K, Verreck G, Peeters J, Brewster M, Van Den Mooter G (2004) Increased physical stability and improved dissolution properties of itraconazole, a class II drug, by solid dispersions that combine fast- and slow-dissolving polymers. J Pharm Sci 93:124–131

Sjögren E, Westergren J, Grant I, Hanisch G, Lindfors L, Lennernäs H, Abrahamsson A, Tannergren C (2013) In silico predictions of gastrointestinal drug absorption in pharmaceutical product development: application of the mechanistic absorption model GI-Sim. Eur J Pharm Sci 49:679–698

Sohnel O, Mullin JW (1988) Interpretation of crystallization induction periods. J Colloid Interface Sci 123:43–50

Somasundaran P, Huang L (2000) Adsorption/aggregation of surfactants and their mixtures at solid–liquid interfaces. Adv Colloid Interface Sci 88:179–208

Somasundaran P, Krishnakumar S (1997) Adsorption of surfactants and polymers at the solid-liquid interface. Colloids Surf A Physicochem Eng Asp 123–124:491–513

Strachan CJ, Windbergs M, Offerhaus HL (2011) Pharmaceutical applications of non-linear imaging. Int J Pharm 417:163–172

Sugano K (2009a) Estimation of effective intestinal membrane permeability considering bile micelle solubilisation. Int J Pharm 368:116–122

Sugano K (2009b) Theoretical investigation of passive intestinal membrane permeability using Monte Carlo method to generate drug-like molecule population. Int J Pharm 373:55–61

Sugano K (2010) Possible reduction of effective thickness of intestinal unstirred water layer by particle drifting effect. Int J Pharm 387:103–109

Sugano K (2012a) Biopharmaceutics modeling and simulations. Wiley, Hoboken, NJ

Sugano K (2012b) Introduction. Biopharmaceutics modeling and simulation, theory, practice, methods and application. Wiley, Hoboken, NJ

Sun Y, Xi H, Ediger MD, Yu L (2008) Diffusionless crystal growth from glass has precursor in equilibrium liquid. J Phys Chem B 112:661–664

Sun Y, Tao J, Zhang GGZ, Yu L (2010) Solubilities of crystalline drugs in polymers: an improved analytical method and comparison of solubilities of indomethacin and nifedipine in PVP, PVP/VA, and PVAC. J Pharm Sci 99:4023–4031

Surana R, Suryanarayanan R (2000) Quantitation of crystallinity in substantially amorphous pharmaceuticals and study of crystallization kinetics by X-ray powder diffractometry. Powder Diffract 15:2–6

Swenson ES, Curatolo WJ (1992) Intestinal permeability enhancement for proteins, peptides and other polar drugs—mechanisms and potential toxicity. Adv Drug Deliv Rev 8:39–92

Szczepanski CR, Pfeifer CS, Stansbury JW (2012) A new approach to network heterogeneity: polymerization induced phase separation in photo-initiated, free-radical methacrylic systems. Polymer 53:4694–4701

Tachibana T, Nakamura A (1965) A method for preparing an aqueous colloidal dispersion of organic materials by using water-soluble polymers: dispersion B-carotene by polyvinylpyrrolidone. Kolloid-Zeitschrift Und Zeitschrift Fur Polymere 203:130–133

Tang XLC, Pikal MJ, Taylor LS (2002a) The effect of temperature on hydrogen bonding in crystalline and amorphous phases in dihydropyrine calcium channel blockers. Pharm Res 19:484–490

Tang XLC, Pikal MJ, Taylor LS (2002b) A spectroscopic investigation of hydrogen bond patterns in crystalline and amorphous phases in dihydropyridine calcium channel blockers. Pharm Res 19:477–483

Tantishaiyakul V, Kaewnopparat N, Ingkatawornwong S (1999) Properties of solid dispersions of piroxicam in polyvinylpyrrolidone. Int J Pharm 181:143–151

Tao J, Sun Y, Zhang GGZ, Yu L (2009) Solubility of small-molecule crystals in polymers: D-mannitol in PVP, indomethacin in PVP/VA, and nifedipine in PVP/VA. Pharm Res 26:855–864

Taupitz T, Dressman JB, Buchanan CM, Klein S (2013) Cyclodextrin-water soluble polymer ternary complexes enhance the solubility and dissolution behaviour of poorly soluble drugs. Case example: itraconazole. Eur J Pharm Biopharm 83:378–387

Taylor LS, Zografi G (1997) Spectroscopic characterization of interactions between PVP and indomethacin in amorphous molecular dispersions. Pharm Res 14:1691–1698

Tho I, Liepold B, Rosenberg J, Maegerlein M, Brandl M, Fricker G (2010) Formation of nano/micro-dispersions with improved dissolution properties upon dispersion of ritonavir melt extrudate in aqueous media. Eur J Pharm Sci 40:25–32

Tian F, Saville DJ, Gordon KC, Strachan CJ, Zeitler JA, Sandler N, Rades T (2007) The influence of various excipients on the conversion kinetics of carbamazepine polymorphs in aqueous suspension. J Pharm Pharmacol 59:193–201

Tian Y, Booth J, Meehan E, Jones DS, Li S, Andrews GP (2013) Construction of drug-polymer thermodynamic phase diagrams using Flory-Huggins interaction theory: identifying the relevance of temperature and drug weight fraction to phase separation within solid dispersions. Mol Pharm 10:236–248

Tishmack PA, Bugay DE, Byrn SR (2003) Solid-state nuclear magnetic resonance spectroscopy—pharmaceutical applications. J Pharm Sci 92:441–474

Tjipangandjara KF, Somasundaran P (1991) Effects of changes in adsorbed polyacrylic acid conformation on alumina flocculation. Colloids Surf 55:245–255

Toth SJ, Madden JT, Taylor LS, Marsac P, Simpson GJ (2012) Selective imaging of active pharmaceutical ingredients in powdered blends with common excipients utilizing two-photon excited ultraviolet-fluorescence and ultraviolet-second order nonlinear optical imaging of chiral crystals. Anal Chem 84:5869–5875

Tsuji H, Ikada Y (1996) Blends of aliphatic polyesters. 1. Physical properties and morphologies of solution-cast blends from poly(DL-lactide) and poly(epsilon-caprolactone). J Appl Polym Sci 60:2367–2375

Van Der Leeden MC, Kashchiev D, Van Rosmalen GM (1993) Effect of additives on nucleation rate, crystal growth rate and induction time in precipitation. J Cryst Growth 130:221–232

Van Eerdenbrugh B, Baird JA, Taylor LS (2010) Crystallization tendency of active pharmaceutical ingredients following rapid solvent evaporation-classification and comparison with crystallization tendency from undercooled melts. J Pharm Sci 99:3826–3838

Vandecruys R, Peeters J, Verreck G, Brewster ME (2007) Use of a screening method to determine excipients which optimize the extent and stability of supersaturated drug solutions and application of this system to solid formulation design. Int J Pharm 342:168–175

Vasconcelos T, Sarmento B, Costa P (2007) Solid dispersions as strategy to improve oral bioavailability of poor water soluble drugs. Drug Discov Today 12:1068–1075

Veesler S, Lafferrere L, Garcia E, Hoff C (2003) Phase transitions in supersaturated drug solution. Org Process Res Dev 7:983–989

Vekilov PG (2004) Dense liquid precursor for the nucleation of ordered solid phases from solution. Cryst Growth Des 4:671–685

Venkatesh S, Lipper RA (2000) Role of the development scientist in compound lead selection and optimization. J Pharm Sci 89:145–154

Verdoes D, Kashchiev D, Van Rosmalen GM (1992) Determination of nucleation and growth rates from induction times in seeded and unseeded precipitation of calcium carbonate. J Cryst Growth 118:401–413

Vyazovkin S, Dranca I (2007) Effect of physical aging on nucleation of amorphous indomethacin. J Phys Chem B 111:7283–7287

Wanapun D, Kestur US, Kissick DJ, Simpson GJ, Taylor LS (2010) Selective detection and quantitation of organic molecule crystallization by second harmonic generation microscopy. Anal Chem 82:5425–5432

Wanapun D, Kestur US, Taylor LS, Simpson GJ (2011) Single particle nonlinear optical imaging of trace crystallinity in an organic powder. Anal Chem 83:4745–4751

Waterman KC, Sutton SC (2003) A computational model for particle size influence on drug absorption during controlled-release colonic delivery. J Control Release 86:293–304

Weeks NE, Karasz FE, Macknight WJ (1977) Enthalpy of mixing of poly(2,6-dimethyl phenylene oxide) and polystyrene. J Appl Phys 48:4068–4071

Wenslow RM (2002) F-19 solid-state NMR spectroscopic investigation of crystalline and amorphous forms of a selective muscarinic M-3 receptor antagonist, in both bulk and pharmaceutical dosage form samples. Drug Dev Ind Pharm 28:555–561

Wesley RD, Cosgrove T, Thompson L (1999) Interactions of star polymers with surfactants. Langmuir 15:8376–8382

Wu T, Yu L (2006) Origin of enhanced crystal growth kinetics near T-G probed with indomethacin polymorphs. J Phys Chem B 110:15694–15699

Yalkowsky SH (1999) Solubility and solubilization in aqueous media. Oxford University Press, New York

Yamada H, Suryanarayanan R (2006) Calculation of the penetration depth of X-rays in intact pharmaceutical film-coated tablets by microdiffractometory. Pharm Res 23:2149–2157

Yamada H, Suryanarayanan R (2007) X-ray powder diffractometry of intact film coated tablets—an approach to monitor the physical form of the active pharmaceutical ingredient during processing and storage. J Pharm Sci 96:2029–2036

Yang Z, Nollenberger K, Albers J, Craig D, Qi S (2013) Microstructure of an immiscible polymer blend and its stabilization effect on amorphous solid dispersions. Mol Pharm 10:2767–2780

Zhang X, Lionberger RA, Davit BM, Yu LX (2011) Utility of physiologically based absorption modeling in implementing quality by design in drug development. AAPS J 13:59–71

Ziller KH, Rupprecht H (1988) Control of crystal growth in drug suspensions: 1. Design of a control unit and application to acetaminophen suspensions. Drug Dev Ind Pharm 14:2341–2370

Zimmermann A, Millqvist-Fureby A, Elema MRB, Hansen T, Müllertz A, Hovgaard L (2009) Adsorption of pharmaceutical excipients onto microcrystals of siramesine hydrochloride: effects on physicochemical properties. Eur J Pharm Biopharm 71:109–116

Chapter 11
Hot-Melt Extrusion: The Process-Product-Performance Interplay

Nathan Boersen, Chad Brown, James DiNunzio, David Johnson, Patrick Marsac, Robert Meyer, and Craig McKelvey

11.1 Introduction

Design and scaling amorphous solid dispersions using the hot-melt extrusion process may be expedited with an understanding of the thermal and rheological properties of the materials involved. Further, a complete understanding of the phase behavior of the materials making up the formulation is required. Avoiding common processing issues requires both engineering and formulation solutions. The hot-melt extrusion unit operation can be broken down into several smaller operations. These include the feeding system(s), conveying of the feed materials into the extruder, mixing and melting of the materials to form a single-phase multicomponent melt, pressurization of the melt to overcome the pressure buildup at the die, and formation of the extruded material via the die. Further, appropriate venting should be provided so as to avoid issues with the expansion of entrapped air or water at the die.

N. Boersen
Celgene Corporation, Summit, NJ, USA

C. Brown • P. Marsac (✉)
Formulation Sciences, Merck & Co., 770 Sumneytown Pike, West Point, PA 19486-0004, USA
e-mail: Patrick_Marsac@merck.com

J. DiNunzio
Merck & Co., Summit, NJ, USA

D. Johnson • R. Meyer
Pharmaceutical Commercialization, Merck & Co., West Point, PA, USA

C. McKelvey
Merck & Co., Inc., Formulation Sciences, West Point, PA, USA

A.C. Templeton et al. (eds.), *Discovering and Developing Molecules with Optimal Drug-Like Properties*, AAPS Advances in the Pharmaceutical Sciences Series 15,
DOI 10.1007/978-1-4939-1399-2_11

11.2 Overview of the Extrusion Process

An appropriate feeding system is required to dispense each material in a consistent and reliable fashion. If materials are individually fed to the extruder, uniformity of the extrudate relies on accurate dispensing and metering by the feeding systems. Each material must be fed at a constant rate to maintain the appropriate ratio of excipients and the API. This approach avoids a pre-blending step, thus eliminating a unit operation from the process train. Alternatively, if a pre-blend is fed into the extruder, it is important to demonstrate homogeneity of the mixture at the beginning of the process and during dispensing (i.e., lack of segregation during dispersion). Most often, twin screw extruders are “starve fed” whereby the throughput is dictated by the feed rate (contrasted by “flood feeding” whereby the throughput is dictated by the screw speed). In the pre-blend starve fed situation, maintaining a constant feed rate is also critical to maintain a constant residence time distribution. Feed systems include gravimetric systems (dispensing by weight) and volumetric systems (dispensing by volume displaced). Most often, solids are introduced via a conveying screw system. Care must be taken to ensure that the solids are introduced at a constant rate. In some instances, pulsatory feeding resulting from compaction during conveying or issues with discharge from the hopper may create problems with the uniformity of feed streams. Further, the addition of liquid streams maybe required and this may be accomplished via liquid feed systems and liquid feed ports. Finally, the location of the feed streams should take into consideration the rheological and mixing properties of each of the materials as well as the thermal liabilities which they may carry. For instance, it is perhaps appropriate to add liquid feeds downstream to powder feeds or to add thermally labile materials downstream to the feed ports of other materials so as to reduce the residence time.

Once the material is fed into the extruder, it is generally transported some distance from the feed port prior to exposing to high temperatures. This conveying section of the extruder is often accompanied by venting ports to allow for release of volatiles such as water which might be introduced with the incoming material. In some instances, the first section(s) of the barrel may require cooling so that heat transfer from downstream cannot induce softening of the incoming material and create problems with clogging in the feed zone. Most often, the material is subsequently exposed to mixing or kneading elements or both. In this region of the extruder, the various materials begin to mix on a particulate level with perhaps some deformation of the particles. Subsequent screw elements are generally selected to provide additional mixing and pressure buildup. Coupled with increased temperature, the screw design should facilitate the formation of a molten phase and a high shear environment—exposing interfaces among the individual materials making up the formulation. Most often, this molten phase is the polymer (i.e., the polymer is above its glass transition temperature) and the discontinuous solid phase is the crystalline drug with other materials added as minor components and introduced as required (i.e., surfactants, plasticizers, pH-modifying agents, and others).

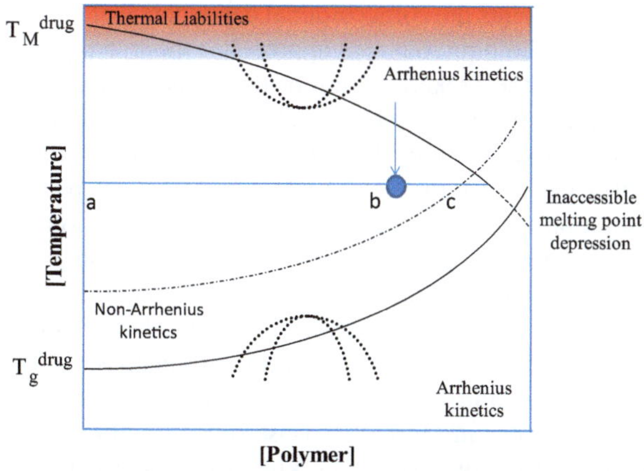

Fig. 11.1 Hypothetical phase diagram for a binary drug-polymer system showing the liquidous line and the glass transition temperature as a function of the polymer composition. Also highlighted are those regions which are often accessible and inaccessible experimentally as well as regions which often follow Arrhenius temperature dependence and regions where there are non-Arrhenius temperature dependence. Hypothetical phase envelops are also displayed. Additional explanation provided in the text

Appropriate downstream processing may include the use of chilled rolls to cool the extrudate (Aitken-Nichol et al. 1996), milling (Albers et al. 2009), and subsequent tablet formulation or direct shaping into final dosage forms (Crowley et al. 2004b), foaming (Verreck et al. 2005), or production of films among others (Low et al. 2013). There are many texts which describe these various processes in detail and this will not be the focus of this publication. Instead, the authors will highlight process-formulation-performance interplay within the framework of a hypothetical phase diagram as outlined in Fig. 11.1. Sufficient time, mechanical input, and thermal energy facilitate thermal homogenization of all of the components in the extrudate. The materials and the process have limitations and a balance between several, often competing, processes must be considered to facilitate a robust process.

11.3 The Phase Diagram as Related to the Hot-Melt Extrusion Process

In the discussion that follows, the HME process will be described within the framework of the phase diagram shown in Fig. 11.1. The *x* and *y* axis represent the polymer composition and temperature respectively. The line originating at the melting point of the API, T_m^{drug}, is referred to as the liquidous line and represents the temperature at which the crystalline API is in equilibrium with the molten API at any polymer composition. The line originating at the glass transition temperature of the API, T_g^{drug}, represents the compositional dependence of the T_g of a mixture of drug and polymer. As the liquidous line approaches the line representing the glass transition temperature (the broken section of the liquidous line), it becomes more difficult to experimentally determine the values of the liquidous line. This is the result of the increased viscosity as the glass transition temperature is approached. Also noted in this phase diagram are the two hypothetical phase envelopes: an upper critical solution temperature and a lower critical solution temperature (the appearance and position are completely arbitrary). In traversing into the temperature and composition space bounded by these phase envelopes, there is a thermodynamic driving force for phase separation to occur. Nevertheless, if the phase envelope resides in a kinetically restricted region of the phase diagram (i.e., below the glass transition temperature), phase separation may not be realized despite there being a thermodynamic driving force for phase separation to occur. Also shown in this phase diagram are regions where the kinetics of physical and chemical processes are expected to obey Arrhenius kinetics and non-Arrhenius kinetics. For instance, well above the glass transition temperature, chemical and physical processes will most often follow Arrhenius kinetics. Further, well below the glass transition temperature, Arrhenius kinetics will be observed. Of course, it is well documented that when approaching the glass transition temperature from above, amorphous solid dispersion systems of API and polymer most often display non-Arrhenius kinetics and can be described by equations such as the Vogel–Tammann–Fulcher equation (Adam and Gibbs 1965; Ferry 1980). Examples of phenomena which display non-Arrhenius kinetics as the T_g is approached from above include viscosity (Andronis and Zografi 1997) and crystallization kinetics (Greco et al. 2012). Also shown on the phase diagram is a hypothetical region for thermal degradation. Thermally induced degradation at temperatures noted in the red-shaded region of the phase diagram must be understood so as to design a robust extrusion process. Finally, we note that the phase diagram is unique for each system.

11.3.1 The Liquidous Curve and Melt Mixing

Production of amorphous solid dispersions which are mixed on a molecular level (i.e., multicomponent single-phase systems) requires an understanding of the liquidous curve. This curve defines the minimum operating temperature in the extrusion process. Introduction of the various components into the extruder and adequate mixing are prerequisites for the formation of a homogeneous multicomponent single-phase system. The materials are sheared and the particle size is often reduced during the conveying and mixing. Introduction of new surfaces (i.e., increasing the surface area via dispersive mixing) expedites the kinetics of mixing. Further, distributive mixing is required so as to create a large number of contact points and surfaces between the various materials. Regardless, formation of a molecular level single-phase system requires a driving force for mixing—often referred to in the extrusion literature as "thermal homogenization." In short, the production of a single-phase system will be expedited by the distributive and dispersive mixing (along with the viscosity of the materials) but it is the thermodynamic driving force for mixing which ultimately describes the achievable endpoint during processing.

Mathematically, this can be described using the traditional solubility and dissolution equations. The Noyes–Whitney equation describes the rate of dissolution as a function of the diffusivity and the solubility (Noyes and Whitney 1897). The dissolution rate or rate of mass transport, as shown in Eq. (11.1), is a function of the difference in the solubility of the drug in the polymer, C_S, and the concentration of the drug already dissolved in the polymer, $C(t)$; the diffusion coefficient, D; the surface area, A; and the boundary layer thickness, h:

$$\frac{\mathrm{d}M}{\mathrm{d}t} = \frac{DA}{h}(C_S - C(t)) \tag{11.1}$$

When working with highly viscous materials, it quickly becomes apparent that the distributive and dispersive mixing acts to increase the dissolution rate by increasing surface area between components, creating new contacts between particles of different type, and decreasing the boundary layer thickness through shearing of the particles. Another important component described in this equation is the diffusivity. As shown in Eq. (11.2), the diffusivity can be described by the Stokes-Einstein equation and is related to the Boltzmann constant, k_B; the temperature, T; the viscosity, η; and the radius of the diffusing species, r:

$$D = \frac{k_B T}{6\pi\eta r} \tag{11.2}$$

From this equation, we note that the diffusivity is increased with temperature. Further, the viscosity of the system is also reduced with increasing temperature. Therefore, the process of reaching the thermodynamic end state at any temperature is expedited as the temperature is increased. It is important to note also that the viscosity will display a shear rate dependence and the interplay in temperature and shear is a strong function of material type and may often define processing limitations.

It is the solubility which dictates the final endpoint of the process. For a pure crystalline drug, the melting point is defined as the point at which the chemical potential of the molten phase is equal to the chemical potential of the crystalline phase or $\mu_{\text{crystal}} = \mu_{\text{drug in molten phase}}$ (Marsac et al. 2009, 2006b). When the molten phase is in contact with a second material (i.e., a polymer), the interactions of the drug with the polymer may be favorable and reduce the chemical potential of the molten drug upon mixing to form a new single-phase multicomponent system or $\mu_{\text{drug in multi-componet molten phase}} < \mu_{\text{drug in molten phase}}$ (Marsac et al. 2006b, 2009). This occurs when the drug and the polymer, or any other material for that matter, form a single phase. Therefore, the temperature at which the chemical potential of the crystalline phase is identical to the chemical potential of the multicomponent molten phase is reduced. Assuming that the extrusion process is designed such that the kinetics of this mixing process are of time scales less than the residence time in the extruder, the depressed melting point can be tabulated as a function of composition so as to give an activity coefficient, γ, in Eq. (11.3):

$$\ln \gamma_{\text{drug}} x_{\text{drug}} = -\frac{\Delta H_{\text{fus}}}{RT}\left[1 - \frac{T}{T_{\text{M}}}\right] - \frac{1}{RT}\int_{T_{\text{M}}}^{T} \Delta C_{\text{P}}^{\text{config}}\,\mathrm{d}T + \frac{1}{R}\int_{T_{\text{M}}}^{T} \frac{\Delta C_{\text{P}}^{\text{config}}}{T}\,\mathrm{d}T \tag{11.3}$$

Here, the melting point depression curve is precisely the liquidous curve shown in Fig. 11.1. In this equation, x is the mole fraction solubility of the drug at the liquidous curve, ΔH_{fus} is the heat of fusion of the pure drug, R is the universal gas constant, and $\Delta C_{\text{p}}^{\text{config}}$ is the configurational heat capacity (i.e., the difference between the heat capacity of the crystalline API and the amorphous counterpart).

11.3.2 The Mixed-Phase Glass Transition Temperature

Strictly speaking, the viscosity, extruder design, and operating conditions will dictate the kinetics of mixing as discussed above. Nevertheless, it is also useful to understand the glass transition temperature of the mixed phase since the material will display solid-like properties below this temperature That is to say, post-melt mixing the operating temperature should be maintained above the glass transition temperature of the mixed phase. Below this temperature, the material will not be easily manipulated in the extruder and drawn from the die. Of course, the glass transition temperature of a miscible system is often different from that of the individual components. As a starting point, the glass transition temperature of the mixture may be estimated from the glass transition temperature of the individual components. For instance, the Gordon-Taylor equation (Eq. 11.4) is derived based on the assumption of ideal mixing, additive free volume, and of course the assumption that the materials are miscible (Gordon and Taylor 1952). As shown in Eq. (11.5), the "*K*" value can be estimated from the glass transition temperature and true density of the individual components (Eq. 11.5):

$$T_{\text{g}}^{\text{mix}} = \frac{w_1 T_{\text{g1}} + K w_2 T_{\text{g2}}}{w_1 + K w_2} \tag{11.4}$$

$$K \approx \frac{T_{\text{g1}}\rho_1}{T_{\text{g2}}\rho_2} \tag{11.5}$$

A similar equation, the Couchman-Karasz equation, is derived after assuming that mixing is ideal and that the mixing is strictly combinatorial but with a K value calculated from the ratio of the change in heat capacity of each material across the glass transition temperature (Eq. 11.6) (Couchman and Karasz 1978):

$$K \approx \frac{\Delta C_{\mathrm{p1}}}{\Delta C_{\mathrm{p2}}} \tag{11.6}$$

Although a useful starting point, these equations should not replace direct measurements since deviations can exist (Khougaz and Clas 2000). In reference to the phase diagram shown in Fig. 11.1, the importance of the mixed-phase glass transition temperature may not be obvious. However, one can envision a similar phase diagram whereby the mixed-phase glass transition temperature is much closer to the liquidous line. For instance, consider a drug substance having a glass transition temperature of 150 °C and no known crystalline forms. Clearly, the processing window will be severely restricted as a result of the high T_g of the drug and presumably the mixed amorphous phase (most often polymers with relatively high T_gs are used in the preparation of amorphous solid dispersions).

11.3.3 Phase Envelops/Miscibility Gaps

Although the pharmaceutical literature has applied many of the concepts which describe melting point depression as a function of composition and the glass transition as a function of composition (Marsac et al. 2006a, 2009; Tian et al. 2013; Sun et al. 2010), relatively little theory has been discussed which describes the possibility of multiple amorphous phases. This may be the result of few examples of amorphous phase separation which is not accompanied by crystallization (Lu and Zografi 1998). Nevertheless, it is useful to review the concepts of the lower critical solution temperature (LCST) and the upper critical solution temperature (UCST) so as not to overlook the manifestation of these phenomena should they occur during extrusion process development. Further, recent advances in understanding the temperature dependence of the free energy of mixing as applied to drug-polymer systems may lead to a better understanding of the phase behavior (Marsac et al. 2012).

The free energy as a function of composition and temperature will determine the phase behavior of the solid dispersion system and only systems with a negative free energy and a single minima over all compositions and over the temperature range experienced in the extruder (and at environmental conditions experienced post extrusion) will have a thermodynamic endpoint which is consistent with a single-phase multicomponent system (i.e., assuming kinetics allow for access) (Gaskell 2003). For instance, consider Fig. 11.2 which shows several possible free energy of mixing curves at an arbitrary temperature. As shown in curve (a), the free energy of mixing is positive over all compositions, and, therefore, the system is "completely" immiscible. Curve (b) shows a system which is miscible over all compositions and

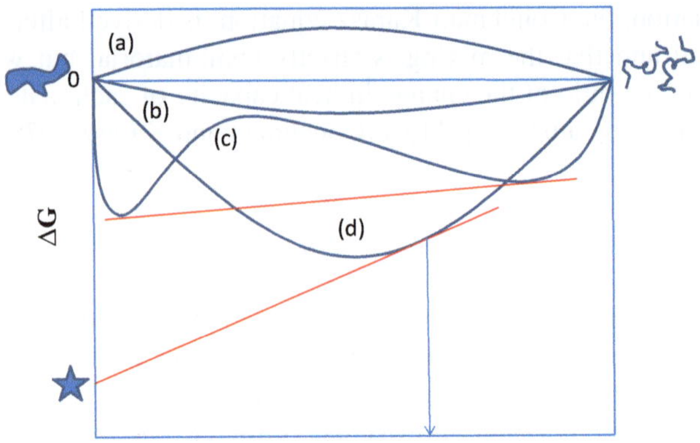

Fig. 11.2 Hypothetical phase diagram at an arbitrary temperature showing an immiscible drug-polymer system (a), a miscible system (b), a partially miscible system (c), and a completely miscible system (d). Although both are miscible, the thermodynamic driving force for mixing is greater in the case of system d as compared with system b. Also shown is the graphical representation of the polymer composition at which the crystalline drug is at equilibrium with the drug is in the amorphous solid dispersion (as discussed in Chap. 10 and noted with the *blue arrow*)

the free energy, although modest, favors mixing. Curve (c) illustrates the situation where two minima exist. Although the free energy of mixing is negative over all compositions, the system will phase separate into the compositions where the chemical potentials of the two phases are identical (as indicated by the red line with the mutual tangent). The mechanism of phase separation will be dictated by the shape of the curve and will include either spinodal decomposition or phase separation via nucleation and growth—with the boundary between the two mechanisms given by the zero points in the second derivative of the curve. Graphically, the distinction in mechanisms of phase separation is shown in Fig. 11.1 by the two curved lines which share a common minimum (in the case of an LCST) or a common maximum (in the case of a UCST) (Gaskell 2003). Nevertheless, it is important to note that realization of any of these processes will depend upon the level of kinetic restriction in the system during processing and storage. As shown in Fig. 11.1, if the UCST exists above the glass transition temperature and if the viscosity allows for access of the two phases over the time scale of cooling from the liquidous to the glassy line and below, two phases may be observed. Alternatively, it is foreseeable that a phase envelope exists in the glassy region (i.e., below the glass transition temperature) or the viscosity of the supercooled liquid is so high and the quench rate is so high that the two-phase region may not be observed. Curve (d) represents a scenario where the free energy of mixing is very favorable and the driving force for mixing is large.

Consider, in more detail, the section of the diagram that shows a phase envelop at the lower temperature (near the T_g). Commonly referred to as a UCST, in this type of behavior, miscibility is favored with increasing temperature. The two lines separating the two mechanisms of phase separation can be described in terms of Flory-Huggins lattice theory. Of course there are several caveats to this approach. Namely, this theory was derived for polymer solutions and here it is extended to describe the behavior of disordered solids, viscous solids, or molten materials—depending on where the critical solution temperature resides relative to the liquidous curve, the glass transition temperature, and also depending on the viscosity temperature profile above the glass transition curve. Strictly speaking, this theory assumes that specific interactions are not significant, the volume change on mixing is negligible, and there are no concentration gradients within the disordered system (Flory 1942). Taking these assumptions into consideration, the Gibbs free energy of mixing can nevertheless be expressed in terms of the volume fractions of the drug and the polymer, Φ_1 and Φ_2, and an interaction parameter, χ, which is meant to represent the sum of an enthalpic term, α/T, and an entropic term, β. Here, the number of moles of each species is given by n_1 and n_2 and k_B represents the Boltzmann constant:

$$\Delta G_{\mathrm{mix}} = k_B T\left(n_1 \ln\Phi_1 + n_2 \ln\Phi_2 + \chi n_1 \Phi_2^2\right) \qquad (11.7)$$

The lines representing the two-phase region can then be drawn according to the following three conditions and considering that phase separation occurs if the sum of the free energy of two separate phases is lower than the free energy of a single phase. Condition 1 explains that phase separation occurs when the chemical potential of each component is identical in each phase. Graphically, this is represented by the common tangent as described above. Condition 2 describes the range of polymer volume fraction where a one-phase system is metastable or unstable. Mathematically, this is given by the points when the second derivative in Eq. (11.7) is equal to zero and graphically represented by the two points where the free energy curve changes from concave up to concave down or vice versa. Between these two points, spontaneous phase separation would occur (i.e., phase separation via spinodal decomposition). Alternatively, between the points identified by the tangent lines and the points which satisfy condition 2, phase separation will occur via nucleation and growth of a second amorphous phase. However, it is important to note again that the distinction between spinodal decomposition and phase separation via nucleation and growth is not necessarily realized in practice since kinetic factors may not allow for access to these thermodynamic endpoints. Finally, a third condition is met which describes the critical point or the temperature, T_c, below which phase separation will occur. This condition is met when the third derivative in Eq. (11.7) is equal to zero.

The application of these equations to pharmaceutical systems requires a complete understanding of the temperature dependence of the thermodynamics of mixing the drug and the polymer as a function of composition and temperature. Of course, this is very challenging given the kinetic and thermal restrictions associated with highly viscous systems. Nevertheless, the concepts may provide insight into the formulation design and identification of modes of failure. For instance, when homogenizing a drug-polymer system at high temperature and high shear, it is possible to achieve a single-phase system at the processing temperature only to access a multiphase system at a lower temperature. If phase separation is realized during the cooling process, this may compromise the performance of the solid dispersion. Also shown in Fig. 11.1 is a hypothetical LCST. Only here, increases in temperature do not favor mixing. In closing, the pharmaceutical literature has, in large part, not recognized CSTs as a failure mode within the scope of preparation of solid dispersions. However, failure modes may have inadvertently been attributed to other mechanisms as a result of lack of recognition of the existence of critical solution temperatures. It is therefore useful to keep these concepts in mind when developing solid dispersion systems.

11.3.4 Polymer Conformation in Solid Dispersions: A Molecular Level Perspective

It is also compelling to note the potential implications of polymer confirmation in solid dispersions and consider extending the models which describe polymer confirmation in solution to the pharmaceutical arena. It is well understood that the confirmation of a polymer will be a function of the environment in which it resides. For instance, when a polymer is placed in a solvent for which favorable interactions exist, it may present as unfolded so as to maximize the strength and extent of interactions with the environment. Alternatively, if the polymer is placed in a solvent for which unfavorable interactions exist, it may present as folded in nature so as to minimize the strength and extent of interactions with the environment. The implications of these polymer confirmations on stability and phase behavior can then be considered. For instance, in the molten state, if we consider a drug-polymer system with incredibly favorable interactions—such as shown in curve (d) of Fig. 11.2—the polymer may be largely unfolded and interact well with the drug. Similarly, if the polymer shows interactions which are more similar to curve (b), more self-interactions may be anticipated. On a molecular level, this may impact the physical stability of the resulting solid dispersion. Consider also the implications on amorphous solid dispersions prepared by spray drying. If a spray-drying solvent is chosen which creates an environment where the polymer is folded vs. unfolded, the resulting dispersion produced may look quite different on a molecule level. The relevance of these concepts is recognized in the polymer literature as demonstrated by the development of various models describing chain confirmation (Rubinstein and Colby 2003; Flory 1942). Ideal chain models do not consider interactions between the repeat units of the polymer and solvent while real chain models consider interactions between repeat units of the polymer and the solvent.

Consider first that the flexibility of a polymer chain may impact the ability of the polymer to interact with the drug molecule it is prepared with. If a polymer chain were completely flexible, calculation of all conformations would average to a total end-to-end distance of zero. Therefore, polymer chain length is often described in solution theory in terms of the root mean square of end-to-end distance. If appropriate restrictions are placed on the bond angles between the backbone atoms, the root mean square end-to-end distance may be further increased. Consideration of steric restrictions and the potential energy of various confirmations may further increase the chain length. Taken together, these factors determine what is often referred to as the "stiffness" of the polymer and summarize "ideal chains" or polymer chains which are not affected by the environment.

However, a polymer may interact with its environment. Similar to regular solution theory, the interaction of polymer with itself can be considered relative to the interaction of the drug with itself and the interaction of the polymer with the drug. Mathematically, this can be described by the potential energy curve associated with placing a polymer in a solvent (or perhaps in this case a molten drug or a spray-dry solvent as the case may be). An excluded volume, V_{excluded}, is defined by the potential energy, U, of bringing two polymer units together, to a distance r, relative to a given thermal energy. Here, as before, k_B and T represent the Boltzmann constant and the temperature, respectively:

$$V_{\text{excluded}} = \int (1 - \exp(-U(r)/k_B T))\mathrm{d}^3 r \tag{11.8}$$

The value of the excluded volume can then be used to describe the "goodness" of the solvent.

If the excluded volume is on the order of the hard core volume of the so-called Kuhn monomer, b (the hypothetical monomer which consists of an equivalent freely jointed chain with the same length of a completely stretched chain and the same root mean square end-to-end distance as a real chain), the solvent is said to be athermal. That is, the interaction between the solvent and the monomer is equal in energy to the interaction between the monomer with another monomer. When the excluded volume is greater than zero but less than the hard core volume, the solvent is good and the attractive interaction between monomers is smaller than the interaction between the solvent and the monomers. When the excluded volume is equal to zero, the solvent is called a theta solvent and the attractive interaction between monomers exactly cancels hard core repulsive interaction. When the excluded volume is less than zero but greater than the negative of the hard core volume, the solvent is termed poor and the attractive interactions between monomers dominate. Finally, if the excluded volume is equal to the negative of the hard core volume, the solvent is termed a non-solvent and the solvent is completely excluded from the coiled polymer (Table 11.1) (Rubinstein and Colby 2003). Indeed, the implications of developing a molecular level understanding of solid dispersions are far reaching and phase behavior in disordered systems in the pharmaceutical arena will continue to borrow from more developed industries.

Table 11.1 Solvent type as a function of excluded volume, v, as defined relative to the Kuhn monomer, b

Value of excluded volume, v	Solvent "type"
$v \approx b^3$	Athermal
$0 < v < b^3$	Good
$v = 0$	Theta
$-b^3 < v < 0$	Poor
$v \approx -b^3$	Non-solvent

11.4 Melt Rheology and Viscous Limitations to Processing

Although the liquidous curve defines the lower operating temperature for which a material can form a single-phase multicomponent system, other properties will determine the kinetics of reaching the thermodynamic end state. Factors such as particle size, residence time, and mechanical input among others will facilitate the mixing process. Viscosity is a critical measurable and an intrinsic parameter that helps to inform the mixing process. Further, it is an important measurable that informs extrusion process modeling. Therefore, this property deserves discussion and is the focus of this section of the chapter. The viscosity of a material (or a system of materials) is a function of the shear rate, temperature, pressure, and composition. The shear rate is equal to the time derivative of the shear strain, γ, or $\mathrm{d}\gamma/\mathrm{d}t$. The corresponding shear stress, τ, is given by the force applied over an area. The shear viscosity is then the ratio of the shear stress and the shear rate as shown in Eq. (11.9):

$$\eta = \frac{\tau}{\mathrm{d}\gamma/\mathrm{d}t} \tag{11.9}$$

This value is critical since, as described above, the path length for diffusion will be a function of the degree of mixing the individual components and the degree of mixing in turn is a function of the viscosity (and differences in viscosity) of the molten component(s). Typically, polymer melts display a less than linear response of shear stress to the shear rate. Therefore, more rigorous screw designs will provide greater than linear impact on the creation of new interfaces and reduction in diffusional path lengths. Typically shear rates within an extruder may range from 1 to 10,000 s^{-1}. Of course, the impact of temperature and pressure should also be considered when interpreting the mixing of various materials in the extruder. A convenient description of the viscosity changes as a function of temperature is given by the Williams-Landel-Ferry equation (Eq. 11.10) (Williams et al. 1955). Here, a_T is known as the shift factor, C_1 and C_2 are constants specific to each material, and the reference temperature, T_S, is taken as some temperature above the glass transition temperature:

$$\log(a_{\mathrm{T}}) = \frac{-C_1(T - T_{\mathrm{S}})}{C_2 + (T - T_{\mathrm{S}})} \tag{11.10}$$

Practically, the utility of the shift factor is more easily recognized when considering how it is used to describe viscosity data. Specifically, the derivative of viscosity with respect to temperature is equal to the viscosity at the reference temperature, T_S, multiplied by the derivative of the shift factor with respect to temperature (Eq. 11.11):

$$\frac{\partial \eta(T)}{\partial T} = \eta(T_{\mathrm{S}}) \frac{\partial a_{\mathrm{T}}}{\partial T} \tag{11.11}$$

Clearly, the non-Arrhenius nature of the viscosity is more pronounced as the glass transition temperature is approached. Practically, this non-Arrhenius behavior near the glass transition temperature will translate into a strong temperature dependence in processability in this region. In viewing the phase diagram, it is noted that in some cases the viscosity of the polymer and the temperature dependence of the viscosity may significantly restrict the processing space. Consider the hypothetical region noted in Fig. 11.1 between the T_g-composition line and the parallel line that lies just above. This region, labeled "non-Arrhenius kinetics," may take on a range of temperatures as shown in the graph or it may extend well into the region noted by the liquidous line. The width of the non-Arrhenius window will be system dependent and will impact the required screw design and processing window. Consider (Fig. 11.3) the viscosity of two commonly used polymers in the production of amorphous solid dispersions, polyvinylpyrrolidone co-vinyl acetate (PVP-VA) and hydroxypropyl methylcellulose acetate succinate (HPMCAS-L). The viscosity as a function of shear rate may show significantly different values and significantly different temperature dependence. For instance, HPMCAS-L displays a very high shear viscosity at low shear rates and shows a strong shear rate dependence.

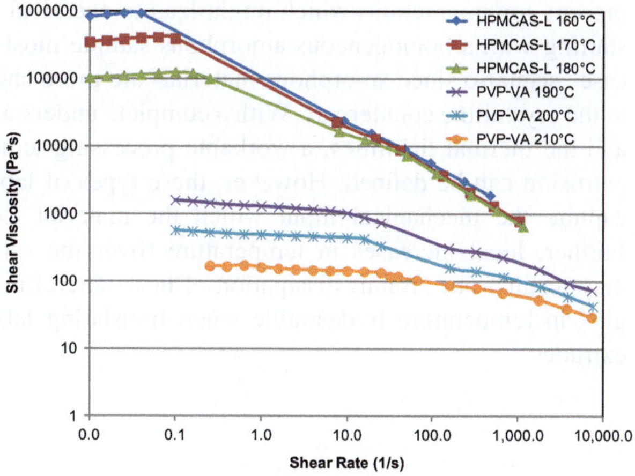

Fig. 11.3 Shear viscosity as a function of shear rate for two polymers commonly used in the development of amorphous solid dispersions. Data courtesy of Polymer Processing Institute (PPI), Newark, NJ

PVP-VA shows a much lower shear viscosity (albeit the temperature range studied is not identical for the two polymers) and a relatively weaker dependence of shear viscosity on shear rate. Clearly, the processing conditions and screw design may be impacted by the material selection. Further, the viscosity will be dynamic as the material mix—highlighting again the process-formulation-performance interplay.

11.5 Thermal Degradation

11.5.1 A Challenge Unique to Extrusion

Avoiding thermal degradation of the API, the polymer, and/or any other functional excipient is often the most difficult challenge in developing an extrusion process. Careful attention to detail is required in this space—especially concerning any degradants which have not been qualified prior to selection of the extrusion operation. As shown in Fig. 11.1, the operating space has a lower limit defined by the liquidous curve and an upper limit defined by the region shown in red. Clearly, the operating space will be a function of the breadth of this temperature range and, since the material passing through the extruder will experience a distribution of shear and thermal history, it is critical that this window accommodate the entire distribution of residence times and processing stresses. As a starting point, thermogravimetric analysis (TGA) may be used early in development to help determine upper bounds in processing temperature. Exposing samples to various residence times and temperatures may help to inform the extrusion process design while exposure to air or a nitrogenous environment may inform oxygen sensitivity at high temperature. Regarding the preparation and presentation of samples for TGA analysis, physical mixtures may not homogenize over the time scale of the experiment. Therefore, it may be useful to prepare small-scale solid dispersions prior to the TGA experiments so that the materials have the greatest opportunity to present any interactions which might lead to chemical degradation issues. Further, starting with a homogeneous amorphous sample most likely provides the "worst-case" scenario since amorphous materials are more chemically labile as compared to the crystalline counterpart. With a complete understanding of the liquidous curve and the thermal liabilities, a workable processing temperature range for hot-melt extrusion can be defined. However, these types of lab-scale experiments will not capture the mechanical input which the material experiences in the extruder. Further, local increases in temperature (over the setpoint product temperature) may result from viscous dissipation of heat. Therefore, a reasonable "safety margin" in temperature is desirable when translating laboratory experiments to the extruder.

11.5.2 Thermal Liabilities of Excipients

Thermal liabilities of the API are critical but the liabilities of other excipients must also be considered. For instance, hydroxypropyl methylcellulose acetate succinate (HPMCAS) has shown promise to enhance supersaturation in GI fluid and inhibit nucleation and crystal growth (Curatolo et al. 2009). However, high temperatures and shear used during the extrusion process may degrade the polymer and influence the release of the final drug product (Schenck et al. 2010; Sarode et al. 2014). For instance, Sarode et al. investigated changes in the physicochemical properties of three grades of HPMCAS as a function of temperature and screw speed (Sarode et al. 2014). A rheological evaluation of the three grades of polymers was performed at 160, 180, and 200 °C. Hot-melt extrusion was also performed at the same temperatures and at three rotational speeds: 100, 200, and 300 rpm. Pure polymer and milled extrudates were evaluated for color, thermal behavior, crystallinity, moisture, liquid viscosity, chemical composition, and solubility at different pH values. The rheological evaluation showed that as temperature increased, zero rate viscosity was reduced. After hot-melt extrusion, an increase in yellowness was observed with temperature, suggesting degradation of the polymer. However, the T_g, amorphous nature, infrared spectra, and moisture content of the pure polymer and milled extrudates were similar. A small reduction in liquid viscosity was observed for the AS-LF and AS-MF milled extrudates as the temperature and screw speed increased. No change in viscosity was observed for AS-HF. The largest change between the pure polymer and milled HME extrudates was observed with an analysis of the substituents of HPMCAS. With an increase in temperature and speed, an increase in free succinic acid was the most significant change observed. An increase in acetic acid and decrease in succinoyl values were also seen. For LF and MF grades, dissolution was not affected by HME processing conditions. However, dissolution of HF was significantly increased and was attributed to the release of free succinic acid. It was found that hot-melt extrusion did not significantly influence the physicochemical properties of milled HPMCAS extrudates. It is hypothesized that HME may not influence dissolution and performance of solid dispersions of the LF or MF grades, but may increase the dissolution time for the HF grade due to the release of succinic acid and reduction in succinoyl content compared to other substituents. Regardless, the thermal liabilities associated with the polymer are noted in this example—highlighting the importance of considering the impact of high-temperature processing not only on the API but also on the excipients used.

11.5.3 Manipulating the Phase Diagram and the Kinetics of Mixing to Avoid Thermal Degradation

Several strategies to minimize or avoid chemical degradation exist. Each aims to decrease the processing temperature and/or mechanical energy input. However, the strategy must not compromise the ability to achieve a single-phase amorphous solid dispersion. One approach is to shift the liquidous curve to lower temperatures. This approach focuses on changing the thermodynamic endpoint of the mixed phase through selection of materials which provide greater melting point depression of the API. Another approach is to shift the glass transition temperature to lower values with the ultimate aim to reduce the viscosity of specific components in the system so as to increase the kinetics of mixing. The development scientist may therefore minimize chemical degradation liabilities by (1) selecting materials which give strong interactions with the drug in the mixed phase, facilitating melting point depression, and/or (2) selecting materials or processing conditions which reduce the viscosity of the mixed phase. Approaches aimed at altering the thermodynamics and kinetics of the system are discussed in this section while processing solutions are discussed in the next section.

One example highlighting the impact of material selection and operating conditions on the resulting thermal degradation is the work of DiNunzio et al. (DiNunzio et al. 2010). Hydrocortisone was used as a model compound to study the impact of formulation and processing conditions on chemical degradation. Physical mixtures of hydrocortisone and hypromellose suggested a melting point depression of hydrocortisone and possible processing at 180 °C. Physical mixtures of hydrocortisone and PVP-VA suggested that the hydrocortisone may be dissolving in the molten polymer at 160 °C. Low-substituted hydroxypropylcellulose was unable to solubilize the active. While mixtures of hypromellose and hydrocortisone could not be processed at temperatures below 180 °C, PVP-VA blends could be extruded with a significantly lower degree of chemical degradation. The authors demonstrated that HME processing parameters, such as processing temperature and residence time, can strongly influence the decomposition of the API. While decomposition of the active was impacted by increased residence times and temperatures, carrier selection was also shown to be an important factor that may help lower chemical degradation of a heat-sensitive API. Thus, manipulating the phase diagram and the ability to access the liquidous curve through appropriate formulation design (i.e., polymer selection) may be a good strategy to avoid chemical degradation.

Several publications have demonstrated that the API itself may inherently plasticize the system by lowering the T_g of the polymer as the materials mix. This is captured, in part, in the cartoon shown in Fig. 11.1. However, in addition to lowering the glass transition temperature, the API will most often reduce the viscosity of the molten phase. Agrawal et al. used mDSC to determine the T_g of PVP-VA64 and amorphous compound X to be 108.5 °C and 77 °C, respectively (Agrawal et al. 2013). An amorphous solid dispersion prepared by hot-melt extrusion produced a single T_g at 95 °C. Compared to the neat polymer, the amorphous solid dispersion had a decreased viscosity and energy consumption during hot-melt extrusion suggesting that compound X is plasticizing the polymer. Crowley et al. also found that the active ingredient could act as a plasticizer (Crowley et al. 2004a). Using either guaifenesin or ketoprofen as the model compound with polyethylene oxide as the polymeric carrier, the APIs were found to act as plasticizers reducing drive load and melt viscosity of the molten polymer. Indeed, these examples highlight the impact of the API on the mixing kinetics. However, it is important to note that the liquidous curve will still ultimately define the lower limit in processing temperature required to achieve a homogeneous amorphous solid dispersion. It is the kinetics of reaching this equilibrium that are ultimately influenced by the viscosity of the molten materials which are being mixed.

Another approach to achieve a single-phase amorphous system is to deliberately add materials which reduce the viscosity of the molten phase and therefore expedite the mixing process (albeit the thermodynamics are also inherently impacted as well). As opposed to the case where the API inherently reduces the viscosity of the system, additives are included with the primary function of acting as plasticizers. The addition of small molecular weight plasticizers into a formulation may increase the free volume, lower glass transition temperature, and reduce the viscosity of the resulting mixed phase. This ultimately may provide an increase in the workable bounds for hot-melt extrusion. One example of this approach is demonstrated by Andrews et al. (Andrews et al. 2010). In this work, the authors used bicalutamide as the model drug and PVP K25 as a model polymer with various plasticizers to reduce processing temperatures. Triacetin, dibutyl sebacate, and triethyl citrate were tested to see how well the liquids plasticized the polymer and API. During early screening, a 13 % mass loss occurred with triacetin and the plasticizer was subsequently dropped from the experimental design. While both dibutyl sebacate and triethyl citrate were able to effectively lower the T_g, the large loss of triacetin highlights that it is important to understand not only the thermal degradation of the polymer and API, but also the volatility and/or thermal degradation of any liquid plasticizes used using extrusion.

While many commonly used plasticizers are liquids, solid-state plasticizers have also been explored to lower the glass transition temperature of polymers during hot-melt extrusion. Wu et al. investigated methylparaben as a solid-state plasticizer for Eudragit RS PO (Wu and McGinity 2003). With increasing concentrations of methylparaben, the T_g of the system and melt viscosity decreased. Moreover, methylparaben was found to be as effective as triethyl citrate at reducing torque during hot-melt extrusion. Andrews et al. used triethyl citrate in combination with citric acid to plasticize Eudragit L100-55 and 5-aminosalicylic acid as the model active (Andrews et al. 2008). Triethyl citrate was shown to effectively plasticize the polymer and drug. However, the addition of citric acid further reduced screw torque and die pressure during hot-melt extrusion.

Surfactants have also been explored to improve the performance during hot-melt extrusion (Ghebremeskel et al. 2007). The processability of the blend, solubilization of the solid dispersion, the microenvironmental effects on dissolution enhancement, and drug release properties were all examined. A poorly soluble active ingredient was hot-melt extruded with various polymers and surfactants. Polymers examined were Plasdone-S630, PVP, HPMC, HPMCAS, and Eudragit L100. Surfactants examined were SLS, docusate sodium, polysorbate 80, poloxamer 188, and polyoxyethylene 40 stearate. Thermal analysis revealed that the combination of API, polymer, and surfactant reduced the melting of the API and, most relevant for this discussion, the combined T_g of the system—as compared to the systems without the surfactant.

In addition to the API or additional excipients plasticizing the polymer, transient plasticizers may be added to reduce T_g. However these plasticizers are removed during or after extrusion and are not present in the final dosage form. Examples of transient plasticizers may include water, hydrates where water is evolved during the extrusion process, or supercritical fluids (Lakshman 2007). In particular, pressurized CO_2 was combined with hot-melt extrusion to act as a plasticizer and to form a foamed extrudate. Itraconazole and PVP-VA were extruded on a co-rotating twin screw Leistritz Micro 18 mm hot-melt extruder (Verreck et al. 2005). The physicochemical properties, dissolution, and morphology of the extrudates with and without pressurized CO_2 were characterized. Pressurized CO_2 was found to act as a plasticizer and was able to lower the processing temperature. However, the addition of pressurized CO_2 did not influence the formation of a solid dispersion. Interestingly, by adjusting processing temperature and pressure, the release rate of the itraconazole was controlled. With pressurized CO_2, the extrudate took on a more foamlike structure due to the expansion of the CO_2 after extrusion. Extrudates with pressurized CO_2 had an increased specific surface area, porosity, hygroscopicity, and milling efficiency. Therefore, transient plasticizers may serve to not only reduce processing temperature but also impart superior processability and performance.

Using Eudragit EPO or PEO and Carvedilol as the model active, a 16 mm co-rotating PrismTM was used to manufacture extrudates with and without supercritical CO_2 (Lyons et al. 2007). As a result of limited pressure buildup during hot-melt extrusion, the screw configuration required optimization before the authors were able to incorporate the CO_2 gas with the polymer and API. Once incorporated, the addition of CO_2 resulted in a reduced viscosity and faster dissolution. However, the addition of CO_2 did lead toward an increased tendency for PEO to crystallize. Pressurized carbon dioxide was also used to hot-melt extrude a thermally liable compound. *p*-Aminosalicylic acid, which decomposes upon melting, was hot-melt extruded with ethylcellulose and pressurized CO_2 to temporarily plasticize the blend (Verreck et al. 2006). CO_2 was able to reduce the processing temperature of the extrusion and degradation of the *p*-aminosalicylic acid was reduced from 17 to less than 5 % with the addition of pressurized CO_2.

11.5.4 Processing Solutions to Avoid Thermal Degradation

High processing temperatures, shear stresses, and melt viscosities may lead to unwanted heat causing the degradation of the API or polymer. While some formulation approaches may be able to lower the melting point, viscosity, and glass transition temperature of the molten dispersion (as discussed above), equipment solutions may also help to prevent degradation. Changing the configuration of the feeding system, water cooling the feed section of the extruder, changing screw design, taking advantage of the shear rate dependence of viscosity, and holding the specific mechanical energy constant all may help control thermal degradation and increase the processing window for hot-melt extrusion.

Changing the feeding location of a thermally liable drug can prevent long residence times in the extruder and reduce degradation. Using a "side stuffer," thermally sensitive materials can be fed downstream and added to the molten polymer, thereby reducing residence time. Bruce and Manning used a side stuffer to feed nicotine bitartrate into an extrusion process (Bruce and Manning 2013). By feeding nicotine bitartrate downstream, the authors were able to avoid water loss and decomposition of the API. However, the impact of this approach on the time allocated for mixing must be considered. That is, feeding the material downstream may cause incomplete mixing of the polymer and API and lead toward inhomogeneity of the polymer and active (Schenck et al. 2010). Examples of this failure mode can be found in the literature and should be kept in mind during process development. Further, as was discussed in Chap. 10, and highlighted above, in some cases it is insufficient to use the glass transition temperature as the sole indicator of homogeneity post-extrusion. The glass transition temperature may not show a strong compositional dependence in which case it may be insensitive to phase separation. Alternatively, the length scale of homogeneity may be such that phase separation is difficult to detect. For instance, in one example, a system showing a single glass transition temperature by DSC displayed a pair distribution function which suggested inhomogeneity (Newman et al. 2008). In another example, compositionally identical extrudates were prepared at two different processing temperatures. Although both gave a single glass transition temperature, the physical stability of the material produced at the lower processing temperature was inferior. Confocal Raman spectroscopy measurements suggested that, despite the single T_g, the two materials in fact showed different degrees of mixing (Qian et al. 2010).

Another engineering solution may include the addition of water cooling to the feed zone. Cooling in the feed zone, adjacent to the heating zone, may reduce thermal degradation of the API by preventing holdup in the feed zone and/or preventing the formation of a plug of material inside feeder. Therefore, this approach may both facilitate feeding and reduce risk of thermal degradation. Forster et al. used four heating zones to hot-melt extrude various APIs and polymers (Forster et al. 2001). In their experimental setup, the first zone was cooled to 20 °C with water to prevent plasticized material from obstructing flow down the barrel. Subsequent heating zones were kept at a constant temperature.

Perhaps a more traditional approach is to explore thermal degradation via a design of experiments. For instance, Ghosh et al. used a statistical design of experiments controlling screw design and assembly, screw RPM, feed rate, and overall process temperature to optimize the degradation of a thermally liable active ingredient (Ghosh et al. 2012). The authors were able to limit chemical degradation without the addition of a plasticizer to lower T_g. From the design of experiments, position of the conveying and mixing elements significantly impacted chemical stability. An optimal position of the screw elements was even more significant since chemical degradation increased with higher processing temperatures and longer residence times. By properly designing the feed screw design and optimizing processing parameters, the authors were able to consistently control the chemical degradation of the API.

Finally, it should be noted that as the product is scaled and optimized, changes to equipment design or processing parameters may result in an unforeseen product temperature increase (global or local). Monitoring the specific mechanical energy (SME) input may help to inform changes in the stress imparted onto the material as a result of changes in equipment and processing parameters. SME is defined as the ratio of the mechanical energy required to process (melt, mix, convey) the extrudate to the feed rate. This parameter is considered crucial to scale-up, quality control, and troubleshooting (Liu et al. 2010). Liu et al. have shown that specific mechanical energy varies with screw configuration, location of kneading block (whether upstream or downstream), and screw speed. The authors also demonstrated that an increased specific mechanical energy conveys more internal energy and increased the melt temperature. If specific mechanical energy is not monitored, different equipment train may impart higher degrees of energy potentially increasing chemical degradation of the API or polymer.

11.6 Glass-Forming Tendency and Quenching

When a material is cooled in the extruder, it necessarily must pass through temperatures well below the liquidous curve as shown by the arrow in Fig. 11.1. Once a single-phase multicomponent amorphous solid dispersion is produced above the liquidous line, the successful isolation of the material will depend on the relative magnitude of the kinetics of crystallization or the kinetics of amorphous–amorphous phase separation vs. the kinetics of quenching. For instance, if an extrudate at the composition noted in the cartoon is held at a temperature equivalent to the indicated point (b), the material will crystallize until the composition of the residual amorphous phase reaches the liquidous line (c) and the relative amounts of each phase will be dictated by the classical tie lines (Gaskell 2003). Specifically, the residual amorphous phase will present in an amount equal to (a–b)/(a–c) whereas the crystalline phase will present in an amount equal to (b–c)/(a–c).

The relative kinetics of cooling vs. crystallization will determine whether the amorphous form is isolated. Several approaches aimed at understanding glass-forming ability have been discussed previously and may be useful to briefly discuss here. The T_g/T_m ~2/3 rule has been used to give a rough estimate of glass-forming ability but some have argued that this rule often only holds well because substances with high values of T_g/T_m do not crystallize and so the melting point, T_m, cannot be measured while substances with low values of T_g/T_m do not form a glass very easily and so the glass transition temperature, T_g, cannot be measured (Alba-Simionesco et al. 1999).

The $T_b/T_m \geq 2$ guideline has also been discussed as a tool to understand glass-forming tendency. This guideline can be rationalized by considering the intermolecular forces involved in glass formation. That is, a material with a high boiling point, T_b, contains strong intermolecular interactions in the liquid state and a material with a high melting point must contain strong intermolecular interactions in the crystalline state. Therefore, materials with high boiling points and low melting points will perhaps have a low driving force for crystallization. Similarly, materials with high melting points and low boiling points will show a high tendency to crystallize. Alba-Simionesco studied *meta*-, *ortho*-, and *para*-isomers of fluorotoluene and found that only *m*-fluorotoluene displayed a T_b/T_m ratio greater than 2 and only *m*-fluorotoluene formed a glass at reasonable cooling rates (Alba-Simionesco et al. 1999). Nevertheless, the guideline failed when applied to some other glass-forming materials exhibiting exceptionally low values of T_b/T_m and when applied to *m*-fluorophenol and *m*-cresol—perhaps due to the presence of hydrogen bonding. Of course, pharmaceutical materials most often degrade prior to boiling. However, it is interesting to consider the link between this rule and the viscosity of a material at the melting temperature (Angell 2002). Low-melting materials, having weak intermolecular interactions in the crystalline state, will generally exhibit a higher viscosity at the melting point. Angell uses this line of reasoning to conclude that it is the crystalline state that determines whether a material will display a high propensity for glass formation. Disubstituted benzene isomers were studied to demonstrate this concept. *Meta*-, *ortho*-, and para-xylene all show a similar boiling point suggesting that the interactions in the liquid state are similar. However, the melting points of the three materials are significantly different with *meta*-xylene having the weakest intermolecular interactions in the crystalline phase and therefore the lowest melting temperature. Indeed, *meta*-xylene displays the highest viscosity at its melting point and forms a glass most readily (Angell 2002).

In terms of the phase diagram, a thermodynamic driving force for crystallization only exists at temperatures below the liquidous curve. As the degree of undercooling is increased, the thermodynamic driving force for crystallization increases. However, the kinetics of crystallization are inhibited as the mobility of the material becomes increasingly restricted with decreasing temperature. A compound with a high tendency to crystallize may require a faster quench rate as compared to a compound with a lower tendency to crystallize. This concept is captured well by the traditional time-temperature-transformation profiles as shown in Fig. 11.4.

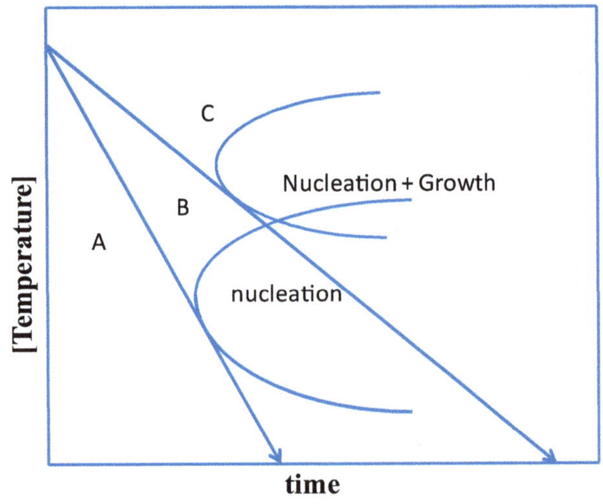

Fig. 11.4 The cooling rate may avoid nucleation and growth altogether (region A), nucleation may occur but bulk crystallization may be avoided (region B), or bulk crystallization may occur (region C)

In this figure, the depressed melting point is displayed on the temperature axis with time = 0. The rate of cooling is given by the slope of the lines shown. Based on this representation of the cooling rate from the liquidous curve through the glass transition temperature, three scenarios may be envisioned. In the first, represented by region A, the quench rate in the extruder is greater than the rate of nucleation or growth and a crystallization is avoided altogether. In the second, represented by region B, the quench rate is such that nucleation occurs but significant crystallization is not observed over the time scale of the quench out of the die of the extruder. In the third, nucleation and growth occur quickly and, although a completely homogeneous phase was produced within the extruder barrel, recrystallization occurs as the material is cooled. Baird et al. used this type of construct to rationalize the various calorimetric behaviors observed for pure APIs as measured by DSC (Baird et al. 2010). In the first scenario, the material may be heated and cooled and no crystallization is observed over the time scale of the DSC experiment (region A). In the second scenario (region B), the crystalline material is heated through the melting point, and cooled, and an exotherm associated with recrystallization is observed on subsequent heating. In the last scenario, the API is most labile to crystallization as reflected by an exotherm associated with recrystallization during the cooling process (region C).

Based on these arguments, the trajectory of the viscosity with cooling will play a key role in successfully isolating a glassy material. It is particularly interesting to note that the addition of a second component will influence the viscosity of the molten phase. That is, the addition of the polymer will facilitate glass formation not only by depressing the melting point and therefore reducing the thermodynamic driving force for crystallization (as described above), but also by increasing the viscosity of the molten phase at any given temperature. This important concept was perhaps first noted by Angell in a publication which examined the glass-forming tendency of a mixture of materials that formed eutectics. The eutectic temperature, representing the lowest temperature at which the binary mixture melts, will have a higher viscosity than any other molten material of the same binary. So long as the second component does not form a more stable crystalline phase with the first component, a larger depression in the melting temperature will lead to a greater increase in the viscosity at the depressed melting point (Angell 1995). Of course, most often in extrusion amorphous polymers are used and so there is no concern over the formation of a more stable crystalline phase. Based on the reasoning presented above, it is clear that polymers which display strong interactions with a particular API may operate in multiple ways to increase glass-forming tendency (of course, here it is assumed that the viscosity of the polymer $\gg$ the viscosity of the API).

The application of these concepts to pharmaceutical systems has been described for pure APIs. Baird et al. evaluated the viscosity of a large set of compounds which displayed various crystallization tendencies associated with the three regions of Fig. 11.4 (Baird et al. 2012). Specifically, steady-state shear viscosity measurements were made near and well below the melting temperature and it was found that materials which crystallize more readily also displayed lower viscosity near the melting temperature (consistent with the arguments above). Generally speaking, when melt viscosities were > 0.01 Pa-s (at a shear rate of 30 s^{-1}), glasses could be successfully isolated (Baird et al. 2012) and this observation was consistent with studies on small organics as outlined by Angell (Angell 2008). Further, the authors note that the temperature dependence of the viscosity also correlated with crystallization tendency. Compounds which showed a greater increase in viscosity with decreasing temperature tended to be more easily isolated as amorphous. It stands to reason that the addition of polymers may operate to increase the glass-forming tendency of APIs which display a strong propensity to crystallize. Further, it is perhaps interesting to note that although highly viscous materials may make mixing difficult during the extrusion process, highly viscous materials may also tend to impart a greater opportunity to form a glass once mixing is achieved.

11.7 Heat, Mass, and Momentum Balances: Modeling the Extrusion Process

With a handle on the phase diagram and an understanding of the extrusion process, the rheological and thermal properties can be used to link the processing, the product (i.e., material properties), and the performance. Modeling the extrusion process, as with any unit operation, starts with a description of the mass, energy, and momentum balance. The first equation, Eq. (11.12), gives the mass balance and can be physically interpreted as the conservation of mass in a finite unit volume (Fig. 11.5):

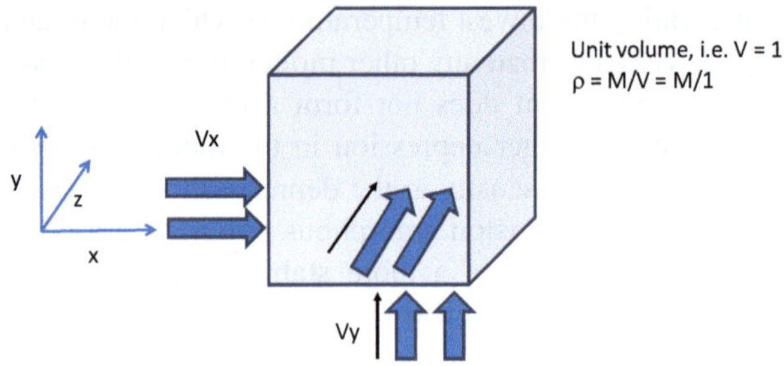

Fig. 11.5 Conservation of mass in a finite unit volume

$$\frac{\partial \rho}{\delta t} + \frac{\partial}{\partial x}(\rho v_x) + \frac{\partial}{\partial y}(\rho v_y) + \frac{\partial}{\partial z}(\rho v_z) = 0 \tag{11.12}$$

Here, the mass balance is written in terms of the density, ρ, and the velocity, v, describes the mass flow rates in each Cartesian direction (x, y, z). In applying this equation to the extruder, it may be most appropriate to consider the system at steady state. That is, the total mass in the extruder is constant with time. Most often no chemical reactions are occurring during the extrusion process and volume changes on mixing the multicomponent system are often ignored. Although shown here in Cartesian coordinates (for simplicity of explanation), the most natural coordinate system may be cylindrical so that the boundary conditions can be captured easily. For instance, if a drag flow boundary condition on the screw and a no-slip condition at the barrel wall is selected, these conditions can be most easily described as a function of the radius, r, from the screw rotational axis with r equal to the outer diameter of the screw and r equal to the barrel inner diameter defining the location of the boundary conditions relatively easily. Although the conservation of mass holds over the entire extruder at steady state, it is important to consider the mass balance over each volume element within the extruder. In the simplest description, the free volume in the extruder can be divided into any number of small-volume elements over which the mass balance can be applied. However, unless a 100 % fill volume is achieved in the extruder, individual volume elements may experience a net accumulation or depletion in mass over any discrete interval in time so long as the sum total of all elements show no accumulation or depletion. This complex process will, in turn, impact the energy and momentum balances and immediately some of the complexities associated with modeling the extrusion process become more obvious.

The next set of equations which must be considered represent the momentum balance (Eqs. 11.13–11.15) or more specifically the Cauchy momentum balance. These equations physically describe the velocity vectors and acceleration resulting from various forces which operate on the system (Fig. 11.6). The first terms in the equation capture the inertial effects and the velocity at the boundaries of any volume element. Acceleration may occur as the result of the action of any external forces—as described on the right-hand side of the equations. The first term on the right-hand side of the equation captures changes in pressure, p, and the last term

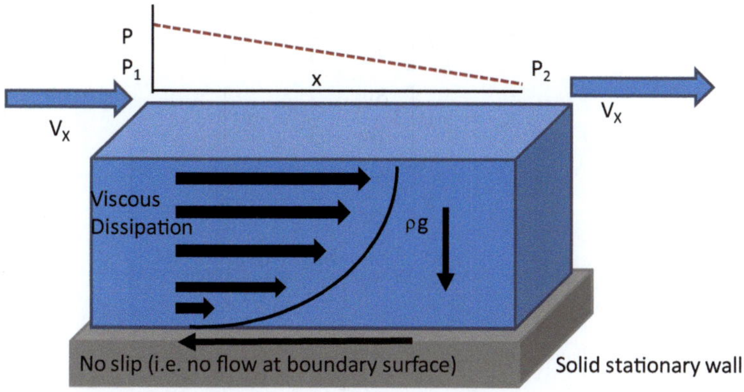

Fig. 11.6 Conservation of momentum in a finite unit volume

captures the gravitational force, g. The pressure term may be important when considering the significant pressure drop across the die or the backpressure which might build up upon formation of a melt plug. Typically, the pressure dependence of viscosity is not significant at typical operating conditions used in pharmaceutical systems. The three remaining terms capture the shear stress tensor, τ, acting on each volume element in each direction. As was described above, the viscosity profile for materials commonly used in the pharmaceutical industry will show a strong functional dependence on the shear rate and temperature. Therefore, these relationships must be built into the model. That is, the viscosity can be measured as a function of temperature and shear rate and modeled accordingly as the material travels through the barrel and experiences the associated shear stress imparted by the screw elements and the temperature profile during processing:

$$\rho\left(\frac{\delta v_x}{dt}+v_x\frac{\delta v_x}{dx}+v_y\frac{\delta v_x}{dy}+v_z\frac{\delta v_x}{dz}\right)=-\frac{\delta p}{dx}-\left(\frac{\delta\tau_{xx}}{\delta x}+\frac{\delta\tau_{yx}}{\delta y}+\frac{\delta\tau_{zx}}{\delta z}\right)+\rho g_x \tag{11.13}$$

$$\rho\left(\frac{\delta v_y}{dt}+v_x\frac{\delta v_y}{dx}+v_y\frac{\delta v_y}{dy}+v_z\frac{\delta v_y}{dz}\right)=-\frac{\delta p}{dy}-\left(\frac{\delta\tau_{xy}}{\delta x}+\frac{\delta\tau_{yy}}{\delta y}+\frac{\delta\tau_{zy}}{\delta z}\right)+\rho g_y \tag{11.14}$$

$$\rho\left(\frac{\delta v_z}{dt}+v_x\frac{\delta v_z}{dx}+v_y\frac{\delta v_z}{dy}+v_z\frac{\delta v_z}{dz}\right)=-\frac{\delta p}{dz}-\left(\frac{\delta\tau_{xz}}{\delta x}+\frac{\delta\tau_{yz}}{\delta y}+\frac{\delta\tau_{zz}}{\delta z}\right)+\rho g_z \tag{11.15}$$

In the case of a simple Newtonian fluid, each of the shear components can be described in terms of the viscosity, μ, and the shear rate (spatial derivatives of velocity) as shown in Eqs. (11.16)–(11.22):

$$\tau_{xx} = -\mu\left[2\frac{\delta v_x}{\delta x} - \frac{2}{3}(\nabla \cdot v)\right] \tag{11.16}$$

$$\tau_{yy} = -\mu\left[2\frac{\delta v_y}{\delta y} - \frac{2}{3}(\nabla \cdot v)\right] \tag{11.17}$$

$$\tau_{zz} = -\mu\left[2\frac{\delta v_z}{\delta z} - \frac{2}{3}(\nabla \cdot v)\right] \tag{11.18}$$

$$\tau_{xy} = \tau_{yx} = -\mu\left[\frac{\delta v_x}{\delta y} + \frac{\delta v_y}{\delta x}\right] \tag{11.19}$$

$$\tau_{yz} = \tau_{zy} = -\mu\left[\frac{\delta v_y}{\delta z} + \frac{\delta v_z}{\delta y}\right] \tag{11.20}$$

$$\tau_{zx} = \tau_{xz} = -\mu\left[\frac{\delta v_z}{\delta x} + \frac{\delta v_x}{\delta z}\right] \tag{11.21}$$

$$(\nabla \cdot v) = \frac{\delta v_x}{\delta x} + \frac{\delta v_y}{\delta y} + \frac{\delta v_z}{\delta z} \tag{11.22}$$

The energy balance is described as shown in Eq. (11.23) (Fig. 11.7). Here, the balance is given again in Cartesian coordinates so as to provide easily interpreted physical descriptions. This equation shows that the change in temperature with time and position (left-hand side of equation) can be described in terms of several external forces acting upon the system. The first group of terms represents the heat flux into the system, the second term shows the change in temperature which accompanies a change in pressure of the flowing media, and the remaining terms capture the frictional heating or viscous dissipation through heating:

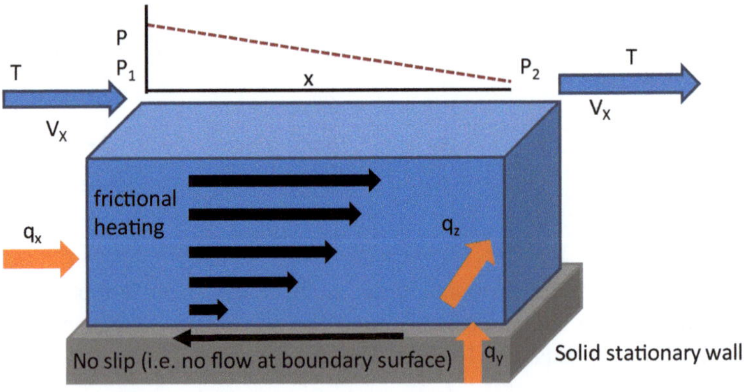

Fig. 11.7 Energy balance in a finite unit volume

$$\rho c_v\left(\frac{\delta T}{dt}+v_x\frac{\delta T}{dx}+v_y\frac{\delta T}{dy}+v_z\frac{\delta T}{dz}\right)=-\left[\frac{\delta q_x}{\delta x}+\frac{\delta q_y}{\delta y}+\frac{\delta q_z}{\delta z}\right]$$
$$-T\left(\frac{\delta p}{\delta T}\right)_\rho\left(\frac{\delta v_x}{dx}+\frac{\delta v_y}{dy}+\frac{\delta v_z}{dz}\right)$$
$$-\left\{\tau_{xx}\frac{\delta v_x}{dx}+\tau_{yy}\frac{\delta v_y}{dy}+\tau_{zz}\frac{\delta v_z}{dz}\right\}$$
$$-\left\{\tau_{xy}\left(\frac{\delta v_x}{\delta y}+\frac{\delta v_y}{\delta x}\right)+\tau_{xz}\left(\frac{\delta v_x}{\delta z}+\frac{\delta v_z}{\delta x}\right)\right.$$
$$\left.+\tau_{yz}\left(\frac{\delta v_y}{\delta z}+\frac{\delta v_z}{\delta y}\right)\right\} \tag{11.23}$$

The conductive heat flux, q, may be described in terms of the heat transfer coefficient, k, and the temperature gradient, $\delta T/\delta x$, as shown in Eqs. (11.24)–(11.26). Similar equations can be written for other directions in space (Eqs. Y and Z):

$$q_x=-k\frac{\delta T}{\delta x} \tag{11.24}$$

$$q_y=-k\frac{\delta T}{\delta y} \tag{11.25}$$

$$q_z=-k\frac{\delta T}{\delta z} \tag{11.26}$$

Together with the definition of appropriate boundary conditions, the equations above can be used to model the processes which occur within the extruder. However, without additional simplification, these equations can become computationally expensive.

In an extrusion process, there are several key factors one must consider in developing these equations and corresponding boundary conditions. Most extruded material has complex rheological behavior. While using a simple single viscosity will help simplify some of the momentum equations, it is highly likely that most materials will require fitting their temperature and shear dependencies using more sophisticated rheology models like the Carreau-Yasuda equation with an Arrhenius temperature effect. Moreover, one needs to account for the fact that the material being fed into the extruder is actually in a solid powder phase, and will be melted to enable the mixing and densification of the material.

As for the equipment itself, there is a rotating volume (i.e., the screw elements) which has a region of overlap for the two rotating elements (Fig. 11.8). It is highly likely that the viscous dissipation due to the rotating screws will generate frictional heat, and depending on what the local extruder elements are you will have different fills around the screw. The barrels have heating elements and liquid-cooling channels which will have their own thermal patterns within the walls of the barrels, and there will be a need for information on the thermal conductivity within the barrels, the screw, the extrusion media, and across each of these mediums.

Fig. 11.8 Schematic showing the presence of a rotating volume (see text)

Most commercially available modeling software incorporates the mass, energy, and momentum balances with some simplifications aiming to increase speed of the simulations. For instance, the Ludovic® software includes a one-dimensional space modeling approach where the radial and azimuthal effects in the extruder barrels are averaged and the primary spatial dimension being tracked is the axial component along the length of the extruder. This software is dedicated to application of co-rotating twin screw extruders. WinTXS is also a one-dimensional modeling approach which is also used for co-rotating twin screw extruders. Alternatively, there are three-dimensional modeling software such as Polyflow, Fluent, or Ximex where the full spatial interactions and effects are tracked during the simulation. These software are designed for evaluation of materials and their associated viscosity–temperature–shear relationships. Polyflow has an extrusion-specific component to the software, and uses a finite element solver where it can track overlapping mesh elements of the screw and extrusion media to determine where material can flow and where solid boundaries exist blocking flow but imparting momentum upon the system. Fluent software is a three-dimensional modeling tool of all states of matter (gas, liquid, solids) but does so using a finite volume approach. This requires subdividing the whole simulation geometry into finite volume elements which track the flow of mass, energy, and momentum into each volume element. Since this system has two rotating bodies with an overlapping region, simulations require remeshing as all motion progresses. Finally, XimeX is a three-dimensional software developed to track mixing and motion in several systems (extruders being one of the key systems). This software has particle-tracking capabilities which try to quantify the degrees of mixing inside the extruder barrels.

11.8 Manufacturing and Scale-Up of Single-Phase Multicomponent Amorphous Materials

The scale-up of a hot-melt extrusion process has been reviewed by many authors (Lowinger 2011; Schenck et al. 2010; Steiner 2003; Dreiblatt 2003, 2012). This is not an all-encompassing review, but a guide for the reader highlighting many aspects of the scale-up process. Prior to scale-up, a fundamental understanding of the thermal liabilities and melt viscosity of the API, polymeric carrier, additional excipient(s), and blends thereof should be acquired. Included in the design is an accounting for how the various inputs to the hot-melt extrusion process impact the critical quality attributes of the formulation. As discussed previously within this text, key quality attributes for consideration include (1) that the API and any other functional components such as the polymeric stabilizer and any surfactants do not degrade, (2) production of an amorphous molecular dispersion, (3) acceptable moisture content of the extrudate to minimize plasticization and reduce physically stability, and (4) stable continuous operation without drift or spikes in the power consumption of the extruder that may be indicative of a non-steady-state process. To accomplish this goal, there are several levers that are available for the design of a successful HME process. First, one can consider the configuration of the extruder itself including the barrel configuration (number of vents, number of feed streams, type of feed zones, feed location, length-to-diameter ratio—*L*/*D*), die configuration (shape, dimension, number of openings), and screw configuration (type, length, and number of conveying and mixing sections). The modular and custom design of the extruder in this regard allows for an infinite number of configurations that can be explored. In addition, the process engineer also has available operating levers including feed rate, screw speed, barrel temperature profile, and cooling temperature which can add significantly to the complexity of the potential design space. It is important to note that all of the inputs discussed to this point are scale and equipment dependent.

In order to garner a more robust fundamental understanding of the process design space one can cast the scale-dependent parameters discussed above in terms of scale-independent parameters such as specific mechanical energy, residence time, number of mixing cycles, maximum shear rate and stress, product temperature, cooling rate, specific volume, degree of fill, and quench rate (Schenck et al. 2010). These scale-independent parameters are defined further in Table 11.2. With a thorough understanding gained through first principles and thoughtfully planned experimentation, the process engineer can gain an appreciation for which critical quality attributes are affected by the scale-independent parameters and their interaction with scale-dependent parameters. As such, one can conduct a design of experiments (DOE) in the scale-independent space and develop a design space where the operation provides a formulation of acceptable quality and a stable and robust process. This information can then be utilized to strategically move between scales. For example, if product performance is shown through experimentation to link to product temperature then as one scales the process to larger equipment the focus can be on maintaining product temperature across scales vs. scale-dependent parameters such as screw rpm and barrel temperature profile. This approach should also lend itself to greater regulatory flexibility in a quality by design paradigm.

The utilization of twin screw extruders to produce amorphous dispersions also lends itself to the utilization of process analytical technologies to monitor key process and product attributes in real time. For example, instrumentation at the die can be utilized to accept or reject material in real time allowing for continuous

Table 11.2 Scale-independent parameters and associated definitions

Scale-independent parameter	Definition
Specific mechanical energy, SME	Amount of power being input by motor into each unit mass of material
	$SME = (P \times \tau \times N_{act}/N_{rated})/Q$
	where P = maximum motor power, τ = motor load or torque (% of Max), N_{act} = actual operating screw speed, N_{rated} = extruder rated screw speed, Q = mass flow rate
Free volume, FV	Approximate volume for 1 L/D of the process section
	$FV = 0.94 \times (D_o^2 - D_i^2) \times D_o$
	where D_o = outer screw diameter, D_i = inner screw diameter
Degree of fill, %Fill	The proportion of the free volume of the extruder that is filled per L/D of the extruder
	$\%Fill = (Q \times t_{mean})/(FV \times \rho) \times 100$
	where Q = mass flow rate, T_{mean} = average residence time, FV = free volume of extruder, ρ = density
Average residence time, t_{mean}	The average amount of time material resides within the process section. In practice residence time is a distribution, $E(t)$
	$t_{mean} = FV \times \rho \times L/D \times \%Fill/Q$
	where FV = free volume per L/D of the extruder, ρ = density, L/D = length to diameter ratio of the extruder, % Fill = proportion of extruder-free volume that is filled, Q = mass flow rate
Mixing cycles, # cycles	The mean number of screw rotations experienced by the material as it passes through the extruder
	$\# \text{ of cycles} = N \times t_{mean}$
	where N = screw speed, t_{mean} = average residence time
Maximum shear rate	Peak shear rate = Pi $\times D \times N/(h)$
	where D = screw diameter, N = screw speed, h = overflight gap $(D_o - D_i)$
Maximum shear stress	Shear stress = peak shear rate × viscosity
Product temperature	Temperature of extrudate as it leaves the die of the extruder
Cooling rate	Rate at which heat is removed from the extrudate to quench from a molten state to a supercooled glass

manufacture of large batches over periods of weeks without jeopardizing quality due to process disturbances. In the simplest form, temperature probes at the die can be utilized to monitor product temperature. Near-infrared (NIR) probes have been utilized to continuously monitor extrudate composition for products containing two to three components (active, polymer, and functional excipient). In addition, this can enable measurement of residence time distributions via the addition of tracers to the feed and their subsequent detection at the die (Gao et al. 1999). Finally, one could envision implementing a Raman probe at the die to continuously monitor the phase state of the extrudate product. The readout from any of these probe configurations could be utilized to control waste gates that would divert material that was outside a known product temperature design space (Schenck et al. 2010).

A number of different approaches can be taken to scale a hot-melt extrusion process, each with its own merits. These include increased batch size, optimization of throughput, and traditional scaling to larger equipment (Dreiblatt 2012). By far, the simplest approach is to scale the process by increasing the batch size. This approach simply uses the same piece of equipment, or multiple clones of the same equipment, operating under the same conditions as the original process. Since twin screw extruders are designed to operate continuously, the process is only limited by operating time (if running an increased batch size on a single extruder), cost of additional capital (if running multiple extruders in parallel), and the ability of downstream operations to handle increased batch sizes. While batch records and regulatory updates may be necessary, scaling by an increase in batch size does not require additional experimentation or material loss resulting in minimal resource and capital investment. As such, this method is ideal for support of early clinical studies where the probability of success of the program is relatively unknown.

If an increase in batch size alone does not yield a time-efficient or a cost-effective option, one may consider redesigning the process to optimize throughput on the same piece of equipment. When increasing the feed rate, several scale-independent parameters, in particular residence time and specific mechanical energy, will be altered from the original process and may potentially have an impact on the critical quality attributes of the final product. An increase in feed rate leads to shorter residence times, thus increasing the likelihood that the extrudate product may contain residual crystalline material or insufficient time for complete homogenization of the components. Often to accommodate the higher feed rate, an increase in screw speed is necessary to avoid flood feeding of the extruder, resulting in higher shear stresses. This may lead to increased product temperatures and an increased risk of degradation of any thermally labile components. Thus, this approach to scale-up will require additional experimentation to ensure that product performance has not been adversely altered by the process changes. One strategy for optimizing throughput in this manner is to ensure that the new process maintains the same specific mechanical energy and product temperature. Specific mechanical energy is the primary factor that influences melt temperature at larger scales (Lowinger 2011). In addition, this approach requires that the die opening be scaled as a function of feed rate. Otherwise, if the die opening is kept constant, higher feed rates will produce an increase in pressure at the die, ultimately resulting in a higher product temperature and an increased risk to thermal degradation (Dreiblatt 2012).

The most common approach to scaling a hot-melt extrusion process is to take advantage of extruders of increased screw diameter. Converse to the previously discussed approaches where capital expense is avoided at the expense of time or additional experimentation, this approach requires capital investment and confidence that the product has significant probability of success of reaching the marketplace. Scaling by increasing the extruder diameter may affect the percent fill, energy delivered, and heat transfer capabilities. If equipment possesses the same ratio of L/D and D_o/D_i, a similarity in the ratio of mass throughput to free volume can be utilized to help maintain residence time distributions (Dreiblatt

2012). The *L*/*D* ratio can also be utilized to maintain the relative location of different screw elements (conveying, kneading, etc.), as the positions of different elements will dictate the residence time distribution of material within the barrel. Scaling to larger extruders will also influence the heat transfer from the barrel to the product. As one increases the diameter of the extruder, heat transfer may become limited as there is a more rapid increase in volume (proportional to D_o^3) compared to surface area (proportional to D_o^2). Based on the inner barrel surface area, heat transfer can be scaled using the ratio of surface areas (Eq. 11.27):

$$Q_{\mathrm{T}} = Q_{\mathrm{M}} \frac{D_{\mathrm{T}}^2}{D_{\mathrm{M}}^2} \tag{11.27}$$

where Q is the throughput, D is the screw diameter, and T and M represent the larger and smaller extruder, respectively (Andersen 1998). Fill volume will influence heat transfer as a larger quantity of material will have greater contact with the surface area of the barrel. Lowering the screw speed may negatively impact heat transfer as material on the surface of the barrel is not wiped away as efficiently. In addition to heat transfer considerations, scaling volumetrically by fill volume allows one to maintain equivalent residence time distribution between scales. Volumetric scale-up of throughput in geometrically similar extruders simplifies to the ratio of the cube of the diameters (Andersen 1998) (Eq. 11.28):

$$Q_{\mathrm{T}} = Q_{\mathrm{M}} \frac{D_{\mathrm{T}}^3}{D_{\mathrm{M}}^3} \tag{11.28}$$

This is the most common and generally accepted approach to scaling throughput on larger machines.

Fukuda et al. have suggested that if shear stress via dispersive mixing is deemed critical to maintaining product quality, a percent drag flow scaling methodology is most appropriate for moving between equipment sizes (Fukuda et al. 2014; Gao et al. 1999). Along a screw profile in an extruder every element has a maximum amount of volume it can convey per revolution. Since twin screw extruders are operated in a starve-fed state and are a partially filled system, only a fraction of the pumping capacity can be utilized which is known at percent drag flow (Eq. 11.29):

$$\text{Percent drag flow} = \frac{\frac{Q}{N_{\text{operating}}}}{\frac{Q}{N_{100\,\%}}} \tag{11.29}$$

Here, Q is the throughput, N is the screw speed, and operating and 100 % refer to the actual operating conditions vs. operation at a fully filled state. The ratio of Q/N is also referred to as the specific throughput. By maintaining percent drag flow as you move between scales, the material can be expected to experience the same residence stress distribution.

Regardless of the scaling approach utilized, it is important to rely on an established fundamental understanding of the process-formulation-performance interplay, particularly how the critical quality attributes of the formulation are influenced by the scale-independent properties. As a final illustration, a scaling case study utilizing an active ingredient with a melting point of approximately 180 °C and a known thermal degradation pathway will be presented to show how a hot-melt extrusion process can be efficiently scaled to larger equipment through a fundamental understanding of scale-independent parameters and their effects on critical quality attributes. This understanding allows for streamlined experiments to verify successful scale-up and a reliance on validated modeling approaches as discussed previously in this chapter. In this example, experiments at the 16 mm scale had demonstrated a link to product temperature and residence time to the critical quality attributes of a single-phase, fully amorphous extrudate and the thermal degradation product. Clinical demand necessitated scale-up to a 27 mm extruder. Volumetric scaling principles were followed and the design of experiments (DOE) in Fig. 11.9 was proposed for confirming the design space for the 27 mm process.

However, due to limited API availability the DOE could not be fully executed. Thus, based on the fundamental understanding acquired linking product temperature and residence time to the critical quality attributes, only the green points were evaluated representing the target operating conditions and the extremes for residual crystallinity (low residence time, low product temperature, low SME) and thermal degradation (high residence times, high product temperature, and high SME). The results shown in Table 11.3 provide confirmation of residual crystallinity at the crystallinity extreme and increased thermal degradation at the degradation extreme. The data from these runs, including measurement of residence time distributions via

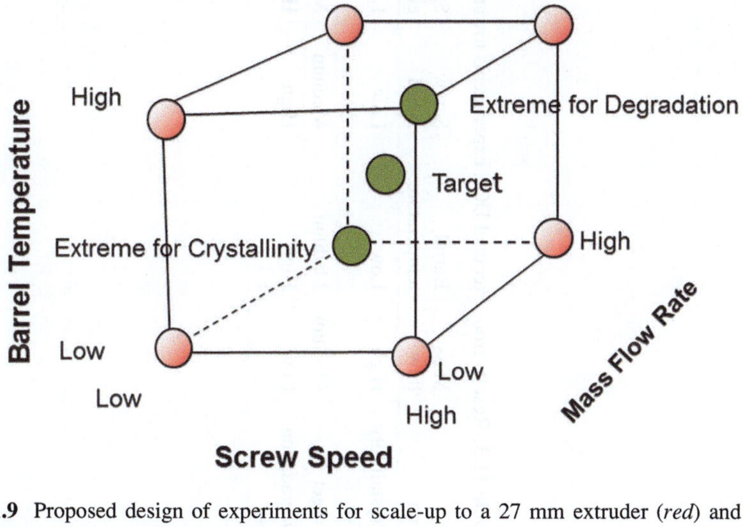

Fig. 11.9 Proposed design of experiments for scale-up to a 27 mm extruder (*red*) and experimentally executed experiments (*green*)

Table 11.3 Results from executed DOE experiments toward scaling up to a 27 mm extruder

Run	Mass flow	Barrel temperature	Screw speed	Specific energy	Residence time	Product temperature (°C)	Phase by XRD	Thermal degradant growth (%)
Crystallinity risk	High	Low	Low	Low	Low	183	Residual crystallinity	0.01
Target	Medium	Medium	Medium	Medium	Medium	197.3	Fully amorphous	0.08
Degradation risk	Low	High	High	High	High	2,115.8	Fully amorphous	0.27

PAT tools, were utilized to validate a first principles model developed using known rheological and heat capacity data of the extrudate. This allowed utilization of the model to predict the outcome of the remaining DOE points and provide confidence in the design space prior to clinical manufacture.

References

Adam G, Gibbs J (1965) On the temperature dependence of cooperative relaxation properties in glass-forming liquids. J Chem Phys 43(1):139–146

Agrawal A, Dudhedia M, Patel A, Raikes M (2013) Characterization and performance assessment of solid dispersions prepared by hot melt extrusion and spray drying process. Int J Pharm 457 (1):71–81

Aitken-Nichol C, Zhang F, McGinity J (1996) Hot melt extrusion of acrylic films. Pharm Res 13 (5):804–808

Alba-Simionesco C, Fan J, Angell C (1999) Thermodynamic aspects of the glass transition phenomenon. II. Molecular liquids with variable interactions. J Chem Phys 110(11):5262–5272

Albers J, Alles R, Matthée K, Knop K, Nahrup J, Kleinebudde P (2009) Mechanism of drug release from polymethacrylate-based extrudates and milled strands prepared by hot-melt extrusion. Eur J Pharm Biopharm 71(2):387–394

Andersen P (1998) The Werner and Pfleiderer twin-screw corotating extruder system. In: Todd D (ed) Plastics compounding: equipment and processing. Hanser Gardner Publications, Cincinnati, pp 71–124

Andrews G, Jones D, Abu Diak O, McCoy C, Watts A, McGinity J (2008) The manufacture and characterisation of hot-melt extruded enteric tablets. Eur J Pharm Biopharm 69(1):264–273

Andrews G, AbuDiak O, Jones D (2010) Physicochemical characterization of hot melt extruded bicalutamide–polyvinylpyrrolidone solid dispersions. J Pharm Sci 99(3):1322–1335

Andronis V, Zografi G (1997) Molecular mobility of supercooled amorphous indomethacin, determined by dynamic mechanical analysis. Pharm Res 14(4):410–414

Angell C (1995) Formation of glasses from liquids and biopolymers. Science 267(5206):1924–1935

Angell C (2002) Origin and control of low-melting behavior in salts, polysalts, salt solvates, and glass formers. In: Gaune-Escard M (ed) Molten salts: from fundamentals to applications. NATO science series, vol 52. Springer, Netherlands, pp 305–320

Angell C (2008) Glass-formers and viscous liquid slowdown since David Turnbull: enduring puzzles and new twists. MRS Bull 33(05):544–555

Baird J, Van Eerdenbrugh B, Taylor L (2010) A classification system to assess the crystallization tendency of organic molecules from undercooled melts. J Pharm Sci 99(9):3787–3806

Baird J, Santiago-Quinonez D, Rinaldi C, Taylor L (2012) Role of viscosity in influencing the glass-forming ability of organic molecules from the undercooled melt state. Pharm Res 29 (1):271–284

Bruce C, Manning M (2009) Melt extruded nicotine thin strips. US Patent WO2011081628, 30 Dec 2009

Couchman P, Karasz F (1978) A classical thermodynamic discussion of the effect of composition on glass-transition temperatures. Macromolecules 11(1):117–119

Crowley M, Fredersdorf A, Schroeder B, Kucera S, Prodduturi S, Repka M, McGinity J (2004a) The influence of guaifenesin and ketoprofen on the properties of hot-melt extruded polyethylene oxide films. Eur J Pharm Sci 22(5):409–418

Crowley M, Schroeder B, Fredersdorf A, Obara S, Talarico M, Kucera S, McGinity J (2004b) Physicochemical properties and mechanism of drug release from ethyl cellulose matrix tablets prepared by direct compression and hot-melt extrusion. Int J Pharm 269(2):509–522

Curatolo W, Nightingale J, Herbig S (2009) Utility of hydroxypropylmethylcellulose acetate succinate (HPMCAS) for initiation and maintenance of drug supersaturation in the GI milieu. Pharm Res 26(6):1419–1431

DiNunzio J, Brough C, Hughey J, Miller D, Williams Iii R, McGinity J (2010) Fusion production of solid dispersions containing a heat-sensitive active ingredient by hot melt extrusion and Kinetisol® dispersing. Eur J Pharm Biopharm 74(2):340–351

Dreiblatt A (2003) Process Design. In: Ghebre-Sellassie I, Martin C (eds) Pharmaceutical extrusion technology, vol. 133. Marcel Dekker Inc, New York, pp 153–169

Dreiblatt A (2012) Technological considerations related to scale-up of hot-melt extrusion processes. In: Hot-melt extrusion: pharmaceutical applications. Wiley, Chichester. pp. 285–300

Ferry J (1980) Viscoelastic properties of polymers. Wiley, New York

Flory P (1942) Thermodynamics of high polymer solutions. J Chem Phys 10(1):51–61

Forster A, Hempenstall J, Tucker I, Rades T (2001) Selection of excipients for melt extrusion with two poorly water-soluble drugs by solubility parameter calculation and thermal analysis. Int J Pharm 226(1–2):147–161

Fukuda G, Chavez D, Bigio D (2014) Investigation of scale-up methodologies in twin-screw compounding. SPE ANTEC. pp. 1138–1146

Gao J, Walsh G, Bigio D, Briber R, Wetzel M (1999) Residence-time distribution model for twin-screw extruders. AICHE J 45(12):2541–2549

Gaskell D (2003) Introduction to the thermodynamics of materials, 5th edn. Taylor & Francis, Washington

Ghebremeskel A, Vernavarapu C, Lodaya M (2007) Use of surfactants as plasticizers in preparing solid dispersions of poorly soluble API: selection of polymer-surfactant combinations using solubility parameters and testing the processability. Int J Pharm 328(2):119–129

Ghosh I, Vippagunta R, Li S, Vippagunta S (2012) Key considerations for optimization of formulation and melt-extrusion process parameters for developing thermosensitive compound. Pharm Dev Technol 17(4):502–510

Gordon M, Taylor J (1952) J Appl Chem 2:493–500

Greco S, Authelin J, Leveder C, Segalini A (2012) A practical method to predict physical stability of amorphous solid dispersions. Pharm Res 29(10):2792–2805

Khougaz K, Clas S (2000) Crystallization inhibition in solid dispersions of MK-0591 and poly (vinylpyrrolidone) polymers. J Pharm Sci 89(10):1325–1334

Lakshman J (2007) Process for making pharmaceutical compositions with a transient plasticizer. WO2007106182 A2

Liu H, Wang P, Zhang X, Shen F, Gogos C (2010) Effects of extrusion process parameters on the dissolution behavior of indomethacin in Eudragit® E PO solid dispersions. Int J Pharm 383(1–2):161–169

Low A, Parmentier J, Khong Y, Chai C, Tun T, Berania J, Liu X, Gokhale R, Chan S (2013) Effect of type and ratio of solubilising polymer on characteristics of hot-melt extruded orodispersible films. Int J Pharm 455(1–2):138–147

Lowinger M (2011) Process development: scaling a melt extrusion process from conception to commercialization. Am Pharmaceut Rev 14(2):80–84

Lu Q, Zografi G (1998) Phase behavior of binary and ternary amorphous mixtures containing indomethacin, citric acid, and PVP. Pharm Res 15(8):1202–1206

Lyons J, Hallinan M, Kennedy J, Devine D, Geever L, Blackie P, Higginbotham C (2007) Preparation of monolithic matrices for oral drug delivery using a supercritical fluid assisted hot melt extrusion process. Int J Pharm 329(1–2):62–71

Marsac P, Shamblin S, Taylor L (2006a) Theoretical and practical approaches for prediction of drug-polymer miscibility and solubility. Pharm Res 23(10):2417–2426

Marsac PJ, Konno H, Taylor LS (2006b) A comparison of the physical stability of amorphous felodipine and nifedipine systems. Pharm Res 23(10):2306–2316

Marsac P, Li T, Taylor L (2009) Estimation of drug-polymer miscibility and solubility in amorphous solid dispersions using experimentally determined interaction parameters. Pharm Res 26(1):139–151

Marsac P, Taylor L, Xi H, Birrell L, Liu Z, Lau H (2012) A novel method for accessing the enthalpy of mixing active pharmaceutical ingredients with polymers. Paper presented at the American Association for Pharmaceutical Scientists National Meeting, Chicago, IL. Nov, 2012

Newman A, Engers D, Bates S, Ivanisevic I, Kelly R, Zografi G (2008) Characterization of amorphous API:polymer mixtures using X-ray powder diffraction. J Pharm Sci 97(11):4840–4856

Noyes A, Whitney W (1897) The rate of solution of solid substances in their own solutions. J Am Chem Soc 19(12):930–934

Qian F, Huang J, Zhu Q, Haddadin R, Gawel J, Garmise R, Hussain M (2010) Is a distinctive single T-g a reliable indicator for the homogeneity of amorphous solid dispersion? Int J Pharm 395(1–2):232–235

Rubinstein M, Colby R (2003) Polymer physics. OUP, Oxford

Sarode A, Obara S, Tanno F, Sandhu H, Iyer R, Shah N (2014) Stability assessment of hypromellose acetate succinate (HPMCAS) NF for application in hot melt extrusion (HME). Carbohydr Polym 101:146–153

Schenck L, Troup G, Lowinger M, Li L, McKelvey C (2010) Achieving a hot melt extrusion design space for the production of solid solutions, chemical engineering in the pharmaceutical industry. Wiley, Rahway, pp 819–836

Steiner R (2003) Extruder Design. In: Ghebre-Sellassie I, Martin C (eds) Pharmaceutical extrusion technology, vol. 133. Marcel Dekker, New York, pp 19–38

Sun Y, Tao J, Zhang G, Yu L (2010) Solubilities of crystalline drugs in polymers: an improved analytical method and comparison of solubilities of indomethacin and nifedipine in PVP, PVP/VA, and PVAc. J Pharm Sci 99(9):4023–4031

Tian Y, Booth J, Meehan E, Jones D, Li S, Andrews G (2013) Construction of drug-polymer thermodynamic phase diagrams using Flory-Huggins interaction theory: identifying the relevance of temperature and drug weight fraction to phase separation within solid dispersions. Mol Pharm 10(1):236–248

Verreck G, Decorte A, Heymans K, Adriaensen J, Cleeren D, Jacobs A, Liu D, Tomasko D, Arien A, Peeters J, Rombaut P, Van den Mooter G, Brewster M (2005) The effect of pressurized carbon dioxide as a temporary plasticizer and foaming agent on the hot stage extrusion process and extrudate properties of solid dispersions of itraconazole with PVP-VA 64. Eur J Pharm Sci 26(3–4):349–358

Verreck G, Decorte A, Heymans K, Adriaensen J, Liu D, Tomasko D, Arien A, Peeters J, Van den Mooter G, Brewster M (2006) Hot stage extrusion of p-amino salicylic acid with EC using CO2 as a temporary plasticizer. Int J Pharm 327(1–2):45–50

Williams M, Landel R, Ferry J (1955) The temperature dependence of relaxation mechanisms in amorphous polymers and other glass-forming liquids. J Am Chem Soc 77(14):3701–3707

Wu C, McGinity J (2003) Influence of methylparaben as a solid-state plasticizer on the physicochemical properties of Eudragit (R) RS PO hot-melt extrudates. Eur J Pharm Biopharm 56 (1):95–100

Chapter 12
Practical Considerations for Spray Dried Formulation and Process Development

Michael Lowinger, John Baumann, David T. Vodak, and Justin Moser

12.1 Spray Drying Process Overview

12.1.1 Spray Drying Introduction

Spray drying is a broadly applicable process for the manufacture of many products in the food, chemical, and pharmaceutical areas. Powders are produced in a continuous process during the drying of liquid feed stocks through an atomization and drying process. The ability to control the final product attributes to maintain the desired physical state, performance, and bulk powder attributes for downstream processing has made spray drying an important unit operation for the manufacture of amorphous solid dispersions. It can enable progression of compounds with poor aqueous solubility throughout the drug development lifecycle from discovery through commercial production.

Efficient development of the process for spray dried amorphous solid dispersions has been demonstrated to ensure the critical quality attributes are maintained during development and scale-up, for example with Vertex Pharmaceuticals' Incivek. To successfully achieve these goals, it is necessary to have a deep understanding of the process and the key formulation attributes along with their potential interaction. By developing this understanding, it is possible to achieve a robust process and formulation that have both been designed to maximize the success during drug development and to achieve the target performance and stability goals.

M. Lowinger (✉)
Discovery Pharmaceutical Sciences, Merck & Co, Rahway, NJ, USA
e-mail: Michael_lowinger@merck.com

J. Baumann • D.T. Vodak
Bend Research, Bend, OR, USA

J. Moser
Formulation Sciences, Merck & Co., West Point, PA, USA

A.C. Templeton et al. (eds.), *Discovering and Developing Molecules with Optimal Drug-Like Properties*, AAPS Advances in the Pharmaceutical Sciences Series 15,
DOI 10.1007/978-1-4939-1399-2_12

The spray drying process allows for a broad range of physicochemical properties of active pharmaceutical ingredients (APIs) and polymers to be used for formulations due to the rapid and controllable process under which the spray dried particles are made. During the manufacture of spray dried amorphous solid dispersions, the API, polymer, and other excipients are completely dissolved in a mutual solvent system to form a homogeneous spray solution. The atomization process produces spray droplets of a defined size distribution, which serves to increase surface area for heat and mass transfer and impacts the final properties of the spray dried particles. Rapid drying and solidification of the particles occurs in the process, resulting in reduced drug and polymer mobility, which in turn improves the likelihood of creating an amorphous, homogenous particle. Use of high glass transition temperature polymers can similarly reduce mobility in order to promote physical stability of the amorphous solid dispersion during shelf life.

12.1.2 *Spray Drying Physical Situation*

The physical situation that occurs during the spray drying process is defined by two primary processes: atomization and drying of the droplets into solid particles. Atomization requires breakup of the liquid feedstock into primary droplets followed by rapid drying and removal of the spray solvent from the spray dried particle. Figure 12.1 shows an overview of the defined control volume and key transitions that occur during the lifecycle of the droplet to final spray dried particle.

Droplet drying is a coupled heat and mass transfer process. Mass transfer during droplet drying can be fundamentally described using Fick's second law of

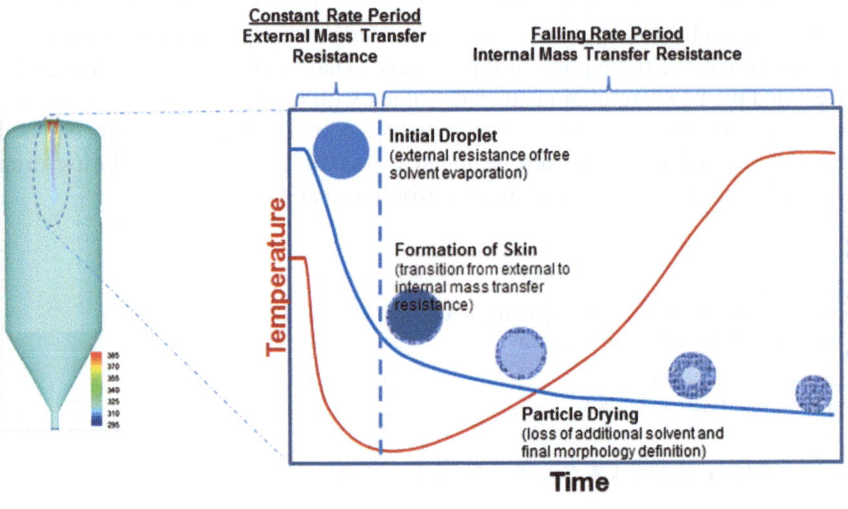

Fig. 12.1 Defined control volume and key transitions during droplet to particle formation

diffusion, as shown in Eq. (12.1) where the time dependent rate of mass transfer is defined by both the diffusion coefficient (D_{AB}) and the concentration gradient between the evaporating droplet surface and the drying gas (Welty et al. 2000).

$$\frac{\partial c}{\partial t} = D_{AB} \nabla^2 c \tag{12.1}$$

Additional simplification of Fick's second law leads to a simple analytical solution that describes the time dependent change of initial droplet diameter (d) based on the starting droplet size (d_0) assuming a constant evaporation rate (κ) (Vehring et al. 2007).

$$d^2(t) = d_0^2 - \kappa t \tag{12.2}$$

Based on the preceding summary of the fundamentals of drying, the key process parameters that dominate the drying process are the concentration driving force (between droplet and drying gas); diffusivity (as a function of drying temperature, or, for constant diffusivity, the final evaporation rate); and droplet size. Assuming adequate heat transfer, initial drying of the droplets is limited by mass transfer of free solvent from the surface of the droplet to the drying gas (evaporation). During this initial constant rate period of drying, a significant drop in temperature of the droplet occurs due to evaporative cooling with the minimum being the wet bulb temperature (Masters 1985; Perry and Green 1997). Solidification of the droplet or "skin" formation occurs after the droplet transitions through a critical viscosity based on the temperature and particle composition. This skin creates additional diffusional resistance for further solvent mass transfer, leading to a falling rate period. This critical point is often reached early in the process, on the order of milliseconds, based on model and experimental outputs (Dobry et al. 2009; Vehring et al. 2007). During the falling rate period, the temperature of the droplet begins to increase with the maximum being defined by the temperature of the drying gas in the chamber, which is equivalent to the outlet temperature of the process. Temperature profiles can be optimized to achieve target morphology or compressibility of the final spray dried powder, as this part of the drying process is responsible for the final particle characteristics. As reported by previous authors, the drying particles can go through further expansion leading to enlarged hollow particles or expansion that leads to buckling or further densification due to shriveling of the particle (Paudel et al. 2013; Vehring 2008; Vicente et al. 2013).

As discussed previously, rapid evaporative cooling occurs immediately as the spray solvent is evaporated from the droplets. Figure 12.4 illustrates a computational fluid dynamics (CFD) model output showing the drying gas temperature contours during typical temperature profiles used in the spray drying process. The heated inlet gas immediately mixes with the atomized droplets, resulting in cooling of the spray plume below the exhaust temperature of the spray dryer. The vast majority of the spray drying chamber remains at the final exhaust temperature of the spray drying process. Manipulation of the drying temperatures can be conducted to change the drying rate, leading to changes in key attributes of the spray dried particles such as bulk density or final residual solvent content.

12.1.3 *Overview of Key Operations in the Spray Drying Process*

The spray drying process can be broken up into multiple operations that occur from preparation of the initial solution feedstock to final collection of the spray dried product. These key operations are shown in Fig. 12.2 and summarized briefly below.

1. *Solution Preparation*: The active pharmaceutical ingredient (API), polymer, and additional components (e.g., surfactants) are dissolved into a homogeneous spray solution in a common solvent system. Amorphous solid dispersions are typically spray dried from organic solvents due to the poor water solubility of the API. Common solvents include acetone, methanol, ethanol, tetrahydrofuran, and mixtures of these with water.
2. *Atomization and Drying*: As the spray solution enters the spray drying chamber, it is passed through an atomizer, breaking it up into high surface area droplets (typical median droplet size range of 20–100 μm) that enable rapid heat and mass transfer. The atomized droplets come into immediate contact with the heated drying gas that provides the driving force for drying of the droplets into final particles. Control of final product attributes is achieved through modification of both the atomization process (e.g., droplet size distribution) and the drying process (e.g., temperatures and feed rates).
3. *Product Collection*: Cyclone separators are the most common collection technique of spray dried amorphous solid dispersions. This is due to the wide range of geometries available that enable high efficiency recovery of the product. The ability for continuous collection and discharge are amenable to the continuous operation of the spray drying process.
4. *Recycling of the Drying Gas*: Pilot and commercial scale spray dryers require the use of a condenser and recycling of the drying gas due to the high drying gas flow rates used at these scales. The condenser outlet temperature defines the amount of solvent vapor that is introduced back into the drying chamber during the process.
5. *Secondary Drying*: Nominal amounts of solvents used for the spray drying process remain after the product has been collected from the cyclone. Therefore, a secondary drying step is usually employed to remove solvent levels down to target levels based on both dosing considerations (International Conference on Harmonisation of Technical Requirements for Registration of Pharmaceuticals for Human Use 2011) and physical stability.

The atomization process is used to form discrete droplets from the feed solution. This serves to control the final particle size of the spray dried amorphous solid

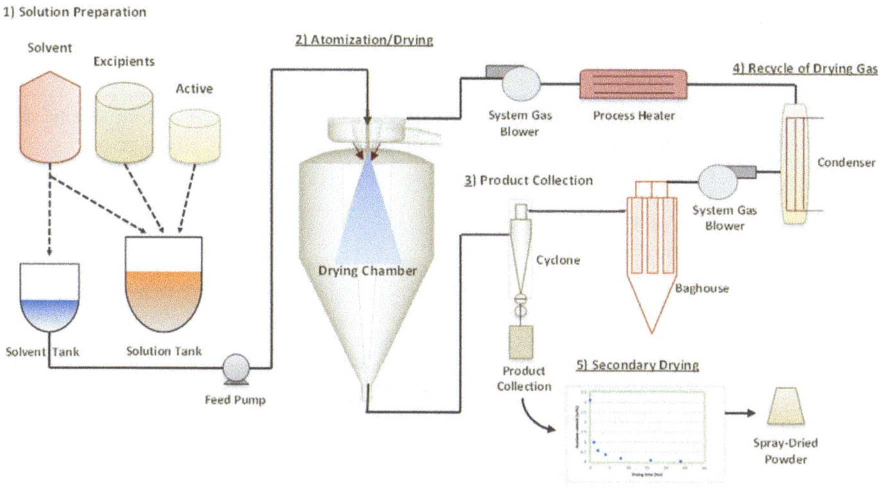

Fig. 12.2 Overview of the spray drying process key operations

dispersion and to increase the surface area for rapid drying of the droplets. The most common atomization techniques used for manufacture of amorphous dispersions are pressure swirl, ultrasonic, and two-fluid atomizers. Based on summaries presented by Bayvel, Lefebvre, and Liu, a comparison of common nozzles is provided in Table 12.1 (Bayvel and Orzechowski 1993; Lefebvre 1988; Liu 2000).

Introduction of the drying gas into the spray drying chamber to ensure contact with the atomized droplets is critical to avoid potential recirculation or product deposition, temperature gradients that would cause recirculation, or risk of physical or chemical instability. The most commonly used gas disperser for introducing the drying gas into the spray drying chamber is the GEA Niro model DPH disperser (Process Engineering A/S 2014). This gas disperser allows for rapid mixing of the heated gas with the atomized droplets for optimal heat and mass transfer and can be effectively employed across laboratory to commercial scale spray dryers (Process Engineering A/S 2014; Hansen and Ullum 2009). Droplet sizes, liquid and gas flow rates, and temperature profiles during the spray drying process are selected to ensure complete drying within the residence time of the spray drying chamber, which is dependent on the scale of spray dryer with laboratory scale spray dryers having much lower residence times than pilot or commercial scale spray dryers.

As shown in Fig. 12.2, closed loop operation where the drying gas is recycled after passing through a condenser is employed at large scale due to the high drying gas flow rate requirements. However, at laboratory and pilot scale, single pass operation is often used where fresh drying gas is continuously fed into the system and the outlet gas exhausted. The key difference is the residual solvent vapor that is present during closed loop operation. A comparison of the outlet relative saturation for single pass operation versus closed loop operation is shown in Fig. 12.3 to demonstrate the impact of closed loop operation and condenser temperature. Closed loop operation primarily reduces the concentration driving

Table 12.1 Comparison of atomizer nozzle types commonly used for pharmaceutical spray drying applications

	Two-fluid	Pressure swirl	Ultrasonic
Advantages	• Ability to handle high viscosity feedstocks • Ability to make small droplet sizes • Low chance of clogging due to relatively large orifice diameter	• Robust and economical • Simple design	• Uniform droplet size • Low velocity
Disadvantages	• High velocity • External source of high pressure atomizing gas required • High energy consumption	• Moderate velocity • High hydraulic pressure required	• Limited throughput makes scale-up challenging
Mechanism for droplet breakup	Pneumatic energy from atomizing gas on liquid core	Hydraulic energy from fluid passing through swirl chambers and final orifice (Bayvel and Orzechowski 1993)	Standing wave imparted through ultrasonic vibration leading to droplet ejection

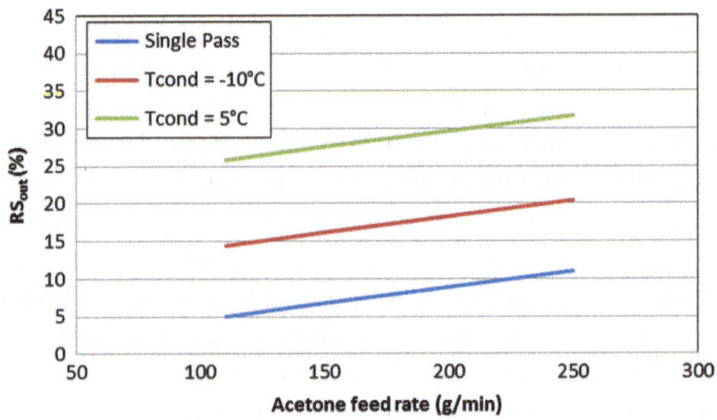

Fig. 12.3 Comparison of Relative Saturation (RS_{out}) as a function of acetone feed rate, mode of operation, and condenser temperature (T_{cond})

force for mass transfer of the droplet to final spray dried particle. Slower drying rates will lead to spray dried products with high bulk densities and higher residual solvents, so an understanding of this is key when considering scale-up and any potential impact this may have to manufacturability of the amorphous solid dispersion or physical stability risks due to the depressed glass transition temperature (T_g).

Collection of the spray dried product is a continuous process enabled through the use of cyclones. Significant work has been published on design and use of high efficiency cyclones enabling a wide variety of optimization to occur for maximizing high collection efficiencies of spray dried products (Azadi and Azadi 2012; Graham

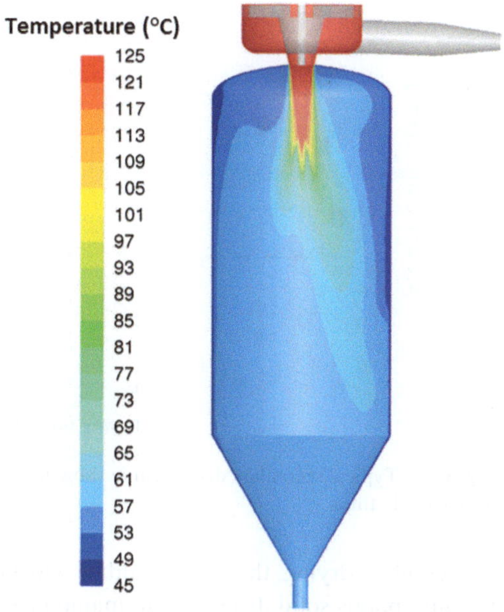

Fig. 12.4 Computational Fluid Dynamic model of drying gas temperature contours in a spray dryer

et al. 2010; Salcedo and Paiva 2010; Wan et al. 2008) Use of computer simulations and numerical models is an effective way to evaluate the impact of changes to cyclone design on collection efficiency of specific spray dried products. Additional consideration should be given to amorphous solid dispersion formulations that might experience attrition or densification during the cyclone collection process.

A secondary drying process is used after spray drying to reduce residual solvents below target limits based on ICH guidelines and potential physical instability (International Conference on Harmonisation of Technical Requirements for Registration of Pharmaceuticals for Human Use 2011). Increased temperature, convection, and relative humidity of the drying gas are all used to increase the kinetics of the drying process. Use of vacuum can also be employed to reduce the overall partial pressure in the vapor phase and decrease drying time. The T_g of the spray dried amorphous solid dispersion while in its "wet" state (prior to secondary drying) should be evaluated and used to select drying conditions that will mitigate risk of crystallization or phase separation throughout the drying process. A representative drying curve for T_g and residual solvent concentration as a function of drying time is shown in Fig. 12.5. Ramping of the temperature during secondary drying is a technique that may be employed to maintain the T_g of the spray dried amorphous solid dispersion at least 10 °C above the process temperature (T). Although this difference of $T_g - T$ is less than generally accepted heuristics to maintain long-term physical stability up to 2 years of storage, this particular approach seeks to mitigate physical stability risk during the relatively short cycle time of the secondary drying process which is on the order of hours to days (Hancock et al. 1995). Consequently, this narrower $T_g - T$ gap may be employed with acceptable risk. Upon completion

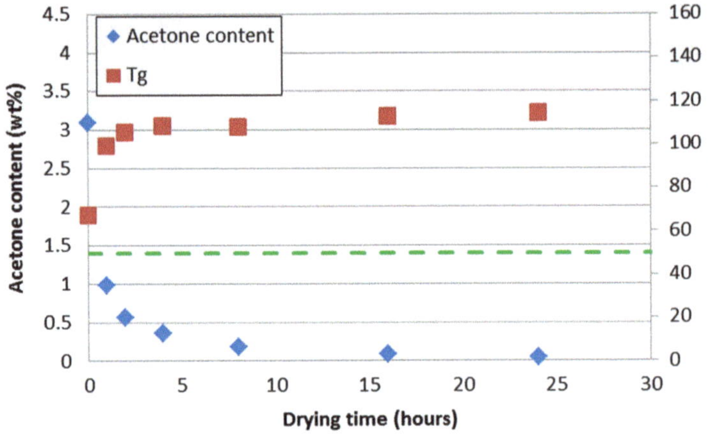

Fig. 12.5 Typical secondary drying curve showing residual solvent content and T_g as a function of drying cycle time

of secondary drying, this gap should be widened to account for long-term storage of the amorphous solid dispersion to maintain adequate physical stability. Commonly used unit operations include convection tray dryers (e.g., Gruenberg), agitated vacuum dryers (e.g., Ekato), or biconical tumble dryers (e.g., Pfaudler).

12.2 Back to Fundamentals

12.2.1 Description of States: Thermodynamic Solubility vs. Kinetic Stability

Binary amorphous solid dispersions comprised of a drug and a polymer may exist in two general cases. In one case, the drug is dissolved to a solubility limit in the polymer matrix and the polymer behaves as a solvent for the drug. In this situation, the drug is considered to be in a thermodynamically stable state (Gupta et al. 2004). To achieve this state, a low drug loading in the amorphous solid dispersion is likely necessary. Further, the impact of environmental perturbations should be considered since variations in temperature and water activity, for instance, will change the equilibrium solubility of the drug in the polymer.

In the other case, the drug is supersaturated with respect to its solubility in the polymer. In this situation, the polymer acts as a stabilizer where its high glass transition temperature and ability to form a glass stabilize the drug due to lack of mobility. Amorphous solid dispersions stabilized in this manner are considered to be kinetically stabilized (Strobl 2007). For this type of dispersion, the glass may phase separate into multiple amorphous phases or the drug may crystallize, as is illustrated in Fig. 12.6 (Serajuddin 1999).

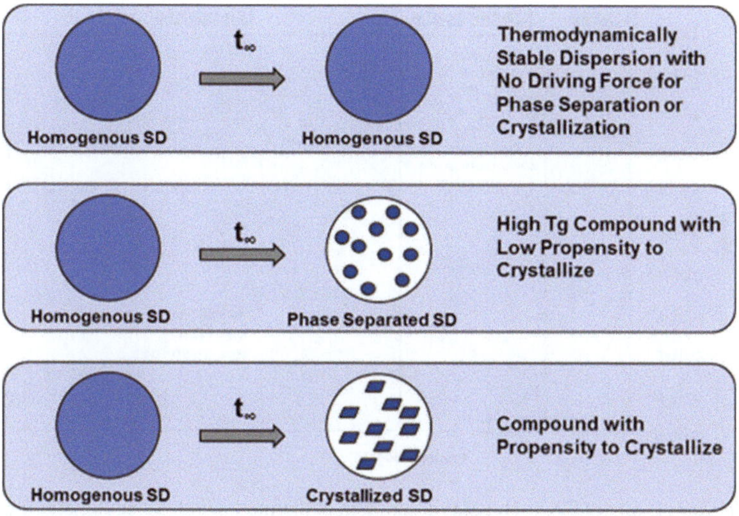

Fig. 12.6 Schematic illustrating the possible final states of amorphous solid dispersions (SD) depending on thermodynamic versus kinetic stability

For kinetically stabilized binary dispersions that undergo amorphous phase separation (i.e., without crystallization), the composition of the resulting phase separated system is a function of the physicochemical properties of the constituents in the mixture and is ultimately determined by the Gibbs free energy as a function of amorphous solid dispersion composition, as it illustrated in Fig. 12.7. The compositional dependence can be represented using the Flory–Huggins relationship developed for polymer blends (Flory 1953). Based on Eq. (12.3) and discussion in Chap. 20, the free energy of mixing is determined based on an entropic term and an enthalpic term. In polymer–drug mixtures, the entropy of mixing is always favored and thus negative. The enthalpic term describes the interaction between the drug and polymer and often dictates the shape of the free energy of mixing curve, as illustrated in Fig. 12.7 (Roe and Zin 1980).

$$\Delta G_{\text{mix}} = \text{RT}\left(\frac{\phi_A}{N_A}\ln\phi_A + \frac{\phi_B}{N_B}\ln\phi_B + \chi n_c \phi_A \phi_B\right) \quad (12.3)$$

A negative free energy of mixing over all compositions displaying a single minimum means that the system is completely miscible. Alternatively, a thermodynamic driving force for amorphous phase separation exists when the curve displays two minima. In this composition regime, the dispersion will phase separate spontaneously, and a homogenous and well dispersed system becomes both more difficult to manufacture and more difficult to stabilize (Strobl 2007).

Based on Fig. 12.7, with increasing drug load, the drug is at greater risk for amorphous phase separation and will require more robust processing conditions and more rapid quenching kinetics to ensure the preparation of a homogeneous,

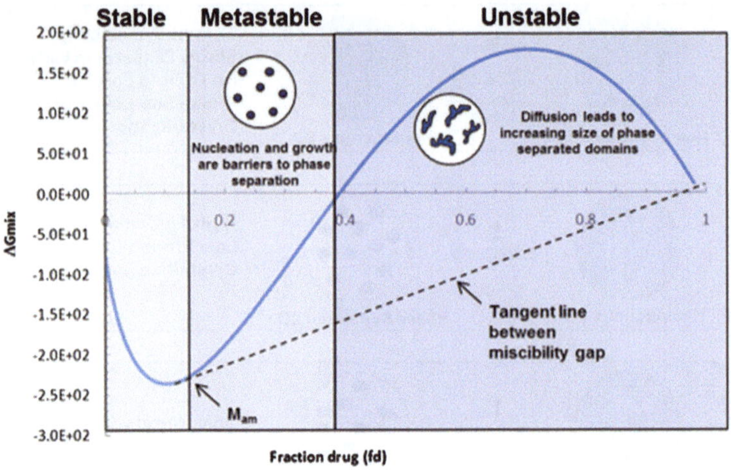

Fig. 12.7 Simple binary Flory–Huggins description of thermodynamic versus kinetic situations in amorphous dispersions, where M_{am} is the miscibility of amorphous active in polymer

dispersed system. Specifically, at compositions to the left of the first vertical line, the drug and polymer will form as a binary single phase system. Between the first and second vertical lines, the drug and polymer will, at equilibrium, exist as two phases, and the mechanism for amorphous phase separation is via nucleation and growth. To the right of the second vertical line, the mechanism for amorphous phase separation is via spinodal decomposition. It is critical in the stabilization of a kinetically stabilized amorphous solid dispersion to begin with a homogeneous system; that is, no phase separation. The focus of this section will cover important aspects of processing to form a homogeneous kinetically stabilized amorphous solid dispersion. It is generally a prerequisite to manufacture a homogeneous amorphous solid dispersion to ensure good physical stability. The physical stability of kinetically stabilized amorphous solid dispersions is also very important, but will be described elsewhere.

Based on the binary phase diagram, the types of applicable processes can be defined. For example, processes in which the compositional trajectory lies entirely in a single-phase region are less reliant on quenching kinetics, and therefore, are typically ideal for processes with inherently slower quenching kinetics (Friesen et al. 2008). As the composition becomes increasingly drug-rich relative to polymer, rapid quenching kinetics become increasingly important and processes such as spray drying, spray freeze-drying, and other precipitation-based techniques (Williams et al. 2013) become more appropriate or even enabling. This section will deal exclusively with how the spray drying process is appropriate for kinetically stabilized amorphous solid dispersions and the important aspects for consideration when developing a process for manufacturing such a product in a robust manner.

12.2.2 Physical Aspects of Composition and the Spray Drying Process

The ternary phase diagram in Fig. 12.8 illustrates compositions of spray solvent, drug, and polymer. This is a simplified yet effective way to describe the thermodynamic compositional history of the spray solution as it progresses from droplet to solid particle. This simplistic representation of the droplet drying history does not accurately portray the true phase diagram as it does not take into account the temperature change of the droplet as evaporative cooling occurs upon solvent removal, it does not reflect compositional changes during drying from mixed solvent systems, and it does not reflect potential compositional heterogeneity introduced by changes to drying kinetics. The temperature can be overlaid but first it is important to clearly understand the composition as it relates to the various phase changes that can be expected as well as the viscosity of the system as it converts from a free flowing liquid to a drying solid.

Following Fig. 12.8, the composition of the droplet transitions from solvent rich to a dried particle, here resulting in a 25 % drug/75 % polymer (w/w) solid dispersion. In this conceptual example, the composition of the droplet will cross several important phase boundaries. Specifically, the composition will cross through the gelation point of the polymer in the organic solution, the API solubility in the organic solvent, and eventually will cross a point where the T_g of the drying droplet is equal to the temperature in the drying chamber. Effectively, the goal of a

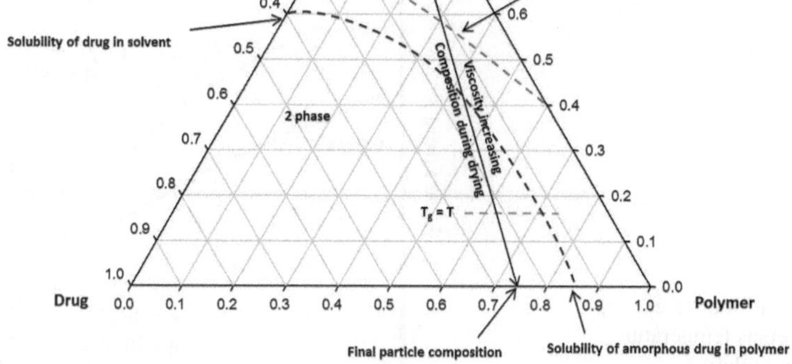

Fig. 12.8 Conceptual ternary phase diagram illustrating overlay of composition during droplet drying to important phase changes, including polymer gelation point, API solubility in spray solvent, and cross-over of T_g to process temperature

successful spray drying process is to transition through these boundaries as rapidly as possible and remove sufficient amount of solvent to arrive at a situation where the T_g of the drying particle rapidly becomes higher than the temperature in the drying chamber. The necessity for rapid solvent removal can be initially explained using the viscosity changes of the components in the droplet as a function of solvent removal. Based on the conceptual ternary phase diagram, it becomes obvious that the system will become exponentially more complex if co-solvent systems or surfactants are added to the composition as required by processing or performance needs. Specifically, when two solvents of differing heat of vaporization are utilized in the evaporation process, the evaporation occurs at different rates for each solvent. The result will be a curved composition history rather than the straight trace represented in Fig. 12.8. Implications of these more complex systems are that the curved drying composition history may be impacted by spray drying process parameters and may have profound effects on both drying kinetics and the final state of the dispersion.

Figure 12.9 conceptually overlays free energy and viscosity changes with decreasing temperature, connecting thermodynamic and kinetic effects. In this schematic, the glass is formed by melting the drug with heat and then upon cooling, trapping the supercooled liquid state at temperatures below the melting point. Cooling continues to form a solid with higher free energy until the temperature cools below the T_g of the glass. Although the glass has a higher free energy and is thermodynamically less stable than the crystalline form, viscosity increases substantially below the T_g slowing the kinetics of phase transitions. As the temperature continues to cool well below the T_g, it is at temperatures in the ranges of 10–20 °C or more below the T_g that the viscosities are sufficiently high that the glass is no longer mobile in pharmaceutically relevant time scales.

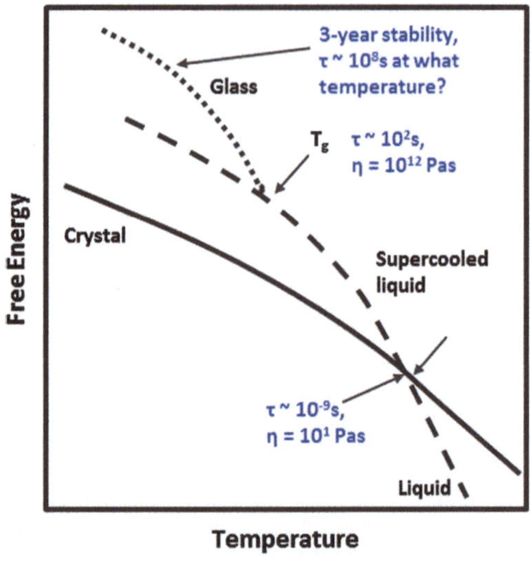

Fig. 12.9 Plot of free energy versus temperature illustrating relative viscosities and relaxation time constants for various phases of the glass

The conceptual plot in Fig. 12.9 illustrates the importance of kinetics on the phase behavior of amorphous solid dispersion formulations. In kinetically stabilized systems, trapping a glass with rapid kinetics is critical. If the glass is held at the supercooled liquid state or near the T_g, viscosities are such that the drug molecules will have mobility. The process must rapidly get to viscosities of $>10^{12}$–10^{14} Pa s to trap the glass without enabling sufficient mobility that can result in phase separation (Meyvis et al. 1999). The higher the drug loading and further into the supersaturated regime of the phase diagram, the faster the process should move from the bottom right to the top left in Fig. 12.9. In the spray drying process, the time constants for solvent removal are generally on the order of milliseconds and are considered enabling for preparing these kinetically stabilized dispersions (Dobry et al. 2009).

12.2.3 Phase State Implications on Kinetics of Solvent Removal from Drying Droplet

The time-dependent viscosity of the glass will primarily be defined by the relationship of droplet temperature (defined by process conditions like droplet size, relative saturation, and dryer outlet temperature) to the glass transition temperature of the drying droplet. The glass transition temperature will be defined by the composition of the amorphous solid dispersion during the process. Figure 12.10 illustrates a conceptual drying curve. Initially, as the droplet passes out of the atomizer and into the hot drying gas inlet of the spray dryer, the organic solvent rapidly flashes off (red zone).

At this point, the droplet has lost much of its solvent but is now physically out of the range of the hot inlet drying gas as it flows through the drying chamber and transitions to equilibration with the spray dryer outlet temperature (yellow to green

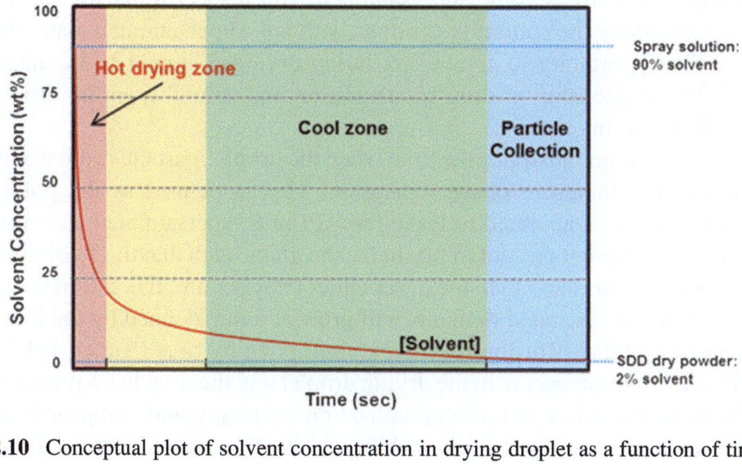

Fig. 12.10 Conceptual plot of solvent concentration in drying droplet as a function of time

zones). If sufficient solvent has not been removed thus far in the drying process to result in a solid particle with a T_g near to or greater than that of the outlet temperature, the process will result in a mobile glass that risks phase separation or worse yet may stick to the drying chamber and lead to spray painting of the amorphous solid dispersion and critical product loss (Dobry et al. 2009). The rate of solvent evaporation will be competing with the rate of solvent diffusion out of the drying droplet due to polymer "skinning" at the surface of the droplet as well as the general viscosity increase during glass formation. Consequently, studies have shown that the spray solvent composition can impact the miscibility and physical stability of spray dried amorphous solid dispersions (Paudel and Van den Mooter 2012). This complexity and its implications on surface composition of the resulting droplet will be discussed in a subsequent section of this chapter.

As the drying process continues, the residual solvent in the particle now plasticizes the dispersion and will lower the T_g according to its concentration in the particle (green to blue zone). The amount of solvent in the particle will be defined by the amount of solvent the formulation adsorbs at a given temperature and relative solvent saturation, as well as the temperature and relative solvent saturation in the drying chamber based on the drying conditions. If the resultant equilibrium conditions—relative saturation of solvent in the dryer and solvent-wet composition of the amorphous solid dispersion formulation—result in a composition with a T_g near or below that of the dryer outlet temperature, the glass will have mobility and the risk of phase separation and potential crystallization is high. The relationship of T_g to the amount of residual solvent in the spray dried formulation can be measured using modulated differential scanning calorimetry (mDSC) after equilibration at varying levels of solvent exposure. Solvent vapor sorption can be used to quantify the amount of solvent the spray dried formulation will adsorb at a given relative solvent saturation condition.

In addition to decreasing the mobility of the glass to kinetically trap the high energy form, the solubility of the API form must be addressed. As solvent leaves the droplet, the drug will rapidly approach its solubility limit in the selected spray solvent and eventually become supersaturated with respect to its solubility in the system. Once above the solubility limit, as with any supersaturated state, the drug will have the opportunity to crystallize during drying. Figure 12.11 conceptually illustrates the relationship of drug concentration and solubility in the droplet as a function of the drying time.

The yellow region illustrates the time when the droplet/particle is most at risk for precipitation and therefore phase separation. This is defined as the point when enough solvent has evaporated to leave the API in a supersaturated state but there is still sufficient solvent present to plasticize the glass, such that the T_g of the drying particle composition is less than the dryer outlet temperature. It is estimated that the size of these phase separated domains will grow at a rate defined by the integral of the square root of the diffusion coefficient multiplied by time (Angell 1985).

If API precipitation occurs in the drying droplet and the drug has a propensity for crystallization, there is a risk of crystallization occurring and compromising performance and/or long term physical stability. Using the drying kinetics of the spray

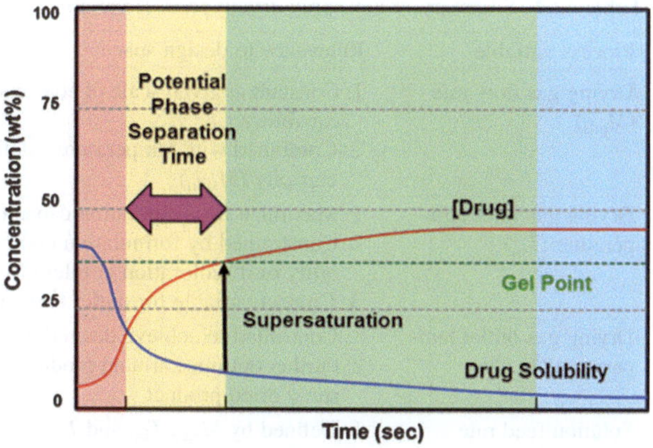

Fig. 12.11 Conceptual plot of solvent (*blue*) and API (*red*) concentration as a function of time

drying process, transition across this zone can be rapid and viscosity increases can inhibit mobility and therefore prevent phase separation or crystallization. While processing can successfully freeze these systems in a homogeneous state, understanding time constants and compositions during the process can be critical. This is especially true when scaling from small scale processes with rapid drying kinetics to larger scale dryers where rapid drying may be compromised somewhat to enable higher throughput. For this reason, fundamental understanding at the early stage can allow for formulation changes during initial development to produce spray dried amorphous solid dispersions that are amenable to scale-up but also meet performance and stability targets.

12.3 Failure Mode: Spray Drying Impact on Particle Physical Properties

12.3.1 Process Determinants of Spray Dried Particle Physical Properties

Definition of a robust spray drying process requires a deep understanding of the process and the potential impact on the spray dried formulation critical quality attributes. Models, experimental tools, and correlations are all used to build this understanding. A primary focus is given to differentiate and study both the atomization and drying droplets separately in an effort to decouple, understand, and optimize the underlying effects from each process.

The relevance of key spray drying process variables that are important for defining a design space are summarized in Table 12.2. Although many of these are dependent variables, each one is critical for a specific aspect of the process and can be optimized using approaches described previously (Dobry et al. 2009; Paudel

Table 12.2 Summary of key spray drying process variables and relevance to design space

Process variable	Relevance to design space
Drying gas flow rate (M_{gas})	1. Constant at given scale of equipment—limited by system pressure capability 2. Combined with temperatures, defines maximum evaporation capacity (M_{soln})
Drying gas inlet temperature (T_{in})	1. Maximize for evaporation capacity 2. Constrained by formulation components—avoid chemical instability or discoloration at inlet of gas disperser 3. Control variable for outlet temperature
Drying gas outlet temperature (T_{out})	1. Controlled to achieve desired particle attributes (bulk density) 2. Further optimize around product yield and physical stability of wet spray dried product
Solution feed rate (M_{soln})	1. Defined by M_{gas}, T_{in}, and T_{out} 2. Key variable to define specific drying ratio (M_{soln}/M_{gas}) and RS_{out}
Relative saturation (RS_{out})	1. With T_{out}, defines drying rate of process 2. Optimized for particle attributes (e.g., bulk density) 3. Defines residual solvent content in particles—physical stability constraint
Condenser temperature (T_{cond})	1. Impacts RS_{out}

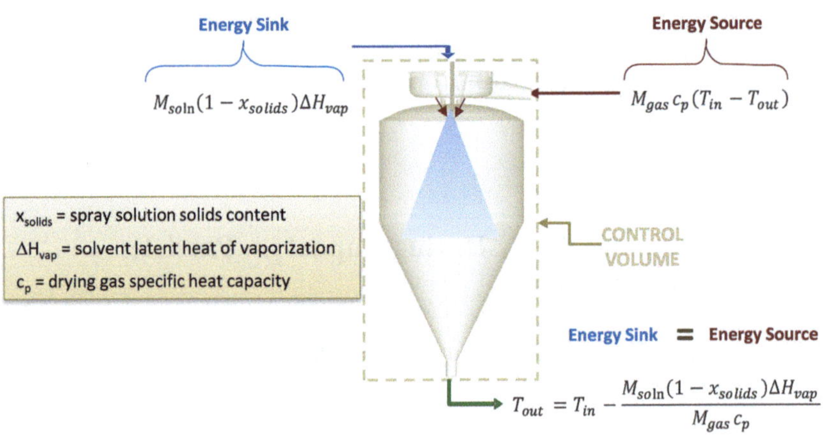

Fig. 12.12 Overview of spray drying mass and energy balance

et al. 2013). This avoids the need for empirical models to describe the impact of spray drying process variables on each other, and, instead, focuses the impact of these process variables on final spray dried amorphous solid dispersion attributes.

The thermodynamic space of the spray drying process can be captured by conducting a direct mass and energy balance on the system using the spray drying chamber as a control volume. Although this approach has been reported previously, it is briefly summarized here for convenience and shown in Fig. 12.12 (Dobry et al. 2009). The energy required to evaporate the solvent, based on the solution

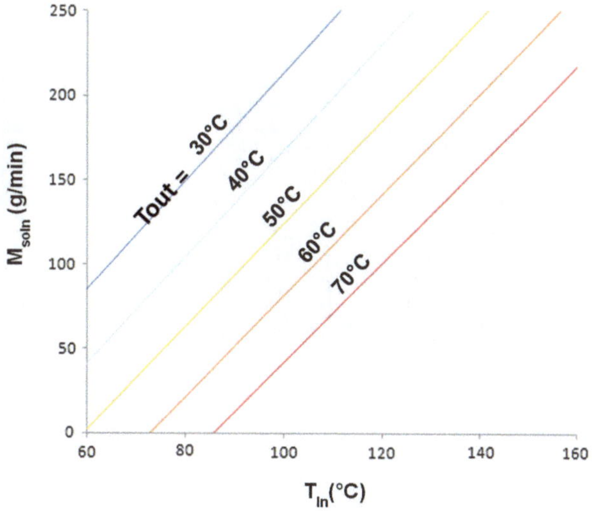

Fig. 12.13 Spray drying thermodynamic space solution for acetone as a spray solvent

flow rate (M_{soln}) and heat of vaporization for the specific solvent, is equivalent to the energy supplied and lost by the drying gas, which is based on the specific heat capacity, drying gas flow rate (M_{gas}), and temperatures of the drying gas (T_{in}, T_{out}). This equation allows for a direct calculation of the outlet temperature based on its relationship to the other parameters and the overall process space across the range of the equipment operability limits can be iteratively calculated using this approach. The specific drying ratio, which is the ratio of the mass flow rates of solution and drying gas, can also be used to replace the liquid flow rate on the *y*-axis and enable the model to be scale independent when accounting for heat loss from the dryer. This is made possible because drying gas flow rate is considered a constant fixed by the scale of spray drying equipment being used. A graphical summary of an example thermodynamic operating space for acetone as a spray solvent is shown in Fig. 12.13. A similar approach and calculation can also be conducted for closed loop operation by including the impact of the condenser on the mass and energy balance of the system, accounting for the additional solvent vapor present in the recycled drying gas.

Although it is considered a dependent variable, T_{out} has the most significant impact on the particle drying rate that ultimately defines final particle morphology and physical properties. This is because the majority of the drying process occurs at this temperature after the initial evaporation and solidification stages as discussed previously. By modifying T_{out}, the morphology and density can be optimized to achieve desired particle attributes. An example of a common correlation observed between T_{out} and changes in bulk density or morphology is shown in Fig. 12.14. As is observed here, the distribution of morphology and particle density is well correlated with the colder, slower drying leading to higher bulk density with primarily buckled sphere morphology. Conversely, at the hotter T_{out}, lower bulk

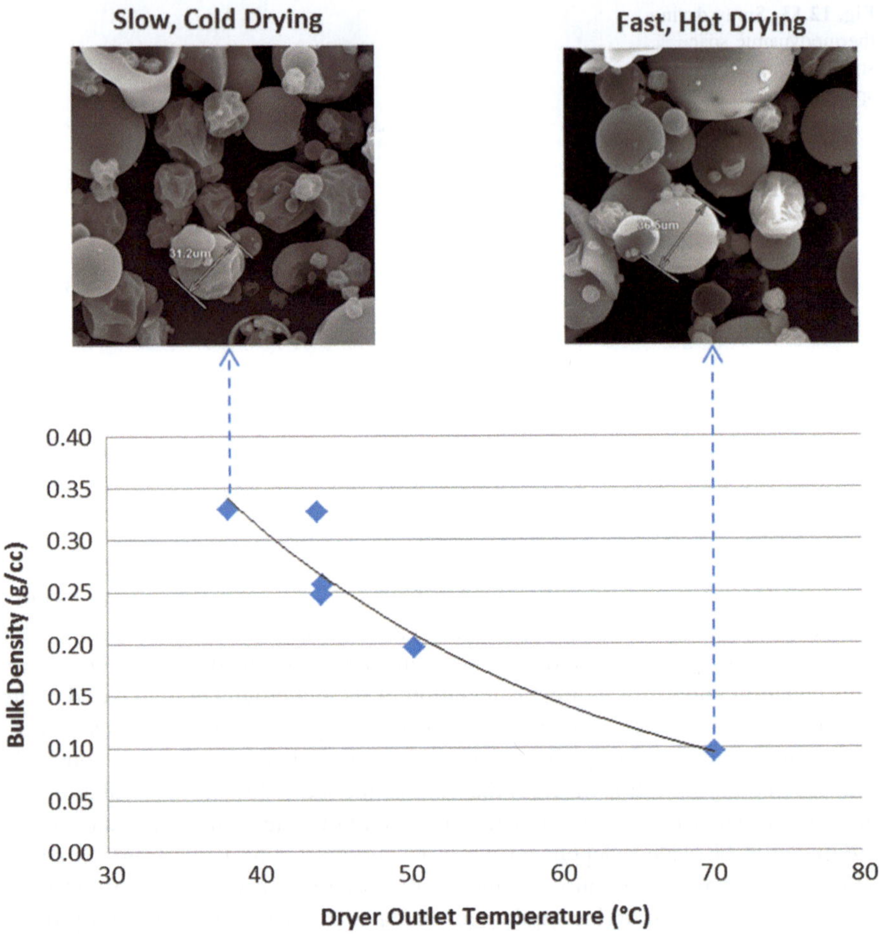

Fig. 12.14 Impact of outlet temperature on bulk density for a spray dried amorphous solid dispersion product. Scanning electron microscopy images of the two extreme outlet temperatures are provided

density is achieved with primarily inflated sphere morphology. In both scenarios, the core of the particle is typically hollow, leading to an internal porous structure.

Although the outlet temperature often describes the process space reasonably well, it cannot describe the impact of changes in other process parameters such as specific drying ratio (M_{soln}/M_{gas}) or condenser temperature. Given this limitation across complex process spaces, additional parameters are required to more completely describe the drying rate of the process. One commonly used parameter is the relative saturation of the solvent vapor at the outlet of the spray dryer. This accounts for both the temperature and the final solvent fraction, which can be adjusted to account for changes in specific drying ratio (i.e., solution to gas feed rate ratio) or condenser temperature. Changes in relative saturation of the incoming

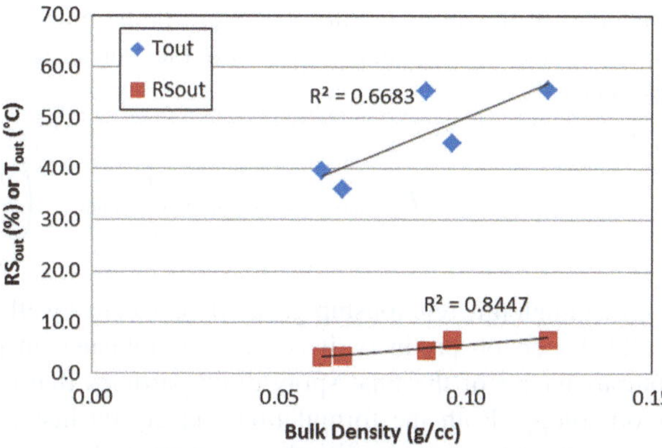

Fig. 12.15 Relationship of bulk density to RS_{out} and T_{out} for an example spray drying process

process gas can impact product morphology at a given outlet temperature. An example data set is shown in Fig. 12.15, illustrating the correlation of T_{out} compared to RS_{out} with particle bulk density for a specific study. In this case, the correlation of RS_{out} to the data set was observed to be a better fit, as this parameter accurately captured the changes in drying rate associated with changes to secondary parameters such as specific drying ratio and condenser temperature.

Additional approaches can be taken to better understand the impact of the key process parameters on final spray dried amorphous solid dispersion product attributes. Statistical analysis is commonly used to assess the impact of secondary variables that are challenging to correlate or predict from first principles analysis. For example, spray solution solids concentration could be studied in this fashion to develop further understanding of how the change in drying rate associated with solids concentration would impact final critical quality attributes, such as compressibility or physical state.

Dimensionless numbers have also been studied and applied successfully for understanding of the process. A commonly used dimensionless number is the Peclet number (Pe), which describes the ratio of convective mass transfer to diffusive mass transfer. This has been effectively demonstrated to characterize spray dried particle drying behavior (Paudel et al. 2013; Vehring et al. 2007). Although these techniques are useful, they can often require significant understanding or experimental data for the particular system to be applied effectively (e.g., measurement of diffusion coefficients). Additional consideration is given in later sections of this chapter that address the impact of drying rate on homogeneity and subsequent physical state of the amorphous solid dispersion after the spray drying process.

Coupling the thermodynamic process with the atomization process is necessary for a holistic understanding of the spray drying operation. The critical attributes of droplet formation relate to the interaction of the high surface area droplets to achieve the rapid drying rate coupled with the definition of final particle size distribution. A simple mass balance approach can be applied to a shrinking sphere physical situation where the final particle size is related to the droplet size as shown in Eq. (12.4), where $D_{particle}$ is the diameter of the spray dried particle, $D_{droplet}$ is the diameter of the atomized droplet, x_{solids} is the mass fraction of solids in the spray solvent, and $\rho_{droplet}$ and $\rho_{particle}$ are the densities of the droplet and particle, respectively.

$$D_{particle} = D_{droplet} \times \sqrt[3]{x_{solids} \times \left(\frac{\rho_{droplet}}{\rho_{particle}}\right)} \tag{12.4}$$

Although this relationship generally correlates well with spray dried amorphous solid dispersion products, it is more complicated in practice. This is due to the porous nature of the final spray dried particles that often exhibit buckled sphere morphology. Both the formulation and drying history of the particle during the spray drying process will also contribute heavily to the final particle size distribution.

Understanding the droplet size distribution can be achieved through the use of experimental tools, models, and empirical correlations. Due to the complexity of the process, it is challenging to directly apply first principles based models to the process. However, extensive research has been conducted on development of predictive models or empirical correlations that are applied for spray drying applications (Bayvel and Orzechowski 1993; Lefebvre 1988; Liu 2000; Masters 1985). Commonly employed experimental techniques give key information regarding the droplet size distribution of the atomized spray plume, velocity, and overall flux of the spray. These techniques include phase Doppler particle analysis (TSI Inc., Shoreview, Minnesota), Spraytec and PVM laser light scattering (Spraytec laser diffraction system; Malvern Instruments Ltd., Malvern, UK), and particle image velocimetry (Micro PIV laser systems, TSI Inc.). Each of these techniques can be successfully used as an offline tool to evaluate and optimize the atomizer geometry and process parameters to achieve the desired droplet size distribution across different scales of operation.

Study of the atomization process encompasses three main focus areas:

1. *Solution Properties*—key properties that impact the process include viscosity, surface tension, and density.
2. *Nozzle Geometry*—both pressure swirl and two-fluid atomizers have a wide variety of geometries that can be selected to provide a range of droplet sizes and flow rates. Ultrasonic nozzles generally have limited geometries available that are constrained by the flow rate and operating frequency.
3. *Nozzle Operating Parameters*—key operating parameters of the nozzle are selected to achieve the target droplet size and flow rate. The energy input into the system controls the final droplet size. In the case of two-fluid or pressure swirl nozzles this is in the form of pneumatic or hydraulic energy, respectively. Ultrasonic nozzles are optimized around the power input for ultrasonic vibration at a specified frequency.

Although not explicitly discussed here, scale-up considerations for spray dried amorphous solid dispersion products should be evaluated early in the development process to ensure the droplet sizes, and thus particle sizes, can be scaled up to production equipment. This is of particular importance as it relates to performance and stability aspects of amorphous solid dispersion formulations. For example, the energy required for single pressure-swirl nozzle atomization will increase with scale, and therefore the minimum achievable droplet size will also increase at larger scale. Additionally, the necessity to develop an efficient, high throughput process at large scale will necessitate the use atomizers with larger effective orifice sizes, further increasing the minimum achievable droplet size. Evaluation of production scale droplets at lab and pilot scale and using the atomized droplet size criteria for scale-up is employed to de-risk potential changes upon scale-up (Thybo et al. 2008; Vicente et al. 2013).

12.3.2 Tying Process Back to Fundamentals for Spray Dried Particle Physical Properties

12.3.2.1 Introduction

Previous discussion of the composition of the droplet during drying illustrated the importance of understanding the phase boundaries that are crossed during the transition from a liquid droplet to a dried dispersion particle (see Fig. 12.8). As described previously, the polymer passes through its gelation point during droplet drying, such that viscosity exponentially increases in the droplet. This viscosity increase should ideally occur before the drug becomes supersaturated in the droplet with respect to its crystalline solubility in the spray solution. Upon viscosity increase in the gelling droplet, the supersaturated drug will have increasingly limited mobility and will finally freeze in a glassy state where it will be unable to collide with other drug molecules to nucleate and grow crystals.

In this preferred situation, prior to gelling of the bulk droplet, the polymer—especially for polymers with good film-forming properties—will undergo the step of "skinning." This phenomenon occurs when the polymer is not able to diffuse toward the center as fast as the drying solvent front around the receding droplet and a thin layer of polymer film forms at the surface of the droplet (Vehring et al. 2007). The skin of the droplet can define the resultant spray dried particle's surface composition and will have important implications for many aspects of the bulk material properties. The following section will describe how the drying process, independent of droplet composition, may alter the surface chemistry and properties of these powders and impact properties from bulk powder flow to physical stability. Several important analytical tools for quantifying the surface compositions of spray dried amorphous solid dispersion particles will also be discussed.

12.3.2.2 Process and Solubility Induced Concentration Gradients Across Droplets/Particles

As a binary drug/polymer amorphous solid dispersion undergoes droplet drying, a small concentration gradient may develop based either on molecular weight or solubility differences between the drug and polymer. In the case of relatively high solubility of both components in the spray solution, the differential diffusion rates of polymer and drug may lead to a polymer skin forming at the surface of the drying droplet and lead to a particle with a relatively polymer-enriched surface. If the API or the polymer has a relatively poor solubility in the spray solution relative to the other component, then the lower solubility component would be expected to precipitate initially and concentrate on the surface due to the slow diffusion coefficient of the larger precipitated material (Boraey and Vehring 2014).

It may be preferable to obtain a particle with a polymer-rich surface. A polymer-rich surface will generally have a higher T_g and have similar flow and dispersability independent of the drug (Sun et al. 2011).

In the situation where both components have high solubility in the spray solution, the kinetics of drying the droplet can impact the surface enrichment of the particle. Figure 12.16 illustrates the concept that at shorter drying time constants, more heterogeneity at the surface—increased surface enrichment of the higher molecular weight species—would be expected instead of a more homogeneous system that one would obtain at longer drying time constants. Based on this model, manipulation of drying kinetics can lead to spray dried powders with different surface compositions, and, therefore, different properties. In the situation where both polymer and drug have similar solubility in the spray solvent, fast drying kinetics will lead to a higher concentration of the larger molecular weight (polymer) species at the surface. Slower drying kinetics will result in less of a concentration gradient throughout the particle as both components will be mobile enough to diffuse homogeneously throughout the droplet. In this way, drying kinetics can partially determine the composition of the surface of the particle.

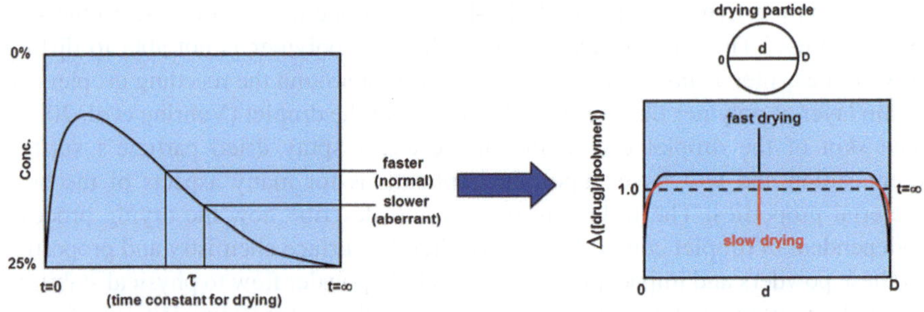

Fig. 12.16 Model relating drug concentration of an atomized droplet composed of 25 % (w/w) API, 75 % (w/w) polymer to drying kinetics and positional concentration gradient across final dry particle

In situations where both drug and polymer are not highly soluble in the spray solvent, it is the kinetics of drying convolved with super saturation that can drive a concentration gradient across a particle. In the ternary phase diagram shown previously (Fig. 12.8), the composition of the droplet first crosses the polymer gelation point and then the drug solubility. If this sequence of events is reversed, then the drug may precipitate out as larger particles in the droplet and diffuse much slower than the solvent and therefore concentrate on the surface of the dried particle. Given very rapid drying kinetics, the liquid can be removed before a precipitation event occurs. However, in the event that the composition crosses the drug solubility line before the polymer gelation point, careful attention must be paid to the process, especially when moving from smaller scale (generally faster drying kinetics) to larger scale (generally slower drying kinetics).

Composition gradients approaching the surface of spray dried amorphous solid dispersion particles can be difficult to detect with common bulk characterization techniques, such as modulated DSC (Wang et al. 2002), solid state nuclear magnetic resonance (NMR) (Pham et al. 2010), or X-ray scattering (Alderborn and Wikberg 1995). Sensitive surface specific techniques can be utilized to correlate process to surface composition, which is especially critical when scaling a process.

12.3.2.3 Particle Surface Characterization Techniques

The preceding discussion outlined the relevance of surface composition of spray dried particles. In this section, specific techniques will be described that can, in a complementary manner, determine the surface composition of the spray dried particles.

Figure 12.17 illustrates three specific techniques that measure the surface composition of spray dried particles using various chemical detection mechanisms as well as evaluating different surface depth profiles. Using each of these techniques alone can be descriptive of the surface composition, but using them together can give a quantitative picture of the concentration gradient that may be established during the spray drying process for a given amorphous solid dispersion formulation.

Depicted in Fig. 12.17a, Time-of-Flight Secondary-Ion-Mass-Spectrometry (TOF SIMS) is a technique used in materials science and surface science to analyze the composition of solid surfaces and thin films by sputtering the surface of the specimen with a focused primary ion beam and collecting and analyzing ejected secondary ions (Benningtoven 1994). The mass–charge ratios of these secondary ions are measured with a mass spectrometer to determine the elemental, isotopic, or molecular composition of the surface to a depth of 1–2 nm (Benningtoven 1994). The depth-dependent composition can be analyzed by sequential sputtering and analysis of the sample. Due to the large variation in ionization probabilities among different materials, TOF SIMS is generally considered to be a qualitative technique, although quantitation is possible with the use of standards (Benningtoven 1994). One of the most effective "standards" is to analyze a sample as a function of depth until the signal reaches its bulk composition. This is the most sensitive surface

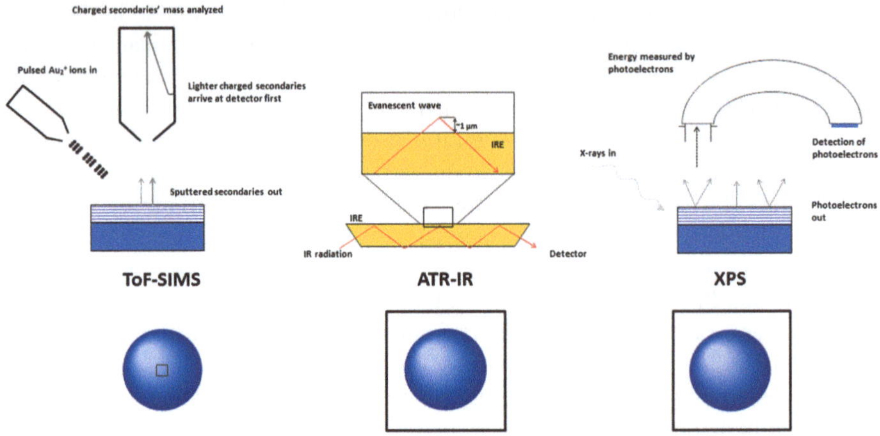

Fig. 12.17 (**a**) TOF-SIMS uses detection of fragmented ions mass spectrometry to generate fingerprint characterization of surface (quantitative) and generate ion mapping images (qualitative); (**b**) ATR-IR monitors ratio of intensities and areas of drug specific vibrations to polymer specific vibrations; (**c**) XPS is a tool for quantitative analysis of elemental surface composition by detection of photoelectrons

analysis technique, with elemental detection limits ranging from parts per million to parts per billion. TOF SIMS is also capable of generating images of particle surfaces based on ion fragment concentrations (Benningtoven 1994).

In pharmaceutical solids, TOF SIMS can be used in depth profiling in solid dosage forms but in a softer mode (via selection of appropriate primary ion type and energy) can sample the surface of spray dried particles and give information about the very outer layer. While this technique can be used in a bulk manner, the beam size is such that if sample preparation is appropriate, single particle surfaces can be evaluated. Drug concentrations can be determined based on the specific mass spectra of the molecules being evaluated.

Attenuated total reflectance (ATR), illustrated in Fig. 12.17b, uses the property of total internal reflection resulting in an evanescent wave. A beam of infrared light is passed through the ATR crystal in such a way that it reflects at least once off the internal surface in contact with the sample. This reflection forms the evanescent wave which extends into the sample. The penetration depth into the sample is typically between 0.3 and 2 μm, with the exact value being determined by the wavelength of light, the angle of incidence, and the indices of refraction for the ATR crystal and the medium being probed (Mirabella 1993).

Practical use of this technique involves evaluation of chemical compositions of the surface of spray dried amorphous solid dispersions based on the vibrational frequencies of bonds specific to the molecule of interest. In this technique it is critical to find a functional group that has a differentiating peak from the polymer matrix in order to be quantitative about the concentration of drug in the sampling depth of the measurement. ATR combined with Fourier Transform Infrared

Spectroscopy (ATR-FTIR) has a much larger depth of penetration than either TOFSIMS or X-ray Photoelectron Spectroscopy (XPS) and therefore is typically not sensitive to surface effects, but is useful in determination of concentration gradients that extend over hundreds of nanometers. Despite its limitations, ATR-FTIR is widely available and therefore is often used as a routine check of samples, while a more detailed understanding requires more sensitive techniques. ATR-FTIR microscopes are also available with spatial resolution of tens of microns.

As illustrated in Fig. 12.17c, X-ray Photoelectron Spectroscopy (XPS) is a photon-in/electron-out technique that involves exposing a sample to X-rays under ultrahigh vacuum and analyzing the emitted electrons as a function of kinetic energy (KE). The binding energy (BE) is calculated as $BE = h\upsilon - KE - \varphi$ where $h\upsilon$ is the energy of the incident X-ray beam and φ is the work function of the analyzer. Because of the low energy of the electrons emitted (100–1,400 eV for a standard XPS system using an Al or Mg anode), the analysis depth for XPS is on the order of 1–10 nm depending on the element analyzed and the characteristics of the material (atomic composition and density). For example, for dense polyethylene, the mean free path escape depth for 1,000 eV electrons (corresponding to N 1 s electrons using an Al anode) is about 3.5 nm and for 400 eV electrons (corresponding to Na 1 s electrons), it is about 1 nm (Nanse et al. 1997).

XPS is useful in sampling bulk surfaces of spray dried particles and is capable of determining not only the elemental composition of the surface but also different bonding environments of those atoms. Drugs or molecules of interest that have atoms with larger cross sections can be particularly appropriate for this type of surface analysis. Compositional depth profiling using XPS is possible either by taking measurements at multiple angles (of limited use for samples that are not atomically smooth), sputtering samples between measurements with an ion beam, or taking measurements as a function of incident X-ray energy at a synchrotron (Nanse et al. 1997).

12.3.2.4 Case Studies from Pharmaceutical Spray Dried Amorphous Dispersion Development Programs

The preceding sections have described the links between the spray drying process, thermodynamic and kinetic phenomena, and their potential impacts on spray dried particle physical properties. The next section delves into several development case studies illustrating scenarios in which the spray drying process may impact particle characteristics.

Impact of Surfactant on Particle Morphology

As discussed previously, an atomized droplet undergoes different stages of drying on the way to formation of the dried particle. The investigation in Fig. 12.18 shows

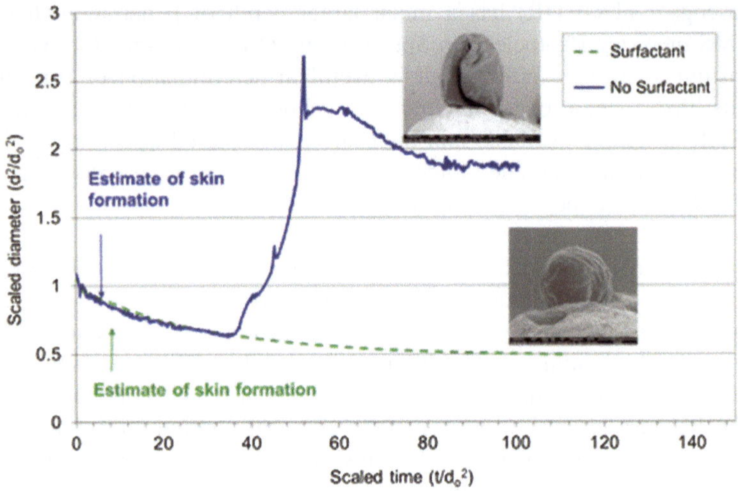

Fig. 12.18 Scaled diameter vs. scaled time data representing drying of levitated droplets containing 20 % (w/w) solids loading in an acetone/water solvent system with 0.4 m/s drying gas at 68 °C. The bulk solids loading at skin formation point was determined to be 24 % (w/w)

the results of a levitated droplet drying study. Several methods of studying single droplet drying have been reported with levitation being a recent method growing in popularity (Sloth et al. 2006). In this study, an approximately 1,000 μm droplet is injected into a chamber and held in position by balanced ultrasonic energy waves. Slow flow of hot gas is passed over the droplet and the change in droplet diameter is observed via high speed camera. Because the initial droplet surface area is significantly different between experiments and this variability will affect drying kinetics, the diameter and time is normalized by the square of the initial droplet diameter. The skinning point is observed by watching the video and visually determining the point at which the surface just begins to wrinkle or show inflection. The total solids loading at the time of skin formation can then be estimated by calculating solvent loss using measurements of both the initial diameter and the diameter at the moment of skin formation, assuming constant droplet density. Since the total mass of solids in a droplet/particle remains the same, mass balance can be used to calculate solids loading using the initial droplet properties.

The initial droplets contained 20 % (w/w) total solids, of which 65 % (w/w) was PVP-PVA polymer. Since droplets containing a high concentration of polymer will generally have a $Pe >> 1$, early skinning of the droplet surface is expected, which is supported by the results. In this study, the impact of surfactant in the amorphous solid dispersion formulation on droplet drying and final particle morphology was investigated. Both formulations began to skin at the droplet surface around the same time and followed similar drying curves during the early, unhindered drying phase. The formulation with surfactant underwent continued drying and shrinking of the particle until it approached the end of the drying curve, when the final particle size formed. However, the droplet without surfactant apparently formed a vapor bubble at a point after skin formation which expanded, causing the overall drying droplet to expand beyond the initial droplet size. The difference in this drying behavior also led to significantly different final particle morphology.

The expanded particle collapsed as the shell strength could not withstand the external forces from final drying and thermal contraction during cooling (Vicente et al. 2013) whereas the particle with surfactant yielded a more spherical particle. Both of these experiments were repeated and approximately the same drying behavior was observed. Since the drying conditions were the same, this case study underscores how surface activity, solubility, and diffusivity of the formulation components play a role in maintaining a certain degree of solvent saturation at the surface and can alter the morphology of the thickening shell during drying (Handscomb 2008; Vehring 2008). These differences may significantly impact downstream performance like suspension uniformity, flow, and compactibility.

Impact of Drying Kinetics on Particle Morphology

Another investigation of droplet drying in the ultrasonic levitator studied the impact of drying gas temperature (as a surrogate for drying kinetics) on skin formation and evolution of dried particle size (Fig. 12.19). For this study, a droplet from a solution of 10 % (w/w) solids containing 80 % (w/w) HPMCAS-L was injected into the levitator and dried with a constant flow of gas at varying temperatures. The time to surface skinning appears to be longer with reduced gas temperature, which may be explained by the extra time needed for solute diffusion and remixing (Vehring 2008). For gas temperatures <55 °C, the droplet continued to shrink down to its final size after skin formation. However, particle ballooning was observed when the gas temperature increased to 65 °C and extreme particle expansion occurred at 99 °C. Given the boiling point of acetone at 56 °C, these observations support the hypothesis that a drying droplet producing a dense, nonporous skin during drying may lead to boiling of trapped solvent that can expand and balloon the particle. The formulation was also studied in a DOE format within a GEA Niro PSD-1 spray dryer. The observed final particle size of the spray dried amorphous dispersion produced at 56 °C outlet temperature was, in fact, larger than the initial droplet size.

12.3.3 Impact of Spray Dried Particle Physical Properties on Oral Suspension Formulations

Amorphous solid dispersions manufactured by spray drying are finding increased use in preclinical toxicology formulations, where sponsors seek to establish extremely wide safety windows through substantial exposure margins using poorly soluble drugs. Figure 12.20 shows a pharmacokinetic profile of a compound dosed

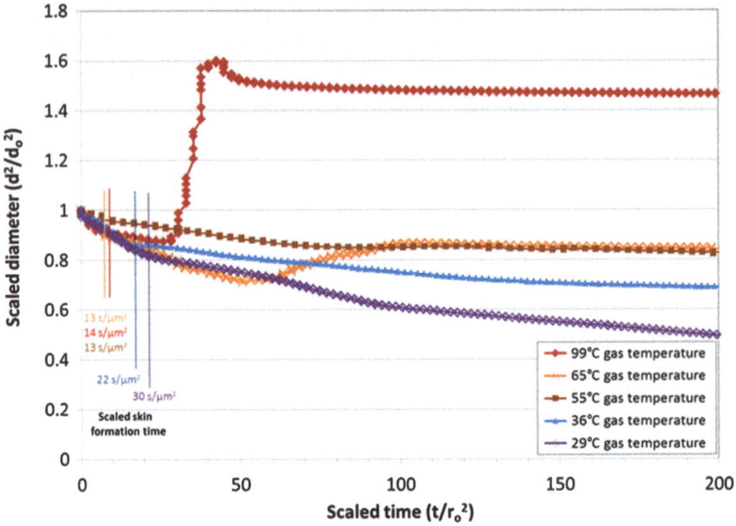

Fig. 12.19 Scaled diameter vs. scaled time data representing drying of levitated droplets containing 10 % (w/w) solids loading in an acetone solvent system with a 0.4 m/s drying gas velocity. The solid formation consisted of 20 % (w/w) drug, 80 % (w/w) HPMCAS

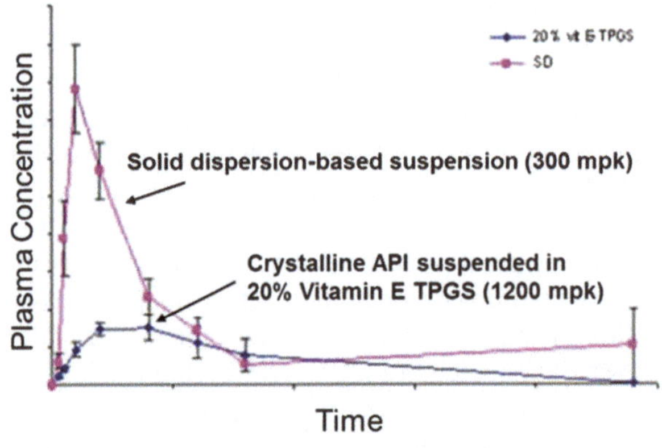

Fig. 12.20 Pharmacokinetic profile showing plasma concentration as a function of time for two formulations dosed to male Sprague-Dawley rats ($n=4$). The spray dried solid dispersion suspension was dosed at 300 mg/kg, whereas the crystalline drug suspension in 20 % Vitamin E TPGS in water was dosed at 1,200 mg/kg

to male Sprague-Dawley rats, comparing a crystalline drug suspended in a vehicle of 20 % (w/v) Vitamin E TPGS in water to a spray dried amorphous solid dispersion suspended in a vehicle of methylcellulose, sodium dodecyl sulfate, and hydrochloric acid in water. The spray dried amorphous solid dispersion is administered at a 75 % lower dose than the crystalline drug suspension, yet it achieves an eightfold higher overall exposure (area under the curve) than the crystalline drug suspension.

Suspension homogeneity is a critical component of effective preparation of oral suspensions. A heterogeneous suspension can result in delivery of substantially more or less dose than desired. The physical properties of spray dried particles may substantially impact the homogeneity of the oral suspension. Quick drying droplets result in hollow spherical particles with significant void space, resulting in low density particles. In theory, particles with a density less than that of the vehicle will float on the surface. In practice, the addition of viscosity modifiers and wetting aids to the vehicle enable an effective dispersion of spray dried powders in aqueous vehicles with powder bulk densities as low as 0.1 g/mL.

Another failure mode of oral suspensions is segregation due to particle settling. Although suspensions are routinely stirred during preparation of preclinical toxicology suspensions, they are typically removed from the stirring apparatus during animal dosing. Consequently, the oral suspension must remain homogeneous for several hours during the course of animal dosing. Stokes' law provides a fundamental basis to model the drag force (F_D) of spherical objects settling in a continuous fluid with laminar flow. Stokes' law can be derived from the Navier–Stokes equation for a spherical particle of diameter D, which is settling in a fluid of viscosity μ at velocity v to give:

$$F_D = 3\pi\mu D v \tag{12.5}$$

Consider a spherical particle of diameter D and density $\rho_{particle}$, which is settling under gravity in a fluid of density ρ_{fluid} and viscosity μ, as depicted in Fig. 12.21. A momentum balance can be constructed, accounting for the downward weight of the particle, the upward buoyant force, the upward drag force, and the downward acceleration of the particle. In order to determine its steady settling velocity, v_s, we assume that the net weight of the particle is counterbalanced by the drag force and there is no acceleration, such that:

$$v_s = \frac{(\rho_{particle} - \rho_{fluid})}{18\mu} g D^2 \tag{12.6}$$

where g is the acceleration due to gravity and other parameters have been defined above. Using Eq. (12.6) and data on the density of the particles as well as the density and viscosity of the vehicle, we can construct an analysis of the settling velocity as a function of spray dried particle size as shown in Fig. 12.22. A similar analysis can be done to assess the impact of spray dried particle density on settling velocity.

The physical properties of spray dried particles may substantially impact the uniformity of oral suspensions. The morphology, density, and size of the particles may be linked directly to process parameters (temperatures, flow rates, atomizer geometry, etc.), solvent systems, and equipment scale. Therefore, an established

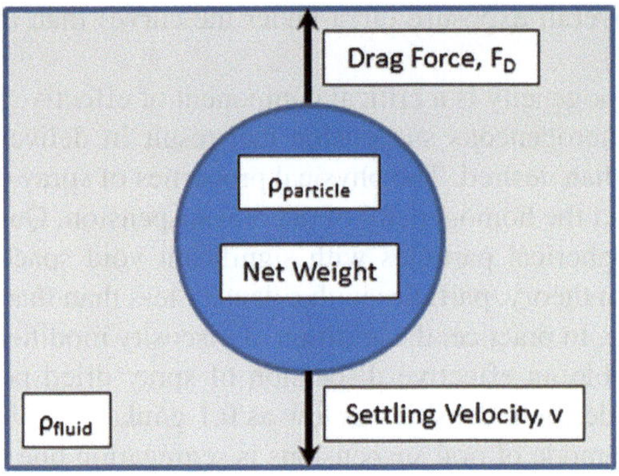

Fig. 12.21 Settling of a spherical particle through a fluid under gravity

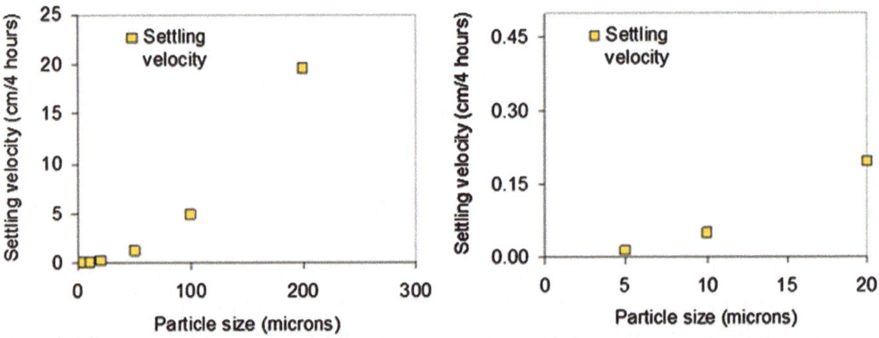

Fig. 12.22 Calculated settling velocity of spray dried particles as a function of particle size

relationship between particle properties and suspension quality can mean the difference between a well-executed toxicology study and an intractable study whose data are so variable that separation of dose groups becomes impossible.

Studies designed to assess the toxicity of new chemical entities in preclinical animal species often necessitate the administration of high (> 400 mg/kg) doses. Since the volume of formulation administered to the animal cannot change dramatically between doses, it is often necessary to develop formulations of high concentration. Random close packing describes the maximum volume fraction of randomly packed solid particles. For the simplified case of monodisperse spherical particles, it has been analytically shown that the volume fraction in random close packing cannot exceed a density limit of 63.4 % (Song et al. 2008). A typical formulation of spray dried hydroxypropyl methylcellulose acetate succinate exhibits a density of 0.3 g/mL. Given the density limit, random close packing would suggest that the maximum achievable concentration of a typical spray dried suspension is 190 mg/mL. Efforts to increase the maximum feasible concentration might be achieved by seeking to alter the density of the spray dried particles, including changes to the drug load, selection of alternate polymers, the incorporation of additional excipients, and the use of different spray drying process parameters.

12.3.4 Impact of Spray Dried Particle Physical Properties on Oral Solid Dosage Forms

Spray dried powder properties can have a profound impact on downstream unit operations including powder and granule flow, compactibility, and tablet disintegration. Polymer deformation properties (e.g., plasticity) have a dominating impact on the compactibility of tablet formulations containing $\geq$30 % (w/w) spray dried amorphous solid dispersion. The porosity and morphology of the spray dried powder also have an impact on compactibility (Alderborn and Wikberg 1995). The particle size, shape, density and surface charge can have a substantial impact on powder flow, both as neat spray dried powder as well as diluted within a formulation (Prescott and Barnum 2000). In addition, the polymer molecular structure and hygroscopicity play a major role in the tablet disintegration properties (Hughey et al. 2013).

As one example, a spray drying design of experiments (DOE) study on a Niro GEA Mobile Minor™ spray dryer evaluating the impact of droplet size and chamber outlet temperature produced a wide array of spray dried particle properties; specifically, bulk density, particle size, and morphology. In this study, no single spray dried particle physical property completely explained the large variation in powder flow observed by studying the unconfined yield strength as a function of major principle stress during shear cell flow measurements (Fig. 12.23). As the chamber outlet temperature is reduced, the particles take on a more wrinkled morphology, particle size and bulk density increase, and the flow is generally negatively impacted. However, as seen by batch 1, very small particle size can produce extreme cohesivity in the blend and severely impact flowability. Samples spray dried using higher outlet temperatures produced smoother surface particle morphology, lower bulk density, and generally exhibited the best flow behavior. These observations are not generalizable and the impact of spray dried particle physical properties on downstream formulation powder and granule flow should be assessed as a component of development because different trends can be observed. Other experience has shown that particles with higher bulk density can impart a more freely flowing tablet formulation, which may be caused by interactions with other physical properties or with the other formulation components. The concentration of spray dried amorphous solid dispersion in the downstream formulation strongly correlates with overall flow ability as well.

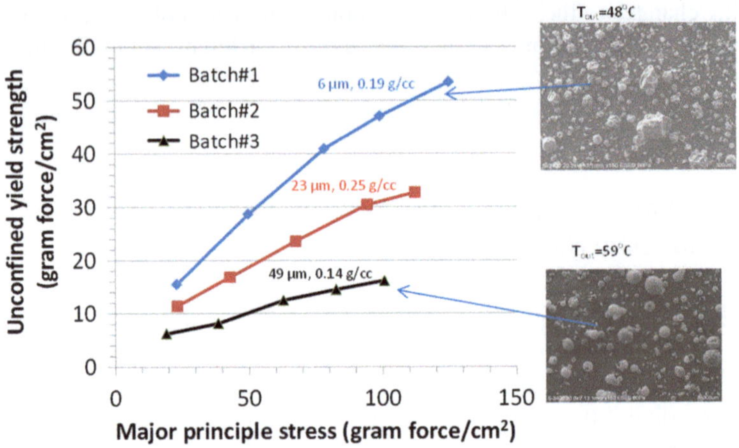

Fig. 12.23 Impact of spray dried powder properties on bulk powder flow measured by shear cell. Concentric ring with rotating top plate setup. 1.2 kgf normal load for preconditioning. Powder bed preconditioned under max load before shear before each test point. Samples analyzed in controlled room environment around 30% RH.80 % (w/w) loading of spray dried powder in Avicel PH102: lactose monohydrate tablet formulation. Spray dried particle size (D_{50}) and bulk density listed by specific data sets, color coded according to the legend

The powder blend compactibility was also significantly impacted by the properties of the spray dried powder (Fig. 12.24), specifically bulk density. Several other powder blend samples were analyzed (not shown) and maximum strength (area under compaction curve) exhibited an approximately linear correlation with spray dried powder bulk density. In addition, the spray dried powder bulk density was predominantly manipulated through changes in the spray dryer chamber outlet temperature. In this work, the particle porosity is defined using the following equation

$$\% \text{ porosity} = \frac{V_{\text{int}}}{\left(V_{\text{int}} + \frac{1}{\rho_{\text{t}}}\right)} \tag{12.7}$$

where V_{int} is the total intruded mercury from 0 to 40,000 psig mercury pressure and ρ_{t} is the true (or skeletal) density of the spray dried intermediate. A common technique called mercury porosimetry was leveraged to measure the amount of mercury intruded into the spray dried powder under a range of pressures. In many cases, a linear correlation can be drawn between particle density and bulk or tapped density unless some significant morphology differences exist (e.g., intact vs. fractured particles).

More specifically, Fig. 12.25 shows a fairly tight correlation between powder and granulated blend tensile strength and the overall porosity of the spray dried particles. The slope for the granulated blend is less steep because some compactibility is lost during dry granulation from the high porosity spray dried

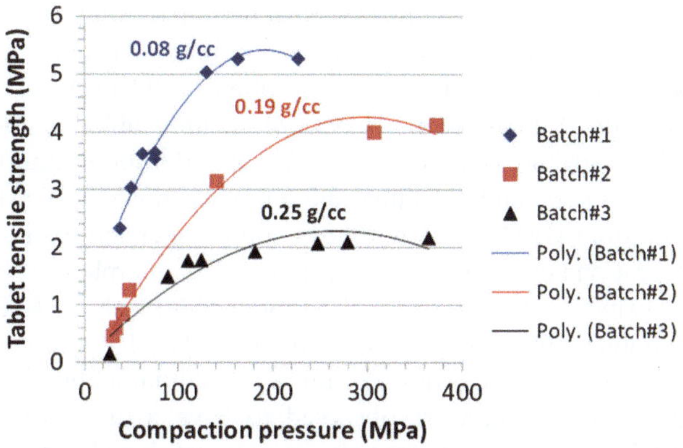

Fig. 12.24 Impact of spray dried powder bulk density on powder blend compactibility. Spray dried powder bulk density indicated. 80 % w/w loading of spray dried powder in Avicel PH102–lactose monohydrate tablet formulation. Spray dried powder is 80:20 w/w HPMCAS-L–drug spray dried from acetone. Compaction conducted on an ESH compaction simulator, single stage linear compaction with displacement control and upper/lower punch force readout. Holland tablet hardness tester

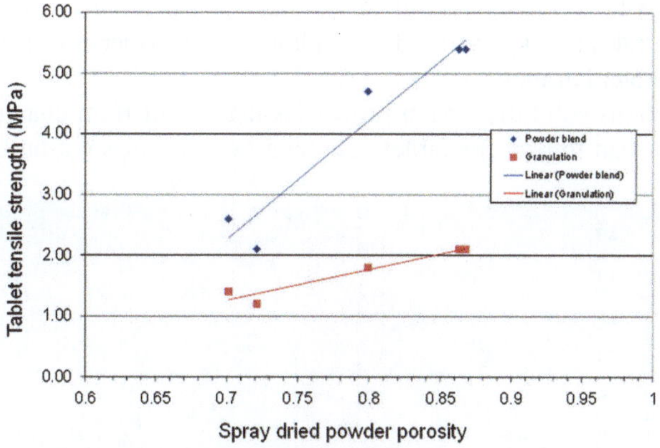

Fig. 12.25 Impact of spray dried powder porosity on tablet tensile strength at 200 MPa compaction pressure, made from powder and granule blend at 80 % w/w loading of spray dried powder in formulation. Compaction conducted on an ESH compaction simulator, single stage linear compaction with displacement control and upper/lower punch force readout. Holland tablet hardness tester. Mercury porosimetry conducted on the spray dried powder using a PoreMaster 60 and 0.5-cc cells. Low pressure intrusion up to 40 psig, high pressure 20–40,000 psig. Data sets merged and overall porosity calculated from total intruded volume

powder. This may be explained by the larger degree of deformation and void filling during compaction creating more surfaces for bonding and leading to higher tablet strength (Alderborn 1995). Fairly good linear correlations of tablet tensile strength or hardness and spray dried particle porosity are observed for many polymer-based spray dried amorphous solid dispersions. In some cases, a weaker effect of particle size or morphology is observed. This may also be impacted by changes in surface area, leading to differences in volume reduction during compaction (Alderborn 1995).

The composition of the spray dried amorphous solid dispersion may play a substantial role in the disintegration behavior of the final tablet. A study comparing the impact of different formulation components (in addition to drug) of the amorphous solid dispersion on tablet disintegration time is shown in Fig. 12.26. Tablets prepared from a drug/HPMCAS amorphous solid dispersion exhibit typical, fast eroding disintegration behavior. However, spray dried amorphous solid dispersions formulated using hydroxypropyl methylcellulose (HPMC) and polyvinylpyrrolidone–polyvinyl acetate copolymer (PVP-PVA) produced a slowly eroding, gelling matrix tablet. After 45 min, a substantial amount of these tablets remained and when broken, were observed to have a dry core. The different behaviors observed between polymer types may be attributed to the differences in hygroscopicity and polymer solubility in the dissolution media (Hughey et al. 2013). Hydrophilic polymers, such as PVP, HPMC, and PVP-PVA, often produce a slowly eroding tablet that quickly forms a thick gel layer near the outer edge of the tablet upon dispersing in aqueous based media (DiNunzio et al. 2012). The gel layer slowly sloughs off during disintegration, allowing for more water penetration and a new layer to propagate slowly into the tablet. Formulation techniques that have proven useful to combat this phenomenon include dilution of the concentration of spray dried amorphous solid dispersion in the tablet and reduction of tablet hardness.

The amorphous solid dispersion properties that result from changing the spray drying process can impact the tablet disintegration and dissolution behavior even

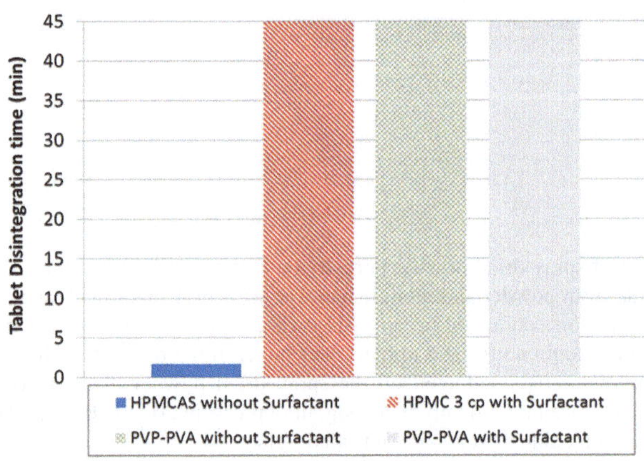

Fig. 12.26 Disintegration time of 500 mg, 15/32 in. round concave tablets containing 50 % (w/w) loading of various types of spray dried amorphous solid dispersions. Spray dried amorphous solid dispersions contained 20 % (w/w) of the same drug. The tablet formulation consisted of the same composition of microcrystalline cellulose, lactose, croscarmellose sodium, glidant, and lubricant. The tablets were compressed to achieve 2 MPa tensile strength. Disintegration times were measured using a Vankel USP disintegration tester filled 900 mL of pH 3.3 phosphate buffer and maintained at 37 °C

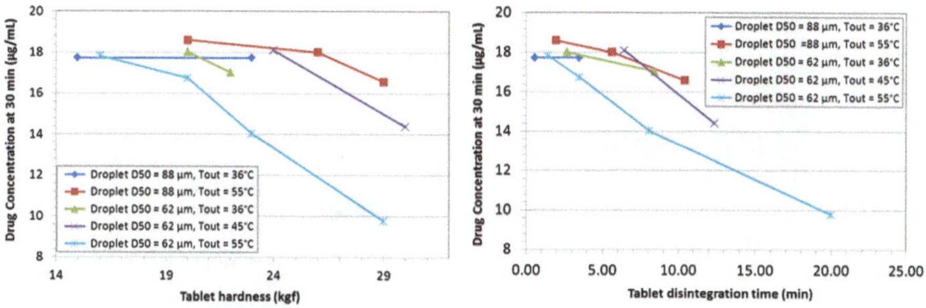

Fig. 12.27 (*Left*) Dissolution of 14/32 in. round concave tablets containing 50 % (w/w) HPMCAS based spray dried amorphous solid dispersion formulation processed at various atomized droplet sizes and gas outlet temperatures. Dissolution was performed with a 20 µg/mL target drug concentration in FaSSIF media in a USP II dissolution vessel at 100 rpm. Samples were pulled and filtered through 1 µm filter before analysis. (*Right*) Same dissolution data set plotted against observed tablet disintegration time in the USP II dissolution bath

when prepared with the same formulation composition and compressed to the same hardness. The left inset of Fig. 12.27 shows the results of a study evaluating the impact of spray drying process conditions on dissolution rate and function of tablet hardness. The spray dried amorphous solid dispersion was blended with excipients and dry granulated at 1 MPa tensile strength, followed by milling, lubrication and tablet compression to various degrees of hardness. The target concentration of drug in the dissolution evaluation was driven by the apparent solubility of amorphous drug in fasted state simulated intestinal fluid (FaSSIF) media (Galia et al. 1998) to provide sufficient sink for full dissolution of drug. The study showed that small droplet size and high outlet gas temperature produced spray dried powder that when processed, yielded much slower dissolution at a given hardness than the rest of the process samples. The sample with the smallest droplet size and highest outlet temperature exhibited the lowest bulk density, 0.06 g/cc, and highest particle porosity, 88 %. The smaller particle size and lower particle density yielded a larger number of particles per tablet. During the compaction event, this larger number of particles may have formed a stronger network of compacted polymer, resulting in slower disintegration (Fig. 12.27, right). Overall, these data illustrate that spray drying process conditions may impact both tablet disintegration and dissolution, even if the tablet formulation and processing are identical.

12.4 Failure Mode: Spray Drying Impact on Chemical Degradation

During the spray drying process, in-process hold times and parameter optimization are conducted to understand the potential impact for chemical interactions and potential impurity formation in the spray dried amorphous solid dispersion formulation. The key areas for evaluation are summarized below:

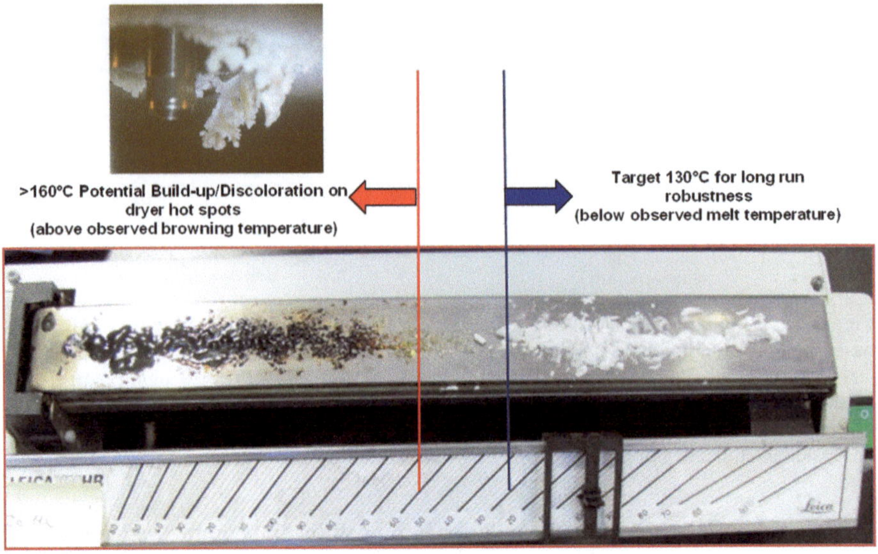

Fig. 12.28 Overview of hot plate experiment used to optimize the drying gas inlet temperature

1. *Spray solution*—evaluate interaction of API with polymer and other excipients in solution over a time scale sufficient to cover the duration of the manufacturing process. Any instability may be mitigated through the use of solution temperature control (e.g., reducing solution temperature) or addition of excipients such as antioxidants that prevent the primary degradation pathway.
2. *Inlet temperature exposure*—given the gas disperser design, recirculation patterns in the drying chamber can allow for accumulation of spray dried material near the drying gas inlet to the chamber. Long drying runs necessitate evaluation of impurity formation at these elevated temperatures. Inlet temperature of the drying gas can be optimized to minimize buildup around the disperser using simple, off-line experiments. A temperature gradient hot plate experiment is shown in Fig. 12.28, where spray dried material is exposed to temperatures similar to the inlet temperature and evaluated for visual transitions such as sticking or discoloration. Chemical analysis may also be performed on these samples. Additionally, active cooling of the disperser is employed in the Niro DPH gas disperser design to further mitigate this potential buildup and high temperature exposure (Process Engineering A/S 2014).
3. *Outlet temperature exposure*—exposure of the spray dried product in the drying chamber and cyclone collector occurs at the outlet temperature of the drying gas. Evaluation of the impact of the spray dried amorphous solid dispersion to this exposure in the presence of solvent can be critical to avoid potential product chemical instability that might occur due to large depositions in the drying chamber or continuous exposure of product during collection. These studies

can be used to define maximum duration of product changeover times or cleaning cycles. Additionally, these experiments can be coupled or used to inform the in-process hold conditions required to maintain purity of the spray dried product prior to secondary drying.

12.5 Failure Mode: Spray Drying Impact on Amorphous Solid Dispersion Phase Changes

12.5.1 Process Determinants of Spray Dried Amorphous Solid Dispersion Phase Changes

Previous discussion has focused on the impact of key spray drying process parameters on final spray dried amorphous solid dispersion bulk particle attributes (e.g., particle size, density, morphology). During process and formulation development, additional consideration must be applied to the final phase state of the spray dried product. This includes understanding the overlap of the process trajectories with potential impacts to product quality. The two key areas are (1) final distribution of components in the dry particle (i.e., particle homogeneity); and (2) residual solvent content in the particle and impact to initial or final physical stability.

Correlation of key process parameters to the final residual solvent content is important for further understanding of the interplay between the process parameters and their potential impact to physical stability of the amorphous sold dispersion. Understanding both the kinetic and thermodynamic aspects of the drying process is useful to facilitate predictive understanding. The former is generally studied using off-line experiments and modeling efforts to avoid interpretation of actual process data, where dependent variables are highly coupled. The latter can also be conducted off-line using experimental tools and correlations that are based on theory.

Other areas of study are focused on understanding the time, temperature, and equilibrium profile of different formulations and the resulting spray drying processes. Key variables that can impact the final product are:

- *Spray solution solids concentration*: this typically contributes to either the viscosity or solubility limit at which the droplet solidifies, changing the temperature profile within the particle.
- *Solvent type*: evaporation rate, viscosity, and equilibrium solubility of the formulation components will be impacted by this. Use of binary and ternary solvent blends may lead to increased complexity of the particle drying trajectories.
- *Spray dryer chamber temperatures*: process temperatures will have a large impact on the drying rate of the spray dried particles. These can be optimized around particle homogeneity or, in more extreme cases, around potential physical instability at the initial spray drying process conditions. The final outlet

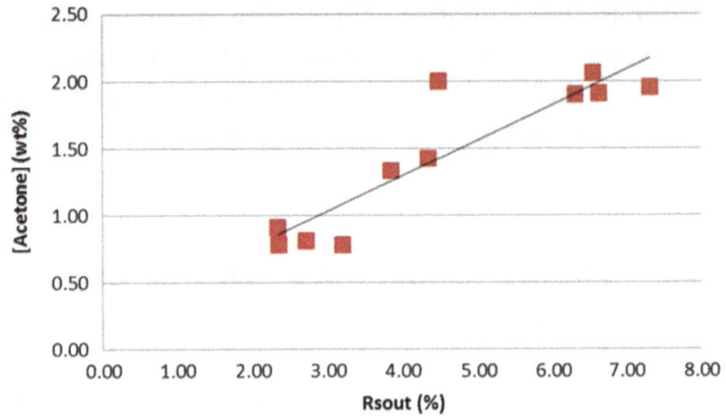

Fig. 12.29 Correlation of residual acetone concentration of a particular spray dried amorphous solid dispersion product with calculated outlet relative saturation

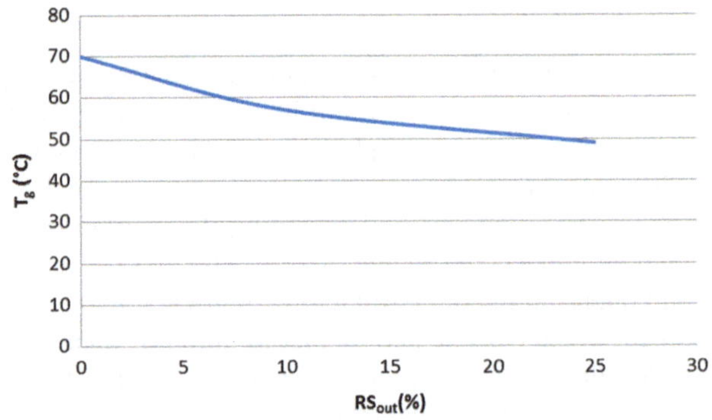

Fig. 12.30 Relationship between amorphous solid dispersion glass transition temperature and calculated outlet relative saturation

temperature should be selected to achieve target powder attributes, while still providing an acceptable margin between the T_g of the amorphous solid dispersion at the final residual solvent content.

- *Spray dryer relative saturation*: outlet relative saturation is defined by the fraction of solvent present in the vapor phase at the spray dryer outlet temperature. This also impacts the drying rate by changing the driving force for mass transfer and will define the equilibrium residual solvent content in the amorphous solid dispersion. Data collected from the spray drying process over a range of outlet relative saturations are shown in Fig. 12.29, demonstrating the impact on final residual solvent concentration.

Assessment of a formulation's equilibrium solvent state can be conducted using dynamic vapor sorption (DVS), where the vapor is the spray solvent of interest rather than the typical moisture sorption study. DVS experiments facilitate understanding of both solvent sorption/desorption kinetics and final equilibrium values. These data can then be coupled with studies to measure the wet T_g of the spray dried product as a function of residual solvent concentration to generate a correlation of the measured physical property to the actual process space. Figure 12.30 illustrates an example of the impact of outlet relative saturation on the T_g by combining the experimental results from equilibrium DVS experiments with measurement of T_g

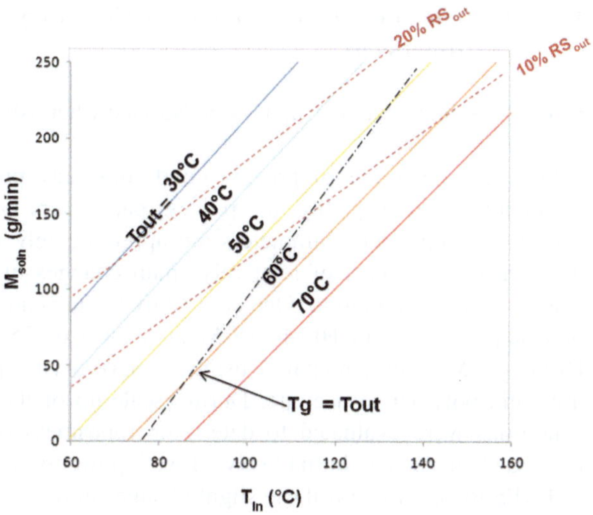

Fig. 12.31 Example of process space employing physical stability constraint

depression of amorphous solid dispersions with a known mass fraction of residual solvent.

Generation of this correlation facilitates evaluation of the quality impact of process parameters by using a thermodynamic model of the process space. A drug product quality constraint operating line for physical stability risk can be used to define the available process space to only regions where risk of physical stability is low based on direct measurements of the amorphous solid dispersion attributes. The T_g can be plotted as a function of outlet relative saturation using the outlet temperature and relative saturation iso-lines that are defined by the thermodynamics of the process space. Additionally, conservative heuristics can be applied (e.g., $T_g - 10\ °C \geq T_{out}$) to ensure that there is low physical stability risk, as illustrated in Fig. 12.31. This approach is most effectively deployed early during process development and formulation selection to ensure that the final process and formulation will be effectively scaled-up to production dryers where closed loop operation will lead to increased outlet relative saturation and increased risk of physical instability.

12.5.2 Tying Process Back to Fundamentals for Spray Dried Amorphous Solid Dispersion Phase Changes

In this section, a case study is used to illustrate how process can impact the surface chemistry and physical stability of a binary drug-polymer spray dried amorphous solid dispersion particle. The "solvent wet" spray dried powder, before secondary drying to remove residual solvent, was placed in stability chambers and the surface of the particle was evaluated to investigate how surface composition may change while the amorphous solid dispersion sits in a mobile condition.

12.5.2.1 Impact of Process on Surface Chemistry of Amorphous Solid Dispersions

Effect of Subtle Process Changes on Surface Chemistry

Although a given set of process conditions may produce an amorphous solid dispersion of good quality, the process parameters typically shift upon scale-up even when dependent parameters are approximately matched. It is important to understand the impact of relatively small changes in process parameters on the quality of the formulation. One case study features an amorphous solid dispersion of Compound X and HPMCAS-M at a ratio of 25 % Compound X to 75 % HPMCAS-M (w/w), prepared using a spray drying process from a solution of drug and polymer in acetone. During scale-up of the material, multiple process conditions were evaluated to determine robustness of the formulation to slight process changes. The initial "solvent wet" powders were evaluated by ATR FTIR and XPS to determine if these slight changes in process conditions had altered the surface chemistries of the resultant spray dried powders.

The structure of the polymer is shown in Fig. 12.32. Compound X is a lipophilic drug molecule that contains an ester group (structure not shown) that could be differentiated by infrared spectrometry as shown by the spectral data illustrated in Fig. 12.33. The ATR FTIR carbonyl region was used to quantify the relative amount of Compound X and HPMCAS at the surface of the particles as a ratio (w/w).

In order to quantify the amount of drug at the surface of the particles, spray dried amorphous solid dispersions comprising various ratios of Compound X and HPMCAS were manufactured at a scale where the particles were sufficiently small that the entire composition was visible to the depth penetration of the particles (~1 μm). ATR FTIR analysis of these samples was used to generate a standard curve (data not shown) of known Compound X and HPMCAS relative to absorbance of the stretches of interest. Based on this standard curve, samples prepared using similar but slightly different process conditions were analyzed (Lots A and B from Fig. 12.35). The plot in Fig. 12.34 shows the relative amounts of Compound X at the surface of the powders from each process. In lot A, approximately 18 % (w/w) of Compound X was measured on the surface. ATR FTIR analysis revealed that lot B had a slightly lower amount of Compound X at the surface, approximately 16 % (w/w).

XPS was used to confirm these relatively subtle differences. Both lots were evaluated as compacts pressed into indium foil and fluorine was used to quantify the

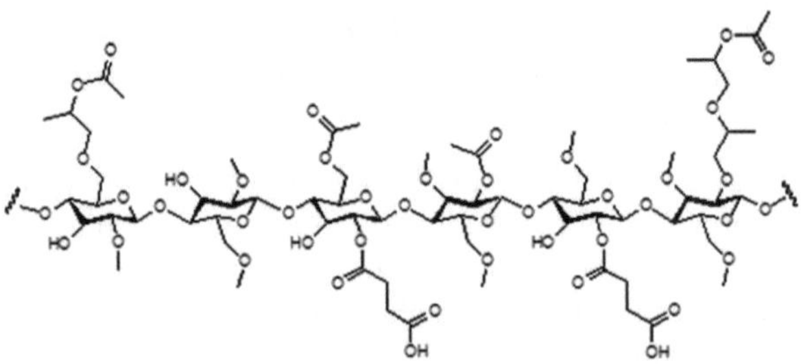

Fig. 12.32 Chemical Structure of hydroxylpropyl methylcellulose acetate succinate (HPMCAS)

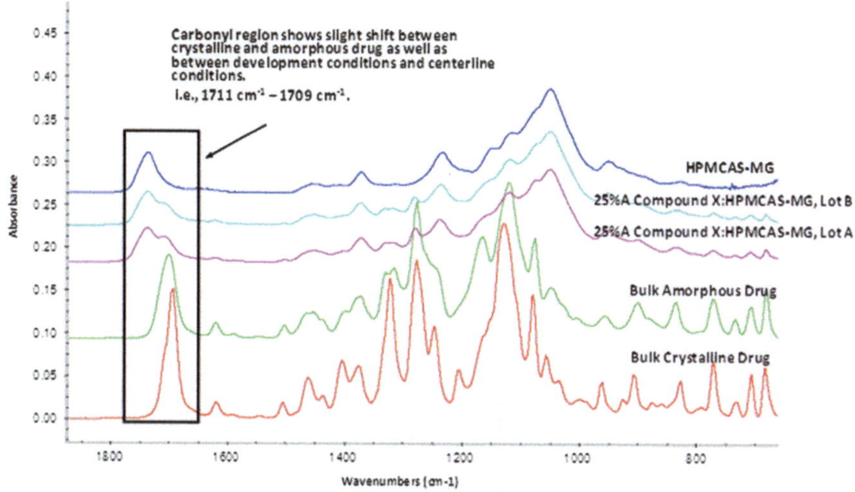

Fig. 12.33 ATR-IR spectra of pure HPMCAS-M, bulk crystalline Compound X, and bulk amorphous Compound X compared with 25 % (w/w) Compound X spray dried amorphous solid dispersions processed using two different sets of processing conditions

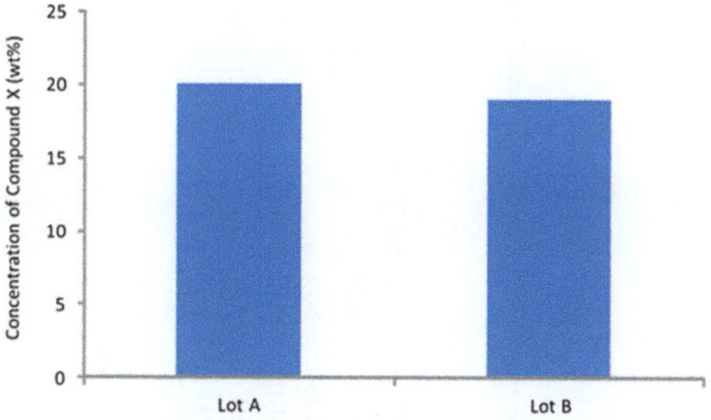

Fig. 12.34 Surface composition of Compound X based on ATR-IR of batches of spray dried amorphous solid dispersions manufactured using different processing conditions

amount of Compound X measured in the experiments. The atomic composition of the surface was determined by fitting a baseline to the spectrum for each transition investigated and determining the area under the curve (A) for each element; then dividing by the radial structure function (RSF) and following the equation, where *i* denotes each of the elements detected:

$$\text{Atomic\%} = \frac{\frac{\text{A}}{\text{RSF}}}{\sum_i \frac{\text{A}(i)}{\text{RSF}(i)}} \tag{12.8}$$

Table **12.3** presents a summary of this analysis. XPS analysis showed that both lots exhibited relatively similar surface concentration gradients. The film-forming

Table 12.3 XPS data summary showing surface elemental composition of two Compound X spray dried amorphous solid dispersions manufactured using different processing conditions

25 % Compound X:75 % HPMCAS Lot A						25 % Compound X:75 % HPMCAS Lot B					
Sample	C/N	C/F	F/N	F/C	% Cmpd X (w/w)	Sample	C/N	C/F	F/N	F/C	% Cmpd X (w/w)
1	131.2	23.4	5.6	0.043	9.82	1	114.3	23.3	4.9	0.043	10.00
2	109.5	26.3	4.2	0.038	8.77	2	134.0	25.8	5.2	0.039	9.12
3	108.3	25.0	4.3	0.040	9.12	3	109.8	23.5	4.7	0.042	9.82
Average	116.3	24.9	4.7	0.04	9.24	Average	119.4	24.2	4.9	0.04	9.65
St. Dev.	12.9	1.5	0.8	0.00	0.53	St. Dev.	12.9	1.4	0.3	0.00	0.46

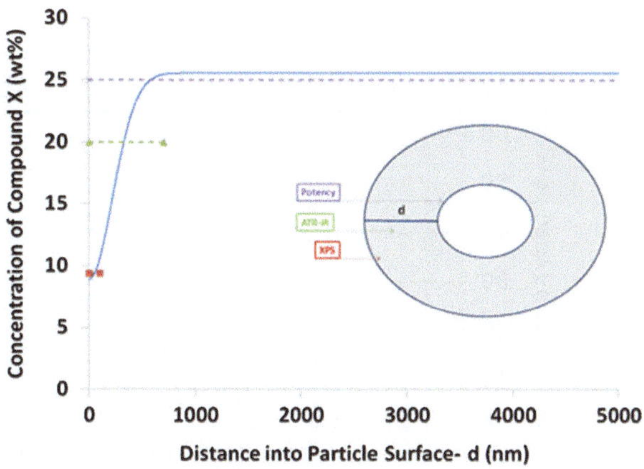

Fig. 12.35 Schematic of Compound X spray dried amorphous solid dispersion particle summarizing characterization data and surface in homogeneity

nature of the polymer and the high active solubility in the spray solution, in this formulation, has led to high polymer content at the surface of these particles. As discussed previously in this chapter, this physical situation is expected to lead to improved physical stability. The similarity of the particle surfaces at the different drying conditions demonstrates the robust nature of the formulation and process. Based on both surface analytical techniques, it was determined that the drug product had demonstrated a process space robust to changes in surface composition. The data was then used to construct a model for the concentration gradient of Compound X across the particle based on the lead formulation and the scaled up process. Figure 12.35 shows a cartoon depiction of the concentration gradient of Compound X at the surface of the spray dried particles based on the depth penetration of ATR FTIR and XPS as complementary analytical techniques.

Effect of Spray Solvent on Surface Chemistry

Another case study evaluates how the water content in the spray drying feed solution impacts the spray dried particle surface concentration using XPS (Fig. 12.36). Scanning electron microscopy (SEM) confirmed that these powders were mainly intact particles, so XPS was probing the outer surface and not a mixture of surfaces. The three spray dried samples in the data set were spray dried from the same solids composition under identical process conditions on a Niro PSD-1 spray dryer. As the drug concentration in the initial feed solution approaches the solubility limit, solvent removal prompts precipitation of the drug on the surface of the spray dried particles (Vehring 2008). The concentration of the surfactant remained unchanged as the water amount varied which initially seems counterintuitive to the solubility limit phenomenon described previously. However, the surfactant concentration was 2–3 times above the target concentration in the feed solution, indicating that the surface activity of the surfactant dominated its

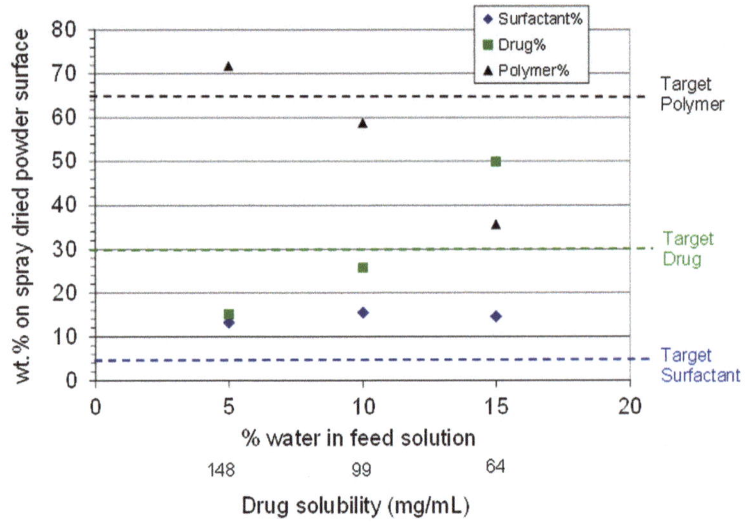

Fig. 12.36 Composition of each component on the surface of bulk spray dried powder as determined by XPS

final concentration gradient within the spray dried particle. The polymer concentration may be a result of mass balance as it has high solubility in all of solvent mixtures studied.

The process can also induce different degrees of heterogeneity within the spray dried particle due to the balance between solute solubility, rate of evaporation, and rate of solute mass transfer (Vehring et al. 2007). Figure 12.37 shows the results of a full factorial design of experiments study on the Niro PSD-1 closed cycle spray drying plant. The independent variables studied were droplet size, solids loading in the feed solution, and outlet gas temperature from the chamber. A statistical correlation including all three main effects was achieved to describe the surfactant concentration on the surface of the spray dried powder. In this case, a clear link between higher surface surfactant concentration and slower drying kinetics was observed. Larger droplet size and lower gas temperature both led to slower drying kinetics because of lower overall surface area per unit volume and smaller temperature differential for heat transfer. Higher solids loading leads to earlier skin formation (data not shown) which will ultimately reduces the drying rate. The surfactant in this case is a small molecule with larger diffusivity than the polymer. In addition, it is a surface active compound, so it is possible that slower drying kinetics will lead to enrichment at the surface when given more time for mass transfer. Although surface enrichment of individual components in spray dried particles may or may not have practical implications on downstream performance or stability these techniques allow for deeper interrogation of the product–process–performance interplay and may be useful in developing a quality by design approach to development.

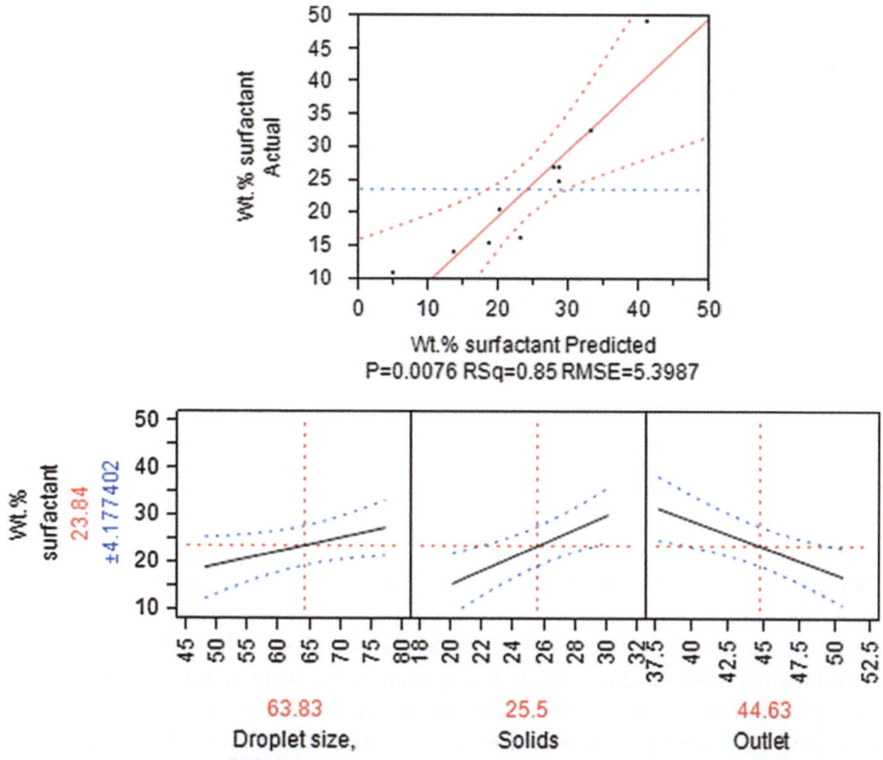

Fig. 12.37 Statistical model output (JMP software) for the surfactant concentration on the surface of the spray dried powder as measured by XPS. (*Upper inset*) Actual vs. predicted data. *Solid red line* is the regression line of the model. *Dashed lines* are the 90 % confidence intervals (CI). (*Lower inset*) Impact of process variables on surfactant concentration at the surface of the spray dried powder. *Solid black line* is the correlation line while the *dotted blue lines* represent the 90 % CI lines. XPS samples were mounted on indium-lined steel washers. PHI 5701 Lsci instrument used with monochromated Alkα 1486.6 eV X-ray source

12.5.2.2 Sources of Phase Change During Storage

Hold time after collection from the spray dryer may substantially impact physical stability of amorphous solid dispersions, given the presence of residual solvent typically at 1–10 % (w/w). As previously discussed, this "solvent wet" material exhibits a reduced T_g and could be prone to phase separation or crystallization. Along with ensuring residual solvent levels in the drug product are aligned with ICH guidelines (International Conference on Harmonisation of Technical Requirements for Registration of Pharmaceuticals for Human Use 2011), secondary drying is often pursued to reduce these solvent levels well below 0.5 % (w/w) to minimize physical stability risk.

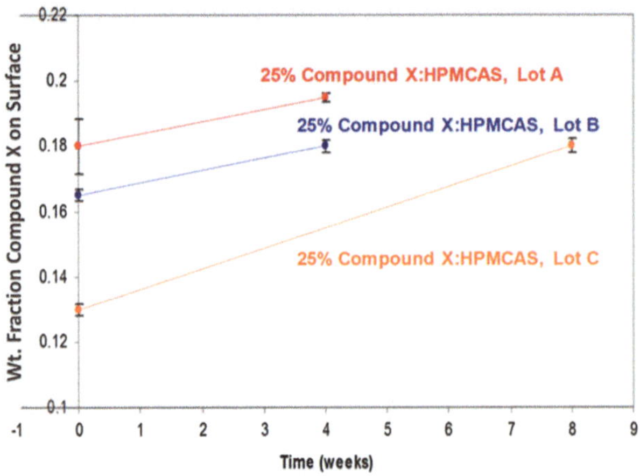

Fig. 12.38 ATR-IR data of Compound X spray dried amorphous solid dispersions manufactured with subtly different process conditions and held at 25 °C/60 % RH under "solvent wet" conditions (before secondary drying), showing increase of surface concentrations as the amorphous solid dispersions relax with time

In order to evaluate the impact of storage of the spray dried amorphous solid dispersion before secondary drying, the "solvent wet" powder was placed in stability chambers and ATR FTIR was used to monitor the increase of Compound X at the surface over time at a given storage condition. Three lots with the composition 25 % Compound X:75 % HPMCAS-M (lots A, B, and C), manufactured with different spray drying conditions, were sampled directly from the cyclone with some discrete level of residual solvent, and put into sealed containers that were placed in a storage chamber at 25 °C/60 % RH. Samples of each lot were pulled, dried by vacuum, and evaluated by ATR FTIR. Samples were evaluated for surface composition of Compound X at initial, 4 week (Lot A and B), and 8 week (Lot C) time points.

The data (Fig. 12.38) illustrate that as the spray dried amorphous solid dispersions are stored in a 'solvent wet' condition (with mobility), Compound X is able to relax toward the equilibrium condition of 25 % Compound X across the particle. Lots A and B have similar surface compositions and show nearly equivalent rates of relaxation given a comparable initial concentration gradient. Lot C was prepared with faster drying kinetics but higher throughput initially and has a lower initial surface concentration of Compound X, consistent with the model. Because a larger concentration gradient has been set up during the process, the mobility enabled by the wet storage drives faster diffusion of Compound X in lot C which increases more rapidly at the surface.

This case study illustrates the potential impact of the spray drying process on the surface composition of spray dried particles. It also underscores that a kinetically stabilized amorphous solid dispersion may relax when given sufficient mobility and time. Implications of these changes are potentially seen in the changes in flow of the spray dried powder depending on wet storage time, but may more profoundly impact longer term physical stability.

In another case study, a stressed stability study was conducted on a 35 % (w/w) drug in HPMCAS-L spray dried amorphous solid dispersion that was immediately

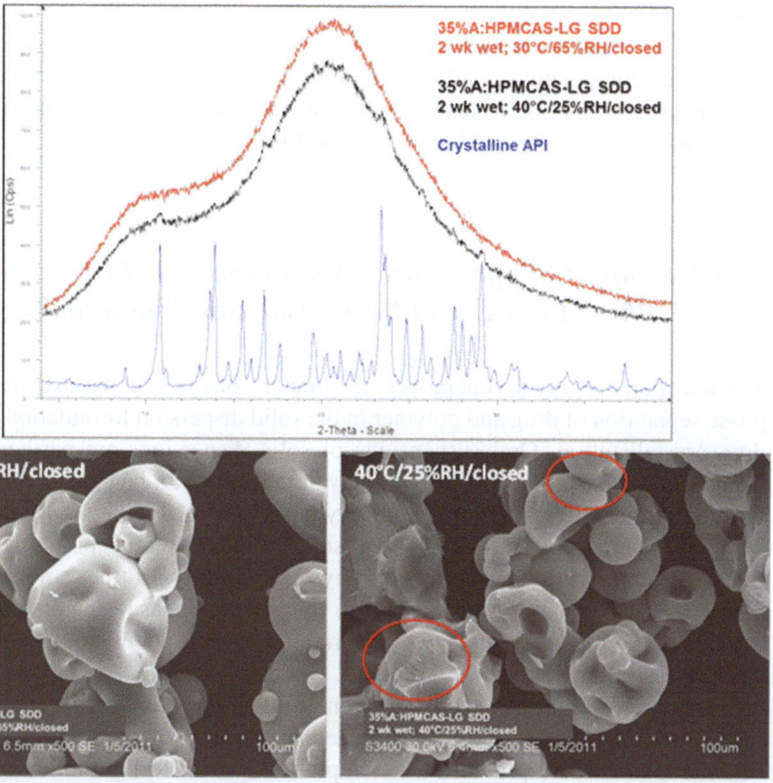

Fig. 12.39 High resolution X-ray powder diffraction (XRPD) on "solvent wet" spray dried amorphous solid dispersion samples stored for 2 weeks at two different conditions (*top*). Scanning electron microscope (SEM) on stored samples, 30 °C/65 % RH (*lower left*), 40 °C/25 % RH (*lower right*). Evidence of particle fusing and potential growth of surface crystals circled in *red*

removed from the spray dryer cyclone (with "solvent wet" residual solvent), transferred to a glass vial capped with an impermeable stopper, and then stored at 30 °C and 40 °C (Fig. 12.39). The dry glass transition of the amorphous solid dispersion was measured to be around 50 °C, so the "solvent wet" T_g was likely even lower. Evidence of crystallization and particle fusing was observed using XRPD and SEM for the sample stored at 40 °C. It is important to consider how the process can impact drug concentration on the particle surface. A melting endothermic peak was also measured using modulated differential scanning calorimetry (data not shown). SEM is a valuable tool to observe subtleties like surface crystal formation that could be at a level below the limit of detection (LOD) of other physical characterization techniques, like XRPD.

It has been demonstrated that crystallization from amorphous systems is more likely to occur at the surface (Sun et al. 2011). It has also been shown that a thin polymer film on the surface can dramatically inhibit crystallization and improve longer term physical stability (Strydom et al. 2009). Therefore, understanding how the surface composition is relaxing to a more drug-rich composition is critical to relating process conditions and how these might impact the initial amorphous solid dispersion surface state and long-term stability.

12.5.3 *Impact of Spray Dried Amorphous Solid Dispersion Phase Changes on Oral Suspension Formulations*

The selection of spray drying process parameters may impact the risk of amorphous phase separation of drug and polymer in the solid dispersion formulation or outright drug crystallization. Oral suspensions of solid dispersions are particularly risky from a physical stability perspective, given that they can be considered to put the formulation in a 100 % relative humidity environment. Figure 12.40 shows polarized light microscopy images of suspensions of two different spray dried formulations, illustrating the challenge of physical stability of the amorphous solid dispersion in aqueous suspensions. The impact of a phase separated or partially crystallized formulation on *in vivo* performance of amorphous solid dispersions has been discussed previously.

Spray dried solid dispersions are prone to hydration in the aqueous suspension media, which may result in heterogeneous swelling of particles in the dosing vehicle. Figure 12.41 provides an example of a polarized light microscopy image of a suspension of a spray dried amorphous solid dispersion. Some particles swell as large as 100 μm, while others retain the original spray dried particle size of approximately 5 μm. This nonuniform swelling phenomenon may impact suspension homogeneity on a macroscopic scale, suspension viscosity, and tendency for creaming and foaming.

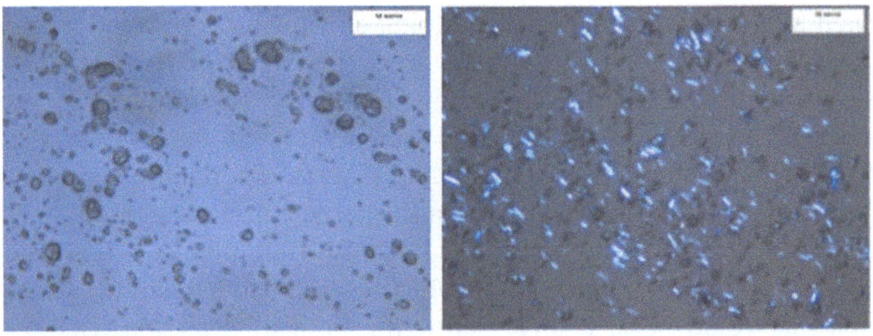

Fig. 12.40 Polarized light microscopy images of aqueous suspensions of two different spray dried formulations. The formulation on the left side (**a**) exhibits isotropic behavior of particles, suggesting amorphous phase state. The formulation on the right side (**b**) shows indications of birefringence, suggesting drug crystallization

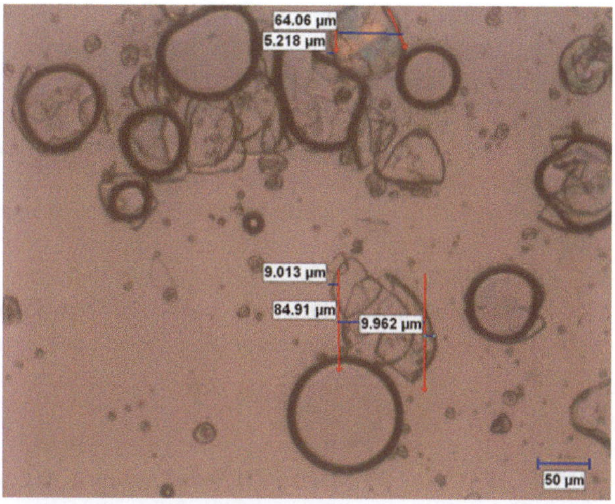

Fig. 12.41 Polarized light microscopy image of an aqueous suspension of a spray dried formulation. Hydration of the spray dried particle results in heterogeneous swelling

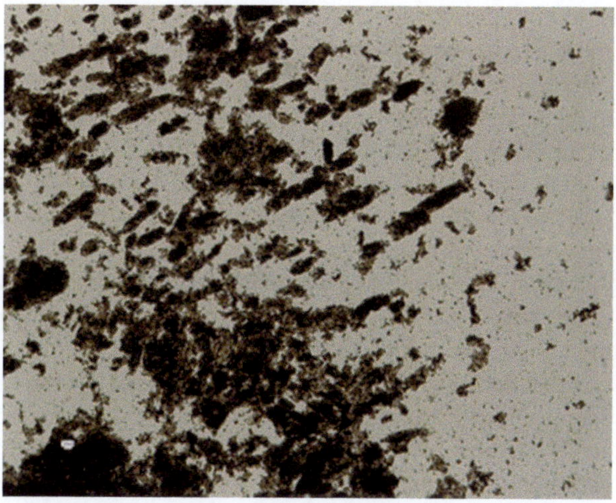

Fig. 12.42 Polarized light microscopy image of an aqueous suspension of a phase separated spray dried formulation. The spray dried particles exhibit agglomeration, resulting in significant gelling and creaming behavior

Hydrophobic drug compounds that are phase separated from polymer may exhibit a tendency to agglomerate in the aqueous suspension media. Figure 12.42 illustrates a polarized light microscopy image of a formulation that emerged from the spray drying process phase separated due to the selection of solvent system and spray dryer process parameters. The image illustrates the tendency of particles to agglomerate, rather than form a fine suspension. At a macroscopic level, the suspension exhibited severe gelling and creaming behavior, substantially reducing maximum feasible concentration and constraining dose uniformity.

The addition of polymers to the aqueous vehicle of oral suspensions has been shown to delay the nucleation and slow the crystal growth of drugs (Simonelli et al. 1970). Previous studies have also shown a correlation between both the concentration and the molecular weight of a polymer and its ability to slow crystal growth rate (Ziller and Rupprecht 1988). Consequently, high molecular weight polymers may be exploited to enhance the physical stability of amorphous solid dispersions in oral suspension formulations. However, the use of high concentrations of high molecular weight polymer may limit maximum feasible concentrations. Surfactants are widely used in suspension formulations as wetting agents and preservatives. However, studies have shown that surfactants, particularly anionic ones, may reduce the inhibiting effect of polymers on drug crystallization (Ziller and Rupprecht 1988). It has been suggested that anionic surfactants, like sodium dodecyl sulfate, aggregate with polyvinylpyrrolidone in the aqueous phase, preventing the polymer from inhibiting growth at crystal faces (Saito et al. 1971).

12.5.4 Impact of Spray Dried Amorphous Solid Dispersion Phase Changes on Oral Solid Dosage Forms

In practice, many spray dried amorphous solid dispersion formulations are kinetically trapped in a metastable state and may phase separate over time. Of relevance to the pharmaceutical world is the time frame over which these phase changes may occur and whether that time frame facilitates sufficient shelf life to maintain desired critical quality attributes. Furthermore, the physical stability challenge is complicated in the drug product, where the amorphous solid dispersion must remain in a single phase amorphous state while mixed with additional excipients and filled into a capsule or compressed into a tablet. Consequently, it is important to investigate drug product stability in both the amorphous solid dispersion powder as well as in the final drug product. Analysis of the neat spray dried amorphous solid dispersion powder decouples several variables that may interfere with interpretation of its stability in the drug product.

Given the kinetic state of the system, environmental factors such as temperature and humidity can impact the stability and phase or physical behavior of the amorphous solid dispersion. The importance of the glass transition temperature has been previously discussed. In addition, prior sections have discussed how the amorphous solid dispersion will approach a liquid-like state at and above the glass transition temperature. A key determinant is mobility of the solutes within the amorphous solid dispersion, which can lead to amorphous-amorphous phase separation or crystallization of drug or other components of the system.

The interaction of the amorphous solid dispersion particles with other excipients in the drug product may also present development challenges. The drug is amorphous and more susceptible to chemical reactivity with other excipients in the final drug product (e.g., capsule or tablet). Chemical or physical interactions of the drug with excipients in the formulation should be considered and studied as part of drug product development. Spectroscopic and other analytical methods to study these interactions should be considered, but are practically challenging to deploy due to interference in the spectra and limit of detection challenges.

In addition to phase stability, particles experiencing conditions above the glass transition temperature can be sticky (Paterson et al. 2007) and begin to adhere to themselves and other surfaces. In extreme cases, this could lead to bulk powder handling issues such as flow and compaction issues or yield losses in process equipment. Particle fusing, leading to an effective surface area decrease, can also reduce the dissolution rate.

The presence of water or residual solvent in the amorphous solid dispersion may have more substantial impact to the product than particle fusing, such as phase separation or drug/excipient crystallization. As discussed earlier, water or residual solvents may reduce the enthalpy of the amorphous system and effectively limit the apparent solubility attainable by the amorphous solid dispersion. Dissolution of spray dried amorphous solid dispersions can present the drug in many forms including true solution (i.e., molecularly dissolved drug), nanoparticles (i.e., drug alone or in composites with excipients), colloids, and precipitation into amorphous or crystalline particles (generally $\geq 1\ \mu m$) (Friesen et al. 2008). These mechanisms and the extent of selectivity of drug into these various forms will very often depend on the homogeneity of the drug and excipients throughout the spray dried amorphous solid dispersion particle. Therefore, phase separation or crystallization due to presence of water or residual solvents can significantly alter the performance of these particles in the drug product.

12.6 Conclusions

Spray drying is a mature process with demonstrated capability for manufacturing amorphous solid dispersions. However, the impact of the drying process on the performance, manufacture, and stability of the drug product is often complex. The process may impact the chemical and physical stability of the formulation, its compatibility with downstream unit operations, and the in vivo performance of the drug product. Although potential failure modes can be conceptually linked to first principles, thoughtful process development leverages a combination of fundamentals, modeling, and empirical experimentation to mitigate risks and ensure a high quality drug product through scale-up and commercialization.

References

Alderborn G (1995) Particle dimensions. In: Alderborn G, Nystrom C (eds) Pharmaceutical powder compaction technology. Marcel Dekker, New York, NY, pp 245–282

Alderborn G, Wikberg M (1995) Granule properties. In: Alderborn G, Nystrom C (eds) Pharmaceutical powder compaction technology. Marcel Dekker, New York, NY, pp 23–24

Angell CA (1985) Strong and fragile liquids. In: Nagai K, Wright GB (eds) Relaxation in complex systems, 1st edn. National Technical Information Service, US Dept Commerce, Springfield, VA

Azadi M, Azadi M (2012) An analytical study of the effect of inlet velocity on the cyclone performance using mathematical models. Powder Tech 217:121–127

Bayvel L, Orzechowski Z (1993) Liquid atomization. Taylor and Francis, Washington, DC

Benningtoven A (1994) Chemical analysis of inorganic and organic surfaces and thin films by static time-of-flight secondary ion mass spectrometry (TOF-SIMS). Angew Chem 33 (10):1023–1043

Boraey MA, Vehring R (2014) Diffusion controlled formation of microparticles. J Aerosol Sci 67:131–143

DiNunzio JC et al (2012) Use of highly compressible Ceolus (TM) microcrystalline cellulose for improved dosage form properties containing a hydrophilic solid dispersion. Drug Dev Ind Pharm 38(2):180–189

Dobry DE et al (2009) A model-based methodology for spray-drying process development. J Pharmaceut Innovat 4(3):133–142

Flory PJ (1953) Principles of polymer chemistry, 1st edn. Cornell University Press, New York, NY

Friesen DT et al (2008) Hydroxypropyl methylcellulose acetate succinate-based spray dried dispersions: an overview. Mol Pharm 5(6):1003–1019

Galia E et al (1998) Evaluation of various dissolution media for predicting in vivo performance of class i and ii drugs. Pharm Res 15(5):698–705

GEA Process Engineering A/S (2014) Specialized GMP spray dryer for pharmaceutical products. http://www.niro.com/niro/cmsdoc.nsf/WebDoc/webb8hde6s?opendocument&q=DPH. Accessed 12 Jan 2014

Graham LJ, Taillon R, Mullin J, Wigle T (2010) Pharmaceutical process/equipment design methodology case study: cyclone design to optimize spray-dried-particle collection efficiency. Comput Chem Eng 34(7):1041–1048

Gupta P, Kakumanu VK, Bansal AK (2004) Stability and solubility of celecoxib-PVP amorphous dispersions: a molecular perspective. Pharm Res 21(10):1762–1769

Hancock BC, Shamblin SL, Zografi G (1995) Molecular mobility of amorphous pharmaceutical solids below their glass transition temperatures. Pharm Res 12(6):799–806

Handscomb C (2008) Simulating droplet drying and particle formation in spray towers. s.l.:s.n

Hansen OE, Ullum TU (2009) Air disperser for a spray dryer and a method for designing an air disperser. US, Patent No. US8157249

Hughey JR et al (2013) The use of inorganic salts to improve the dissolution characteristics of tablets containing Soluplus(R)—based solid dispersions. Eur J Pharm Sci 48:758–766

International Conference on Harmonisation of Technical Requirements for Registration of Pharmaceuticals for Human Use (2011) Impurities: guideline for residual solvents Q3C(R5), s.l.:s.n

Lefebvre AH (1988) Atomization and sprays. CRC, New York, NY

Liu H (2000) Science and engineering of droplets: fundamentals and applications, 1st edn. Noyes, Norwich

Masters K (1985) Spray drying handbook, 4th edn. George Godwin, London

Meyvis TK et al (1999) Fluorescence after photobleaching: a versatile tool for mobility and interaction monitoring in pharamceutical research. Pharm Res 16(8):1153–1162

Mirabella FM (1993) Practical spectroscopy series; internal reflection spectroscopy: theory and applications, 1st edn. Marcel Decker, Inc., New York, NY

Nanse G et al (1997) Fluorination of carbon blacks: an X-ray photoelectron spectroscopy study: I. Aliterature review of XPS studies of fluorinated carbons. Carbon 35(2):175–194

Paterson AH, Zuo JY, Bronlund JE, Chatterjee R (2007) Stickiness curves of high fat dairy powders using the particle gun. Int Dairy J 17:998–1005

Paudel A, Van den Mooter G (2012) Influence of solvent composition on the miscibility and physical stability of naproxen/PVP K 25 solid dispersions prepared by cosolvent spray-drying. Pharm Res 29:251–270

Paudel A et al (2013) Manufacturing of solid dispersions of poorly water soluble drugs by spray drying: formulation and process considerations. Int J Pharm 453(1):253–284

Perry RH, Green DW (1997) Perry's chemical engineer's handbook, 7th edn. McGraw Hill, New York, NY

Pham TN et al (2010) Analysis of amorphous solid dispersions using 2D solid-state NMR and 1H T1 relaxation measurements. Mol Pharm 7(5):1667–1691

Prescott JK, Barnum RA (2000) On powder flowability. Pharmaceut Tech 60–84:236

Roe RJ, Zin WC (1980) Determination of polymer–polymer interaction parameter for polystyrene–polybutadiene pair. Macromolecules 13:1221–1228

Saito S, Taniguchi T, Kitamura K (1971) Interactions of anionic surfactants with nonionic polymers. Comparison of guanidinium, tetraalkylammonium, and alkali metal ions as counterions. J Colloid Interface Sci 37(1):154–164

Salcedo R, Paiva J (2010) Pharmaceuticals: efficient cyclone systems for fine particle collection. Filtrat Separ 47(1):36–39

Serajuddin A (1999) Solid dispersions of poorly water soluble drugs: early promises, subsequent problems and recent breakthroughs. J Pharm Sci 88:1058–1066

Simonelli AP, Mehta SC, Higuchi WI (1970) Inhibition of sulfathiazole crystal growth by polyvinylpyrrolidone. J Pharm Sci 59(5):633–638

Sloth J et al (2006) Model based analysis of the drying of a single solution droplet in an ultrasonic levitator. Chem Eng Sci 61:2701–2709

Song C, Wang P, Makse H (2008) A phase diagram for jammed matter. Nature 453:629–632

Strobl GR (2007) The physics of polymers, 3rd edn. Springer, New York, NY

Strydom S, Zhu L, Yu L, de Villiers M (2009) Ultra-thin coatings for stabilizing amorphous drugs against surface-enhanced crystallization. Scientia

Sun Y et al (2011) Glasses crystallize rapidly at surfaces by growing crystals upward. Proc Natl Acad Sci U S A 108:5990–5995

Thybo P et al (2008) Scaling up the spray drying process from pilot to production scale using an atomized droplet size criterion.Pharmaceutical. Research 25(7):1610–1620

Vehring R (2008) Pharmaceutical particle engineering via spray drying. Pharm Res 25(5):999–1021

Vehring R, Foss WR, Lechuga-Ballesteros D (2007) Particle formation in spray drying. Aerosol Sci 38:728–746

Vicente J, Pinto J, Menezes J, Gaspar F (2013) Fundamental analysis of particle formation in spray drying. Powder Tech 247:1–7

Wang L, Velikov V, Angell CA (2002) Direct determination of kinetic fragility indices of glass forming liquids by differential scanning calorimetry. J Phys Chem 117(22):10184–10192

Wan G, Sun G, Xue X, Shi M (2008) Solids concentration simulation of different size particles in a cyclone separator. Powder Tech 183(1):94–104

Welty JR, Wicks CE, Rorrer GL, Wilson RE (2000) Fundamentals of momentum, heat, and mass transfer, 4th edn. Wiley, New York, NY

Williams H et al (2013) Strategies to address low solubility in discovery and development. Pharmacol Rev 65:315–499

Wu T, Sun Y, Li N, de Villiers M, Yu L (2007) Inhibiting surface crystallization of amorphous indomethacin by nanocoating. Langmuir 23(9):5148–5153

Ziller KH, Rupprecht H (1988) Control of crystal growth in drug suspensions. Drug Dev Ind Pharm 14:2341–2370

Chapter 13
Nanosizing: "End-to-End" Formulation Strategy for Poorly Water-Soluble Molecules

Elaine Merisko-Liversidge

13.1 Nanonization/Nanosizing Poorly Water-Soluble Compounds

Nanonization of poorly water-soluble compounds can be accomplished through a "bottom-up" or "top-down approach" (Fig. 13.1). The "bottom-up" approach includes processes ranging from controlled precipitation (Chan and Kwok 2011), solvent evaporation (D'Addio and Prud'homme 2011), supercritical fluid technology (Pathak et al. 2005), and spray drying (Nandiyanto and Okuyama 2011), to recently explored template printing technology (Kolakovic et al. 2013). All of these approaches have been used to successfully generate drug nanoparticles (Williams et al. 2013; Douroumis and Fahr 2013; Krishnaiah 2010). The "top-down" strategies (Van Eerdenburgh et al. 2008) are generally aqueous-based formulations that do not require solvents for processing (Fig. 13.2) and employ a number of crystal fracturing technologies such as high-pressure homogenization (Keck and Muller 2006), microfluidization (Bosch et al. 1996), sonication (Ige et al. 2013), and wet media milling (Merisko-Liversidge and Liversidge 2011; Peltonen and Hirvonen 2010; Mende and Schwedes 2006; Knieke et al. 2010). The advantages and disadvantages of each method have been extensively reviewed (Merisko-Liversidge et al. 2003; Timpe 2010; Chen 2011; Kesisoglou et al. 2007a, b). Since the launch of the first NanoCrystal® product, Rapamune® in 1999, there has been a significant amount of academic and pharmaceutical research focused on this topic (Fig. 13.3). For most

The terms nanosuspension, nanocrystalline drug particles, nanoparticles, and nanosized formulation are used interchangeably throughout this chapter when referring to a dosage form wherein the active pharmaceutical ingredient (API) has a mean particle size less than one micron.

E. Merisko-Liversidge (✉)
Pharmaceutical Research and Development, Alkermes, INC., 852 Winter Street, Waltham, MA, USA
e-mail: Elaine.Liversidge@alkermes.com

A.C. Templeton et al. (eds.), *Discovering and Developing Molecules with Optimal Drug-Like Properties*, AAPS Advances in the Pharmaceutical Sciences Series 15,
DOI 10.1007/978-1-4939-1399-2_13

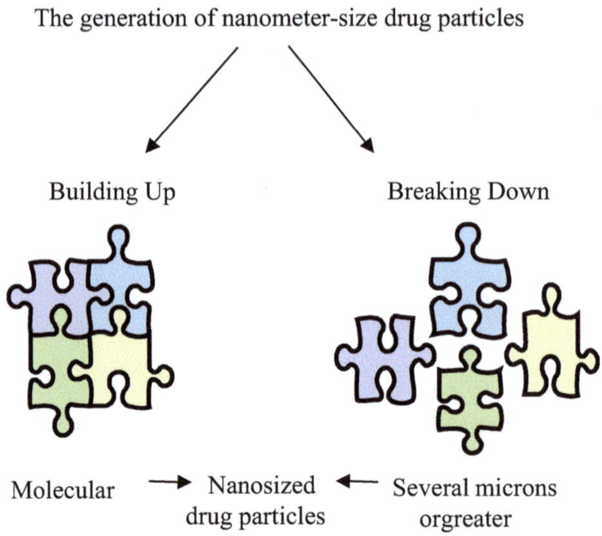

Fig. 13.1 A pictorial image of the two strategies employed for generating nanometer-sized drug crystals; the "bottom-up," approach which involves molecular assembly of nanometer-sized crystals typically using controlled precipitation methodology and the "top-down" strategy which is a crystal-fracturing strategy

nanonization processes, the intended goal is to increase the surface area of a poorly water-soluble crystalline compound. Dissolution rate of a compound is directly proportional to its surface area (Noyes and Whitney 1897) and reducing the crystal size of a compound to the nanometer-size range can significantly increase the dissolution rate. In comparing nanonization to micronization, a nanosized crystalline compound has ~50-fold increase in surface area in contrast to a micron-sized drug crystal (Fig. 13.4). However, increased surface area with increased interfacial energy can lead to colloidal instability. Most nanosized formulations require a surface modifier to dampen the surface energy of the fractured crystal. Without a surface modifier, most nanosized formulations would aggregate or agglomerate. For ease of development, the surface modifiers used in drug nanosuspensions are commonly used excipients that are capable of wetting the surface of a poorly water-soluble molecule while providing steric and/or ionic surface stabilization. This stabilization is needed for shelf-life stability, compatibility, and optimal performance after dosing.

Nanocrystalline dosage forms also benefit from the superior kinetic solubility and surface coverage provided by nanometer-sized drug crystals as opposed to a conventional formulation. Both of these properties can help improve the bioavailability of the molecule. To further explain, consideration must be given to the interplay between particle size, the boundary layer as described by (Galli 2006), and absorption surface for orally administered molecules. The boundary layer can be loosely described as the area between the absorptive surface and the dissolving particle. A number of mathematical models (Higuchi and Hiestand 1963; Shefter

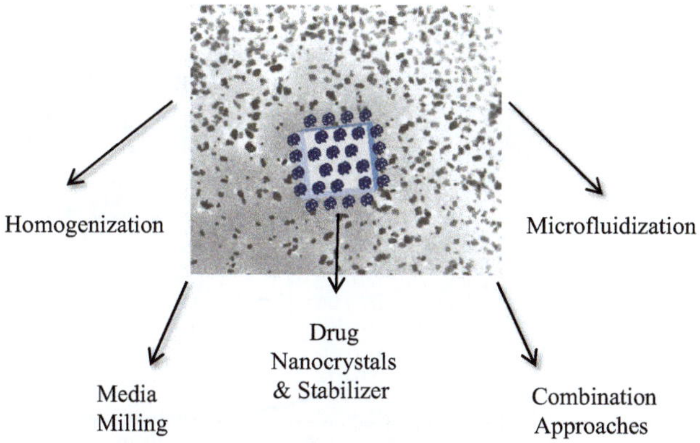

Fig. 13.2 Employing a "top-down" strategy to generate nanometer-sized drug crystals can be accomplished using a variety of approaches. Most popular of which are homogenization, wet milling and microfluidization. The transmission electron micrograph shows a nanosized diagnostic imaging agent WIN 8883 processed using wet milling (Magnification 100 K). The diagram of the fractured crystals demonstrates the nanocrystalline-fractured compound with a surface modifier that is generally required for maintain the physical stability of the nanosized dispersion

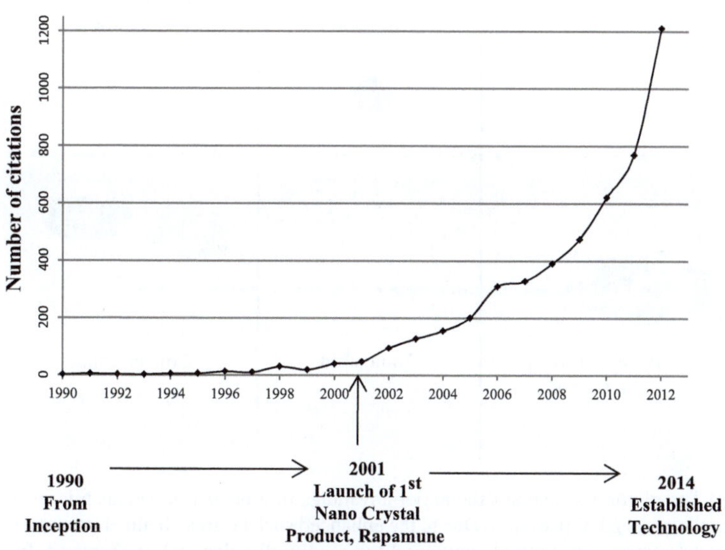

Fig. 13.3 The plot was obtained using Google Scholar searching for the term drug "nano-suspension" from 1990 through 2013. The search includes publications, patents, and patent applications. After 2001 and, the launch of the first nanocrystalline product, the publication rate has logarithmically increased

and Higuchi 1963; Wu and Nancollas 1998) attempt to provide a comprehensive correlation between dissolution rate and bioavailability by viewing the boundary layer as being dynamic and decreasing proportionately to the particle size of the dissolving drug. This effect further enhances the performance of a nanosized drug

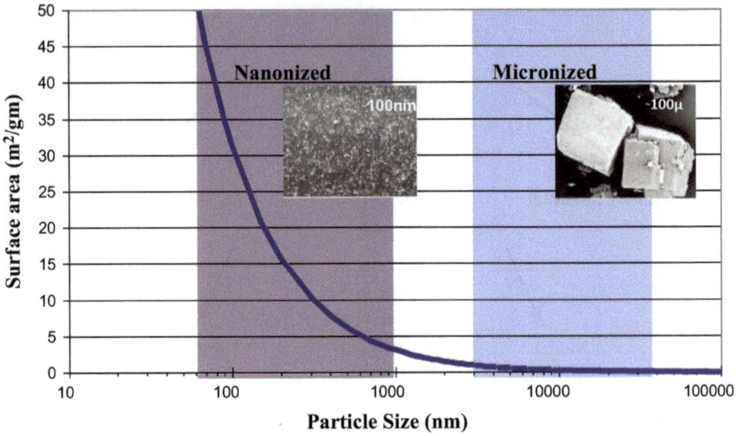

Fig. 13.4 The graph compares the surface area of nanosized and micronized drug crystals. The scanning electron micrographs (Magnification = 10 K) are images of a poorly water-soluble compound before and after nanonization

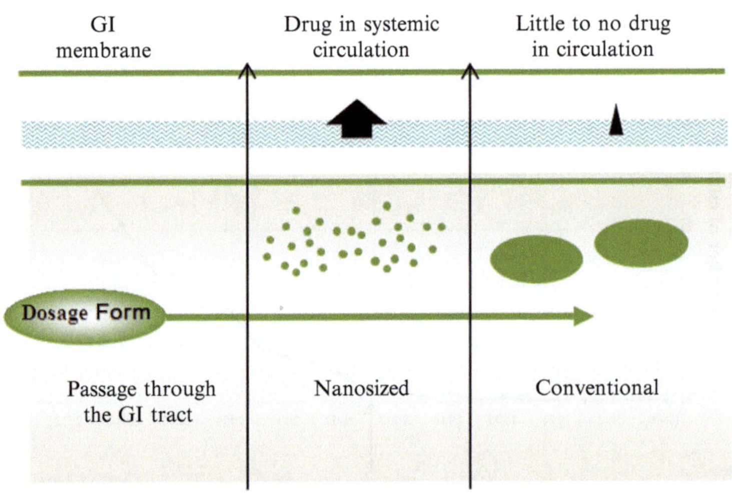

Fig. 13.5 The pictorial illustrates the major advantage of a nanosized versus micronized dosage form following oral administration. Due to the enhanced surface area obtained from the nanosizing process, a dissolution rate limited compound can readily dissolve and be absorbed. In contrast a more conventional dosage form cannot readily dissolve and hence cannot be absorbed. For compounds affected by regional absorption, this difference between a nanocrystalline formulation and a more conventional dosage form is further enhanced

particle versus its micron sized counterpart (Fig. 13.5). Junemann and Dressman (2012) demonstrate that in comparing 5- and 0.4-μm drug crystals, the latter has 12.5 times the surface area of the former. If one then considers the boundary layer being inversely proportional to particle size, theoretically, the dissolution rate of a nanosuspension can be up to 156 times faster than the non-nanosized formulation.

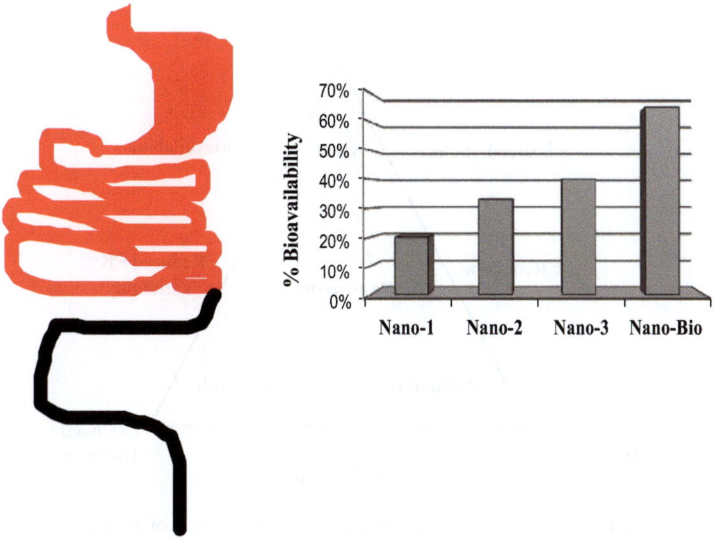

Fig. 13.6 The bar graphs compare the bioavailability of four nanosized formulations in a rat model following oral gavage. The four formulations were identical with respects to particle size (Mean ~200 nm). For each formulation, a different surface stabilizer was used. The formulation with the greatest bioavailability was stabilized with a bioadhesive polymer. The diagram further illustrates the potential for selective coating of the gastrointestinal tract which can affect bioavailability

In addition, the effect of nanosizing on the solubility of poorly water-soluble compounds has been studied (Junghanns and Muller 2008). A number of investigators have demonstrated positive effects on kinetic solubility of nanosized compounds (Jinno et al. 2006; Muller et al. 2006; Kesisoglou et al. 2007b). However, effects on intrinsic saturation solubility have been difficult to consistently demonstrate. This may be related to current limitations in reproducibly generating nanocrystalline drug particles that have a significant percentage of particles below 100 nm, which is the size range that may affect the thermodynamic solubility of a compound (Van Eerdenburgh et al. 2010). This area continues to be a subject of intense research and debate and can be heavily influenced by the analytical methods used to determine the solubility of the compound in a nanosized formulation (Van Eerdenburgh et al. 2010). A final consideration for the enhanced performance of nanosuspensions is the interaction or sequestration of the nanometer-sized particles in the mucin-coated microvilli of the duodenal/jejunal region. Extensive coverage and retention of nanocrystalline particles over the mucosal surface can enhance absorption (Moschwitzer and Muller 2006; Bosch et al. 2002). As shown in Fig. 13.6, enhanced retention of a nanosuspension has the potential to further improve bioavailability after oral dosing. In this particular example, adding a bioadhesive component to the stabilizer system increased the compound's bioavailability. In addition to oral dosing, Muller et al. (2011) have studied the effects of topical application of nanosized-rutin crystals and have demonstrated that

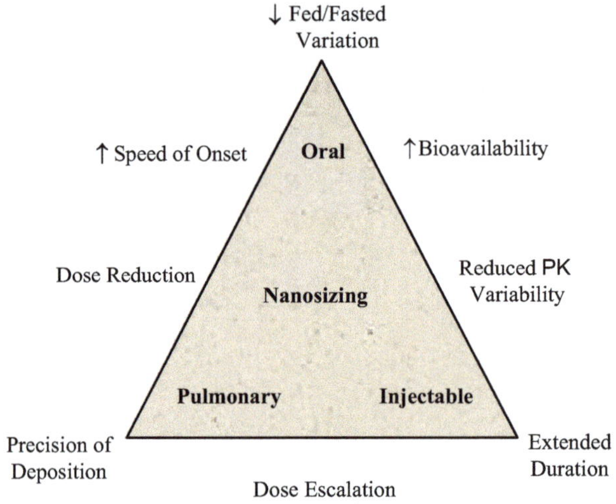

Fig. 13.7 This pictorial summarizes the demonstrated advantages of nanosized dosage forms. Nanonization can be used for the most commonly used routes of administration, including, but not limited to: oral, injectable, and pulmonary delivery

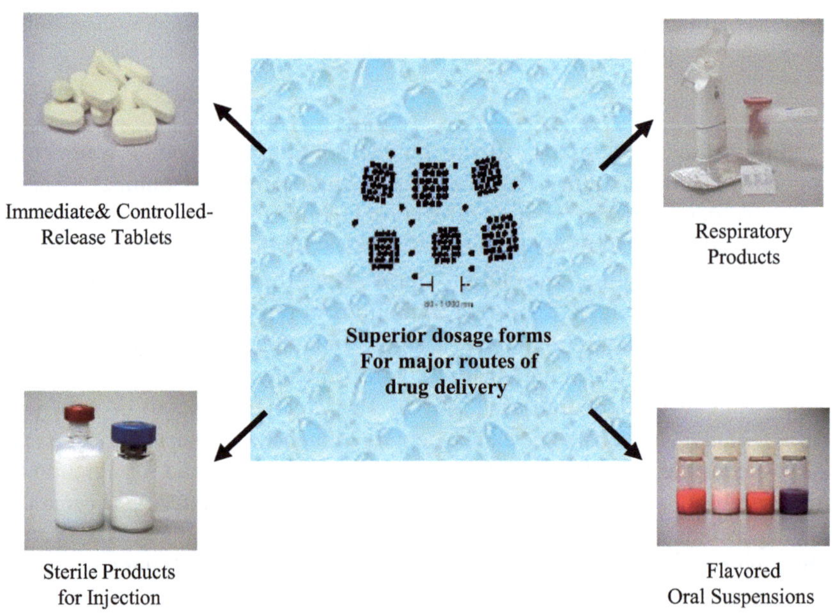

Fig. 13.8 The image shows various nanocrystalline dosage forms. Nanosized crystals can be post-processed using conventional pharmaceutical processes to produce tablets or lyophilized cakes. In addition, the liquid nanocrystalline dispersion is physically stable and compatible with flavoring agents and devices

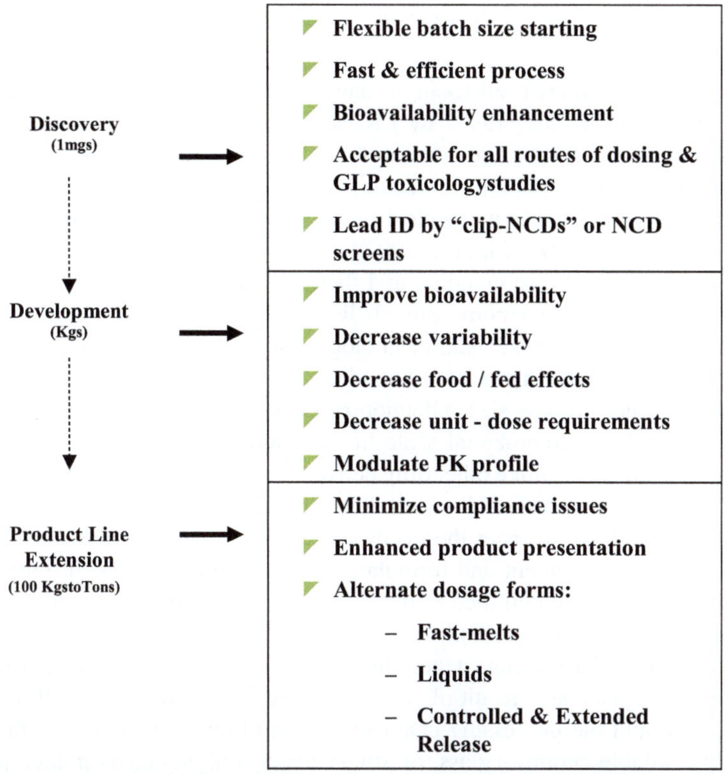

Fig. 13.9 Nanonization/nanosizing has a role to play through the drug development process. The diagram elaborates how this approach for poorly water-soluble compounds has been used in discovery, development and as a product line extension tool

improved surface coverage and enhanced kinetic solubility provide sufficient driving force which improves the efficacy of a nanocrystalline dispersion as compared to a more conventional formulation.

The benefits of formulating poorly water-soluble molecules as nanosized dosage forms are described in Fig. 13.7. Of primary importance is the versatility of the approach. Nanosized dosage forms are suitable for all routes of administration (Fig. 13.8). Nanocrystalline dosage forms can be orally administered as a water-based liquid dispersion or post-processed into tablets, capsules or fast melts using standard pharmaceutical equipment following the fracturing process. Nanosizing has been used in discovery and development to improve bioavailability and help identify and advance lead development candidates. In development and as a line extension strategy, the approach has provided opportunities to not only improve bioavailability but to provide patient friendly dosage forms (Fig. 13.9). The most successful nanosizing technology, in terms of products and development candidates, is Alkermes' NanoCrystal® Technology, which will be discussed in detail in this chapter

13.2 NanoCrystal® Technology

NanoCrystal® technology is all about the nanocrystalline drug particle. The approach was developed in the early1990s by Liversidge et al. (1992) and Liversidge and Cundy 1995) in an effort to find a better, faster, more commercializable approach to formulate poorly water-soluble molecules. At the time, the options available for advancing a poorly soluble molecule from discovery to development and beyond were limited (Lipper 1999; Lipinski 2002). The development path for poorly water-soluble molecules was discouraging, and there was a recognized need in the pharmaceutical industry to overcome this challenge (Di et al. 2012; Stegemann et al. 2012; Waring 2010). NanoCrystal technology primarily utilizes wet media milling technology to generate nanocrystalline drug particles; although the technology encompasses other particle size reduction processes as well. Wet-media milling is used on a very large commercial scale in other industries (e.g., paints and photographic film industry). Unlike other drug delivery technologies, that start at the bench and either experience difficulties or failures during scale-up, wet-media milling required scaling down to meet the needs of the pharmaceutical industry. Over the years, processing equipment and formulation "know-how" have been implemented such that a formulation identified in discovery can be readily optimized, developed and commercialized (Fig. 13.10).

During wet-media milling, depending on the design of the mill, particle-size reduction takes place as a result of shear forces or impact between milling media, drug crystals, and the processing chamber. The milling media can be either commercially available ceramic, glass, plastic beads or a highly cross-pinked polystyrene referred to as PolyMill® 500 milling media (Quinn et al. 2012). Choice of media is dependent on accessibility, target particle size, efficiency of processing, and chemical compatibility. Generally, the smaller the media, the finer the nanosuspension and the more effective and efficient is the processing. There are reports of using 50-μm media (Czekai and Seaman 1999; Rosenflanz et al. 2011) but more typically, 100–1,000 μm-sized milling media are used for product concepts targeting a formulation with a particle size that is <1 μm. Processing issues, such as harvesting and equipment design plus the expense and energy input associated with very fine milling media, oftentimes outweighs the advantage gained in fracturing and/or fracturing efficiency. Figure 13.11 shows an example of the in-processing particle-size distribution profile obtained during a wet media milling process using PolyMill 500 media for a poorly water-soluble crystalline compound in a water-based stabilizer solution. Having wet-media milled a multitude of poorly water-soluble compounds over the years; it is amazing how forgiving and universal the process is. Another advantage of wet-media milling in comparison to other "top-down" approaches is that the particle-size distribution profiles are generally tighter and more unimodal than those reported in publications for homogenization and microfluidization processes. A common misperception cited in the literature as a disadvantage of the approach is how difficult it is to separate the nanosuspension from the milling media (harvesting). Harvesting the nanosuspension from the media

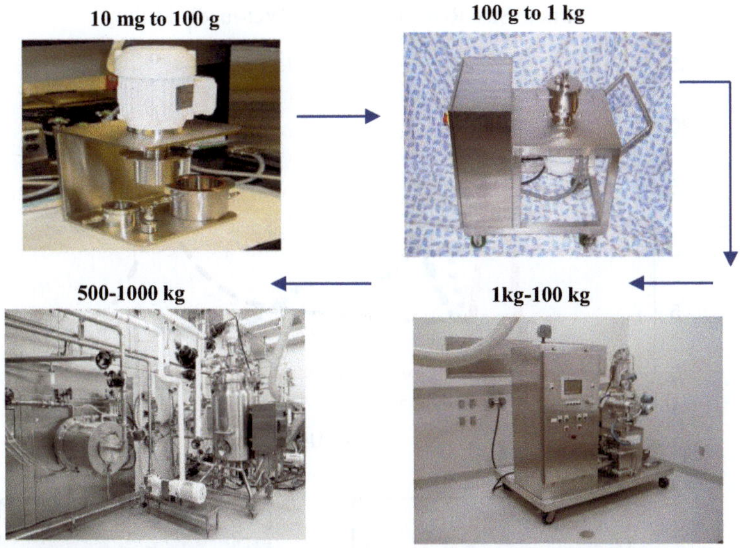

Fig. 13.10 The images are of the NanoMill series of wet milling instrumentation that is used to generate NanoCrystal dispersions and used to manufacture most of the NanoCrystal products that have been launched. As shown, the instrumentation can be used from the 1 mg to the 1,000 kg scale. The 10 mg–1 kg units are operated in a batch mode, and the processing vessel and shaft are in a vertical configuration. The larger production mills, 1–1,000 kg, are operated in a recirculation mode and are configured horizontally. For all operations, the milling vessel is charged with milling media, water, stabilizer, and compound. Particle-size reduction occurs as a result of shear forces between milling media contact points, drug crystals, and the vessel walls

is not an issue at discovery or commercial scales. During discovery, harvesting can be accomplished using a syringe with an appropriately sized needle (gauges 23–26). Alternatively, vacuum filtration or a peristaltic pump with in-line filter can be used. At the development and commercial scales, the processing mills are designed such that the milling media is retained in the processing chamber. As the liquid dispersion exits the processing chamber, the media is retained via an appropriately sized Johnson screen. In addition, as a final precaution, the nanosuspension can be filtered using a depth filter during harvesting. Another question related to the milling media is the potential for generating in process impurities related to the "breakage" of the media during milling. PolyMill polymeric milling media was designed and engineered to be able to withstand the shear forces of the high-energy NanoMill® series of media milling systems. Ceramic media are also commonly utilized. Recently, Juhnke et al. (2012) provided an in-depth study of various types of ceramic media and in-process metal impurities generated during processing. The study demonstrated the importance of selecting milling media and processing conditions to optimize efficiency and minimize impurity levels resulting from the media and the mill chamber. The use of either ceramic or plastic milling media or any new material requires extensive validation of the process and extensive characterization of the product to verify quality. The fact that there are five NanoCrystal

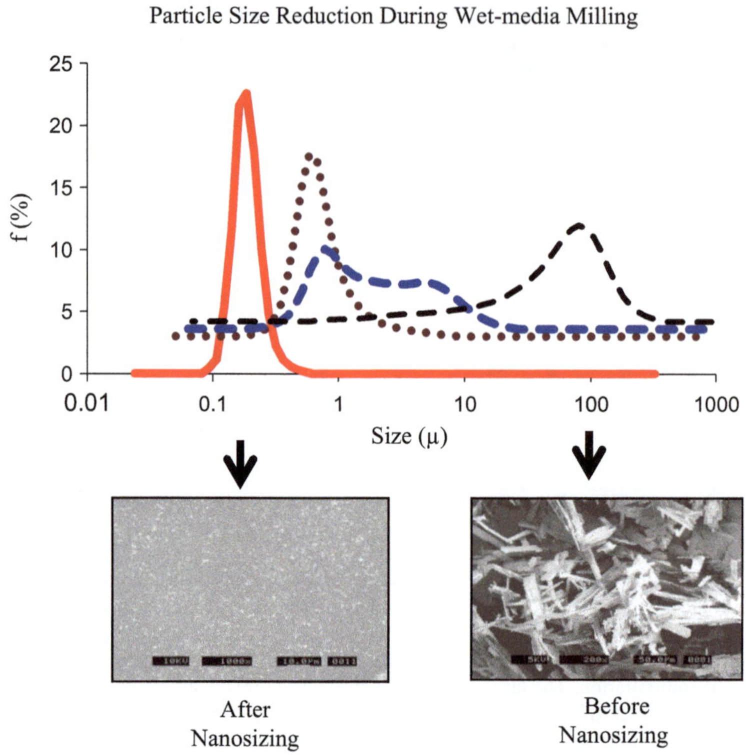

Fig. 13.11 This depicts the particle-size distribution curves of a poorly water-soluble compound before, during and after processing in a NanoMill. Prior to processing, the compound exhibited a mean particle size of approximately 100 μm (*black dashes*). During processing, the particle-size distribution curve becomes increasingly unimodal. The final nanosized dispersion (*red*) has a mean particle size of approximately 200 nm, D90 < 400 nm, D50—150 nm, and D10—100 nm. The scanning electron micrographs (Magnification = 10 K) shown are images of the crystals before and after processing

technology products commercially available worldwide and a number in late stage development lends validation to the acceptability of the process and safety of the product.

The biopharmaceutical performance value that particle size reduction can impart on a product is best illustrated by reference to the NanoCrystal products in the marketplace. The first NanoCrystal product that was launched was Rapamune, an immunosuppressive medication in tablet form, indicated for renal transplant patients (Shen and Wu 2007). The active ingredient, rapamycin (sirolimus), is a poorly water-soluble fermentation product (MW = 914.17 g/mol). In accordance with the Biopharmaceutical Classification System (BCS) (Benet 2013), Rapamycin is a Class IV molecule, i.e., poorly soluble and poorly permeable. The product was initially launched as a liquid-lipid solution containing phosphatidylcholine, propylene glycol, monoglycerides, ethanol, soy fatty acids, ascorbyl palmitate and polysorbate 80, i.e.,

Phosal ®50® PS. This solution is stored refrigerated and has a rather elaborate procedure for dispensing and dosing. Conventional tableting approaches proved to be problematic and incompatible with the chemical stability of the molecule. However, with wet-media milling, rapamycin is fractured as a crystalline solid. Maintaining the molecule in a crystalline solid state during processing, improved chemical stability (Liversidge and Wei 2007) and provided opportunity for the product, by launching a tablet after liquid solution. In addition, the nanocrystalline tablet formulation prepared from a nanosuspension of rapamycin exhibited a 27 % improvement in bioavailability (Shen and Wu 2007). Although improved bioavailability was not a goal for the project, it was interesting to observe that when formulated using NanoCrystal technology, the nanosized formulation out-performed the liquid-lipid solution in terms of bioavailability, stability, and ease of administration.

The second NanoCrystal product introduced to the market was Emend, an antiemetic. The active ingredient, aprepitant, is a poorly water-soluble substance P antagonist having a molecular weight of 534.43 (g/mol) and a melting point of 254 °C. The molecule is a free base but the solubility between pH 2 and 10 is poor (<1 μg/mL). The use of NanoCrystal technology for aprepitant is a good example of how an enabling drug delivery approach can "rescue" a product concept. Clinical data indicated that the molecule exhibited an unacceptable fed/fasted variability in bioavailability. For the conventional formulation approach used in early clinical studies, food was required to achieve the target blood levels. Being an antiemetic, the need to consume food prior to dosing was jeopardizing further development of the molecule. The use of a NanoCrystal formulation to eliminate the need for food and to increase bioavailability in the fasted state is one of the hallmark studies describing the value of the approach. The preclinical bioavailability studies performed by Wu et al. (2004) are frequently cited and provide a good demonstration of regional absorption, effects of particle-size reduction on bioavailability, and the potential for using particle-size reduction to decrease or eliminate food effects associated with the behavior of this molecule. The product, Emend, was launched as a capsule containing sugar beads spray dried with an aprepitant nanosuspension with dosage strengths ranging from 40 to 125 mg. Details of the solid dosage development for Emend are described by Oliver et al. (2007).

TriCor 145, (fenofibrate) indicated for hypercholesterolemia, was the first blockbuster NanoCrystal product achieving sales in excess of $ 1 billion. Incorporating NanoCrystal technology into the TriCor family was a line extension strategy to improve product performance. Fenofibrate is a prodrug that is converted to fenofibric acid, which is the active form of the molecule. Fenofibrate has a molecular weight of 250 g/mol and a melting point of approximately 75 °C. TriCor 145 provides an interesting example of the applicability of using wet-media milling technology for a crystalline compound having a relatively low melting point. The physically and chemically stabile nanosuspension formulation of fenofibrate was post-processed into a tablet. The tablet demonstrated improved bioavailability such that the dose was lowered and a food effect observed with the previous dosage form was eliminated (Sauron 2006). Similar to Emend, eliminating the food effect should not be taken lightly, especially in cases were increased bioavailability leads to unwanted side effects and in cases like Emend, TriCor 145, and Megace ES (to be described below), where a high-fat diet is contraindicative of the medical need.

Megace ES is indicated for anorexia, cachexia, or unexplained significant weight loss in patients with AIDS. The active ingredient, megesterol acetate, is a steroid with a melting point of 220 °C, molecular weight of 384.5 g/mol and solubility in water at approximately 2μg/mL. The molecule is a classic BCS Class II product, poorly water-soluble and highly permeable. The mechanism of action as an appetite stimulant is unclear but it is believed not to be related to the antiandrogenic effects of megesterol acetate, but to its downstream effects on the hypothalmus and inhibition of proinflammatory cytokines, all of which are known to affect appetite. During development of the nanosuspension formulation, the original product (an oral suspension) was discovered to have a substantial fed/fasted variability in bioavailability. In order to maximize bioavailability and efficacy with the original formulation, patients were required to take it with food. However, this patient population has difficulties with eating. Formulating megesterol acetate as a ready-to-use liquid nanosuspension improved bioavailability in the fasted state, reduced the fed/fasted variability (Deschamps et al. 2009), and the reduced viscosity of the medication provided a more patient friendly and easy to ingest product. The liquid nanosuspension is taken with a teaspoon (5 mL) rather than a 20 mL dosing cup which was required for the original formulation. From a technical advancement, Megace ES demonstrates that if properly formulated, a liquid nanosuspension can be extremely stable. There are no storage restrictions on the medication, and formulated, at 125 mg/mL, Megace ES, the flavored-liquid nanosuspension.

The most recently approved NanoCrystal Technology product is an injectable, Invega Sustenna. This antipsychotic is a once-monthly intramuscularly administered dosage form of paliperidone palmitate available in a ready-to-use prefilled syringe (Gopal et al. 2010). Paliperidone palmitate has a molecular weight of 664.89 g/mol and is poorly soluble in water (<1 ug/mL). Successful development of paliperidone palmitate as a nanosuspension product suitable for intramuscular injection reflects the versatility provided by a nanocrystalline dosage form. For the previous products described, nanocrystalline drug particles have been described as a means to formulate poorly water-soluble molecules so that dissolution rate is rapid and thereby bioavailability is positively affected. The strategy behind Invega Sustenna is a one-eighty reversal in product design, i.e., making a soluble drug (paliperidone) less soluble by the addition of palmitic acid to provide sustained slow release after dosing intramuscularly. Invega Sustenna is an excellent example of combining two drug delivery platforms to generate a novel concept and a product that has proved beneficial to the intended patient population as, a once-monthly medication for a patient population that tends to be noncompliant with regard to a daily dosing regimen. A similar concept has been applied to TMC278, a non-nucleoside reverse transcriptase inhibitor (NNRTI) to treat HIV. TMC 278 is not a prodrug but a very poorly soluble small molecule. The NanoCrystal concept for TMC 278 is to provide a long-acting injectable to be administered intramuscularly (Baert et al. 2009). The concept is currently being evaluated in clinical trials (Verloes et al. 2008).

It is worth mentioning a few other clinical and preclinical studies that further demonstrate the versatility of using a particle size reduction approach as a formulation tool (Figs. 13.12, 13.13, and 13.14). A NanoCrystal formulation of itraconazole, a poorly water-soluble antifungal agent, was evaluated in the clinic in a Phase I study designed to compare pharmacokinetic properties of a nanosuspension to a cyclodextrin solution of itraconazole following intravenous dosing (Mouton et al. 2006). The study demonstrated that nanocrystalline drug particles can be safe and effective when dosed intravenously. There have been many publications wherein nanoparticles have been dosed intravenously using preclinical animal models. Similar to the work on itraconazole, the formulations were well tolerated, provided means to dose escalate, improve efficacy, and achieved the absolute bioavailability of a poorly water-soluble molecule. In addition to intravenously dosing, Kraft et.al. (2004) reported results from a Phase I clinical trial for pulmonary delivery of a nanocrystalline budesonide formulation. The formulation was delivered using a nebulizer, was well-tolerated and exhibited linear pharmacokinetics after inhalation. In comparison to Pulmicort® Respules®, the nanocrystalline formulation required significantly shorter nebulization time which was cited as being advantageous, especially, for pediatric and geriatric patients. In summary, this section has focused primarily on products incorporating NanoCrystal technology and development candidates produced by wet milling, other drug-particulate products are described below.

13.3 Additional Drug-Particulate Products

There are a number of drug-particulate products, in addition to those mentioned above, that have been launched. These include Triglide, Abraxane, Juvedical, and the recently approved Zorvolex™. Triglide is a tablet formulation of fenofibrate marketed by First Horizon and launched in 2005. As reviewed by Tziomalos (Tziomalos and Athyros 2006), this tablet formulation utilizes Insoluble-Drug Delivery (IDD®) high-pressure homogenization approach to generate small particles of fenofibrate stabilized with a phospholipid surface-modifying agent. As previously discussed, the increased surface area of the homogenized drug crystals improves oral bioavailability. The IDD-P fenofibrate 160 mg tablets are bioequivalent in the fed and fasted state, similar to TriCor 145, and administering the tablets, independently of food, provides greater convenience for the patient, thereby, improving compliance.

Abraxane is the first and only drug nanoparticulate product that is dosed intravenously. Abraxane, marketed by Celgene Corporation, is a cremophor-free, albumin-bound nanoparticle formulation of paclitaxel referred to as a *nab* technology product and as described by Hawkins et.al. (2008), the reduced toxicity profile of the product has provided an avenue for dose escalation compared to the more

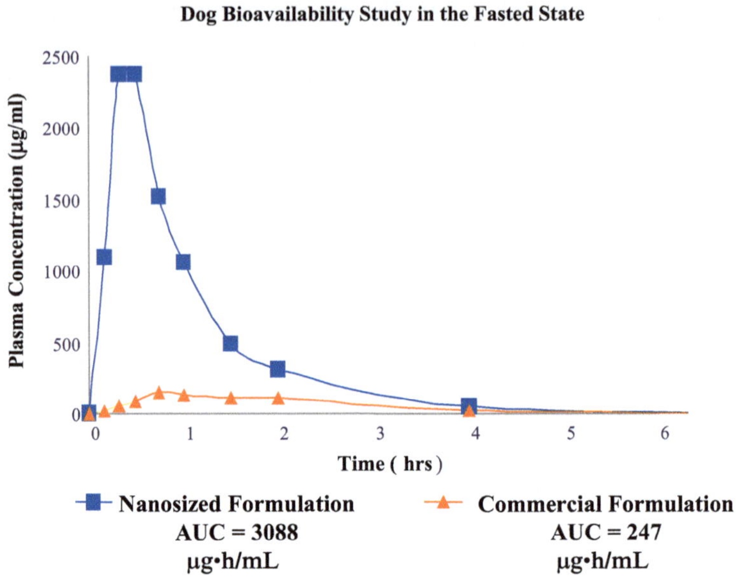

Fig. 13.12 A dog bioavailability study was performed in the fasted state comparing the marketed product, (a conventional dosage form), to a nanoparticle formulation. In the fasted state, the AUC of the nanocrystalline formulation is 12.5 times greater than the commercial product

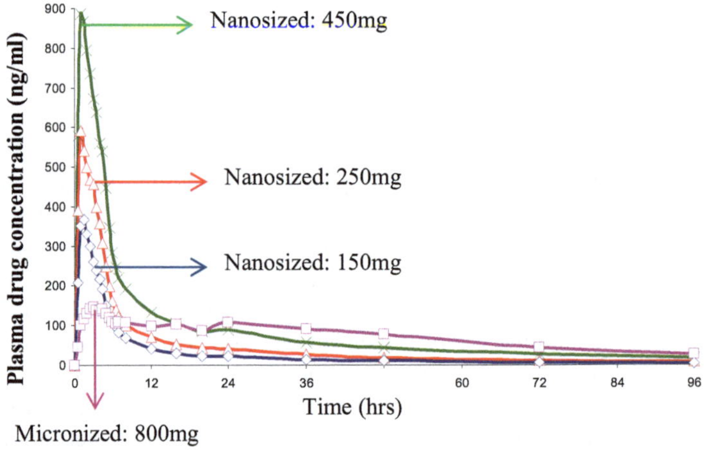

Fig. 13.13 Compound A, a poorly water-soluble compound, was formulated as a nanocrystalline or a micronized dosage form. The formulations were evaluated in the clinic using healthy volunteers in the fasted state. The data demonstrate (1) improved bioavailability of the nanocrystalline dosage forms and (2) the exquisite dose proportionality that is frequently observed with nanosized formulations

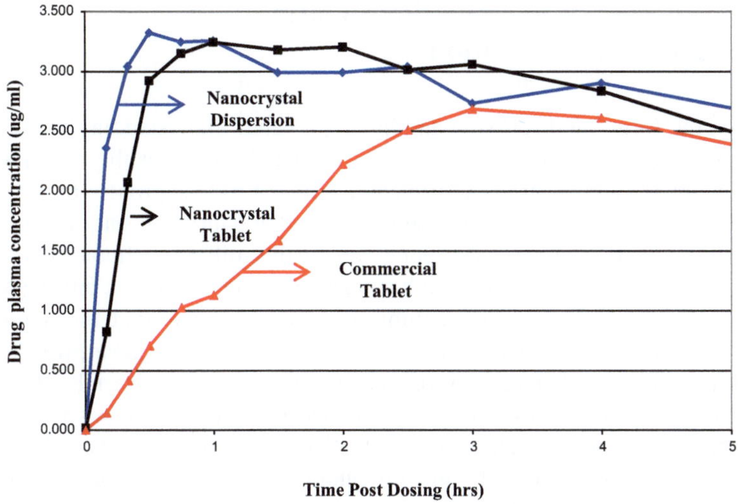

Fig. 13.14 Compound B, a poorly water-soluble compound, was evaluated in the clinic using healthy volunteers to compare the bioavailability of a nanocrystalline liquid dispersion, a nanocrystalline solid dosage form, and the commercial tablet. The data demonstrate (1) improved bioavailability of the nanosized dosage forms and (2) that the conversion of the nanocrystalline liquid into a solid dosage form does not compromise the bioavailability observed

conventional solvent-based paclitaxel formulation, Taxol®. Abraxane was launched in 2005 for the treatment of breast cancer in patients. Since then it has gained FDA approval for non-small-cell lung cancer, and, more recently, has been approved to treat patients with late-stage pancreatic cancer (Patel et al. 2014).

There are now a few cosmetic products that employ a nanonization processing to improve the effectiveness of the active ingredient. As was previously mentioned, nanosizing the active ingredient in a topical cream enhances surface coverage and kinetic solubility resulting in improved performance (Muller et al. 2011). The first cosmetic product line uses smartCrystal® technology from PharmaSol GmbH which covers various processes to more effectively generate nanocrystalline particles. Juvedical, launched in 2007 by Juvena, is a nanoparticulate formulation of rutin, a natural flavonoid found in many plants and associated with antioxidant activity. Since then a number of companies have launched additional nano-rutin products for various cosmetic applications (Keck et al. 2013).

Finally, the most recent nanosized product approval was announced in January 2014 by Iroko Pharmaceuticals. The product, Zorvolex, is a nanoparticulate capsule formulation of diclofenac indicated for the treatment of osteoarthritis pain. This is the first and only FDA-approved nonsteroidal product using SoluMatrix Fine Particle Technology™, a dry milling approach (Atkinson et al. 2013), which provides the benefits of a nanosized oral dosage form that have been previously discussed, i.e., improved bioavailability, faster onset of action, lower doses, and improved patient compliance.

13.4 Nanosizing: Formulation Identification and Characterization

Based on the forgoing discussion and having a better understanding of the advantages associated with nanosizing (Fig. 13.7), it is hoped that the reader is sufficiently interested and eager to experimentally evaluate the approach. In this section, a few operational pointers for identifying, characterizing and developing a nanosuspension will be provided using wet media milling technology as a basic tool. As previously mentioned, the advantages of using wet media milling technology to nanosize a poorly water-soluble drug candidate is that the approach is forgiving, broadly applicable, and scalable. At the discovery level, the challenge is working with small quantities of poorly characterized material. There are a number of publications describing particle size reduction processes for screening nanosized formulations that use as little as "1 mg" of drug substance. Some of the approaches rely on proprietary instrumentation such as the NanoMill units of Alkermes Pharma Ireland Limited and the multi-screening, small-volume media mill described by Brewster et al. (2012). However, there are simple cost effective ways to get started. The use of a low energy roller mill for small scale screening nanosuspensions is an option (Juhnke et al. 2010; George and Ghosh 2013). For low-energy processing, ceramic media, ranging in size from 400 to 800 μm, is ideal for most applications and is commercially available (Liversidge et al. 1992; Juhnke et al. 2012; Merisko-Liversidge et al. 1996). In getting started, the first question that must be addressed is the amount of API available for screening and the route of administration being targeted. Since wet media milling is dependent on shear and the impact of the media against the crystals of API and the vessel walls, at a minimum, it is recommended that processing be initiated with at least 0.5 mg API per gram of crude slurry. The crude slurry consists of API, water and stabilizer. In selecting a stabilizer, it is crucial to be cognizant of the eventual route of dosing. For wet media milling, some of the commonly reported stabilizers that act as surface modifiers have been previously discussed (Gao et al. 2013; Merisko-Liversidge and Liversidge 2011; Van Eerdenburgh et al. 2008; Liversidge et al. 1992). A good stabilizer is an excipient with structural domains having hydrophobic and hydrophilic properties. A stabilizer must be able to wet the surface of the crystal and at the same time, through either steric and/or ionic interactions, stabilize the fractured crystals in the aqueous fluid phase of the dispersion. For oral applications, there are a number of commonly used excipients that have performed well during wet milling. These are the povidones, cellulosics, pluronics, polysorbates, and TPGS (tocopheryl polyethylene glycol) with and without an anionic surfactant like sodium lauryl sulfate (SLS) or docusate sodium (Merisko-Liversidge and Liversidge 2011; Liversidge et al. 1992). For injectables, the approved excipient list is very limited but success has been achieved using low molecular weight povidones, pluronics, polysorbates, solutol, and cremophor. To generate a stable nanosuspension, the starting drug-to-stabilizer ratio on a weight basis, should be 4:1–10:1. Too little or too much stabilizer produces stability issues. A stabilizer "starved" formulation will aggregate and too much stabilizer can induce Ostwald ripening, i.e., crystal growth. If the amount of compound is very limited, 10 μL of a 1–5 % stabilizer, the solution can be placed on the slide and 1–2 μg of the compound added to the liquid. After gently stirring, one can generally rank a series of stabilizers regarding the ability to wet the surface of the compound being evaluated. If the compound readily partitions into the liquid phase, this is a good indication that the stabilizer will be suitable surface modifier for nanosizing the compound in question.

Prior to processing, one should define the goals of the study. Is the goal of the study to achieve the smallest size possible? The goals will be driven by the amount of compound available for screening. For first studies, most often the intention is to determine how well the stabilizer is wetting the crystal surface and processing to completion is not required. If the availability of compound is limited and the goal is to determine if particle size reduction would improve bioavailability, limited processing may be sufficient to address this question. Once addressed, the formulation and process can be optimized to further improve performance and/or stability. For processing, when using a low energy mill, it typically will take 24–96 h to achieve a mean particle size of <1 μm. For higher-energy systems, particle size reduction can be completed within 30–60 min when processing milligram to gram quantities of active. It is advisable to periodically sample during processing. To evaluate the quality of the nanosuspension during and after processing, there are a number of very straightforward observations that should be considered. When working with small amounts of API, visual nondestructive analysis is extremely important. It is easy to detect signs of aggregation, solubilization, and fluidity just from visual inspection of the dispersion. If the fluid is transparent and clear, the "pH" of the fluid should be recorded, and this may provide an avenue for dosing a solubilized liquid formulation. If there are signs of aggregation, foaming, or creaming, typically an alternate stabilizer should be tried or continue processing after the addition of a secondary stabilizer. Addition of an anionic or cationic surfactant may remedy the situation. If this is being tried, the anionic/cationic should be added at no more than one-tenth the concentration of the primary stabilizer, and processing should be continued for a few minutes to hours depending on whether a low-or high-energy process is being used. The goal is to generate a free flowing dispersion with no signs of aggregation and/or large micron-sized crystals when analyzed using light microscopy.

For particle-size determination, laser light scattering instrumentation is most frequently used and is generally readily available in most pharmaceutical laboratories. A concern regarding the approach is that the dispersion does have to be significantly diluted. Diluents can range from deionized distilled water to a dilute-stabilizer solution. Data generated should be verified using an orthogonal approach which has been extensively discussed in a number of publications (Tinke et al. 2006). A very straightforward verification method is the use of light microscopy equipped with phase optics. The Brownian motion of submicron particles should be readily detectable. The advantage of light microscopy is that the dispersion does not have to be diluted or dried prior to visualization. In essence, the dispersion is evaluated as it exists in the processing chamber or on the shelf. All light-scattering measurements should reflect what is being visually observed. If aggregation, crystal growth, heterogeneity or, most desirably, homogeneity is observed, this should be reflected in the particle-size distribution data obtained from light scattering instrumentation. Both static (laser light diffraction) and dynamic laser light scattering (Photon Correlation Spectroscopy) are the primary analytical tools used for particle size analysis of nanosuspensions. Laser light diffraction has the advantage of detecting a very broad particle-size range from approximately 10 nm to hundreds of microns based on the various manufacturers' specifications. A potential disadvantage is that, for particles below 1 μm, the refractive index (real and imaginary) component should be determined for accuracy (Keck and Muller 2008). Dynamic light scattering or photon correlation spectroscopy has a much narrower particle size detection range than laser light diffraction. It is used for particulate formulations having a mean particle size <6 μm and is extremely useful for ultrafine particles less than 50 nm. Dynamic light scattering or photon correlation spectroscopy does not require prior knowledge of the refractive index of the compound being studied. The measurement is based on the autocorrelation constant obtained during analysis. For high-quality nanosuspensions, data obtained from both approaches will vary, but, only to a modest extent. If there is a wide discrepancy between the readings there may be an issue with the dispersion which can be further analyzed using microscopy, as described above. Based on the data, the dispersion may require additional processing, additional stabilizer, or the dispersion may be acceptable but the particle-size method maybe the issue, i.e., changing the diluents used for particle size analysis should be considered.

Additional analyses that should be performed include quantifying the amount of soluble drug after nanosizing. This has to be performed very carefully, and it is recommended that the dispersion be ultracentrifuged at 50,000 × *g*-forces for 30 min. The supernatant should be carefully collected and filtered via a 0.1μm filter. This can be defined as the soluble fraction. In most cases, nanosizing will not affect the saturated solubility of the compound but due to the presence of stabilizers that have surface active properties and the nanometer size of the processed drug crystals, this should be confirmed. In addition, verification of crystallinity using powder X-Ray Diffraction is recommended. Ideally, processing should not affect the solid-state properties of the molecule. If there is a solid state change, processing conditions can be optimized to maintain crystal integrity. This is generally accomplished by changing stabilizer, changing processing time and/or temperature. Most analytical methods used for characterization of product quality, performance, and integrity can be applied and/or revalidated for a nanocrystalline formulation. Although oftentimes overlooked, it is also important to obtain an understanding of the compatibility of the formulation in the biological fluid that the formulation will encounter on dosing, i.e., simulated gastric, intestinal fluid, salt solution, or plasma. If a physical instability is observed, effort should be placed in performing additional excipient selection studies to determine if the stability of the dispersion can be improved. The addition of a new stabilizer, and/or an excipient (e.g., a carbohydrate or salt) may rectify the situation. If the stability cannot be improved, then further analysis or investigations regarding the benefits of the nanosuspension should be pursued, realizing this particular caveat. Finally, care should be given to nanosuspensions of actives that are free bases. In these cases, it is important to be aware of the solubility of the compound in the low "pH" environment of the stomach and the potential for uncontrolled precipitation while in transit to the intestine. If this occurs, results obtained using a nanosuspension will be compromised. In comparison to a conventional formulation, the nanometer-size crystals will dissolve faster and may be subject to more extensive precipitation leading to a loss rather than an improvement in bioavailability.

Dissolution testing of nanosized dosage forms is generally not performed in the early to late discovery phase. However, if resources are available, dissolution testing in various simulated fluids can be an invaluable tool for prioritizing formulations to be screened in vivo. For nanosized formulations, dissolution and subsequent solubilization should be rapid, unless the dosage form is specifically designed to be a delayed or sustained release solid dosage form. Since a classical approach to dissolution testing is to filter solubilized versus unsolubilized active using a 0.2–0.45 μm filter, this may not be an optimal separation approach for nanosized formulations. For many nanosized formulations, particle-size distribution curves indicate that 10 % or more of the particles are less than 100 nm, and this size range may pass through a typical 0.2 μm syringe filter resulting in an overestimate of dissolved material. More recently, it has been suggested that smaller-bore filters, centrifugation, ultracentrifugation or a combination approach may be more appropriate separation methods for nanosuspensions (Junemann and Dressman 2012; Liu et al. 2013; Crisp et al. 2007). As more products incorporating nanotechnology move to the marketplace, regulatory guidance regarding appropriate dissolution methodology and other formulation characterization requirements for nanomaterials will evolve (Narang et al. 2013).

13.5 Preclinical In Vivo Evaluation of Nanosuspensions

Nanosuspensions have been used for a wide variety of applications. However, the majority of preclinical studies involve oral, intravenous, subcutaneous or intramuscular dosing to obtain pharmacokinetic and/or pharmcodynamic data. For the most commonly used dosing regimens, the aqueous liquid dispersion can be readily dosed. For oral dosing, the formulation can be administered via gavage, intraduodenally, or placed in a capsule immediately prior to dosing. If the capsule option is employed, it is best to make sure that the capsule components are compatible with the nanosuspension. It may be desirable to add a viscosity-thickening agent to the aqueous nanosuspension. Most nanosuspensions should have a viscosity reading close to water, especially, if formulated at a drug concentration of 200 mg/mL or less. For dosing the liquid dispersion, there is literature suggesting that adding a viscosity-modifying agent may also improve the

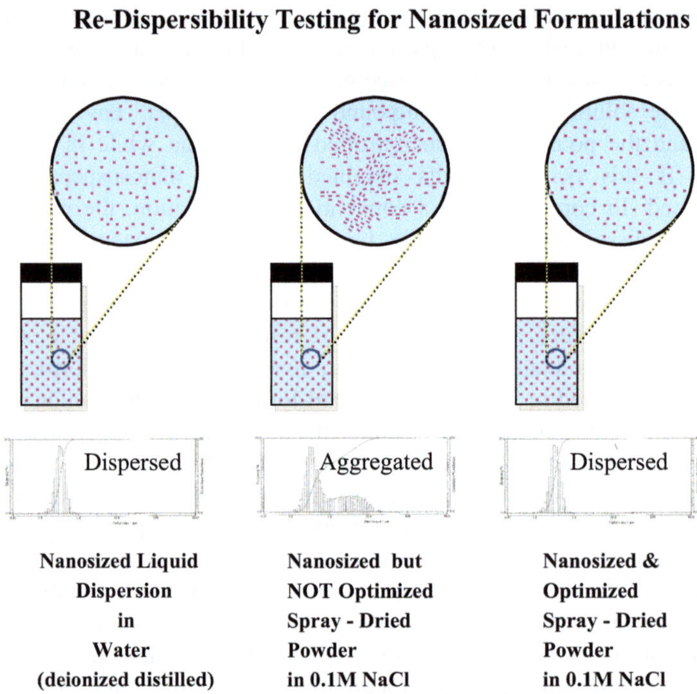

Fig. 13.15 The image demonstrates the importance of monitoring the physical stability of the nanosuspension whether in liquid or dried-solid dosage form in biorelevant fluids. If the particle size of the nanosuspension is not stable in the biorelevant fluid, performance after dosing will be compromised

bioavailability (Carver et al. 1999). As long as the viscosity modifier is compatible with water and does not adversely affect the colloidal stability of the nanosuspension, there should be no problem in adjusting the viscosity prior to dosing. Enhanced bioavailability of orally dosed nanosuspensions has not typically required a viscosity modifier. However, if the viscosity of the comparator formulation is greater than that of the nanosuspension, it would be of value to test the nanosuspension with and without a viscosity modifier. If a solid dosage formulation is desired, for early studies when API and formulation are available in limited quantities, the nanosuspension can be spray-dried, film-dried or granulated using small-scale, lab-top units that are now commercially available. Typically, the water-based nanosuspension will require the addition of a matrix forming reagent so that after drying, the formulation re-disperses into nanometer-sized particles (Cooper et al. 2010). Figure 13.15 shows why re-dispersibility testing is needed after a nanosuspension is dried. To re-disperse the dried nanosuspension, add water or simulated gastric or intestinal fluids and then analyze using laser light scattering instrumentation and/or microscopy. The particle-size distribution curve of the re-dispersed formulation should be very similar or identical to the original

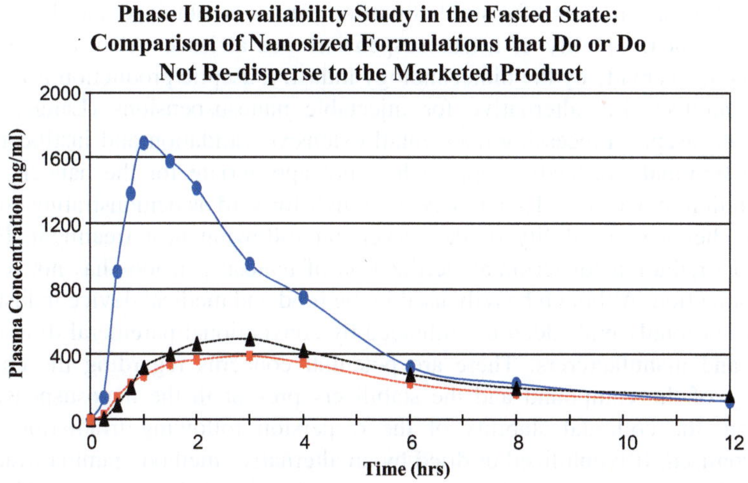

Fig. 13.16 Compound C, a poorly water-soluble compound, was evaluated in the clinic using healthy volunteers to compare two nanosized dosage forms to the marketed product. As discussed in Fig. 13.16, the first-generation tablet (*black triangles*) did not properly re-disperse in simulated fluids, and similar to the marketed product (*red circles*), bioavailability was poor. In contrast, the second-generation tablet (*blue circles*) was designed to completely re-disperse in simulated fluids and bioavailability in the fasted state was significantly improved

pre-dried formulation. Figure 13.16 provides an excellent example of the benefits of performing re-dispersibility testing on nanocrystalline solid dosage forms. If not properly dried, the performance of the nanocrystalline dosage form will be compromised.

For injectables, care should be taken to design preclinical formulations using excipients that are approved for the intended route of administration and are generally recognized as safe (GRAS reagents). In addition, there is the sterility issue that must be considered. If the compound has a high melting point and with the aid of cloud point modifiers, nanosuspensions have been successfully sterilized using heat (Na et al. 1999). During heat treatment, there may be a slight growth in particle size but if the formulation is optimized to withstand the heat, the slight change in particle size may be acceptable. Formulations that are not stable to heat sterilization can be processed by aseptic processing and/or sterile filtration. For sterile filtration, the nanosuspension must have an average mean particle size <100 nm with 100 % of the particles being <200 nm. Although nanosuspensions can have a mean particle size of approximately 200 nm, since the nanoparticles are nonelastic solid crystals, sterile filtration is not always possible. On occasion, this ultrafine particle-size range has been reported (Zhang and Bosch 1997) but more routinely nanosuspensions are in a size range that is not suitable for sterile filtration. Aseptic processing should be considered if terminal heat and filtration are not possible. Aseptic processing would be a more forgiving approach, especially if the ultrafine size required for sterile filtration cannot be obtained and the chemical and/or physical stability of a nanosuspension was at issue. Also, as a result of advancements made by the biotechnology industry, aseptic production is a reasonable manufacturing alternative for injectable nanosuspensions (Langer 2013). However, aseptic processing does entail extensive validation and justification on why a terminal sterilization approach is not appropriate for the nanocrystalline formulation in question. This may be as straightforward as demonstrating colloidal and/or chemical instability of the dispersion following heat treatment. Finally, gamma irradiation for terminal sterilization of nanosuspensions has not received much attention. Although heavily used in the food and medical device industries, it is still not totally embedded or embraced by conventional parenteral drug formulators and manufacturers. There are potential concerns regarding the chemical stability of the compound and the stabilizers present in the nanosuspension. In addition, the colloidal stability of the dispersion following irradiation maybe compromised. If lyophilized or dried by an alternative method, gamma irradiation for sterility assurance of nanosuspensions may be a workable approach. However, similar to the discussion regarding other drying techniques for solid dosage forms, care must be taken to ensure that the lyophilized dosage form re-disperses to a particle size similar to the "neat" dispersion. Typically, this entails adding matrix forming reagents, such as a sugar, to the formulation prior to drying (Liversidge et al. 1994; Beirowski et al. 2011a, b).

For injectable applications, an advantage that a nanosuspension has over other particulate formulation options is that at high concentrations approaching 200–300 mg/mL, they are very fluid and very syringeable. For dosing intravenously, it is recommended that the nanosuspension be compatible with plasma, filterable via a 5 μm syringe filter and dosing be performed at a rate of 10 mg/min to ensure safe dosing (De Garavilla et al. 1996). Since nanosuspensions are water based, for intravenous dosing, there should be complete compatibility with blood or plasma. Formulation related issues associated with injectables have been thoroughly reviewed by Yalkowsky et al. (1998). The review discusses issues and provides in vitro/in vivo studies that can be performed to evaluate and screen formulations to optimize compatibility with blood and tissue components. Although published prior to the use of nanosizing in the pharmaceutical industry, the approaches described are relevant and can be readily applied to nanocrystalline formulations. In addition, there is a tremendous amount of activity in the area of using micro and nanofluidics for establishing in vitro–in vivo correlations. These newer approaches may provide future opportunities to screen and optimize nanosized dosage forms (Huh et al. 2012).

As previously discussed, although a nanocrystalline formulation of itraconazole exhibited a pharmacokinetic profile similar to a cyclodextrin solution in man following intravenous dosing (Mouton et al. 2006), this may not always be the case. For nanocrystalline formulations pharmacokinetic and tissue distribution are dependent on the solubility and dissolution rate of nanocrystalline drug particles in blood and potential interaction with blood components (Merisko-Liversidge and

Liversidge 2011; Rabinow 2004). In essence depending on the goals of a program, a nanocrystalline formulation, to an extent, can be designed to modulate the pharmacokinetic profile of a poorly water-soluble compound. Since pure drug nanoparticles are not encapsulated, following intravenous dosing, interaction of the particles with blood components along with dissolution rate is a very dynamic event. Particle size and the adsorbed surface modifier are known to affect pharmacokinetic and pharmacodynamic properties of the molecule when formulated as a nanosuspension (Dufort et al. 2012; Moghimi et al. 2012). If the nanoparticle does not immediately dissolve in the blood pool, its size and surface interactions can impact performance. The interaction of drug nanoparticles and blood components is a very active area of research and a number of academic groups are focused on characterizing protein–particle interactions in healthy and in diseased animal models (Aggarwal et al. 2009; Monopoli et al. 2012). For intramuscular and subcutaneous dosing, the nanosuspension is dosed similar to a solution and, once again, size and surface coating may impact pharmacokinetics and provide an opportunity for enhancing or prolonging release at the site of dosing.

13.6 Scale-Up and Commercialization

There are a number of marketed products that incorporate a nanonization step during manufacturing and it is anticipated that nanosizing will be a widely established processing tool for poorly water-soluble compounds in the very near future. To advance a nanosuspension from early discovery to late-stage development, Alkermes plc's NanoCrystal technology provides a good example of how this can be accomplished. In Figure 13.10, the nanosizing equipment used, and the results for a processing study (Figs. 13.11 and 13.17), are shown. A nanosuspension identified with a few milligrams of compound can be scaled to production-level capacity, i.e., hundreds of kilograms. In addition to Alkermes' NanoMill media milling systems, there are a number of commercial vendors who offer similar equipment to support discovery, development, and commercial manufacturing. During scale-up, wet-milling operations are generally converted from a batch mode to a recirculation configuration. In the recirculation configuration, the milling media is generally retained in the processing chamber with a Johnson screen that allows free passage of the slurry from processing vessels through the mill and back. For nanosizing, in order to achieve a fine, homogeneous dispersion of nanoparticles, the size of milling media is typically <1 mm.

Potential issues that can be encountered during scale-up are similar to those encountered in most conventional scale-up operations. A change in manufacturer for the active can affect milling performance as a result of change in specifications for residual solvents used during processing or the final particle size of the active as received. For wet milling, starting with micronized material is not a requirement. In

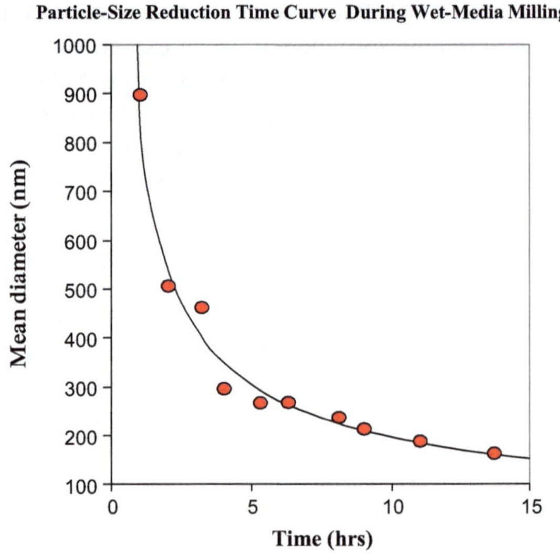

Fig. 13.17 A particle-size reduction time curve is shown for a poorly water-soluble compound processed using a commercial manufacturing NanoMill and 500 μm plastic milling media. Processing was performed in a recirculation mode configuration to manufacture approximately 10 kg of active

fact, actives with a particle size ranging from 50 to 100 μm are typically easier to wet and handle than micronized material. Scale-up efforts typically involve maximizing drug loading to improve the efficiency of the process. As the concentration of the active increases, potential issues related to increased viscosity of the initial slurry or increased viscosity of the milled active can pose issues. For example, charging the processing chamber prior to milling and harvesting the final processed dispersion may be difficult. These types of issues can be addressed by decreasing the drug load, decreasing media load, adjusting the drug to primary stabilizer concentration, or adding a secondary stabilizer. Similar to all pharmaceutical manufacturing processes, a quality-by-design program can be readily implemented. Using an experimental design strategy, operating parameters that are crucial for the process and quality of the product can be identified and optimized. For in-process testing, particle-size analysis can be performed by manually sampling, or more recently, online particle-size methods have been described (Higgins et al. 2003). During validation, the processing time required to obtain the target particle size or particle-size distribution profile is established. Commercial manufacturing batches are typically processed for the established time period. Once optimal processing conditions are identified, the wet-milling process is very reproducible (Merisko-Liversidge and Liversidge 2011).

Although not always appreciated, one of the value-added features of nanosizing is that all downstream processing utilizes classical conventional pharmaceutical operations. Figure 13.18, shows the equipment line wherein the nanosized dispersion is processed as a granulation feed for tableting. For drying the nanosuspension, wet granulation is only one option. Spray drying, spray coating and lyophilization are also options. In essence, the media-milling operation aligns nicely with the more established processing equipment required to manufacture high-quality products for parenteral and non-parenteral applications.

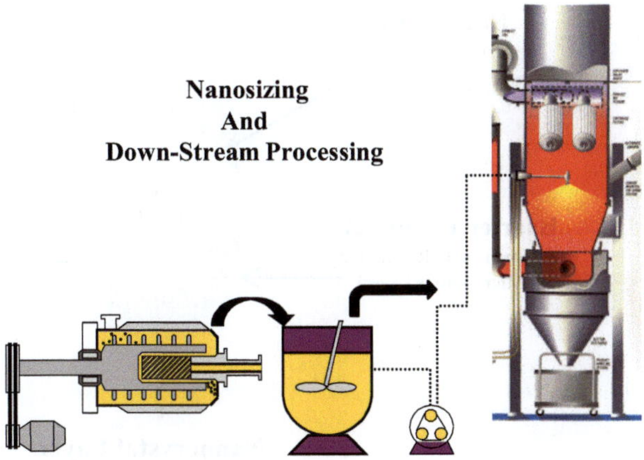

Fig. 13.18 The drawing demonstrates that during nanosizing using wet-media milling technology, with the exception of the milling unit, all post-processes use conventional standard pharmaceutical equipment

13.7 Future Opportunities

As for the future, there are a number of exciting opportunities in the field of nanodelivery. Nanosized formulations are being combined with classical controlled-release technologies to provide more patient-friendly dosage forms. An example is the NanOsmotic® technology approach (Ruddy et al. 2010), wherein a nanosized dispersion is spray coated onto an inert bead in combination with various modifying agents. Modifying agents can be permeation enhancers, precipitation inhibitors, or other reagents required to increase bioavailability of a poorly water-soluble compound (Fig. 13.19). The approach is a good illustration of the potential utility of combining various formulation strategies to enhance performance. Also, it is foreseeable that complex disease states that require a cocktail of actives to be effective could be formulated as an easy-to-administer nanosuspension or nanosized dosage form. Currently, co-injection of GSK1265744 and TMC278, two poorly water-soluble development candidates, is being evaluated in the clinic as long-acting parenteral nanosuspensions for HIV infection (Spreen et al. 2013). In the future, this concept may be extended and nanosized cocktail formulations maybe available and tailored to meet the individual patient's need. In addition, there are reports regarding the generation of stable-amorphous nanosuspensions produced using classical nanosizing instrumentation (Auweter et al. 1998; Dai 2012). The approach would provide enhanced solubility from both the particle size reduction process and presenting the compound in a high energy amorphous state. The long-term stability of such formulations is questionable and will be have to be established.

Since we now have formulation tools to handle poorly water-soluble compounds, it is foreseeable that compound selection strategies may change. More emphasis will be placed on efficacy and safety and less on the aqueous solubility of the molecule. Screening protocols may also change. Nanosizing in a multiplate format has already been documented and there are a number of promising approaches on the horizon (Cunningham et al. 2004) that could be adapted. Finally,

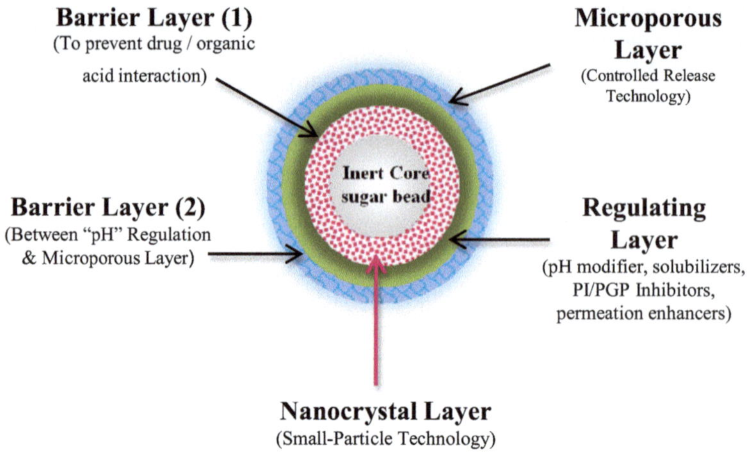

Fig. 13.19 The image depicts the design of an oral dosage form using NanOsmotic technology. By combining nanonization with controlled-release technologies, in addition to improving performance by increasing surface area, it is also possible to enhance solubility, inhibit undesirable precipitation, and potentially incorporate additional reagents that could impact bioavailability

the surface modifiers used to stabilize nanocrystalline formulations are conventional commercially available excipients. The future holds great promise for incorporating tissue-specific or disease-specific biomarkers into stabilizer design (Chan et al. 2004; Wang et al. 2012; Cheng et al. 2012; Xiao et al. 2012; Fuhrmann et al. 2014). It's visionary, but the goal would be to have a library of "smart" surface modifiers that could be incorporated into the design of a nanocrystalline formulation to further improve product performance, tolerability, and patient compliance.

13.8 Conclusions

In this chapter, nanosizing has been described as an end-to-end formulation approach for poorly water-soluble compounds. Nanosizing is now an established technology and will continue to be incorporated into many product design concepts for future applications. Being suitable for sterile and non-sterile applications provides an added advantage over other formulation platforms. Today, the pharmaceutical scientist has an array of good robust approaches for handling and developing poorly water-soluble molecules. With the many formulation tools that are available, it is tempting to state that the formulation issues previously associated with poorly water-soluble compounds have been addressed. It is now time to aggressively incorporate these tools into early discovery programs so that decision-making regarding compound selection, lead identification, and early formulation development strategies can be streamlined and executed in an expeditious and cost-effective manner.

Acknowledgements Special acknowledgement is given to Gary G. Liversidge, Ph.D. for his many contributions to NanoCrystal technology and its applications and for his invaluable encouragement and input in writing this chapter. In addition, I thank Tarek Zeidan, Ph.D., Chris Hencken, Ph.D., Jules Remenar, Ph.D., Fidelma Callahan, Eva Stroymowski, and Amy Trevvett for their critical reviews, to all past colleagues of Nanosystems LLC, and Elan Drug Technologies, and to the current management of Alkermes for their support.

References

Aggarwal P, Hall JB, McLeland CB, Dobrovolskaia MA, McNeil SE (2009) Nanoparticle interaction with plasma proteins as it relates to particle biodistribution, biocompatibility and therapeutic efficacy. Adv Drug Deliv Rev 61:428–437

Atkinson TJ, Fudin J, Jahn HL, Kubotera N, Rennick AL, Rhorer M (2013) What's new in NSAID pharmacotherapy: oral agents to injectables. Pain Med 14:S11–S17

Auweter AV, Horn D, Luddecke E (1998) The function of gelatin in controlled precipitation processes of nanosize particles. J Disp Sci Tech 19:163–168

Baert L, van't Klooster G, Dries W, Francois M, Wouters A, Basstanie E, Iterbeke K, Strappers F, Stevens P, Schueller L, van Remoortere P (2009) Development of a long-acting injectable formulation with nano particles of rilpirivine (TMC278) for HIV treatment. Eur J Pharm Biopharm 72:502–508

Beirowski J, Inghelbrecht S, Arien A, Gieseler H (2011a) Freeze drying of nanosuspensions, 1: freezing rate versus formulation design as critical factors to preserve the original particle size distribution. J Pharm Sci 100:1958–1968

Beirowski J, Inghelbrecht S, Arien A, Gieseler H (2011b) Freeze-drying of nanosuspensions, part 3: investigation of factors compromising storage stability of highly concentrated drug nano suspensions. J Pharm Sci 101:354–362

Benet LZ (2013) The Role of BCS (biopharmaceutics classification system) & BDDCS (biopharmaceutics drug disposition classification system) in drug development. J Pharm Sci 102(1):34–42

Bosch HW, Marcera DM, Mueller RI, Swanson JR, Mishra DS (1996) Process for preparing therapeutic compositions containing nanoparticles. US Patent 5,510,118, 23 Apr 1996

Bosch HW, Cooper ER, McGurk SL (2002) Bioadhesive nanoparticulate compositions having cationic surface stabilizers. USA Patent 6,428,814, 6 Aug 2002

Brewster M, Swinney K, Herman J, Arien T, Verreck G, Li B, Vanhoutte F, Stokbroekx S, Van den Mooter G, Van Gelder J (2012) Use of a down scaled multiplexed mill to assess the effect of milling time on itraconazole nanosuspensions. American Association of Pharmaceutical Science, Chicago, IL

Carver P, Fleisher D, Zhou SY, Kaul D, Kazanjiian P, Cheng L (1999) Meal composition effects on the oral bioavailability of indinavir in HIV-infected patients. Pharm Res 16:718–724

Chan HK, Kwok PCL (2011) Production methods for nanodrug particles using the bottom-up approach. Adv Drug Deliv 63:406–416

Chan CP-Y, Bruemmel Y, Seydack M, Sin K-K, Wong L-W, Merisko-Liversidge E, Trau D, Renneberg R (2004) Nanocrystal biolabels with releasable fluorophores for immunoassays. Anal Chem 76:3638–3645

Chen H, Khemtong X, Chang X, Gao J (2011) Nanonization strategies for poorly water-soluble drugs. Drug Discov Today 16:354–360

Cheng Z, Zaki AA, Hui JZ, Muzykantov VR, Tsourkas A (2012) Multifunctional nanoparticles: cost versus benefit of adding targeting and imaging capabilities. Science 338:903–910

Cooper ER, Bullock JA, Chippari JR, Schaefer JL, Patel RA, Jain R, Straster J, Ryde N (2010) US Patent 7,695,739, 13 Apr 2010

Crisp MT, Tucker CJ, Rogers TL, Williams RO III, Johnston KP (2007) Turbidimetric measurement and prediction of dissolution rates of poorly soluble drug nanocrystals. J Control Release 117:351–359

Cunningham J, Liversidge E, Cooper ER, Liversidge GG (2004) Milling microgram quantities of nanoparticulate candidate compounds. US Patent Application Publications No.2044/0173696

Czekai DA, Seaman LP (1999) Method of grinding pharmaceutical substances. US Patent No. 5,862,999, 26 Jan 1999

D'Addio SM, Prud'homme RK (2011) Controlling drug nanoparticle formation by rapid precipitation. Adv Drug Deliv Rev 53:417–426

Dai W (2012) Why amorphous nanoparticles. American Association of Pharmaceutical Sciences, Chicago, IL, 2012

De Garavilla L, Peltier N, Merisko-Liversidge E (1996) Controlling the acute hemodynamic effects associated with IV administration of particulate drug dispersions in dogs. Drug Dev Res 37:86–96

Deschamps B, Musaji N, Gillespie JA (2009) Food effect on the bioavailability of two distinct formulations of megesterol acetate oral suspension. Int J Nanomedicine 4:185–192

Di L, Fish PV, Mano T (2012) Bridging solubility between drug discovery and development. Drug Discov Today 17(9/10):486–495

Douroumis D, Fahr A (2013) Drug delivery strategies for poorly water-soluble drugs. Wiley, New York, NY

Dufort S, Sancey L, Coll J-L (2012) Physico-chemical parameters that govern nanoparticles fate also dictate rules for their molecular evolution. Adv Drug Deliv Rev 64:179–189

Fuhrmann K, Gauthier MA, Leroux J-C (2014) Targeting of injectable drug nanocrystals. Mol Pharm 11:1762–1771. doi:10.1021/mp5001247

Galli C (2006) Experimental determination of the diffusion boundary layer width of micron and sub micron particles. Int J Pharm 313:114–122

Gao L, Liu G, Wang X, Zhou L, Li X, Wang F (2013) Application of drug nanocrystal technologies on drug delivery of poorly soluble drugs. Pharm Res 30:307–324

George M, Ghosh I (2013) Identifying the correlation between drug/stabilizer properties and critical quality attributes (CQA) of nanosuspension formulation prepared by wet media technology. Eur J Pharm Sci 48:142–152

Gopal S, Gassmann-Mayer C, Palumbo J, Samtani MN, Shiwach R, Alphs L (2010) Practical guidance for dosing and switching paliperidone palmitate treatment in patients with schizophrenia. Curr Med Res Opin 26:377–387

Hawkins MJ, Soon-Shiong P, Desai N (2008) Protein nanoparticles as drug carriers in clinical medicine. Adv Drug Deliv Rev 60:876–885

Higgins JP, Arrivo SM, Thurau G, Green RL, Bowen W, Lange A, Templeton AC, Thomas DL, Reed RA (2003) Spectroscopic approach for on-line monitoring of particle size during the processing of pharmaceutical nanoparticles. Anal Chem 75:1777–1785

Higuchi WI, Hiestand EN (1963) Dissolution rates of finely divided drug powders.1. Effect of a distribution of particle sizes in a diffusion-controlled process. J Pharm Sci 52:67–71

Huh D, Torisawa Y, Kim HJ, Ingber DE (2012) Mircoengineered physiological biomimicry: organs-on-a chip. Lab Chip 12:2156–2164

Ige PP, Baria RK, Gattani SG (2013) Fabrication of fenofibrate nanocrystals by probe sonication method for enhancement of dissolution rate and oral bioavailability. Colloids Surf B Biointerfaces 108:366–373

Jinno J, Kamada N, Miyake M, Yamada K, Mukai T, Odomi M, Toguchi H, Liversidge GG (2006) Effect of particle size reduction on dissolution and oral absorption of a poorly water-soluble drug, cilostazol, in Beagle dogs. J Control Release 111:56–64

Juhnke M, Berghausen J, Timpe C (2010) Accelerated formulation development for nano milled active pharmaceutical ingredients using a screening approach. Chem Eng Technol 33:1412–1418

Juhnke M, Martin D, John E (2012) Generation of wear during the production of drug nano suspensions by wet media milling. Eur J Pharm Biopharm 81:214–222

Junemann D, Dressman J (2012) Analytical methods for dissolution testing of nanosized drugs. J Pharm Pharmacol 64:931–943

Junghanns J-UAH, Muller RH (2008) Nanocrystal technology. Drug delivery and clinical applications. Int J Nanomedicine 3:295–309

Keck CM, Muller RH (2006) Drug nanocrystals of poorly soluble drugs produced by high pressure homogenization. Eur J Pharm Biopharm 62:3–16

Keck CM, Muller RH (2008) Size analysis of submicron particles by laser diffractometry-90 % of the published measurements are false. Int J Pharm 366:150–163

Keck C, Chen R, Muller R (2013) SmartCrystals for consumer care & cosmetics: enhanced dermal delivery of poorly soluble plant actives. H&PC Today 8:18–24

Kesisoglou F, Panmai S, Wu Y (2007a) Nanosizing-oral formulation development and biopharmaceutical evaluation. Adv Drug Deliv Rev 59:631–644

Kesisoglou F, Panmai S, Wu Y (2007b) Application of nanoparticles in oral delivery of immediate release formulations. Curr Nanoscience 3:183–190

Knieke C, Steinborn C, Romeis S, Peukert W, Breitung-Faes S, Kwade A (2010) Nanoparticle production with stirred-media mills: opportunities and limits. Chem Eng Technol 33:1401–1411

Kolakovic R, Viitala T, Ihalainen P, Genina G, Peltonen J, Sandler N (2013) Printing technologies in fabrication of drug delivery systems. Expert Opin Drug Deliv 10:1711–1723

Kraft WK, Steiger B, Beussink D, Quiring JN, Fitzgerald N, Greenberg MD, Waldman SA (2004) The pharmacokinetics of nebulized nanocrystal budesonide in healthy volunteers. J Clin Pharmacol 44:67–72

Krishnaiah YSR (2010) Pharmaceutical technologies for enhancing oral bioavailability of poorly soluble compounds. J Bioeq Bioavail 2:28–36

Langer ES (2013) Seeking innovation to solve persistent problems. Pharmaceut Manuf 12:8–1479

Lipinski CA (2002) Poor aqueous solubility: an industry wide problem C in drug discovery. Am Pharm Rev 5:82–85

Lipper RA (1999) E Pluribus product. Mod Drug Discov 2:55–60

Liu P, Wulf D, Laru J, Heikkila T, van Veen B, Kiesvaara J, Hirvonrn J, Peltonen L, Laaksonen T (2013) Dissolution studies of poorly soluble drug nanosuspensions in non-sink conditions. AAPS PharmSciTech 14:748–756

Liversidge GG, Cundy KC (1995) Particle size reduction for the improvement of oral bioavailability of hydrophobic drugs: 1. Absolute bioavailability of nano crystalline danazol in Beagle dogs. Int J Pharm 125:91–97

Liversidge E, Wei L (2007) Stabilization of chemical compounds using nano particulate formulations. US Patent Application 2007/0224279 A1. 27 Sept 2007

Liversidge GG, Cundy KC, Bishop JF, Czekai DA (1992) Surface modified drug nano particles. US Patent 5,145,684, 8 Sept 1992

Liversidge GG, Phillips CP, Cundy KC (1994) Method to reduce particle size growth during lyophilization. US Patent 5,302,401, 12 Apr 1994

Mende S, Schwedes J (2006) Mechanical production and stabilization of nano particles by wet comminution in stirred media mills. Powder Handl 18:366–373

Merisko-Liversidge E, Liversidge GG (2011) Nanosizing for oral and parenteral drug delivery: a perspective on formulating poorly-water soluble compounds using wet media milling technology. Adv Drug Deliv Rev 63(1):427–440

Merisko-Liversidge E, Sarpotdar P, Bruno J, Hajj S, Wei L, Peltier N, Rake J, Shaw MJ, Pugh L, Polin L, Jones J, Corbett T, Cooper ER, Liversidge GG (1996) Formulation and antitumor activity of nanocrystalline suspensions of poorly soluble anticancer drugs. Pharm Res 13:272–278

Merisko-Liversidge E, Liversidge GG, Cooper ER (2003) Nanosizing: a formulation approach for poorly water soluble compounds. Eur J Pharm Sci 18:113–120

Moghimi SM, Hunter AC, Andresen TL (2012) Factors controlling nanoparticle pharmacokinetics: an integrated analysis and perspective. Annu Rev Pharmacol Toxicol 52:481–503

Monopoli M, Aberg C, Salvati A, Dawson KA (2012) Biomolecular coronas provide the biological identity of nanosized materials. Nat Nanotechnol 7:779–786

Moschwitzer J, Muller RH (2006) Spray coated pellets as a carrier system for mucoadhesive drug nano crystals. Eur J Pharm Biopharm 62:282–287

Mouton JW, Van Peer A, deBeule K, Vliet V, Donnelly JP, Soons PA (2006) Pharmacokinetics of itraconazole and hydroxyitraconazole in healthy subjects after single and multiple doses of a novel formulation. Antimicrob Agents Chemother 50:4096–4102

Muller RH, Moschwitzer J, Bushrab FN (2006) Manufacturing of nanoparticles by milling and homogenization techniques. In: Gupta RB, Kompella UB (eds) Nanoparticle technology for drug delivery, drugs and the pharmaceutical sciences, vol 159. Taylor & Francis Group, LLC, New York, NY, pp 21–51

Muller RH, Gohla S, Keck CM (2011) State of the art of nanocrystals—special features, production, nanotoxicology aspects and intracellular delivery. Eur J Pharm Biopharm 78:1–9

Na GC, Stevens HJ, Yuan BO, Rajagopalan N (1999) Physical stability of ethyl diatrizoate nanocrystalline suspension in steam sterilization. Pharm Res 16:569–574

Nandiyanto ABD, Okuyama K (2011) Progress in developing spray-drying methods for the production of controlled morphology particles: from the nanometer to the submicrometer size ranges. Adv Powder Tech 2:1–19

Narang AS, Chang R-K, Hussain MA (2013) Pharmaceutical development and regulatory considerations for nanoparticles and nano particulate drug delivery systems. J Pharm Sci 102:3867–3882. doi:10.1002/jps.23691

Noyes AA, Whitney WR (1897) The rate of solution of solid substances in their own solutions. J Am Chem Soc 19:930–934

Oliver I, Shelukar S, Thompson KC (2007) Nanomedicines in the treatment of emesis during chemotherapy: focus on aprepitant. Int J Nanomedicine 2:12–18

Patel JD, Krilov L, Adams S, Aghajanian C, Basch E, Brose MS, Carroll WL, deLima M, Gilbert MR, Kris MG, Marshall JL, Masters GA, O'Day SJ, Polite B, Schwartz GK, Sharma S, Thompson I, Vogelzang NJ, Roth BJ (2014) Clicical cancer advances 2013: annual report on the progress against cancer from the American society of clinical oncology. J Clin Oncol 32:129–161

Pathak P, Meziani MJ, Sun Y-P (2005) Supercritical fluid technology for enhanced drug delivery. Expert Opin Drug Deliv 2:747–761

Peltonen L, Hirvonen J (2010) Pharmaceutical nanocrystals by nanomilling: critical process parameters, particle fracturing and stabilization methods. J Pharm Pharmacol 62:1569–1579

Quinn K, Gullapalli RP, Merisko-Liversidge E, Goldbach E, Wong A, Liversidge GG, Hoffman J-M, Bullock J, Tonn G (2012) A formulation strategy for gamma secretase inhibitor ELN006, a BCS Class II compound: development of a nanosuspension formulation with improved oral bioavailability and reduced food effects in dogs. J Pharm Sci 101:1462–1474

Rabinow BE (2004) Nanosuspension drug delivery. Nat Rev Drug Deliv 3:785–796

Rosenflanz HZ, Barnes AS, Wallace JT (2011) Micro-Milling Media. Ceramin Industry. www.3M.com/zgc

Ruddy SB, McGurk SL, Patel R, Bullock J, Kewalramani R (2010) Drug delivery compositions. US Patent Application Publications No. 2010/0260858 A1, 14 Oct 2010

Sauron R (2006) Absence of a food effect with a 145 mg nanoparticulate fenofibrate tablet formulation. Int J Clin Pharmcol Ther 44:64–70

Shefter E, Higuchi WI (1963) Dissolution rates of finely divided drug powders. 2. Micronized methylprednisolone. J Pharm Sci 52:162–164

Shen LJ, Wu F-L (2007) Nanomedicines in renal transplant rejection—focus on sirolimus. Int J Nanomedicine 2:25–32

Spreen W, Williams P, Margolis D, Ford S, Crauwels H, Lou Y, Gould E, Stevens M, Piscitelli S (2013) First study of repeat dose co-administration of GSK1265744 and TMC278 long-acting parenteral nano suspensions: pharmacokinetics, safety, and tolerability in healthy adults. 7th IAS Conference on HIV Pathogenesis, Treatment and Prevention, 30 June–3 July, 2013, Kuala Lumpur, Malaysia

Stegemann S, Leveiller F, Franchi D, DeJong H, Linden H (2012) When poor solubility becomes an issue: from early stage to proof of concept. Eur J Pharm Sci 31(5):249–261

Timpe C (2010) Drug solubilization strategies applying nano particulate formulation and solid dispersion approaches in drug development. Pharm Rev 13:12–21

Tinke AP, Govoreanu R, Vanhoutte K (2006) Particle size and shape characterization of nano and submicron liquid dispersions. Am Pharmaceut Rev 9:33–37

Tziomalos K, Athyros VG (2006) Fenofibrate: a novel formulation (Triglide™) in the treatment of lipid disorders: a review. Int J Nanomedicine 1:129–147

Van Eerdenburgh B, Van den Moorter G, Augustijins P (2008) Top-down production of drug nanocrystals: nanosuspension stabilization, miniaturization and transformation into solid products. Int J Pharm 364:64–75

Van Eerdenburgh B, Vermant J, Martens JA, Froyen L, Van Humbeeck J, Van den Mooter G, Augustijins P (2010) Solubility increases associated with crystalline drug nanoparticles: methodologies and significance. Mol Pharm 7:1858–1870

Verloes R, van't Klooster G, Baert L, van Velsen F, Bouche M-P, Spittaels K, Leempoels J, Williams P, Kraus G, Wigerinck P (2008) TMC 278 long acting—a parenteral nanosuspension formulation that provides sustained clinically relevant plasma concentrations in HIV-negative volunteers. Paper presented at XVIIth international AIDS conference, Mexico City, Mexico, 3–8 Aug 2088

Wang L, Li M, Zhang N (2012) Folate-targeted docetaxel-lipid-based-nanosuspensions for active-targeted cancer therapy. Int J Nanomedicine 7:3281–3294

Waring MJ (2010) Lipophilicity in drug discovery. Expert Opin Drug Discov 5(3):235–248

Williams HD, Trevaskis NL, Charman SA, Shanker RM, Charman WN, Pouton CW, Porter CJH (2013) Strategies to address low drug solubility in discovery and development. Pharmacol Rev 65(1):315–499

Wu W, Nancollas GN (1998) A new understanding of the relationship between solubility and particle size. J Solution Chem 27:521–531

Wu Y, Loper E, Landis E, Hettrick L, Novak L, Lynn K, Chen C, Thompson K (2004) The role of biopharmaceutics in the development of a clinical nano particle formulation of MK-0869: a Beagle dog model predicts improved bioavailability and diminished food effect on absorption in human. Int J Pharm 285:135–146

Xiao Z, Levy-Nissenbaum E, Alexis F, Luptak A, Teply BA, Chan JM, Shi J, Digga E, Cheng J, Langer R, Farokhzad OC (2012) Engineering of targeted nanoparticles for cancer therapy using internalizing aptamers isolated by cell uptake selection. ACS Nano 6:696–704

Yalkowsky SH, Krzyzaniak JF, Ward GH (1998) Formulation-related problems associated with intravenous drug delivery. J Pharm Sci 87:787–796

Zhang JY, Bosch HW (1997) Sterile filtration of NanoCrystal drug formulations. Drug Dev Ind Pharm 23:1087–1093

Chapter 14
Leveraging Solid State Form and Physiochemical Properties for Early Clinical Formulation Efforts: Opportunities and Challenges During Telcagepant Liquid Capsule Development

Dan Zhang, Allen C. Templeton, William Marinaro, Alfred C.F. Rumondor, Filippos Kesisoglou, Brett Duersch, Karen Thompson, Joyce Stellabott, and Michael H. Kress

14.1 Lipid Based Formulations

14.1.1 Introduction to Lipid Based Formulations

Lipid based formulations (LBFs) have been used as an effective tool to improve the oral bioavailability of poorly water soluble compounds (Narang and Srivastava 2002; Porter et al. 2007; Khoo et al. 2001). By dissolving the drug in liquid vehicles, this approach avoids the solid/liquid barrier for drug release and facilitates the transfer of drug to the gastrointestinal (GI) fluid for absorption. This delivery system also includes the encapsulation of drugs in self-emulsifying vehicles, otherwise, known as self-emulsifying drug delivery systems (SEDDS) (Charman and Stella 1986) and self-microemulsifying drug delivery systems (SMEDDS) (Lespine et al. 2006). Well known marketed LBF products include Sandimmune® or Neoral® (cyclosporine), Norvir® (ritonavir), Fortovase® (saquinavir), Avodart®

D. Zhang (✉) • W. Marinaro • A.C.F. Rumondor • K. Thompson • J. Stellabott
Formulation Sciences, Merck & Co., Summit, NJ, USA
e-mail: dina_zhang@merck.com

A.C. Templeton
Analytical Sciences, Merck & Co., Summit, NJ, USA

F. Kesisoglou
Biopharmaceutics, Merck & Co., West Point, PA, USA

B. Duersch
Analytical Development-Commercialization, Merck & Co., West Point, PA, USA

M.H. Kress
Process Chemistry, Merck & Co., Rahway, NJ, USA

A.C. Templeton et al. (eds.), *Discovering and Developing Molecules with Optimal Drug-Like Properties*, AAPS Advances in the Pharmaceutical Sciences Series 15,
DOI 10.1007/978-1-4939-1399-2_14

(dutasteride), and Advil® Liqui-Gels® (ibuprofen). LBFs represent a wide diversity in formulation strategy from semisolids such as clofazimine, in which micronized active pharmaceutical ingredient (API) is suspended in a waxy matrix (Narang and Srivastava 2002), to Tessalon® (benzonatate) Perles, which contains the active ingredient dissolved in glycerin with preservatives added. Due to relatively straightforward processing steps, this dosage type offers a number of advantages, such as rapid development timeline, low requirements for drug usage, and high tolerability of variability in API physical and processing attributes (particle size, morphology, flow, etc.), making it an attractive option to support early clinical studies (Ichihashi et al. 1992; Kwei et al. 1998). However, like other delivery systems, LBFs face a number of challenges. The first obvious issue is the potential for insufficient drug solubilities in the liquid vehicles, leading to difficulty in achieving high unit dose strengths often required for competitive marketing and patient convenience. Another central problem is capsule shell cross-linking which could potentially delay the drug release and alter pharmacokinetics (PK) profiles (Ueda et al. 1983; Hugger et al. 2002). Since the drug is in solution with excipients that have the potential for reactivity, drug and formulation stability is also a major challenge during development (Rege et al. 2002; Feinle et al. 2003).

In general, formulation research activities focus on providing stable, bioavailable, processible, and marketable dosage forms to support clinical trials, registration and eventual product commercialization. Balancing amongst these competing goals, e.g., achieving necessary bioavailability objectives with a market-competitive dosage unit, can be challenging to achieve due to the specific pharmaceutical properties of the drug substance and commercial constraints around the drug product. One significant hurdle can be a change in drug phase and/or dosage forms and the requisite pharmacokinetic bridging of the formulations throughout the different development stages. Since drug development is often interdependent across many functional areas, the true attributes and performance of the formulation are often only understood after the availability of human PK data and substantial stability data. The delay in clarity around these important parameters can also present challenges and risks. This chapter will address these development considerations for LBFs in greater depth, focus on key challenges, and then illustrate a case study in which these practices were applied and challenges were encountered and addressed.

14.1.2 Fundamentals of Lipid-Based Formulation Systems

In addition to the obvious capacity of LBFs to solubilize lipophilic compounds, there is a growing body of knowledge around the unique and advantageous downstream effects of LBFs on in vivo drug absorption and distribution. Although this is described in depth in a previous chapter, it is helpful to note the comprehensive review by Porter et al., which describes and classifies these effects into three major areas: “alteration of the character and composition of the intestinal milieu, the recruitment of intestinal lymphatic drug transport, and the interaction with enterocyte-based transport processes" (Porter et al. 2007).

The first mechanism refers to the ability for even relatively small quantities of lipid in typical LBFs to initiate a series of digestive processes which benefit drug delivery. Of most obvious impact would be the stimulation of biliary secretions, bile salts and phospholipids. These endogenously secreted components directly impart solubilizing capacity to the intestinal milieu. Besides this well-known and exploited phenomenon, the role of lipase secretion in the GI tract can have a profound and complicated effect on LBF performance. Gastric and pancreatic lipases digest diglycerides and triglycerides to their constituent free fatty acids, thus altering their performance with respect to solubilizing capacity. Long-chain lipids generally maintain increased solubilization capacity upon digestion and incorporation into colloidal structures compared to medium-chain lipids. Conversely, drug solubility in these long-chain lipids is typically worse than drug solubility in medium-chain lipids. This contradiction sets up a formulation balancing act between drug loading (formulation solubility) and product performance (in vivo solubility).

The second unique mechanism by which LBFs may affect drug absorption is through the promotion of intestinal lymphatic transport. Briefly, the anatomy of the intestinal (mesenteric) lymph system is such that it will deposit fluid and materials from the gut into circulation without passing through the liver. Thus, lymphatic transport is a possible route directly from oral administration to systemic circulation while avoiding first pass-liver metabolism. Within intestinal enterocytes, materials from the intestinal milieu may be transported into the mesenteric lymph after incorporation into lipoproteins (primarily chylomicrons and very low density lipoproteins). These lipoproteins are then transported into the interstitial space, via exocytosis, where they are available to the mesenteric lymph. Lipophilic drugs may take advantage of this route by being incorporated into the highly lipophilic cores of these lipoproteins. More importantly, LBFs may also enhance this drug route by stimulation of the lipoprotein formation. Several examples have shown significant increases in lymphatic drug transport through co-administration of lipids (Khoo et al. 2001). Through modeling the mesenteric lymph, Charman and Stella sought to quantify the utility of this phenomenon in relation to drug delivery (Charman and Stella 1986). They proposed that significant lymphatic transport of drugs would only occur for molecules with a $\text{Log}\,D > 5$ and a solubility in long-chain triglycerides of 50 mg/g. This rule of thumb has been subsequently borne out through several examples (Ichihashi et al. 1992; Lespine et al. 2006; Kwei et al. 1998; Ueda et al. 1983).

Porter et al. describes the various ways in which lipids and lipidic excipients interact with enterocyte biology to affect drug delivery. The mechanisms are varied and include interactions with apical membrane lipid transporters and efflux proteins (Hugger et al. 2002; Rege et al. 2002), as well as a complex system of intracellular trafficking mechanisms and lipid pooling. Some of these interactions are controversial and this field is likely not yet mature enough to be elegantly exploited a priori during LBF development. However, consideration of these systems is important in the design and interpretation of in vitro and in vivo experimentation.

Lastly, it should also be noted that lipids generally slow GI motility and gastric emptying. The effects of these changes will be compound specific and are the subject of numerous work elsewhere (Feinle et al. 2003). However, the opposite effect has been noted through administration of PEG 400 enriched formulations (Basit et al. 2001), and the latter may negatively impact drug absorption (Basit et al. 2001). This is particularly important to note since PEG 400 enriched vehicles are often highly leveraged in early compound screening, including the present case study.

Taken as a whole, these mechanisms suggest several delivery advantages including potential faster T_{max} and higher C_{max} due to the elimination of the dissolution step of a conventional oral solid, decreased food effect by stimulation of digestive processes to in part mimic the fed state, increased absorption by increasing drug solubility in the gastric milieu, and decreased first pass metabolism by promoting intestinal lymphatic transport. Of particular importance to the case study presented later is the application of lipid based systems to impart a fast onset PK profile and enhance bioavailability. This has been demonstrated via several examples including one by Yuksel et al. in which the greatly enhanced C_{max} of piroxicam was demonstrated through an LBF strategy (Yuksel et al. 2003).

14.1.3 Designing a Lipid-Based Formulation System

In designing a lipid-based formulation (LBF) system, the interplay between different physical, chemical, and biological factors need to be considered. While a number of approaches have been discussed (Humberstone and Charman 1997; Charman 2000; Wasan 2001; Mullertz et al. 2010; Pouton 2000; Pouton and Porter 2008; Singh et al. 2009; Kuentz 2012; Jannin et al. 2008), the major steps in LBF development can be summarized as illustrated in Fig. 14.1. Just as with other dosage forms, formulation design for LBF often begins with excipient selection. Two factors directly contribute to the selection of LBF excipients. The first factor is chemical compatibility between the excipients and the active pharmaceutical ingredient (API) of interest. The second factor is the solubility of the API in the excipients, many of which are liquids at room temperature. Combined with the target dose, the solubility of the API in the excipients affects the amount of API that can be delivered in a typical fill volume of a soft or hard gelatin capsule (e.g., 600 μL), which in turn dictates the number of dosage units required. The type of LBF system desired should also be determined as early as possible, which can be guided by the use of the Lipid Formulation Classification System (Pouton 2000, 2006).

Armed with the solubility values of the API in the excipients and the type of LBF system desired, potential formulation combinations can be designed. The behavior of the potential formulations can then be assessed, which is commonly

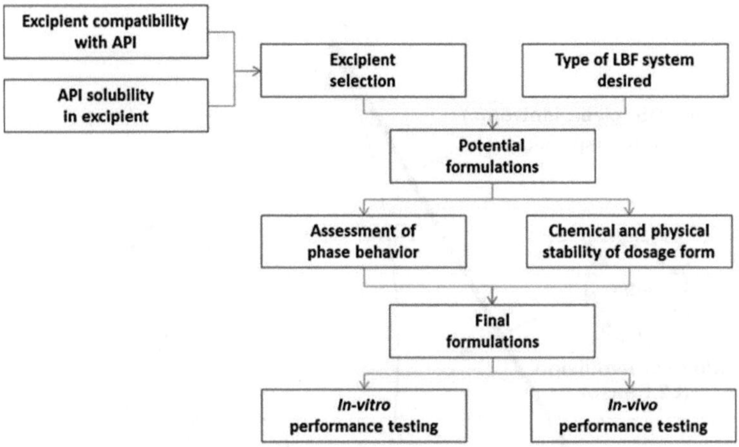

Fig. 14.1 Typical major steps in lipid-based formulation (LBF) development

accomplished through the construction of ternary phase diagrams, as shown in Fig. 14.2. At the same time, the compatibility of potential formulations with soft and hard gelatin capsule ingredients should be verified. Through this selection process, the most promising formulations can be identified. Finally, optimization can be done by evaluating the performance of the formulations either in vitro or in vivo, or both. Each of these steps is discussed in further details in the following sections.

One concept that will help in subsequent discussions is the concept of hydrophilic and lipophilic (or hydrophobic) balance. A hydrophilic compound is a compound that is "water-loving," or has strong affinity for water. When a hydrophilic liquid is added into water, it forms a one-phase solution for the whole range of composition (complete miscibility). On the other hand, a lipophilic (or hydrophobic) compound has strong affinity to lipid, and lacks affinity to water. When added into water, a lipophilic liquid tends to form a discrete phase or a separate layer. Many compounds, especially surfactants, are amphiphilic in nature, possessing both hydrophilic (*water-loving*, polar) and lipophilic (*fat-loving*) properties. To quantify the affinity of a surfactant to water, Griffin proposed a classification system termed "HLB," which stands for "hydrophile–lipophile balance" (Griffin 1949). In this classification system, the HLB value of an amphiphilic compound can be calculated according to the following formula (Griffin 1954):

$$\mathrm{HLB} = 20 \cdot \frac{M_{\mathrm{h}}}{M}$$

where M_{h} is the molecular mass of the hydrophilic portion of the molecule and M is the molecular mass of the whole molecule. When applied against this classification system, every surfactant can be described using an HLB value between 0 and 20, with 0 corresponding to a completely lipophilic molecule, and 20 corresponding

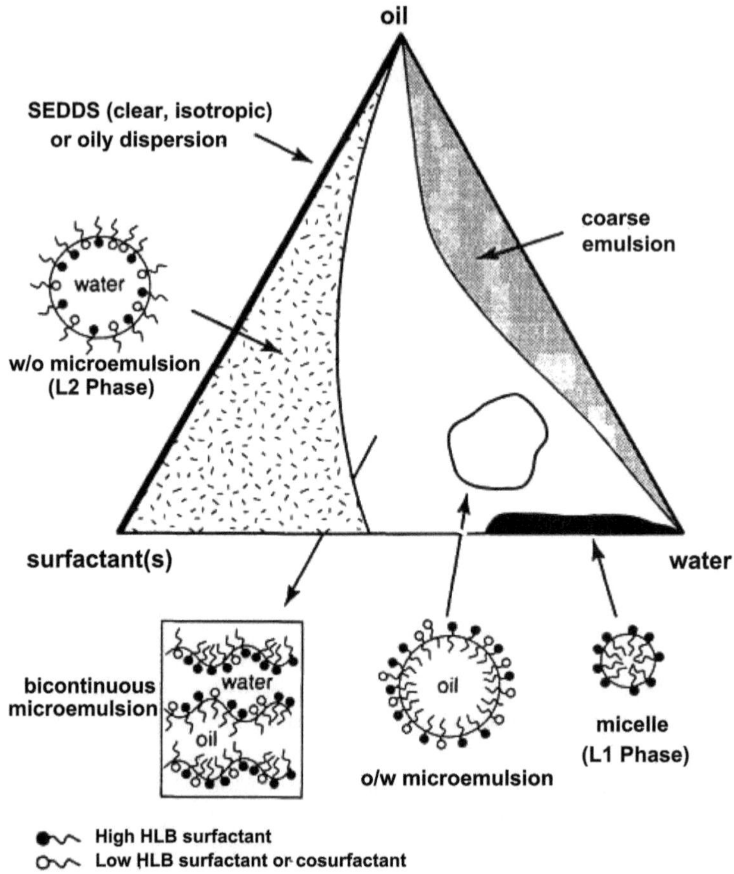

Fig. 14.2 Hypothetical ternary phase diagram of an oil/surfactant/water system depicting the types of systems that can be generated upon dispersion of an LBF in a water–micellar solution, self-emulsifying drug delivery system (SEDDS), and a coarse emulsion. Adopted from Constantinides (1995)

to a completely hydrophilic molecule. For example, the HLB values of Tween 80 and Span 80 are 15 and 4, respectively. Tween 80 can play many roles, including oil-in-water emulsifier, wetting agent, dispersant, and solubilizer. On the other hand, Span 80 is more hydrophobic and acts as a water-in-oil emulsifier.

14.1.4 Commonly Used Excipients in LBFs

Different excipients encompassing a wide range of chemistry can be considered in designing an LBF system (Strickley 2004; Hauss 2007). For the sake of discussion, LBF excipients are divided into the following major classes: medium- and long-chain triglycerides (MCTs and LCTs) and their derivatives, chemical combinations of MCT- or LCT-derived products and co-solvents, and other excipients such as phospholipids, antioxidants, and P-glycoproteins (P-gp) efflux inhibitors.

The chemical structures of MCTs and LCTs are relatively similar; both have a glycerol backbone with all three hydroxyl moieties esterified to fatty acids. The difference lies in the length of the fatty acid chains: 6–10 carbon length for MCTs, and 16 or more carbon length for LCTs. MCTs and LCTs can be extracted from food-grade ingredients (such as coconut oil, corn oil, and sesame oil), or synthetically produced. MCTs and LCTs can be liquid or semisolid at room temperature, depending on the length of the fatty acid chain as well as the degree of saturation. The presence of unsaturated bonds, especially in *cis*-configuration, introduces a "kink" in the chain, which disrupts chemical packing during solidification, resulting in a lower melting temperature (Nelson and Cox 2012). Although LCTs and MCTs are chemically similar, their rates of digestion are different, which can lead to different API solubilization and uptake behavior (Kaukonen et al. 2004).

Due to their hydrophobic nature, MCTs and LCTs have limited dispersibility in water. On the other hand, the partial glycerides, monoglycerides and diglycerides (e.g., glycerol monooleate and glycerol monostearate), where only one or two of the hydroxyl moieties on the glycerol are esterified with fatty acids, have better dispersibility in water and are more amenable to forming emulsions (Pouton and Porter 2008). Despite the name, monoglycerides typically also contains diglycerides and triglycerides (Gibson 2007); so care should be taken in selecting the proper ingredient. Pure free fatty acids (FFAs), such as oleic acid, can also be used as a formulation vehicle. Since FFAs are formed as the hydrolysis or digestion products of mono, di, and triglycerides, it is often prudent to measure the solubility of the API in these vehicles even if they are not directly added in the final dosage form.

The next class of excipients used in lipid-based formulations is co-solvents, which include propylene glycol (PG), polyethylene glycols (PEGs), ethanol, glycerol, and isopropanol. They are hydrophilic in nature and are completely miscible with water. Propylene glycol is a colorless, clear, viscous liquid at room temperature, with two hydroxyl moieties on a three-carbon chain (O'Neil 2013). Polyethylene glycols are an oligomer or a polymer of ethylene oxide. Traditionally, PEGs are named according to their average molecular weight; for example, the MW of PEG 300 is approximately 300 g/mol, which corresponds to six repeating units of ethylene glycol. Depending on the chain length, PEGs can either be liquid (e.g., PEG 300 and PEG 400) or solid (e.g., PEG 1000) at room temperature.

Ingredients from the first class are often chemically combined with ingredients from the second class to impart specific functionalities. For example, polyglycerol fatty acid esters (trade names Labrasol®, Labrafil®, and Gelucire®) are composed of a chain of glycerol molecules, linked together by ether linkages, and esterified with one or more fatty acid molecules. The hydrophilic nature of the molecule increases as the polyglycerol chain length or the number of free hydroxyl moieties increases, and decreases with increasing number or chain length of the esterified fatty acids. PEGs can also be combined with castor oil, which is the case for polyethoxylated castor oil (trade name Kolliphor® EL). In a similar fashion, PG, ethanol, and isopropanol can be esterified with different fatty acids.

Also related to this class of ingredients are sorbitan derivates and polysorbates. Sorbitan is a mixture of isomeric organic compounds derived from the dehydration of sorbitol and usually consists of 1,4-anhydrosorbitol, 1,5-anhydrosorbitol and 1,4,3,6-dianhydrosorbitol (O'Neil 2013). Sorbitan monostearate (trade name Span®) is an ester of sorbitan with stearic acid. Polysorbates are PEG-ylated sorbitans esterified with fatty acids. For example, polysorbate 80 (trade name Tween®) is an ester of PEG-ylated sorbitan and oleic acid.

Another class of lipidic formulation ingredients is phospholipids, which is a major component of cell membranes. Phospholipids are comprised of a hydrophilic head which contains a negatively charged phosphate group and a glycerol backbone and the hydrophobic tail, which usually contains two long fatty acid hydrocarbon chains (Fricker et al. 2010). An example of a phospholipid commonly used in LBF systems is lecithin, which can be extracted from egg yolk or soybean. While there are differences in the composition of fatty acids in phospholipids extracted from egg or soybean, the two have similar surface activities (Palacios and Wang 2005).

Other ingredients can also be included in LBF systems to address special formulation needs. For example, antioxidants can be added for APIs susceptible to oxidation. Commonly used oil-soluble antioxidants include butylated hydroxytoluene (BHT), butylated hydroxyanisole (BHA), and propyl gallate. P-gp inhibitors such as vitamin E TPGS (D-α-tocopherol polyethylene glycol succinate) can also be added if reducing the API efflux rate is desired, for example in the case of BCS class IV compounds. Vitamin E TPGS is an ester of vitamin E succinate with polyethylene glycol. In addition to its role as a surfactant, vitamin E TPGS has been shown to overcome the inhibition of P-gp transporters, which can further increase the bioavailability of the API (Dintaman and Silverman 1999; Varma and Panchagnula 2005; Collnot et al. 2010; Zhang et al. 2012).

14.1.5 Determining the Solubility of APIs in Lipidic Excipients

Since the primary objective of using an LBF is bioavailability enhancement through solubility improvement, it is important to assess the solubility of the API in different excipients. The most rigorous approach in assessing the solubility of a solid API in a liquid vehicle is by using the shake-flask method, where an excess amount of API is dispensed into the vehicle and the container is mixed at controlled speed and temperature. Samples are collected regularly, e.g., after 24, 48, and 72 h. The liquid is then separated from the solid, and the amount of API dissolved is quantified. Comparison of values obtained at different time points can provide insight into whether the solubility values acquired are close to the thermodynamic solubility value. Although different protocols have been described by different authors (Avdeef 2001, 2007; Blasko et al. 2001; Kerns 2001; Bergstrom et al. 2002; Chen and Venkatesh 2004; Glomme et al. 2005; Zhou et al. 2007), the principles of the shake-flask solubility measurement method are well-established.

Owing to the large number of lipidic excipients available, quantitative measurement of API solubility can be time consuming and labor intensive. In order to minimize the time and effort spent in generating solubility data, a qualitative assessment can first be done by dispensing the API in different amounts of liquid excipients. Visual inspection of the samples can provide an estimate of the maximum and minimum solubility of the API in each excipient. Once a smaller number of excipients with high API solubility have been identified, a more rigorous solubility assessment can be undertaken. Solubility measurement also lends itself to miniaturization and automation. Commonly, kinetic solubility of the API is assessed in place of the thermodynamic solubility to maximize the number of samples analyzed in a relatively short time, especially during early stages of the drug development process (Alsenz and Kansy 2007). Different techniques such as nephelometry or UV spectroscopy can be adopted to perform this work (Bevan and Lloyd 2000; Dehring et al. 2004; Hoelke et al. 2009; Chen et al. 2002; Bard et al. 2008; Mansky et al. 2007; Lipinski et al. 1997). In contrast, Alsenz et al. described a quantitative, partially automated screening approach that can generate thermodynamic solubility data faster than the shake-flask method with good correlation to the more rigorous technique (Alsenz et al. 2007). Note that the experimental approaches described above are all quite amenable to automation and are scale independent. For these reasons LBFs are often leveraged for enabling solubility early in development.

In addition to experimental approaches, models to predict the solubility of an API in lipidic excipients a priori have been attempted. For example, through the application of Partial Least Squares (PLS) modeling, Persson et al. were able to correlate the solubility of different compounds in soybean oil (LCT) and Captex 355 (MCT) with molecular descriptors such as API melting temperature, polar surface area, number of nitrogen atoms, number of double bonds, eccentricity, topological charge, size and shape (Persson et al. 2013). This approach can also reduce the number of solubility experiments that have to be conducted. However, a complete mathematical model to accurately predict the solubility of small molecule APIs in lipidic excipients with wide-varying chemical properties and hydrophilicity has proved to be challenging, in part due to the complexities of the role of absorbed water and its ability to form hydrogen bond structures (Rane and Anderson 2008).

14.1.6 Designing the Final LBF System and Evaluation of Performance

Once excipients with acceptable chemical compatibility and relatively high API solubility have been identified, potential formulations can be designed by combining excipients from different classes. In 2000, Pouton proposed the Lipid

Table 14.1 Classification of lipid-based formulations (LBFs) according to composition with relative advantages and disadvantages

LBF type	Composition	Advantages	Disadvantages
Type I	Pure oils (MCTs, LCTs, or mixed monoglycerides and diglycerides)	Relatively simple formulations, compatible with capsule ingredients	Non-dispersing, requires digestion, poor solvent capacity for many APIs
Type II	Oils and water-insoluble surfactants (HLB < 12, 20–60 % w/w)	Forms SEDDS (0.25–2 μm), unlikely to lose solvent capacity upon dispersion	Turbid dispersion
Type IIIa	Oils, water-soluble surfactants (HLB > 12, 20–40 % w/w), and co-solvents (up to 40 % w/w)	Clear or almost clear SMEDDS (100–250 nm), absorption without digestion	Possible loss of solvent capacity upon dispersion, less easily digested
Type IIIb	Oils, water-soluble surfactants (HLB > 12, 20–50 % w/w), and co-solvents	Clear SMEDDS (<100 nm), absorption without digestion	Likely loss of solvent capacity on dispersion
Type IV	Surfactants and co-solvents (up to 50 %)	Good solvent capacity for many APIs, disperses to micellar solution	Loss of solvent capacity on dispersion, may not be digestible

Adapted from Pouton (2000, 2006), Pouton and Porter (2008)

Formulation Classification System (LFCS) to delineate between the different types of lipidic formulations according to their characteristics (Pouton 2000, 2006). Each LFCS formulation type has its own strengths and challenges (see Table 14.1). For example, a Type I formulation, which is comprised mostly of pure MCTs or LCTs, must rely on digestion to accommodate the formation of emulsified species. On the other hand, a Type IV formulation rapidly mixes with water to form a micellar solution with small droplets, but is susceptible to drug precipitation upon dilution in the gastrointestinal tract. Selection of the appropriate LBF type to deliver the API can help streamline formulation development efforts while avoiding known potential problems. In designing potential formulations, the HLB value of the mixed systems should be adjusted to match the desired HLB value of the final formulation. This can be done by varying the composition of ingredients and the HLB value of the final system is the weight-averaged contributions of the HLB of each ingredient (Griffin 1949).

The behavior of the potential formulations upon dispersion in water is typically evaluated through the construction of ternary phase diagrams. The formation of a microemulsion, a coarse emulsion, or phase separation into distinct liquid phases should be carefully examined. In addition, the ability of the system to maintain API super-saturation upon dispersion should be evaluated, which has been shown to be crucial in enhancing API absorption in selected cases (Gao et al. 2003, 2004, 2009). At the same time, the physical and chemical stability of the formulation should be evaluated, leveraging studies at elevated temperatures when appropriate. The compatibility of the formulation with capsule ingredients such as gelatin or hypromellose should also be included as a part of this evaluation.

In an oral tablet development program, dissolution testing is commonly employed to provide insight into the performance of the final dosage form. However, selection of the appropriate in vitro testing for an LBF is still a subject of active research due to the complexities of biological and physiological processes such as lipid digestion and uptake by lymphatic transport. Nevertheless, various in vitro methods have been used to evaluate the performance of LBFs. The most widely accepted test is to monitor the dispersibility of the LBF when diluted into dissolution or biorelevant media. The goal is to achieve molecular or nanoparticle dispersions. In addition to in vitro testing, the appropriate use of animal models is often unavoidable in evaluating the performance of the final LBF dosage form (Williams et al. 2012a, b).

14.1.7 LBF Manufacture

LBFs are typically presented either as a soft gelatin (softgel) or a hard-shell capsule. Softgels are single piece gel-encapsuled oral dosage forms that are formed, filled and sealed in a single operation. In the manufacturing process, the gelatin is first melted with water and plasticizer. This material is cooled on the encapsulation machine into two ribbons which ultimately form the shell of the capsule. The fill material enters from the top of the encapsulation machine and is injected in-between the ribbons, which are sealed immediately after filling. After they are formed, softgels are usually dried in drying tunnels. On the other hand, in the manufacturing of liquid-filled hard-shell capsules, the two pieces of capsules must be pre-made before the fill material is added. After filling, the two pieces of the capsule are sealed through banding or microspray sealing. Gelatin is often preferred as capsule shell material; however, other materials such as hypromellose can also be used to manufacture hard-shell capsules.

Both softgels and hard-shell capsules can accommodate a wide range of formulations; however, there are inherent differences between them. In softgels, plasticizers are commonly used, which can lead to relatively higher oxygen permeability. In addition, the amount of moisture adsorbed by soft gelatin shells containing different plasticizers has been shown to be higher than hard gelatin capsules (Cole et al. 2008). On the other hand, a significant risk of hard gelatin capsules is their tendency to leak between the joint of body and cap. Even with banding or microspray technologies, this remains a major liability. As a result, hard gelatin capsules are often employed in early clinical studies due to their low resource requirements, but a switch to an SGC is a common practice to ensure commercial success. Regardless of the type of final dosage form chosen, the API is usually dissolved before placement into the fill material during the manufacturing of LBFs. As a result, the manufacturing process of LBFs is more amenable for highly toxic and highly potent compounds.

14.1.8 Stability Challenges in Developing LBFs

As highlighted above, the development of a lipid-based formulation requires a thorough understanding of the interplay between different physical, chemical, and biological factors. In most cases, drug degradation will occur more rapidly in solution than in the solid state, thus an obvious concern in LBF development is chemical stability. Moreover, key excipients common to LBF formulations are mixtures of compounds with reactive functionalities and are thus prone to chemical reaction through a number of mechanisms. Stella has described the literature and experiences of many drug development scientists regarding these issues (Stella 2013). Typical routes of drug degradation observed in lipidic vehicles include: oxidation, esterification, transesterification, amidation, transamidation, acylation, transacylation, epimerization, and impurities resulting from peroxide, formaldehyde, and formic acid addition. These reactions are either the result of the functional groups of the excipients themselves, as in transesterification seen with triglycerides, or due to impurities common in the excipients, as in oxidation reactions observed with polyethylene oxides. Table 14.2 highlights many of the common reactions seen in LBFs and links them with the excipient root cause. While the table is not comprehensive, it is highly instructive in describing the route of most LBF specific stability concerns.

In 1998, concerns over the lipid-based formula of Norvir® (ritonavir, semisolid capsules) brought physical stability risks of LBFs to the forefront of the pharmaceutical sciences world. In this case, a previously undiscovered and less soluble form of the API began appearing in commercial supplies. This caused batch release failures due to out-of-specification dissolution values and prompted an immediate reformulation effort (Bauer et al. 2001). Interestingly, there are a few other

Table 14.2 Chemical stability liabilities of some common lipid-based formulation (LBF) excipients

Excipient	Function	Chemical stability liabilities
Triglycerides	Solubilizer	Hydrolyze to free fatty acids and glycerol, losing functionality and yielding reactive moieties. Direct participation in transesterification reactions (Ballard et al. 2007).
Free fatty acids (e.g., oleic acid)	Solubilizer	Esterification reactions with alcohol containing drugs (Stella 2013).
Propylene glycol	Cosolvent	Transesterification with ester containing drugs (Ma et al. 2002).
Polyethylene oxides (esp. polyethylene glycols, PEGs)	Cosolvent	Contain high levels of hydroperoxides (Wasylaschuk et al. 2007), peroxy radicals (Yang et al. 1996), and formaldehyde (Waterman et al. 2008).
Polysorbates (Tween), vitamin E TPGS	Surfactant	These excipients contain PEG and have thus demonstrated the liabilities associated with polyethylene oxides.

examples in the literature of API physical stability challenges in LBF systems. The authors speculate that this is likely due to work emphasizing phase separation and precipitation risks early in development. As discussed above, the earliest steps in LBF development include screening API solubility in excipients and target vehicles, followed by an understanding of the vehicle phase and constructing a phase diagram of the composition. In addition, early development often includes stressing prototypes at extreme temperatures and temperature cycling specifically searching for physical stability liabilities. It is likely that because of this intensive upfront effort, physical stability concerns often do not appear during late development of LBFs. Thus, the case study later in this chapter offers an interesting and instructive example.

A final factor to consider regarding stability is the actual capsule shell itself. Cross-linking of the capsule shell is a well-known difficulty in formulating capsules generally, but especially for liquid-filled soft gelatin capsules. Degenis et al. provides a succinct review of the chemistry of cross-linking and considerations for formulation and storage (Digenis et al. 1994). Briefly, the capsule shell is a mixture of water soluble proteins derived primarily from collagen. Capsule shell cross-linking arises from gelatin polymerization. The chemistry of cross-linking has been well documented and is facilitated by temperature, humidity, light and exposure to aldehydes (Meyer et al. 2001). Cross-linking results in the formation of a thin, water-insoluble membrane called a pellicle, which can act as a barrier to cause dissolution slowdown. Studies have showed that these effects primarily impact the in vitro testing results rather than the in vivo performance (Basit et al. 2001), as digestive enzymes in the body are able to break down the cross-linked gelatin. Literature studies suggest that a two-tiered in vitro dissolution test may constitute a better indication for the in vivo behavior of gelatin-encapsulated formulations than single tier testing (Basit et al. 2002). In addition, several excipients (or their impurities) utilized in formulations have been implicated in gelatin cross-linking and dissolution slowdown. For example, aldose sugars and aldehydes directly catalyze gelatin cross-linking, while starch and rayon in packaging contain impurities which may directly, or after further degradation, catalyze cross-linking. As discussed, many of the excipients used in LBF formulations are well known for forming aldehydes after autoxidation including PEGs, polyethylene glycol, non-ionic surfactants, and plasticizers, among others.

Since LBFs can introduce the drug to the gastrointestinal tract in a solubilized state, this novel delivery system has emerged as an effective tool to enhance bioavailability and reduce both interpatient and intrapatient variability and food effects. However, it also accompanies a number of challenges as presented before, including difficulties in accommodating high unit dose strengths, capsule cross-linking, and risks in chemistry and physical stabilities. We would like to elucidate these points further through the following case study.

14.2 Opportunities and Challenges During Telcagepant Liquid Capsule Development

Telcagepant (Table 14.3) is a novel calcitonin gene related peptide (CGRP) receptor antagonist studied for the treatment of acute migraine. As a potent vasodilatory neurotransmitter, CGRP is believed to play a key role in the pathophysiology of migraine. CGRP levels in cranial circulation are increased during a migraine attack and CGRP itself has been shown to trigger migrainous pain with concomitant dilation of the middle cerebral artery. Thus, inhibition of CGRP-driven pathophysiological processes has been pursued as a novel therapeutic approach for the treatment of migraine-associated pain (Ceruti et al. 2011; Ho et al. 2010; Tepper and Millson 2003). In order to have clinical utilities as a migraine therapy, the drug must rapidly reach therapeutic plasma levels (i.e., levels predicted to result in >90 % blockade of CGRP receptor activity) and sustain them for at least 2 h, ideally 4 h, to reduce the chance of symptom recurrence. To help achieve clinical targets, the critical requirements for the formulation include fast onset of pharmacokinetics and sustainment of exposures within the therapeutic range for the required duration. Besides pain, migraines can often cause nausea and vomiting, thus a single small dosage form was strongly desired to enhance patient convenience and compliance. As migraine attack can occur under any daily

Table 14.3 Physicochemical properties of telcagepant crystalline neutral form

Properties	Data for crystalline neutral form
Structure	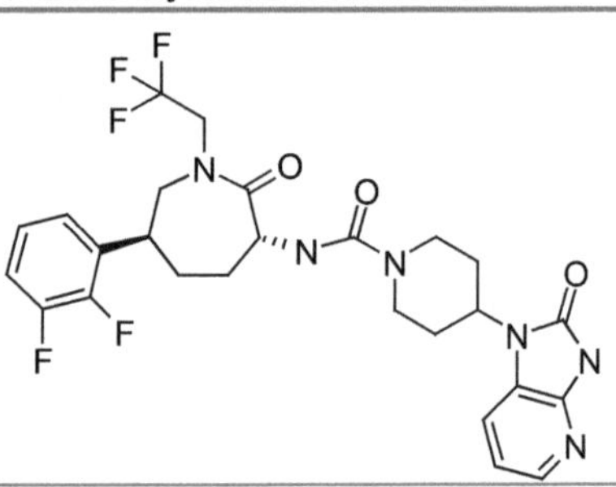
Molecular weight	566.54
Permeability	Papp = 11.7×10^{-6} cm/s (Caco-2)
p*K*a	2.2 and 9.5
Melting point (°C)	223.5
Aqueous Solubility (mg/mL)	0.03 in water (pH 7.3)
	0.03 at pH 6
	0.04 at pH 4
	0.08 at pH 2
Stability	Stable in pure solid state and aqueous solutions
Projected Efficacious/Top Phase I dose (mg)	300 mg/1,800 mg

circumstances, stability to support a room temperature product is also critical so as not to limit patient access to the medicine.

From a physiochemical property standpoint, the crystalline neutral form of telcagepant presented a number of significant challenges. As shown in Table 14.3, telcagepant has low solubility across the physiologically relevant pH range. Based on the high projected doses, it was preliminarily classified as a BCS class II compound (low solubility, high permeability). The key risk for developing this type of compounds is inadequate exposures and/or loss of dose proportionality. To make an insoluble compound orally bioavailable, a critical step is to achieve and maintain a high solubility in the gastrointestinal milieu (Basit et al. 2004; Desai et al. 2006; Patil et al. 2011). Over the years, various approaches have been developed, such as salt formation (Damayanthi et al. 2008), nanoformulations (Kesisoglou et al. 2007), liquid formulations (Benza and Munyendo 2011), and amorphous formulations (neat or dispersed in polymer matrix most often via spray drying or hot melt extrusion) (Vasconcelos et al. 2007). Each of the formulation tools has its advantages and limitations. In addition, the formulation development strategy depends not only on drug physiochemical properties, and target product characteristics, but also on its relative position in the development continuum. In the early stage of development, factors such as uncertainty in the projected efficacious dose, project timelines and API availability can play key roles in the decision making process. Based on the physiochemical properties of the compound, drug supply, timeline and clinical objectives of the Phase I study, a simple formulation of the crystalline neutral form in PEG 400, encapsulated in hard gelatin capsules (HGC) was developed. The stability risk associated with the known incompatibility of PEG 400 and the HGC matrix was mitigated by special packaging and cold storage (at 5 °C). In the formulation redesign for the Phase IIB study, various approaches were explored to improve stability, enhance bioavailability and reduce unit size. A soft-gelatin capsule (SGC) formulation using telcagepant potassium salt ethanolate with the incorporation of surfactants was developed. The formulation was successful in resolving the incompatibility issue between vehicle and capsule shell while doubling the exposure. However, on shelf storage, cross-linking of the capsule shell was observed. In addition, chemical reactivity between telcagepant potassium salt ethanolate and vehicle led to surfactant hydrolysis, salt disproportionation and finally loss of solubilization of the drug within the vehicle matrix. Both in vivo and in vitro studies were performed to help understand capsule cross-linking and the impact of drug crystallization.

14.3 Experimental Section

14.3.1 Materials

The following chemicals were obtained from commercial sources and used as received: potassium hydroxide and sodium hydroxide from Mallinckrodt Baker; polyethylene glycol and propylene glycol from Dow; polysorbate 80 (Tween®) and Imwitor 742® from Croda; polyoxyl 35 castor oil (Cremophor® EL, now Kolliphor® EL), Cremophor® RH40 (now Kolliphor® RH40), Eudragit® E, poloxamer 407 (Pluronic®) and sodium lauryl sulfate from BASF, hydroxypropylcellulose from Ashland; lactose anhydrous from Kerry; lactose monohydrate from Foremost Farms; magnesium stearate from Covidien, calcium phosphate dibasic from Innophos; sodium citrate from Archer Daniels Midland; Labrafil® from Gattefosse; and hard gelatin capsules from Capsugel. Two types of soft gelatin capsules were also used: acid bone-based supplied by Cardinal Health, and lime bone-based supplied by Banner Pharmacaps, Inc. The active pharmaceutical ingredients (telcagepant neutral form and telcagepant monopotassium salt ethanolate) were synthesized internally at Merck & Co., Inc. The milled API was achieved via either pin milling using a 100 UPZ from Alpine or dyno-milling by using a Dynomill KDL-A from Glen Mills.

14.3.2 Experimental Methods

14.3.2.1 Polymorph and Salt Screening

Both polymorph and salt screens of telcagepant were extensively conducted. The polymorph screens included slurries, anti-solvent, wet grinding, and thermal heating/cooling. Counter ions for forming both acidic and basic salts were considered. The screens were performed by dissolving appropriate amounts of free base and counter-ions in selected solvents and evaporating the solvents. The resultant solids were characterized by microscopy, X-ray powder diffraction (XRPD) and differential scanning calorimetry (DSC).

14.3.2.2 Phase Conversion Diagram

A phase diagram was obtained by studying the physical stability of each polymorph. The stability samples were stationed at various temperatures and humidity conditions and the API phase was monitored by a number of different techniques.

14.3.2.3 Additional Measurements

(a) X-ray powder diffraction (XRPD). XRPD was conducted using a Philips model 1710 X-ray diffractometer. The diffraction profile was measured from 4° to 40° using a 2D step size of 0.05° and dwell time of 1 s.
(b) Differential Scanning Calorimetry (DSC). DSC was performed using a TA Instruments Model 2920 DSC and data analyzed using TA Universal Analysis 2000 software. Testing was performed at a ramp rate of 10 °C/min from 25 to 250 °C under nitrogen purge at a flow rate of 40 mL/min.
(c) Solubility Measurements. Samples were continuously stirred on a stir plate at 400 rpm at 25 °C during the course of the study. At 24 h, an aliquot was added to a centrifuge filter device (filter pore 0.45 μm) and centrifuged for 2–5 min at ~13,000 rpm at 25 °C. The filtrates were analyzed by HPLC. The solid residue of each sample was collected and analyzed by X-ray powder diffraction.
(d) p*Ka*. Potentiometric titration was performed using a GLpKa/D-PAS by Sirius Analytical.
(e) Dissolution Measurements. Dissolution measurements were performed using a DISTEK 2100C Apparatus equipped with Fiber-optic probes and a UV photo diode array detection system (OPT-DISS Fiber-optics system). The bath temperature was set at 37 °C with a paddle speed of 50 rpm.
(f) Stability studies. Stabilities of API and formulations were evaluated by storing stability samples at various stations with controlled temperature and humidity. The standard stations included: 5 °C/ambient humidity (as reference), 25 ºC/60 % RH, 30 ºC/65 % RH, 40 ºC/ambient humidity and 40 ºC/75 % RH. Stability samples were monitored at regular time intervals for chemical and physical changes.
(g) Permeability. The permeability of telcagepant was determined by the Caco-2 cell monolayer approach. A donor solution was prepared by adding a small amount of C^{14} telcagepant to 20 mL of Hank's Balanced Salt Solution (HBSS) at pH 7. Permeability was measured across Caco-2 cell monolayers grown to confluence on Transwell® filter membranes via a bidirectional transport experiment. Samples were taken from the receiver compartment at 0, 15, 30, 45, and 60 min and from the donor compartment at 0 and 60 min. The receiver solutions were replaced with fresh HBSS after each sampling timepoint. The samples were analyzed using a liquid scintillation counter immediately following the experiment. A mannitol flux experiment using ^{3}H-mannitol was conducted concurrently with telcagepant to monitor the integrity of the monolayers used. Mannitol transport was found to be less than 1 % after 60 min, indicative of good monolayer integrity.
(h) Preparation of liquid-filled capsule formulations. The liquid vehicle was prepared via w/w percentage and the drug was added to the vehicle at the target concentration. The mixture was then stirred until a homogenous solution was formed. The formulation was encapsulated in size 00 hard gelatin capsules for stability or animal dosing studies.

(i) Preparation of dry-filled capsule formulations. Dry-filled capsules were made by encapsulating either dry blends or granulations of specified compositions.
(j) Preclinical PK studies. Capsules were dosed to fasted Beagle dogs or Yucatan Minipigs to assess the comparative bioperformance of different formulations. Animals were housed in an AAALAC-accredited facility in accordance with appropriate guidelines. All appropriate internal and external animal welfare regulations were followed in the conduct of the studies. All formulation dosing studies were conducted under a protocol approved by the Merck IACUC. The plasma concentration of telcagepant over time following oral administration was sampled and analyzed. Area under the curve (AUC), observed maximum plasma concentration (C_{max}), and time of C_{max} (T_{max}) were calculated in WinNonLin v5.01.
(k) Clinical Relative Bioavailability Studies. PN005 (a formulation comparison study) was an open-label, randomized, five-period crossover study to evaluate the comparative bioavailability of five formulations of telcagepant administered orally as a single 200 mg dose in 20 healthy male human subjects. In each period, pharmacokinetic samples were collected at specific time points for 48 h after administration of telcagepant. Each treatment was separated by a minimum 72 h washout period between administrations. PN017 and PN038 were open-label, randomized, three-period crossover studies to understand the impact of storage on the relative bioavailability of selected LBF formulations. All studies were conducted following the necessary approvals by the appropriate Institutional Review Boards and Regulatory Agencies.
(l) Assay results were obtained through High Performance Liquid Chromatography (HPLC). Quantitation was achieved against an authentic reference standard of telcagepant drug substance. Methods used were validated to have acceptable sample preparation and chromatographic performance. Filtration of the individual capsules in the precipitate screening experiments was performed using centrifuge filters.
(m) Titration results were obtained using an automated pH titrator. 0.01 N HCl was used as the titrant.

14.4 Results and Discussion

14.4.1 Solid State Form and Physicochemical Properties of Telcagepant

For the Phase I study, a crystalline neutral form was identified as the single most stable form following both polymorph and salt screens. The physicochemical properties of the telcagepant crystalline neutral form are listed in Table 14.3.

The crystalline neutral form is non-hygroscopic and shows a single melting endotherm at 223.5 °C. Although the neutral form is physically and chemically stable, it has poor solubility in water and across the physiological pH range. The permeability is moderate to high (Papp = 11.7×10^{-6} cm/s), making the compound preliminarily BCS Class II at the projected clinical doses. The primary objectives of the Phase I study were to assess the pharmacokinetics of telcagepant and demonstrate acceptable safety and tolerability in single and multiple dose administrations. Investigated for the treatment of migraine, the drug needed to rapidly reach (<1 h postdose) the pharmacokinetic target of 1 μM associated with 90 % CGRP receptor blockade, and maintain this target for at least 2 h. The projected top Phase I dose was as high as 1,800 mg. To support these clinical goals, a formulation with a fast onset PK and high unit dose strength was required. Based on the pharmaceutical properties of the compound, streamlined approaches for developing both solid and liquid dosage forms were conducted in parallel, balancing the short term need of the early clinical study and the long term goals of commercial product development.

14.4.2 Formulation Development of the Crystalline Neutral Form

Research toward a solid formulation was focused on optimization of bioavailability and fast dissolution/absorption by seeking to improve the rate and extent of drug solubilization. The approaches included particle size reduction of the drug substance, incorporation of various wetting and solubilizing agents, and conversion of the crystalline drug into an amorphous form. The underlining principles are summarized in part by the Noyes–Whitney Equation (Noyes and Whitney 1897):

$$\frac{\mathrm{d}m}{\mathrm{d}t} = A\frac{D}{d}(C_\mathrm{s} - C_\mathrm{b})$$

where m, mass of dissolved material; t, time; A, surface area of the interface between the dissolving substance and the solvent; D, diffusion coefficient; d, thickness of the boundary layer of the solvent at the surface of the dissolving substance; C_s, mass concentration of the substance on the surface; C_b, mass concentration of the substance in the bulk of the solvent.

The aforementioned approaches were aimed at increasing surface area and solubility while reducing the boundary layer. Furthermore, the addition of surfactants was to aid wetting and to maintain supersaturation, thereby facilitating absorption. Simple dry-blend formulations containing milled drug substance and surfactants were evaluated. Both ionic and nonionic surfactants were considered, including sodium lauryl sulfate (SLS), Pluronics, and Cremophor® RH40. The lead solid formulation of telcagepant with 2 % SLS/10 % Cremophor® in a dry-blend filled capsule (DFC) was tested in dogs versus a self-emulsifying system, namely an Imwitor®/Tween® 80 (1:1) LBF. Plasma exposure following oral dosing with the solid formulation was significantly lower than that with the Imwitor®/Tween® 80 formulation, as shown in Table 14.4. In addition, the conversion of the crystalline neutral form into an amorphous dispersion via co-precipitation of the drug with

Table 14.4 Mean pharmacokinetic parameters after oral administration of telcagepant to male beagle dogs (mean ± SD)

	Imwitor/Tween LBF (1:1)	Crystalline API/surfactant DFC	Amorphous API/surfactant DFC	In-situ salt in PEG 400 LBF
Composition	Neutral form (5 %)	Neutral form (5 %)	Neutral form (5 %)	In situ K salt (13.5 %)
	Imwitor® (47.5 %)	SLS (2 %)	PEG 600 (85 %)	PEG 400 (86.5 %)
	Tween® (47.5 %)	Cremophor® RH40 (10 %)	Poloxamer 407 (10 %)	
		Lactose (83 %)		
AUC_{0-24} (μM h)	3.66 ± 2.52	0.41 ± 0.52	1.64 ± 0.86	3.80 ± 1.48
C_{max} (μM)	1.12 ± 0.67	0.09 ± 0.10	0.59 ± 0.28	1.28 ± 0.54
T_{max} (h)	2.00 ± 0.00	2.20 ± 1.80	1.50 ± 0.87	1.17 ± 0.76

a polymer or a polymeric surfactant system was also explored. Although significant enhancement in bioavailability was obtained for a formulation containing amorphous drug/surfactant in comparison to the simple dry-blend formulation, the amorphous drug formulation still had only ~40 % bioavailability of the liquid solution formulation (Table 14.4).

To develop liquid formulations, as discussed in the introduction, the first key task is to find the appropriate vehicle. A variety of liquid vehicle systems were screened, including single, binary, and ternary systems that contained oils, monoglycerides, diglycerides, triglycerides, polymers, and various surfactants. The key challenge found was inadequate solubility in the liquid vehicles, making it difficult to achieve the high unit dose strength required by the clinical studies. As shown in Fig. 14.3, none of the liquid systems were able to provide the solubility for the crystalline neutral form to support a 100 mg-potency unit dose, which was required for Phase I studies. To increase the drug solubility, a novel in situ salt was prepared by adding one molar equivalent of base (KOH or NaOH) to the liquid formulation. This approach resulted in a dramatic increase in the solubility of the drug in vehicle, especially in PEG systems. The ionization of the drug in the presence of a strong base makes it much more soluble in the polar polymeric vehicles, yielding a solubility >300 mg/mL (Fig. 14.3).

More significantly, the in situ salt/PEG formulation achieved comparable bioavailability to that of the LBF of a self-emulsifying system (Table 14.4) and was found to be chemically and physically stable. Even though the drug phase-separated upon dispersion in aqueous media to form a milky suspension, the precipitate was of small drug particles in the amorphous state which presumably facilitated fast dissolution and subsequently absorption in vivo. However, it is well-known that PEG 400 is not compatible with hard gelatin capsule (HGC) shells due to its hygroscopic nature (Bhutkar et al. 2008; Stein and Bindra 2007). The vehicle can extract moisture from the capsule shell and make it brittle, leading to leakage and/or

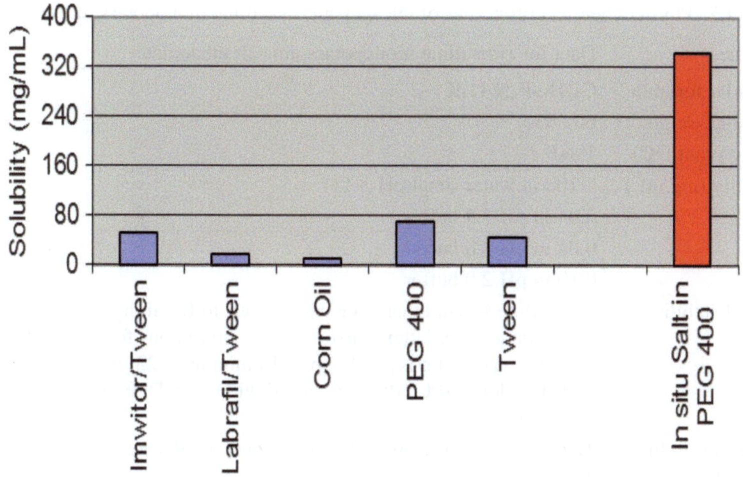

Fig. 14.3 Solubility of Telcagepant in various liquid vehicles

breakage problems. In practice, this issue is commonly resolved with a switch to a soft gelatin capsule. Given the excellent bioperformance observed, the in situ salt/PEG formulation in hard gelatin capsules was selected as a simple initial formulation for the Phase I studies based on the balance of risk, clinical objectives, drug availability, and timeline. The risk of physical instability was mitigated by packing the clinical supplies in small bottles with rayon coils and storing the product at 5 °C prior to clinical evaluation. The planned strategy was to switch to soft gelatin capsules (SGCs) for later phase development and further research additional formulation options. The in situ salt/PEG 400 liquid capsule formulation significantly enhanced the bioavailability of telcagepant, achieved a fast onset PK and met the unit strength requirement. It overcame the key challenges of a BCS II compound and allowed the program to move into Phase I successfully with limited API supply and a rapid development timeline.

14.4.3 Research with Crystalline Potassium Salt

Although the in situ salt/PEG 400 formulation successfully supported the Phase I studies, the limitations of this formulation made it clear that further research was needed towards identifying a room temperature stable formulation to support later phase clinical studies and commercialization. The early Phase I clinical data provided additional guidance for the formulation with respect to the required exposures. Although the observed exposure was adequate for Phase I studies, it became evident that further improvement in the bioavailability and increase in the unit dose strength would be important to increase the market competitiveness of the product. A new effort with extensive manual salt screens yielded a crystalline

Table 14.5 Physicochemical properties of telcagepant crystalline monopotassium salt ethanolate

Properties	Data for crystalline monopotassium salt ethanolate
Molecular formula	$C_{28}H_{32}F_5N_6O_4K$
Molecular weight	604.630
Melting point (°C)	196.0
Solubility (mg/mL)	>100 in water (final pH >11)
	0.04 in pH 7.5 buffer
	0.04 in pH 4.0 buffer
	0.10 in pH 2.0 buffer
Physical stability	Crystalline K-salt ethanolate can convert to K-salt hydrate at ambient condition and the latter converts to the amorphous form at high relative humidity. The drug is stable up to 18 months at 25 °C/60 % RH when stored in doubled LDPE bags stored within a HDPE container with desiccant.
Chemical stability	The drug is chemically stable in the pure solid state
Projected efficacious dose	300 mg

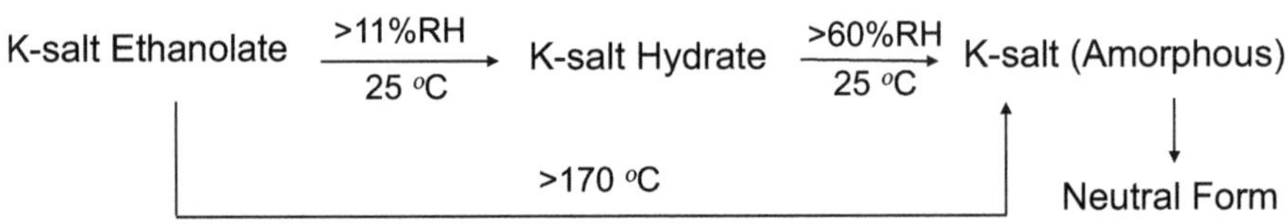

Fig. 14.4 Phase conversion diagram for K-salt ethanolate

potassium salt (K-salt), isolated as an ethanolate solvate. The properties of the K-salt ethanolate and its phase stability diagram are shown in Table 14.5 and Fig. 14.4.

The K-salt ethanolate provided a new opportunity for formulation research. However, efforts to utilize the new drug form were accompanied with challenges. As shown in Fig. 14.4, the K-salt ethanolate is not physically stable and can convert to K-salt hydrate under ambient conditions and the latter can further turn over to the amorphous form at high humidity. Moreover, it was found that the amorphous K-salt can disproportionate, resulting in the neutral form that is less bioavailable. The solvent molecule (ethanol) is tightly bound into the crystal. Solvent evaporation commenced only at around 170 °C and K-salt ethanolate converted to amorphous form upon removal of ethanol. The solubility of the potassium salt in water is much improved relative to the neutral form; however, it also results in a high native solution pH of 11. In buffers across the physiological pH range, the solubility is very low with little improvement over the crystalline neutral form. Thus, the newly discovered salt form faced similar challenges to those of the neutral form for formulation research.

As noted previously, the efficacy of a molecule for migraine treatment depends heavily on the pharmacokinetic profile achieved. Thus, a formulation that provides rapid absorption is a key success factor. In addition, to avoid bridging between formulations after the onset of pivotal clinical trials, it was desired to have a final market image formulation with unit dose strength of 300 mg for Phase IIB studies. With the new salt form in hand, all of the available formulation options were reexamined with the primary strategy to develop a solid dosage form to help streamline and integrate the formulation development for both the monotherapy and potential fixed dose combinations. The formulations evaluated included the surfactant-modified solid formulation of the K-salt ethanolate, nanoformulation of the neutral form, amorphous solid dispersion, as well as liquid capsule formulations. The main challenges for the surfactant-modified solid formulation of the K-salt ethanolate were physical instability and a lack of significant improvement in bioavailability. In aqueous media at physiological pH, the K-salt has the potential to self-associate, resulting in gelation. Furthermore, the initial dissolved drug disproportionated. The resultant neutral form rapidly created an impervious shell over the remaining undissolved drug, slowing drug release and yielding low exposure. Various buffering agents were also explored without success, such as sodium carbonate and sodium bicarbonate. The dyno-milled neutral form nanoformulation was chemically stable but its bioperformance was low due to the poor re-dispersion of nanoparticles from the solid dosage form to the solution phase. Both of these solid formulations were considered for a clinical biocomparision study due to the lack of correlation between the animal and the human model.

For the liquid capsule formulation research, a new approach was explored by adding surfactants to the K-salt/PEG 400 formulation to prevent drug precipitation during formulation dissolution in aqueous media. Although the K-salt has a high solubility in PEG 400, phase change occurred during dissolution due to rapid disproportionation, resulting in the insoluble neutral form. Surfactants, such as Cremophor® and Tween®, are well known for their multiple functions as wetting agents, dispersant, as well as solubilizers (Sarpal et al. 2010). In our investigation, it was found that upon contact with the dissolution medium, the new liquid formulation spontaneously formed a translucent solution. Cremophor® EL and Tween® were found to effectively maintain drug supersaturation during dissolution by delaying nucleation and retarding crystal growth, possibly through micelle encapsulation. The significant improvement in in vitro performance was confirmed by an in vivo study which demonstrated a twofold increase in exposure in dogs.

Four formulations were selected for the relative bioavailability comparison study based on the criteria of animal PK, stability, and processability. The key attributes of these formulation candidates are summarized in Table 14.6.

These formulations were tested in an open-label, randomized, five-period crossover study to evaluate their comparative bioavailability. As showed in Fig. 14.5, the data indicated that the PEG/PG/Cremophor® EL/Tween® formulation had the most favorable PK profile. It achieved a fast onset of absorption and much improved exposure relative to the other formulations. The geometric mean ratio (GMR) of

Table 14.6 Key attributes of formulation candidates for the relative bioavailability study PN005

	B	C	D	E
	LBF K-salt/ PEG/PG	LBF K-salt/ PEG/surfactants	DFC Dyno-milled neutral	DFC K-salt ethanolate
Major components	K salt ethanolate (30 %)	K salt ethanolate (30 %)	Dyno-milled neutral form (60 %)	K salt ethanolate (57.5 %)
		PEG 400 (24.5 %)	Lactose (29.5 %)	Lactose (19.2 %), poloxamer 407 (3.3 %), sodium lauryl sulfate (1.3 %)
	PEG 400 (62.5 %)	Propylene Glycol (7.5 %)	A-Tab® (6 %)	Eudragit® E (13.3 %)
		Tween® 80 (19 %)	Poloxamer 407 (4 %)	
	Propylene glycol (7.5 %)	Cremophor® EL (19 %)	MgSt (.5 %)	MgSt (0.8 %), others (4.6 %)
Projected potency	300 mg/dose	300 mg/dose	300 mg/dose	Up to 400 mg/dose
Process	Soft gel	Soft gel	Conventional WG	Conventional RC
Bioavailability (ratio to Ph I LBF in dogs)	1.0	2.6 (a K-salt/ PEG/Tween® system was tested)	0.28	1.2
Stability	Epimerization	No issue	No issue	Dissolution slow down upon storage at 40 °C/75 % RH
Additional Issues/Risks	Not amenable to FDC	Not amenable to FDC	Need to scale up dyno-milled neutral form under narrow particle size range	Physical stability of the drug substance and drug product

this formulation's C_{max} to that of the Phase I formulation was 2.18 while the GMR for $AUC_{0-\infty}$ was 1.54. The enhancement of bioavailability was attributed to the addition of surfactants which minimized drug precipitation owing to disproportionation. This formulation significantly improved bioavailability, met the desired unit dose strength of 300 mg, and displayed acceptable stability based on the accelerated stability data. To mitigate the incompatibility between PEG 400 and the hard gelatin capsule shell, the PEG/PG/Cremophor® EL/Tween® formulation was manufactured as soft gelatin capsules (SGCs). Supported by this formulation, the Phase IIB study was conducted to understand efficacy and required dose range (Table 14.7).

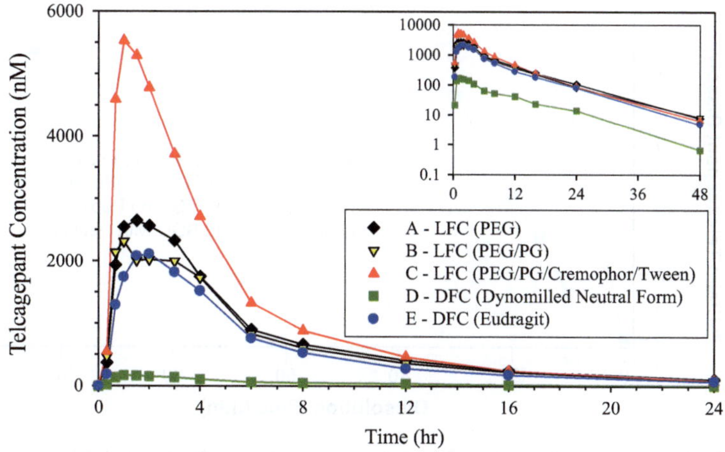

Fig. 14.5 Mean plasma concentration of Telcagepant following administration of a 400 mg oral dose of the test formulations in healthy human volunteers (Study PN005)

Table 14.7 Mean pharmacokinetic values following single doses of 400 mg of various telcagepant formulations (Study PN005)

Formulation	AUC_∞ (μM h)	C_{max} (μM)	T_{max}	$t_{1/2}$ (h)
A: LBF (PEG)—Phase I	17.64	3.15	1.5	6.2
B: LBF (PEG/PG)	15.92	2.97	1.75	6.4
C: LBF (PEG/PG/Crem/Tween)	26.35	6.68	1.5	5.7
D: DFC (Dyno-milled Neutral)	1.45	0.21	2.0	7.8
E: DFC (Eudragit)	14.38	2.65	1.75	6.4

14.4.4 Capsule Cross-Linking

Soft gelatin capsule (SGC) supplies, made with acid-bone gelatin showed dissolution slow-down after 20 weeks storage at 25 °C/60 % RH, which was largely attributed to gelatin cross-linking, a common phenomenon for gelatin capsules. As noted in the introduction, capsule shell cross-linking has been extensively studied. As a reminder, cross-linking is facilitated by temperature, humidity, light, and exposure to aldehydes (Meyer et al. 2001). Based on numerous studies, it was widely accepted that that these effects primarily impact the in vitro testing results rather than the in vivo performance (Basit et al. 2001) as digestive enzymes in the body are thought to break down the cross-linked gelatin. A two-tier in vitro dissolution test was thus developed using enzymes (pepsin or pancreatin), according to USP <711>. The test was conducted first in a dissolution medium without enzymes, followed by retesting in medium containing appropriate levels of biorelevant enzymes. Literature studies suggest that the two-tier in vitro dissolution test may constitute a better indication for the in vivo behavior of gelatin-encapsulated formulations than single tier testing (Basit et al. 2002). As

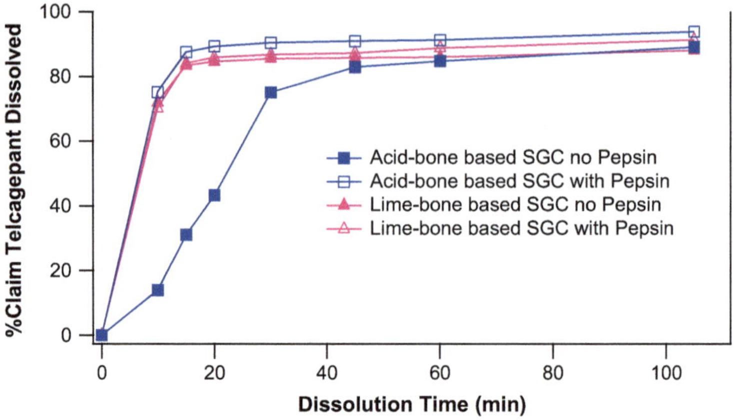

Fig. 14.6 Dissolution profiles of acid-bone based capsules (at 25 °C/60 % RH/20 weeks) and lime-bone based capsules (at ambient condition/9 months) with Pepsin and without Pepsin

demonstrated in several studies, typically cross-linked capsule formulations suffer little in vivo exposure loss even though drug release was significantly delayed in the standard in vitro dissolution test. For example, the Phase II SGC made with acid-bone gelatin had a dissolution slow-down where only 75 % of the drug was released in 30 min. With the addition of pepsin, the cross-linked capsule released >90 % in 30 min. This indicated that the cross-linked capsules might be comparable to the unstressed capsules in vivo. Dog studies were also conducted with fresh and "aged" supplies and also did not indicate a reduction in bioavailability.

To have a better understanding of the physiological relevance of the cross-linking observed with the acid-bone gelatin capsules, a clinical relative bioavailability study of cross-linked and freshly made capsules was conducted at the 300 mg dose. Contrary to the results from the two-tier dissolution test, the clinical data showed that the capsules on storage resulted in a moderate decrease in exposure (~25 %) relative to fresh capsules. In this case, the dissolution testing with the addition of pepsin was not able to discriminate between stressed and the fresh capsules. While the cross-linked capsules in vitro demonstrated 86 % release in 60 min, a larger bioavailability difference was observed in the clinic. However, it should be acknowledged that the reduction in exposure in the clinic was modest and it is inherently difficult to quantitatively link dissolution and clinical data. In addition, the incomplete release observed for the cross-linked capsules at 60 min could be indicative of additional formulation changes other than cross-linking that affect the dispersability of the formulation. For example, it was later found that surfactant hydrolysis occurred and resulted in lower apparent solubilization after formulation dispersion in an aqueous environment.

In order to reduce cross-linking, different types of gelatin were evaluated. The acid-bone gelatin was derived from acid-catalyzed hydrolysis, whereas the lime-bone gelatin was from base-catalyzed hydrolysis. It was found that the lime-bone gelatin capsules could significantly reduce cross-linking, although not completely eliminate this phenomenon. As shown in Fig. 14.6, the drug release profile of the

Table 14.8 Mean pharmacokinetic parameters after oral administration of telcagepant to male miniature pigs (mean ± SD)

	$AUC_{0\text{–}24h}$ (μM h)	C_{max} (μM)	$AUC_{0\text{–}24\ h}$ ratio
Lime-bone SGCs, initial ($n = 7$)	8.7 ± 1.2	1.5 ± 0.21	1.0
Lime-bone SGCs, stored for 18 months ($n = 7$)	7.9 ± 2.2	1.4 ± 0.50	0.9

Table 14.9 Potency-adjusted bioequivalence assessment for telcagepant soft gelatin capsules ($N = 47$)

	$AUC_{0\text{–}\infty}$ GMR	C_{max} GMR
12 months 5 °C supplies vs. 5 months 5 °C supplies	1.13 (1.06–1.22)	0.90 (0.84–0.97)
12 months RT supplies vs. 5 months 5 °C supplies	1.12 (1.02–1.24)	0.91 (0.82–1.00)

lime-bone based SGCs stored after 9 months showed little difference with or without pepsin.

Given the gaps in in vitro–in vivo correlation and the lack of discrimination for the effect of cross-linking in the dog model, efforts were directed to explore an alternative animal model to predict clinical performance. It was found that the miniature pig model showed a good correlation with the human data; aged SGCs similar to those tested in the clinic resulted in approximately 20 % reduction in bioavailability compared to fresh HGCs. The lime-bone based capsule was thus also tested in the miniature pig model to assess the effect of storage on bioavailability. The superior performance of the lime-bone capsule in reducing cross-linking was further confirmed by the PK study in miniature pigs. As shown in Table 14.8, comparable exposures were observed for the newly made SGC and the SGC stored for 18 months at 25 °C/60 % RH.

The acceptable performance on storage of the developed SGC was finally confirmed in the clinic where capsules stored for 12 months at room temperature were shown to be bioequivalent to capsules stored for 12 months at 5 °C and to full scale supplies stored at 5 °C for 5 months (Table 14.9).

14.4.5 In Vitro to Preclinical to Clinical Studies Correlations

It is a common practice during formulation development to use preclinical tools such as solubility/dissolution assays and animal models to evaluate the performance of new formulations. This is typically done in an iterative fashion, i.e., formulations are initially qualified by dissolution and subsequently tested in vivo (Wu and Kesisoglou 2010). Ultimately, the final selected formulations proceed to a clinical study. These clinical data serve as the "validation" of the preclinical tests and ideally allow for establishment of either in vitro–in vivo correlations (even a rank order relationship would be considered useful) or help with a more quantitative translation of preclinical animal data to humans. The clinical data further drive refinement of the preclinical tests, for example, a change in dissolution conditions. However, the nature of the formulation and specific drug properties may affect the suitability of some of these tests. In this section, the above concepts are discussed in relation to the early telcagepant formulation development data as presented above.

Early formulation development efforts for telcagepant focused on maximizing bioavailability by enhancing the effective drug solubility. As seen in Table 14.6, formulations that provide pre-solubilized drug in liquid resulted in significantly improved exposure over solid formulations. Within the solid formulations, a rank order relationship between amorphous and crystalline API is observed, as expected. These early dog model data clearly indicate the need for maintaining supersaturation in the intestine to maximize telcagepant absorption. However, clearly establishing an IVIVC for liquid capsule formulations is not an easy task. The process of in vivo solubilization for LBFs is more complex and varies based on the specific excipients employed. For telcagepant, a higher solubility in the dosing vehicle (Fig. 14.3—in situ salt in PEG400) did not necessarily result in a significant bioavailability increase compared to the Imwitor®/Tween® system. The in vivo absorption process for these systems is different, with the former being a dispersable co-solvent system while the Imwitor®/Tween® system will likely undergo digestion. At the early stages of formulation development for telcagepant and in the absence of clinical data, formulation selection relied primarily on animal data as opposed to in vitro testing.

The next iteration of formulation efforts focused on improving the LBF formulation. Given the reasonable performance in Phase I of the initial LBF system, the dog model was considered a suitable screening tool. Furthermore, given the previously observed impact of solubility on exposure, in vitro testing was also performed. The focus of the formulation research was thus on producing co-solvent/surfactant systems that were expected to undergo very limited digestion in vivo. The desirable characteristics were the formation of stable micellar dispersions upon contact with the GI fluid, which was obtained with the PEG/Cremophor®/Tween® system. Preclinical dog studies (Table 14.4) indeed confirmed the potentially favorable effect of surfactant inclusion where a K-salt/PEG/Tween® system resulted in 2.6-fold higher exposure compared to the Phase I formulation. Additional studies with the solid formulations again confirmed the importance of supersaturation with exposures significantly lower compared to the co-solvent/surfactant system. However, at least one formulation with surfactants and polymer (poloxamer/SLS/Eudragit®) appeared to potentially reach the exposure seen with the Phase I formulation.

The results of the clinical relative bioavailability study served as a validation of the preclinical dog model. The rank order seen in clinical human studies largely reflected that seen in the dogs: PEG/surfactant > PEG (Phase I) ~ Eudragit®/surfactant solid formulation > dyno-milled neutral API. However, the relative exposures observed between formulations were somewhat different with a more pronounced difference for the neutral API formulation wherein absorption in human was <10 %

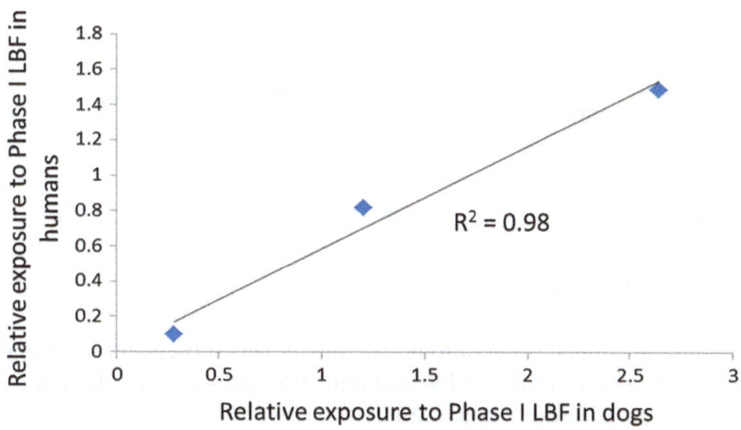

Fig. 14.7 Relative performance of Telcagepant formulations in dogs and in humans

of the exposure of the Phase I formulation. The dog studies were run at 40 mg to allow for a comparison with the 300 mg dose in the clinic (on mg/kg basis) and also accounted for similar dosed water volumes of ~3.5 mL/kg. Thus, one could postulate that the expected higher bile salt concentration in dogs may have facilitated a somewhat better absorption in the dog model compared to human. For the PEG/surfactant formulation, the increase in exposure was not as pronounced as seen in the dog model at least for AUC; only a 50 % increase in AUC was found compared to 2.5-fold in the dog model; the C_{max} increase (~2-fold), on the other hand, was more in line with the dog data. Figure 14.7 provides a graphical depiction of the performance of the formulations in both dog and human.

With the selection of the formulation composition for subsequent studies, the focus of biopharmaceutical evaluation shifted toward understanding the effect of cross-linking on bioavailability. The two-tier in vitro dissolution test using enzymes (pepsin or pancreatin), commonly employed to assess cross-linking, suggested that full release could be obtained within 30 min. A dog study conducted with freshly prepared capsules against aged capsules either 14-month or 19-month in storage did not indicate any significant differences in bioavailability (AUC 4.66 ± 0.848 μM h vs. 5.93 ± 1.37 μM h vs. 4.85 ± 0.848 μM h respectively, mean $\pm$ SE, $n = 5$). However, in human studies (Study PN017), a moderate bioavailability reduction was observed for aged capsules; the 13-month acid-bone based SGC underperformed compared to hard gelatin capsules with an $AUC_{0-\infty}$ GMR 0.74 (0.65–0.83) and C_{max} GMR 0.78 (0.65–0.92). The high dose in dogs may have contributed to the discrepancy; since clinical supplies were used, the intact 300 mg capsule in dogs is a much higher dose on a mg/kg basis compared to the human clinical studies. In order to circumvent this limitation, a miniature pig study was conducted where seven Yucatan miniature pigs of ~50 kg were dosed with 300 mg capsules. On average, an approximately 20 % reduction in exposure was seen. It is difficult to draw too much from studies with the small number of animals employed in a conclusive manner; retrospectively, one could state that the miniature pig data

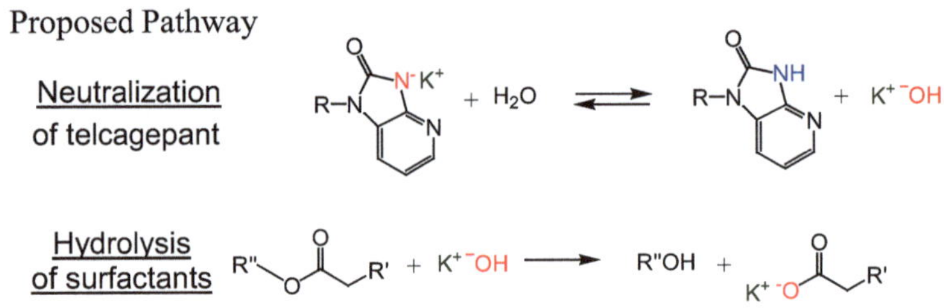

Fig. 14.8 Proposed pathway for drug crystallization

trended with the clinical observations. The model was subsequently utilized to test capsules as described in Table 14.4; and miniature pig data were later found to be similar to human in further clinical trials.

In summary, animal studies played a major role in preclinical formulation development. The dog model was shown to be reasonably predictive in assessing the relative performance of different formulations when dogs were dosed at 40 mg potency (~4 mg/kg). However, the dog model was not suitable in assessing subtle differences between the same LBF to assess the impact of cross-linking. An alternate animal model was utilized to allow for testing of intact clinical supplies. The experience with telcagepant highlights the difficulties faced by formulation and biopharmaceutical scientists in early formulation development where understanding of bioperformance has to rely on limited data sets. Whenever clinical data are generated, they should serve as validation sets for previous assumptions and for refinement or change of preclinical models. For telcagepant, although perfect "correlations" were not established, the preclinical data were very helpful to identify a suitable formulation for the clinic.

14.4.6 Chemical and Physical Instability Challenges with the Liquid Capsule Formulation

To achieve room temperature storage for the soft gelatin formulation as a commercial product, stability data was generated by stressing the capsule at 30 °C/65 % RH (an extreme condition for an SGC). A more serious issue emerged as drug crystals were observed in the SGC capsule supplies stored at 30 °C/65 % RH for 1 year. Analysis of the crystals by XRPD and Raman spectroscopy identified the crystals as a new crystalline neutral form of the drug. The proposed mechanism of the neutralization is given in Fig. 14.8, where water promotes the protonation of the drug molecule. This reversible reaction is driven to neutralization when the hydroxide created in that reaction is consumed by a subsequent hydrolysis of ester groups which are present in both of the surfactants used in the formulation, Cremophor® and Tween®. This proposed mechanism was supported by the presence of oleic acid

Table 14.10 Percentage of potassium salt remaining as a function of age, temperature, and humidity stress conditions

			% relative humidity						
Temperature (ºC)	Time (months)	LBF Lot	20	30	38	45	52.5	60	65
5	6	1			61				
25	6	1		51	47	45	40	37	
	12	2			53			34	
30	6	1	42			38			
	12	2			43				24

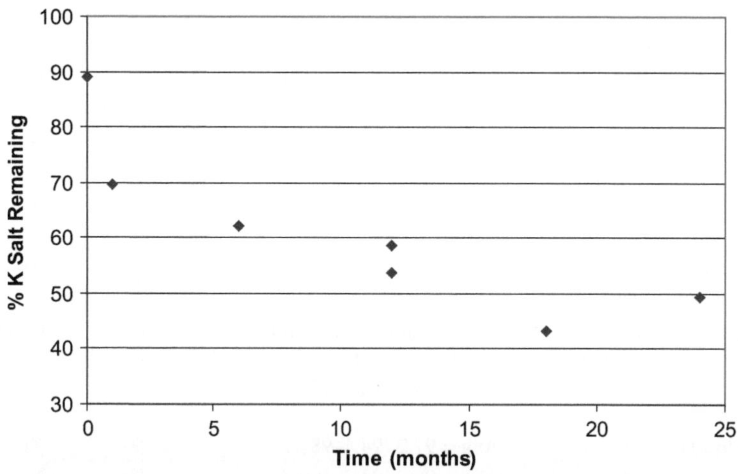

Fig. 14.9 % K salt remaining as a function of time and storage conditions

(one of the products of the hydrolysis reaction) at elevated levels in stressed samples.

To assess the risk to the formulation, several studies were conducted. First, the extent of neutralization was studied through pH titration experiments. The results in Table 14.10 give the amount of potassium salt still remaining in the liquid vehicle for samples of various ages, temperature, and humidity stress conditions. The results indicate that both temperature and humidity increase the extent of conversion. This finding is consistent with the mechanism proposed in Fig. 14.8. These experiments also indicate that there was a significant amount of conversion occurring in the soft gelatin formulation.

The titration experiments were also used to explore the rate of conversion. In Fig. 14.9, titration results of samples stored at room temperature for different times are shown. Samples tested are from different lots manufactured over time. These results showed that the initial conversion occurred very quickly. Within 1 month, 30 % of the potassium salt had converted to the neutral form. Given this rapid conversion to the neutral form, it was then important to determine if the formulation

Table 14.11 Evaluation of saturation solubility of telcagepant K-salt ethanolate in degraded lipid-based formulation vehicle

	mg drug/g LBF vehicle	
	Before spike	After spike
LBF vehicle from capsules stored 1 year at room temperature/ambient humidity	239	227
LBF vehicle from capsules stored 1 year at 30 °C/ambient humidity	237	196

Table 14.12 Evaluation of telcagepant concentration in samples stored under a variety of storage and time conditions

Warehouse samples				
Storage conditions	Total age (months)	Telcagepant concentration (% target)	# capsules	Crystals observed
RT—2 mo 5 °C—3 mo	5	Avg = 97.1 (94.8–98.4)	300	No
RT—3 mo 5 °C—9 mo	12	Avg = 97.7 (95.2–99.2)	301	No
RT—12 mo	12	Avg = 97.1 (96.0–100.8)	301	No
Clinical return samples				
Clinical site location	Total age (months)	Telcagepant concentration post-filtration (% target)	# capsules	Crystals observed
USA, Colorado	12	Avg = 98.1 (96.4–99.2)	50	No
USA, Florida	12	Avg = 97.2 (94.4–98.4)	50	No
USA, Pennsylvania	12	Avg = 98.2 (97.2–100.4)	50	No
Columbia	12	Avg = 98.2 (96.0–99.6)	74	No
Denmark	12	Avg = 97.5 (94.8–98.6)	75	No

was still adequate to solubilize the drug under normal storage conditions. One challenge was that the absolute solubility was difficult to define, since the solubility decreased as the neutralization progressed. In the deprotonated form, the drug itself formed micelles that promoted self-solubilization. As the drug was neutralized, the micelle formation decreased and the overall solubility was reduced. Additionally, as part of the neutralization reaction (Fig. 14.8), surfactants added as solubility aids were chemically degraded and led to a further loss of solubility. Finally, the addition of water, which changed with storage conditions, was also shown to impact the solubility of the vehicle. To evaluate the solubility of the compound in this dynamic system, excess neutral drug crystals were spiked into the liquid fill that was extracted from aged capsules and the combination was allowed to stir for 24 h to obtain equilibration. Excess crystals were removed by centrifugation and the total telcagepant concentration in the supernatant was determined by HPLC. The results given in Table 14.11 indicate that the samples stored at room temperature and ambient humidity had reached saturation after storage for 1 year. The sample stored at 30 °C and ambient humidity was already supersaturated (no visible crystals were present in the fill solution extracted from the soft gelatin capsules).

At the time of these observations, the clinical program with the liquid filled capsules was discontinued. However, there was a significant body of knowledge that had been obtained with the liquid capsule formulation. Since the solubility experiments indicated that it was possible that supplies used in the clinical program could have been supersaturated, a study was designed to evaluate the likelihood that samples that had been used had insoluble material. To evaluate this probability a statistical sampling plan was developed to detect a ≤1 % defect rate (at the 95 % Confidence Interval). This required 300 capsules to be evaluated from four different populations. The first three populations were from the newest and the oldest capsules used in the clinical trials that had been stored in the Merck Clinical Supplies Warehouse. The fourth population was from samples that were returned from the clinical trial sites. The distribution of the samples from clinical trials sites was taken to simulate a distribution of hot/cold climates and domestic and international sites.

Each of the 1,201 capsules was individually opened and the contents were passed through a 0.45 μm filter and the filtrate was analyzed by HPLC. The capsule fill and the empty capsule shells were visually inspected for the presence of visual particles. As the data in Table 14.12 indicate, there was no evidence of precipitated material found in any of the samples. From this it was concluded that there was not more than a 1 % chance that any patient that had participated in the clinical trial had received as capsule containing precipitated material. These data were reviewed by regulatory agencies and agreement was obtained to use the clinical data from those early trials.

While it was felt that the surfactant degradation and drug crystallization were unlikely to alter the safety profile of the formulation, observation of crystalline neutral form raised bioavailability concerns as crystals of the telcagepant neutral form are minimally bioavailable. Two major hurdles were presented to the continuation of the ongoing clinical program: (1) to demonstrate bioequivalence between fresh and aged SGC at expiry to support the integrity of data collected to date in the study; and (2) to bridge the Phase III SGC formulation to an alternative formulation that is amenable for commercialization. The first hurdle was overcame by a successful BE study. The second challenge will be the subject of a separate research report discussing the development of the solid dosage form.

14.5 Conclusions

Lipid based formulations (LBFs) represent an excellent formulation option to deliver enhanced bioavailability for poorly soluble compounds, but pose a number of unique technical challenges that must be addressed during development and commercialization. A LBF consisting of a hard gelatin capsule (HGC) formulation of the crystalline neutral form of telcagepant in PEG 400 was developed for Phase I clinical studies. The stability risk associated with the incompatibility of PEG 400 and HGC was mitigated by cold storage. This strategy successfully brought the drug into the Phase I study rapidly with limited drug supplies. In the formulation redesign for the Phase II studies, various approaches were explored to improve stability, enhance bioavailability and increase unit dose strength. A soft-gelatin capsule (SGC) formulation using telcagepant potassium salt ethanolate was identified and used. The addition of surfactants enhanced and sustained drug supersaturation, resulting in a twofold increase in exposure relative to the Phase I formulation. However, formulation instability involving surfactant hydrolysis and salt disproportionation resulted from telecagepant and excipient chemistry and led to ultimate loss of solubilization of the drug within the vehicle matrix. Both in vivo and in vitro studies were performed to help preserve the integrity of the clinical data. A solid oral dosage form was developed and PK data collected to bridge from the LBF for further exploration in Phase III clinical studies and is the subject of a separate published paper (Mahjour et al. 2014). The clinical development program for telcagepant was put on hold during Phase III owing to an insufficient risk/benefit balance derived from a careful evaluation of clinical efficacy and safety data.

Acknowledgements The authors would like to acknowledge the telcagepant development team and a number of current and former Merck scientists who contributed to the research presented herein or provided helpful comments on the chapter composition: D. Breslin, Y.-H. Kiang, N. Pourkavoos, R. Gokhale, Y. Liu, Y. Cheng, T. Gibson, E. Nelson, K. Hetick, G. Ma, H. Flanagan, W. Bowen, D. Ellison, B. Hill, S. Ghosh, L. Alani, G. Szpunar, and D. Camposano.

References

Alsenz J, Kansy M (2007) High throughput solubility measurement in drug discovery and development. Adv Drug Deliv Rev 59(7):546–567. doi:10.1016/j.addr.2007.05.007

Alsenz J, Meister E, Haenel E (2007) Development of a partially automated solubility screening (PASS) assay for early drug development. J Pharm Sci 96(7):1748–1762. doi:10.1002/jps.20814

Avdeef A (2001) Physicochemical profiling (solubility, permeability and charge state). Curr Top Med Chem 1(4):277–351. doi:10.2174/1568026013395100

Avdeef A (2007) Solubility of sparingly-soluble ionizable drugs. Adv Drug Deliv Rev 59(7):568–590. doi:10.1016/j.addr.2007.05.008

Ballard JM, Zhu L, Nelson ED, Seburg RA (2007) Degradation of vitamin D3 in a stressed formulation: the identification of esters of vitamin D3 formed by a transesterification with triglycerides. J Pharm Biomed Anal 43(1):142–150

Bard B, Martel S, Carrupt P-A (2008) High throughput UV method for the estimation of thermodynamic solubility and the determination of the solubility in biorelevant media. Eur J Pharm Sci 33(3):230–240. doi:10.1016/j.ejps.2007.12.002

Basit AW, Newton JM, Short MD, Waddington WA, Ell PJ, Lacey LF (2001) The effect of polyethylene glycol 400 on gastrointestinal transit: implications for the formulation of poorly-water soluble drugs. Pharm Res 18(8):1146–1150. doi:10.1023/a:1010927026837

Basit AW, Podczeck F, Newton JM, Waddington WA, Ell PJ, Lacey LF (2002) Influence of polyethylene glycol 400 on the gastrointestinal absorption of ranitidine. Pharm Res 19 (9):1368–1374. doi:10.1023/a:1020315228237

Basit AW, Podczeck F, Newton JM, Waddington WA, Ell PJ, Lacey LF (2004) The use of formulation technology to assess regional gastrointestinal drug absorption in humans. Eur J Pharm Sci 21(2–3):179–189. doi:10.1016/j.ejps.2003.10.003

Bauer J, Spanton S, Henry R, Quick J, Dziki W, Porter W, Morris J (2001) Ritonavir: an extraordinary example of conformational polymorphism. Pharm Res 18(6):859–866. doi:10.1023/a:1011052932607

Benza HI, Munyendo WLL (2011) A review of progress and challenges in soft gelatin capsules formulations for oral administrations. Int J Pharm Sci Res 10:20–24

Bergstrom CAS, Norinder U, Luthman K, Artursson P (2002) Experimental and computational screening models for prediction of aqueous drug solubility. Pharm Res 19(2):182–188. doi:10.1023/a:1014224900524

Bevan CD, Lloyd RS (2000) A high-throughput screening method for the determination of aqueous drug solubility using laser nephelometry in microtiter plates. Anal Chem 72 (8):1781–1787. doi:10.1021/ac9912247

Bhutkar MA, Yadav AV, Sevukarajan M, Sonawane RO (2008) Hard capsules: a suitable alternative to soft gelatin capsules for delivering liquid drugs. Res J Pharm Technol 1 (3):280–282

Blasko A, Leahy-Dios A, Nelson WO, Austin SA, Killion RB, Visor GC, Massey IJ (2001) Revisiting the solubility concept of pharmaceutical compounds. Monatshefte Fur Chem 132 (7):789–798. doi:10.1007/s007060170065

Ceruti S, Villa G, Fumagalli M, Colombo L, Magni G, Zanardelli M, Fabbretti E, Verderio C, van den Maagdenberg AMJM, Nistri A, Abbracchio MP (2011) Calcitonin gene-related peptide-mediated enhancement of purinergic neuron/glia communication by the algogenic factor bradykinin in mouse trigeminal ganglia from wild-type and R192Q Ca(v)2.1 knock-in mice: implications for basic mechanisms of migraine pain. J Neurosci 31(10):3638–3649. doi:10.1523/jneurosci.6440-10.2011

Charman WN (2000) Lipids, lipophilic drugs, and oral drug delivery—some emerging concepts. J Pharm Sci 89(8):967–978. doi:10.1002/1520-6017(200008)89:8<967::aid-jps1>3.0.co;2-r

Charman WNA, Stella VJ (1986) Estimating the maximal potential for intestinal lymphatic transport of lipophilic drug molecules. Int J Pharm 34(1–2):175–178. doi:10.1016/0378-5173(86)90027-x

Chen TM, Shen H, Zhu CY (2002) Evaluation of a method for high throughput solubility determination using a multi-wavelength UV plate reader. Comb Chem High Throughput Screen 5(7):575–581

Chen XQ, Venkatesh S (2004) Miniature device for aqueous and non-aqueous solubility measurements during drug discovery. Pharm Res 21(10):1758–1761. doi:10.1023/b:pham.0000045225.76650.53

Cole ET, Cad D, Benameur H (2008) Challenges and opportunities in the encapsulation of liquid and semi-solid formulations into capsules for oral administration. Adv Drug Deliv Rev 60 (6):747–756. doi:10.1016/j.addr.2007.09.009

Collnot E-M, Baldes C, Schaefer UF, Edgar KJ, Wempe MF, Lehr C-M (2010) Vitamin E TPGS P-glycoprotein inhibition mechanism: influence on conformational Flexibility, intracellular ATP levels, and role of time and site of access. Mol Pharm 7(3):642–651. doi:10.1021/mp900191s

Constantinides PP (1995) Lipid microemulsions for improving drug dissolution and oral absorption—physical and biopharmaceutical aspects. Pharm Res 12(11):1561–1572. doi:10.1023/a:1016268311867

Damayanthi RD, Narayanan N, Elango K, Jayshree N, Reddy PN, Sadhwani NC (2008) Soft Gelatin Capsules - A Review. Pharm Rev 6(32):42–45

Dehring KA, Workman HL, Miller KD, Mandagere A, Poole SK (2004) Automated robotic liquid handling/laser-based nephelometry system for high throughput measurement of kinetic aqueous solubility. J Pharm Biomed Anal 36(3):447–456. doi:10.1016/j.jpba.2004.07.022

Desai J, Alexander K, Riga A (2006) Characterization of polymeric dispersions of dimenhydrinate in ethyl cellulose for controlled release. Int J Pharm 308(1–2):115–123. doi:10.1016/j.ijpharm.2005.10.034

Digenis GA, Gold TB, Shah VP (1994) Cross-linking of gelatin capsules and its relevance to their in-vitro in-vivo performance. J Pharm Sci 83(7):915–921. doi:10.1002/jps.2600830702

Dintaman JM, Silverman JA (1999) Inhibition of P-glycoprotein by D-alpha-tocopheryl polyethylene glycol 1000 succinate (TPGS). Pharm Res 16(10):1550–1556. doi:10.1023/a:1015000503629

Feinle C, O'Donovan D, Doran S, Andrews JM, Wishart J, Chapman I, Horowitz M (2003) Effects of fat digestion on appetite, APD motility, and gut hormones in response to duodenal fat infusion in humans. Am J Physiol Gastrointest Liver Physiol 284(5):G798–G807. doi:10.1152/ajpgi.00512.2002

Fricker G, Kromp T, Wendel A, Blume A, Zirkel J, Rebmann H, Setzer C, Quinkert R-O, Martin F, Mueller-Goymann C (2010) Phospholipids and lipid-based formulations in oral drug delivery. Pharm Res 27(8):1469–1486. doi:10.1007/s11095-010-0130-x

Gao P, Akrami A, Alvarez F, Hu J, Li L, Ma C, Surapaneni S (2009) Characterization and optimization of AMG 517 supersaturable self-emulsifying drug delivery system (S-SEDDS) for improved oral absorption. J Pharm Sci 98(2):516–528

Gao P, Guyton ME, Huang T, Bauer JM, Stefanski KJ, Lu Q (2004) Enhanced oral bioavailability of a poorly water soluble drug PNU-91325 by supersaturable formulations. Drug Dev Ind Pharm 30(2):221–229

Gao P, Rush BD, Pfund WP, Huang TH, Bauer JM, Morozowich W, Kuo MS, Hageman MJ (2003) Development of a supersaturable SEDDS (S-SEDDS) formulation of paclitaxel with improved oral bioavailability. J Pharm Sci 92(12):2386–2398

Gibson L (2007) Lipid-based excipients for oral drug delivery. In: Hauss DJ (ed) Oral lipid-based formulations: enhancing the bioavailability of poorly water-soluble drugs. CRC, New York, pp 33–62

Glomme A, Marz J, Dressman JB (2005) Comparison of a miniaturized shake-flask solubility method with automated potentiometric acid/base titrations and calculated solubilities. J Pharm Sci 94(1):1–16. doi:10.1002/jps.20212

Griffin WC (1949) Classification of surface-active agents by 'HLB'. J Soc Cosmet Chem 1 (5):311–326

Griffin WC (1954) Calculation of HLB values of non-ionic surfactants. J Soc Cosmet Chem 5 (4):249–256

Hauss DJ (ed) (2007) Oral lipid-based formulations: enhancing the bioavailability of poorly water-soluble drugs. Drugs and the pharmaceutical sciences. CRC, New York

Ho TW, Edvinsson L, Goadsby PJ (2010) CGRP and its receptors provide new insights into migraine pathophysiology. Nat Rev Neurol 6(10):573–582. doi:10.1038/nrneurol.2010.127

Hoelke B, Gieringer S, Arlt M, Saal C (2009) Comparison of nephelometric, UV-spectroscopic, and HPLC methods for high-throughput determination of aqueous drug solubility in microtiter plates. Anal Chem 81(8):3165–3172. doi:10.1021/ac9000089

Hugger ED, Novak BL, Burton PS, Audus KL, Borchardt RT (2002) A comparison of commonly used polyethoxylated pharmaceutical excipients on their ability to inhibit P-glycoprotein activity in vitro. J Pharm Sci 91(9):1991–2002. doi:10.1002/jps.10176

Humberstone AJ, Charman WN (1997) Lipid-based vehicles for the oral delivery of poorly water soluble drugs. Adv Drug Deliv Rev 25(1):103–128. doi:10.1016/s0169-409x(96)00494-2

Ichihashi T, Takagishi Y, Yamada H (1992) Factors determining the intrinsic lymphatic partition rate of epitiostanol and mepitiostane. Pharm Res 9(12):1617–1621. doi:10.1023/a:1015824710957

Jannin V, Musakhanian J, Marchaud D (2008) Approaches for the development of solid and semi-solid lipid-based formulations. Adv Drug Deliv Rev 60(6):734–746. doi:10.1016/j.addr.2007.09.006

Kaukonen AM, Boyd BJ, Porter CJH, Charman WN (2004) Drug solubilization behavior during in vitro digestion of simple triglyceride lipid solution formulations. Pharm Res 21(2):245–253. doi:10.1023/B:PHAM.0000016282.77887.1f

Kerns EH (2001) High throughput physicochemical profiling for drug discovery. J Pharm Sci 90 (11):1838–1858. doi:10.1002/jps.1134

Kesisoglou F, Panmai S, Wu Y (2007) Nanosizing—oral formulation development and biopharmaceutical evaluation. Adv Drug Deliv Rev 59(7):631–644. doi:10.1016/j.addr.2007.05.003

Khoo SM, Edwards GA, Porter CJH, Charman WN (2001) A conscious dog model for assessing the absorption, enterocyte-based metabolism, and intestinal lymphatic transport of halofantrine. J Pharm Sci 90(10):1599–1607. doi:10.1002/jps.1110

Kuentz M (2012) Lipid-based formulations for oral delivery of lipophilic drugs. Drug Discov Today Technol 9(2):e97–e104

Kwei GY, Novak LB, Hettrick LH, Reiss ER, Fong EK, Olah TV, Loper AE (1998) Lymphatic uptake of MK-386, a sterol 5 alpha-reductase inhibitor, from aqueous and lipid formulations. Int J Pharm 164(1–2):37–44. doi:10.1016/s0378-5173(97)00392-x

Lespine A, Chanoit G, Bousquet-Melou A, Lallemand E, Bassissi FM, Alvinerie M, Toutain PL (2006) Contribution of lymphatic transport to the systemic exposure of orally administered moxidectin in conscious lymph duct-cannulated dogs. Eur J Pharm Sci 27(1):37–43. doi:10.1016/j.ejps.2005.08.003

Lipinski CA, Lombardo F, Dominy BW, Feeney PJ (1997) Experimental and computational approaches to estimate solubility and permeability in drug discovery and development settings. Adv Drug Deliv Rev 23(1–3):3–25. doi:10.1016/s0169-409x(96)00423-1

Ma MH, DiLollo A, Mercuri R, Lee T, Bundang M, Kwong E (2002) HPLC and LC-MS studies of the transesterification reaction of methylparaben with twelve 3- to 6-carbon sugar alcohols and propylene glycol and the isomerization of the reaction products by acyl migration. J Chromatogr Sci 40(3):170–177

Mahjour M, Kesisoglou F, Cruanes M, Xu W, Zhang D, Maguire T, Rege B, Rosen L, Templeton AC, Kress MH (2014) Effect of added alkalizer and surfactant on dissolution and absorption of the potassium salt of a weakly basic poorly water-soluble drug. J Pharm Sci 103(6):1811–1818. doi:10.1002/jps.23985

Mansky P, Dai W-G, Li S, Pollock-Dove C, Daehne K, Dong H, Eichenbaum G (2007) Screening method to identify preclinical liquid and semi-solid formulations for low solubility compounds: Miniaturization and automation of solvent casting and dissolution testing. J Pharm Sci 96(6):1548–1563. doi:10.1002/jps.20799

Meyer MC, Straughn AB, Mhatre RM, Hussain A, Shah VP, Bottom CB, Cole ET, Lesko LL, Mallinowski H, Williams RL (2001) The effect of gelatin cross-linking on the bioequivalence of hard and soft gelatin acetaminophen capsules. Pharm Res 18(5):718

Mullertz A, Ogbonna A, Ren S, Rades T (2010) New perspectives on lipid and surfactant based drug delivery systems for oral delivery of poorly soluble drugs. J Pharm Pharmacol 62 (11):1622–1636. doi:10.1111/j.2042-7158.2010.01107.x

Narang AS, Srivastava AK (2002) Evaluation of solid dispersions of clofazimine. Drug Dev Ind Pharm 28(8):1001–1013. doi:10.1081/ddc-120006431

Nelson DL, Cox MM (2012) Lehninger principles of biochemistry. W. H. Freeman, New York

Noyes AA, Whitney WR (1897) The rate of solution of solid substances in their own solutions. J Am Chem Soc 19(12):930–934

O'Neil MJ (ed) (2013) The Merck Index: an encyclopedia of chemicals, drugs, and biologicals, 15th edn. Royal Society of Chemistry, London

Palacios LE, Wang T (2005) Egg-yolk lipid fractionation and lecithin characterization. J Am Oil Chem Soc 82(8):571–578. doi:10.1007/s11746-005-1111-4

Patil SK, Wagh KS, Parik VB, Akarte AM, Baviskar DT (2011) Strategies for solubility enhancement of poorly soluble drugs. Int J Pharm Sci Rev Res 8(2):74–80

Persson LC, Porter CJH, Charman WN, Bergstrom CAS (2013) Computational prediction of drug solubility in lipid based formulation excipients. Pharm Res 30(12):3225–3237. doi:10.1007/s11095-013-1083-7

Porter CJH, Trevaskis NL, Charman WN (2007) Lipids and lipid-based formulations: optimizing the oral delivery of lipophilic drugs. Nat Rev Drug Discov 6(3):231–248. doi:10.1038/nrd2197

Pouton CW (2000) Lipid formulations for oral administration of drugs: non-emulsifying, self-emulsifying and 'self-microemulsifying' drug delivery systems. Eur J Pharm Sci 11:S93–S98. doi:10.1016/s0928-0987(00)00167-6

Pouton CW (2006) Formulation of poorly water-soluble drugs for oral administration: physicochemical and physiological issues and the lipid formulation classification system. Eur J Pharm Sci 29(3–4):278–287. doi:10.1016/j.ejps.2006.04.016

Pouton CW, Porter CJH (2008) Formulation of lipid-based delivery systems for oral administration: materials, methods and strategies. Adv Drug Deliv Rev 60(6):625–637. doi:10.1016/j.addr.2007.10.010

Rane SS, Anderson BD (2008) What determines drug solubility in lipid vehicles: is it predictable? Adv Drug Deliv Rev 60(6):638–656. doi:10.1016/j.addr.2007.10.015

Rege BD, Kao JPY, Polli JE (2002) Effects of nonionic surfactants on membrane transporters in Caco-2 cell monolayers. Eur J Pharm Sci 16(4–5):237–246. doi:10.1016/s0928-0987(02)00055-6

Sarpal K, Pawar Y, Bansal A (2010) Self-emulsifying drug delivery systems: a strategy to improve oral bioavailability. Curr Res Inform Pharm Sci 11:42–49

Singh B, Bandopadhyay S, Kapil R, Singh R, Katare OP (2009) Self-emulsifying drug delivery systems (SEDDS): formulation development, characterization, and applications. Crit Rev Ther Drug Carrier Syst 26(5):427–521

Stein D, Bindra DS (2007) Stabilization of hard gelatin capsule shells filled with polyethylene glycol matrices. Pharm Dev Technol 12(1):71–77. doi:10.1080/10837450601166627

Stella VJ (2013) Chemical drug stability in lipids, modified lipids, and polyethylene oxide-containing formulations. Pharm Res 30(12):3018–3028. doi:10.1007/s11095-013-1051-2

Strickley RG (2004) Solubilizing excipients in oral and injectable formulations. Pharm Res 21(2):201–230. doi:10.1023/B:PHAM.0000016235.32639.23

Tepper SJ, Millson D (2003) Safety profile of the triptans. Expert Opin Drug Saf 2(2):123–132. doi:10.1517/eods.2.2.123.21388

Ueda CT, Lemaire M, Gsell G, Nussbaumer K (1983) Intestinal lymphatic absorption of cyclosporin a following oral-administration in an olive oil solution in rats. Biopharm Drug Dispos 4(2):113–124. doi:10.1002/bdd.2510040203

Varma MVS, Panchagnula R (2005) Enhanced oral paclitaxel absorption with vitamin E-TPGS: effect on solubility and permeability in vitro, in situ and in vivo. Eur J Pharm Sci 25(4–5):445–453. doi:10.1016/j.ejps.2005.04.003

Vasconcelos T, Sarmento B, Costa P (2007) Solid dispersions as strategy to improve oral bioavailability of poor water soluble drugs. Drug Discov Today 12(23–24):1068–1075. doi:10.1016/j.drudis.2007.09.005

Wasan KM (2001) Formulation and physiological and biopharmaceutical issues in the development of oral lipid-based drug delivery systems. Drug Dev Ind Pharm 27(4):267–276. doi:10.1081/ddc-100103726

Wasylaschuk WR, Harmon PA, Wagner G, Harman AB, Templeton AC, Xu H, Reed RA (2007) Evaluation of hydroperoxides in common pharmaceutical excipients. J Pharm Sci 96(1):106–116. doi:10.1002/jps.20726

Waterman KC, Arikpo WB, Fergione MB, Graul TW, Johnson BA, Macdonald BC, Roy MC, Timpano RJ (2008) N-methylation and N-formylation of a secondary amine drug (Varenicline) in an osmotic tablet. J Pharm Sci 97(4):1499–1507. doi:10.1002/jps.21119

Williams HD, Anby MU, Sassene P, Kleberg K, Bakala-N'Goma J-C, Calderone M, Jannin V, Igonin A, Partheil A, Marchaud D, Jule E, Vertommen J, Maio M, Blundell R, Benameur H, Carriere F, Muellertz A, Pouton CW, Porter CJH (2012a) Toward the establishment of standardized in vitro tests for lipid-based formulations. 2. The effect of bile salt concentration and drug loading on the performance of type I, II, IIIA, IIIB, and IV formulations during in vitro digestion. Mol Pharm 9(11):3286–3300. doi:10.1021/mp300331z

Williams HD, Sassene P, Kleberg K, Bakala-N'Goma J-C, Calderone M, Jannin V, Igonin A, Partheil A, Marchaud D, Jule E, Vertommen J, Maio M, Blundell R, Benameur H, Carriere F, Mullertz A, Porter CJH, Pouton CW (2012b) Toward the establishment of standardized in vitro tests for lipid-based formulations, part 1: method parameterization and comparison of in vitro digestion profiles across a range of representative formulations. J Pharm Sci 101(9):3360–3380. doi:10.1002/jps.23205

Wu Y, Kesisoglou F (2010) Immediate release oral dosage forms: formulation screening in the pharmaceutical industry. In: Dressman JB, Reppas C (eds) Oral drug absorption: prediction and assessment, 2nd edn. Informa Healthcare, New York, pp 296–337

Yang L, Heatley F, Blease TG, Thompson RIG (1996) A study of the mechanism of the oxidative thermal degradation of poly(ethylene oxide) and poly(propylene oxide) using H-1- and C-13-NMR. Eur Polym J 32(5):535–547. doi:10.1016/0014-3057(95)00182-4

Yuksel N, Karatas A, Ozkan Y, Savaser A, Ozkan SA, Baykara T (2003) Enhanced bioavailability of piroxicam using Gelucire 44/14 and Labrasol: in vitro and in vivo evaluation. Eur J Pharm Biopharm 56(3):453–459. doi:10.1016/s0939-6411(03)00142-5

Zhang Z, Tan S, Feng S-S (2012) Vitamin E TPGS as a molecular biomaterial for drug delivery. Biomaterials 33(19):4889–4906. doi:10.1016/j.biomaterials.2012.03.046

Zhou L, Yang L, Tilton S, Wang J (2007) Development of a high throughput equilibrium solubility assay using miniaturized shake-flask method in early drug discovery. J Pharm Sci 96 (11):3052–3071. doi:10.1002/jps.20913

Index

A.C. Templeton et al. (eds.), *Discovering and Developing Molecules with Optimal Drug-Like Properties,* AAPS Advances in the Pharmaceutical Sciences Series 15,
DOI 10.1007/978-1-4939-1399-2

MIX
Papier aus verantwortungsvollen Quellen
Paper from responsible sources
FSC® C105338

If you have any concerns about our products,
you can contact us on
ProductSafety@springernature.com

In case Publisher is established outside the EU,
the EU authorized representative is:
Springer Nature Customer Service Center GmbH
Europaplatz 3, 69115 Heidelberg, Germany

Printed by Libri Plureos GmbH
in Hamburg, Germany